ENGINEERING MATHEMATICS-I

ENGINEERING MATHEMATICS-I

DR. REENA GARG
M.Sc. (Goldmedalist), M.Phil, Ph.D
Asstt Professor (Mathematics)
YMCA University of Science & Technology,
Faridabad – 121006 (Haryana)

KHANNA BOOK PUBLISHING CO. (P) LTD.
Publisher of Engineering and Computer Books
4C/4344, Ansari Road, Darya Ganj, New Delhi-110002
Phone : 011-23244447-48 **Mobile:** +91-9910909320
E-mail : contact@khannabooks.com
Website : www.khannabooks.com

Price: ₹ 465.00

ISBN: 978-93-82609-55-1

Engineering Mathematics-I
by Dr. Reena Garg

First Edition: 2015
Revised Edition: 2018
Reprint Edition: 2023, **2025**

Published by:
Khanna Book Publishing Co. (P) Ltd.
CIN: U22110DL1998PTC095547

Visit us at: **www.khannabooks.com**
Write us at: **contact@khannabooks.com**

Printed in Bharat.

PREFACE

We, the authors feel pleasure in presenting the book as a text-book of Maths for the first year students of Science and Engineering. This book is designed in such a way that the students feeel comfortable in understanding the subject with out the help of teachers. It meets all the needs of the students of Various Universities in the country.

The long experience of authors in teaching the subjects has formed the background to bring out the book. The book is based entirely on the class notes and the class lectures of the authors.

In order to make the studeents familiar with the recent style which the Universities adopt in setting the question paper, a number of problems from University papers have been included in the book. The problems which the students feel difficult in solving have been solved in this book to understand the subject easily.

We are sure that the book will be useful to the students and will earn them warm receptions by the users.

We proudly express our gratitude to our friends and well- wishers for their valuable guidance and suggestions to write the book.

We are highly thankful to our Publisher's Sh. Puneet Khanna, to bring out the book in a short time.

Though every effort has been taken to keep the book away from errors, we shall be highly obliged if any error that may have crept in may be brought to the notice of the authors.

Any suggestions for improvement of the book shall be highly appreciated and thankfully acknowledged.

ABOUT THE AUTHOR

Dr. Reena Garg, M.Sc. Mathematics (Gold Medalist), M.Phil, Ph.D is Assistant Professor (Mathematics) in YMCA University of Science & Technology, Faridabad (Haryana). She also taught in C.I.T.M. Faridabad (presently known as Manav Rachna International University, Faridabad). Her teaching experience of more than a dacade has made this book more valuable for the knowledge seekers. She has published more than 10 research papers in various International Journals. She is a life-time member of Forum of Inter-disciplinary mathematics in India. She is a member of Editorial Board in International Journal of Engineering & Research. She is a member of Reviewer Board in IJRET, Bangalore.

Prof. R.S. Goel is Retired Principal of Aggarwal College, Ballabgarh, Haryana (India) and Retired Professor & HOD (Deptt. of Mathematics) Manav Rachna International University, Faridabad Haryana. He has a teaching experience of more than 45 years.

Dr. Deepankar Sharma is Director in D.J. College of Engineering, Modi Nagar (UP) since last 10 years. He has a total teaching experience of 15 years. He has supervised & guided more than 20 Researcher & Ph.D scholars. He has published more than 80 research papers in various International Journals.

CONTENTS AT A GLANCE

DETAILED TABLE OF CONTENTS

CHAPTER 1

Infinite Series

1.1 Introduction

We are familiar with arithmetic and geometric series. In an arithmetic series each term after the first term is obtained by adding a constant fixed number to the preceeding term. In geometric series, each term after the first is obtained by multiplying a constant fixed number to the preceeding terms. Thus a series can also be formed by applying some other rule to its terms. For example the series $1^3 + 2^3 + 3^3 + ...10^3$ is formed by the cubes of the first ten natural numbers. In some series there is no last term. Such a series is called infinite series.

Before the study of series we have to know about sequence.

1.2 Sequence

A sequence of real numbers (or a real sequence) is defined as a function $f : N \to R$ where N is set of natural numbers and R is the set of real numbers.

If $f : N \to R$ is sequence, then for each $n \varepsilon N$, $f(n)$ is real number.

$f(n)$ can be denoted as f_n. A sequence may be written as

$< f_1, f_2, f_3, ... f_n, ... >$ or $< f_i >$ or $\{f_n\}$. the real numbers $f_1, f_2, f_3, ... f_n$ ......are called the terms or elements of the sequence. f_1 is called first term, f_2 is second term and f_n is the nth term of the sequence $< f_n >$. We can write all the terms of a sequence in its nth term.

> **Note :**
>
> **(i) The term of a sequence occuring at different positions is treated as distinct term even if they have the same value.**
>
> **(ii) The number of terms of a sequence is always infinite. However its range may be finite.**

Illustrative examples of sequence

(i) If $f_n = n!$, then the sequence defined is $<1, 2, 6, 24, 120 ......>$

(ii) If $f_n = \dfrac{1}{2n-1}$, then the sequence defined is $<1, \dfrac{1}{3}, \dfrac{1}{5}, \dfrac{1}{7}...... >$.

1.3 Constant Sequence

A sequence $<f_n>$ defined by $f_n = C$ for all $n \varepsilon N$ is calle a constant sequence.

For example

$<f_n> = <C, C, C, ...>$ is a constant sequence with range $= \{C\}$.

1.4 Bounded Sequence

(a) **Bounded above :** A sequence $<f_n>$ is said to be bounded above if there exist a real number K such that $f_n \leq K \; \forall \; n \varepsilon N$. Here K is called an upper bound of sequence $<f_n>$.

(b) **Bounded below :** A sequence $<f_n>$ is said to be bounded below if there exist a real number k such that $k \leq f_n$ (i.e., $f_n \geq k$) $\forall \; n \varepsilon N$. Here k is called lower bound.

A sequence $<f_n>$ is said to be bounded if there exist real numbers k and K such that $k \leq f_n \leq K \; \forall \; n \in N$.

OR

we can say a sequence is said to be bounded if it is both bounded above and bounded below. Otherwise it is unbounded.

<u>Illustrative Examples</u> :

1. $<n^2> = <1, 4, 9, 16 ......>$ is bounded below by 1 but not bounded above.
2. $<1 + (-1)^n> = <0, 2, 0, 2 ...>$ is bounded, 0 is lower bound and 2 is upper bound.
3. $\left\langle \dfrac{(-1)^{n-1}}{n!} \right\rangle = \left\langle 1, \dfrac{-1}{2!}, \dfrac{1}{3!}, \dfrac{-1}{4!} ... \right\rangle$ is bounded since $-\dfrac{1}{2} \leq f_n \leq 1 \; \forall \; n \in N$

1.5 Convergent, Divergent and Oscillatory Sequence

(a) **Convergent Sequence :** A sequence $<f_n>$ is said to be convergent to a real number l if for any given $\varepsilon > 0$, however small it may be, there exists a positive integer m (depending upon ε) such that $|f_n - l| < \epsilon \; \forall \; n \geq m$.

The real number l is called limit of the sequence $<f_n>$.

OR

A sequence $<f_n>$ is said to be convergent if $\lim_{n\to\infty} f_n$ is finite say 'l'.

The real number l is called the limit of sequence $<f_n>$ and is written as

$\lim_{n\to\infty} f_n = l$ or $f_n \to l$ as $n \to \infty$.

(b) **Divergent Sequence :** Any sequence is said to be divergent if it is not convergent.

i.e., $\lim_{n\to\infty} f_n = \infty$ or $\lim_{n\to\infty} f_n = -\infty$

(c) **Oscillatory Sequence :** If a sequence $<f_n>$ neither converges to a finite number nor diverges to $+\infty$ or $-\infty$ then it is called oscillating sequence.

Oscillatory sequence are of two types :

(a) Oscillate Finitely : A sequence is said to oscillate finitely if
- (i) it is bounded.
- (ii) it neither converges nor diverges.

(b) Oscillate Infinitely :
- (i) it is unbounded.
- (ii) it neither converges nor diverges.

Illustrative Examples :

1. $<(-1)^n> = <-1, 1, -1, 1 \ldots>$ oscillates finitely.
2. $<1 + (-1)^n> = <0, 2, 0, 2 \ldots>$ oscillates finitely.
3. $<(-1)^n \cdot n> = <-1, 2, -3, 4, -5, 6 \ldots>$ oscillates infinitely.
4. $\left\langle 1, 2, \frac{1}{2}, 3, \frac{1}{3} \ldots \right\rangle$ oscillates infinitely.

Remark :

A bounded sequence either converges or oscillates finitely. An unbounded sequence either diverges to $+\infty$ or $-\infty$ or oscillates infinitely.

Important Observation :

1. *A sequence cannot converge to more than one limit.*
2. *Every convergent sequence is bounded but the converse is not true.*

1.6 Monotonic Sequence

A sequence is said to be monotonic if it is either monotonically increasing or decreasing.

(a) A sequence $<f_n>$ is said to be monotonically increasing if $f_{n+1} \geq f_n \ \forall \ n \in N$.

i.e., $f_1 \leq f_2 \ \leq f_3 \leq ... \leq f_n \leq f_{n+1} \leq ..$

(b) A sequence $<f_n>$ is said to be monotonically decreasing if $f_{n+1} \leq f_n \ \forall \ n \in N$.

i.e., $f_1 \geq f_2 \geq ... \geq f_n \geq f_{n+1} \geq ..$

(c) A sequence $<f_n>$ is said to be strictly monotonically increasing if $f_{n+1} > f_n \ \forall \ n \in N$.

(d) A sequence $<f_n>$ is said to be strictly monotonically decreasing if $f_{n+1} < f_n \ \forall \ n \in N$

Illustrative Examples :

1. The sequence $<1, 1, 2, 2, 3, 3, 4, 4 ...>$ is a monotonically increasing sequence.
2. The sequence $<n>, <n^2>, <3^n>$ are strictly monotonic increasing.
3. The sequence $\left\langle 1, 1, \frac{1}{2}, \frac{1}{2}, \frac{1}{3}, \frac{1}{3} ... \right\rangle$ is a monotonically decreasing sequence.
4. The sequence $\left\langle \frac{1}{n} \right\rangle, \left\langle \frac{1}{n^2} \right\rangle$ are strictly monotonically decreasing sequences.

1.7 Limit of a Sequence

A sequence $<f_n>$ is said to approach the limit l when $n \to \infty$, if for each $\varepsilon > 0$ there exist a positive integer m (depending upon ε) such that $|f_n - l| < \varepsilon \ \forall \ n \geq m$.

Symbolically we can write $\lim\limits_{n \to \infty} f_n = l$.

For example :

1. $\left\langle \frac{1}{n} \right\rangle = \left\langle 1, \frac{1}{2}, \frac{1}{3} ... \right\rangle$ has only 0 as its limit point.
2. $<n^2> = <1, 4, 9 ...>$ has no limit point.

1.8 Some Important Results

(a) Sum of two convergent sequence is convergent.
If $\lim a_n = A$ and $\lim b_n = B$
then $\lim (a_n + b_n) = A + B$

(b) If $<a_n>$ and $<b_n>$ are two convergent sequences such that $\lim a_n = A$ and $\lim b_n = B$ then $\lim (a_n + b_n) = A + B$

(i) Sequence $<a_n b_n>$ is also convergent and converges to AB.

(ii) Sequence $\left\langle \frac{a_n}{b_n} \right\rangle$ is also convergent and converges to $\left\langle \frac{A}{B} \right\rangle$.

(c) A sequence $<|a_n|>$ converges to zero iff the sequence $<a_n>$ converges to zero.

(d) If a sequence $<a_n>$ converges to 'a' and $a_n \geq 0 \; \forall \; n$ then $a \geq 0$.

(e) If $a_n \to a,\; b_n \to b$ and $a_n \leq b_n \; \forall \; n$ then $a \leq b$.

(f) If $a_n \to l,\; b_n \to l$ and $a_n \leq c_n \leq b_n \; \forall \; n$ then $c_n \to l$.

(g) The necessary and sufficient condition for the convergence of a monotonic sequence is that it is bounded.

(i) A monotonic increasing sequence which is bounded above converges.

(ii) A monotonic decreasing sequence which is bounded below converges.

(h) If a monotonic increasing sequence is not bounded above, it diverges to $+\infty$.

(i) If a monotonic decreasing sequence is not bounded below, it diverges to $-\infty$.

Illustrative Examples

Example 1:

Prove that the sequence whose nth term given is monotonic. Is the sequence increasing or decreasing ?

$$\frac{1}{2}+\frac{1}{2^2}+\frac{1}{2^3}+\ldots+\frac{1}{2^n}$$

Solution :

$$f_n = \frac{1}{2}+\frac{1}{2^2}+\frac{1}{2^3}+\ldots+\frac{1}{2^n}$$

$$f_{n+1} = \frac{1}{2}+\frac{1}{2^2}+\frac{1}{2^3}+\ldots+\frac{1}{2^n}+\frac{1}{2^{n+1}}$$

$$f_{n+1}-f_n = \frac{1}{2^{n+1}} > 0 \; \forall \; n$$

$f_{n+1} > f_n$ for all n

Hence, sequence $<f_n>$ is monotonically increasing. Proved

Example 2:

Discuss the convergence of sequence $<f_n>$ where

$$f_n = \frac{2n-7}{3n+2}$$

Solution :

$$f_n = \frac{2n-7}{3n+2}, \quad f_{n+1} = \frac{2n-5}{3n+5}$$

$$f_{n+1} - f_n = \frac{2n-5}{3n+5} - \frac{2n-7}{3n+2}$$

$$= \frac{(2n-5)(3n+2)-(2n-7)(3n+5)}{(3n+5)(3n+2)}$$

$$= \frac{(6n^2+4n-15n-10)-(6n^2+10n-21n-35)}{(3n+5)(3n+2)}$$

$$= \frac{25}{(3n+5)(3n+2)} > 0$$

$f_{n+1} > f_n$ for all n

Hence, sequence $<f_n>$ is monotonically increasing sequence.

Also, $$<f_n> = \left\langle -1, -\frac{3}{8}, -\frac{1}{11}, \frac{1}{14} \ldots \right\rangle$$

$f_n \geq -1$ for all n.

$$f_n - 1 = \frac{2n-7}{3n+2} - 1 = \frac{2n-7-3n-2}{3n+2}$$

$$= \frac{-n-9}{3n+2} = \frac{-(n+9)}{3n+2} < 0 \text{ for all } n$$

$f_n < 1$ for all n

Thus we have $-1 \leq f_n < 1$ for all n.

$\Rightarrow$ $<f_n>$ is a bounded sequence.

Since, $<f_n>$ is monotonically increasing and bounded above.

$\Rightarrow$ $<f_n>$ is convergent.

1.9 Infinite Series

A series is the sum of the terms of a sequence. If $u_1, u_2, u_3 \ldots$ is a sequence then the sum $u_1 + u_2 + u_3 \ldots$ of all the terms is called an infinite series and is denoted by $\sum_{n=1}^{\infty} u_n$ or $\sum u_n$.

If $\sum u_n = u_1 + u_2 + u_3 + \ldots + u_n + \ldots$ is an infinite series, then

$$s_n = u_1 + u_2 + \ldots + u_n \forall n$$

The sequence $\{s_n\}$ is called the sequence of partial sums of the series and

$$s_1 = u_1$$
$$s_2 = u_1 + u_2$$
$$s_3 = u_1 + u_2 + u_3$$

To every infinite series $\sum u_n$ there corresponds a sequence $\{s_n\}$ of its partial sums.

Note:

An infinite series is said to converge, diverge or oscillate according as its sequence of partial sum $\{s_n\}$ converges, diverges or oscillates.

1.10 Alternating Series

A series in which the terms are alternatively positive and negative is called an alternating series. Thus the series

$$\sum(-1)^{n-1} = u_1 - u_2 + u_3 \ldots (-1)^{n-1} u_n + \ldots$$

where $u_n > 0 \forall n$ is an alternating series.

Remarks:

(a) The series $\sum u_n$ is said to be convergent if the sequence $\{s_n\}$of its partial sums is convergent.

i.e., $\sum u_n$ is convergent if $\lim_{n \to \infty} s_n$ = a finite quantity.

(b) The series $\sum u_n$ diverges if the sequence $\{s_n\}$ of its partial sums diverges.

i.e., $\sum u_n$ is divergent if $\lim_{n \to \infty} s_n = +\infty$ or $-\infty$

(c) The series $\sum u_n$ oscillates finitely if $\{s_n\}$ is bounded and neither converges nor diverges.

(d) The series $\sum u_n$ oscillates infinitely if $\{s_n\}$ is unbounded and neither converges nor diverges.

(e) Convergence or divergence of an infinite series does not get affected if we add or ommit some terms in series.

(f) Nature of an infinite series does not change on multiplying the given series by a non zero constant.

(g) (i) If $\sum a_n$ is convergent then $\lim_{n\to\infty} a_n = 0$

Since $\sum a_n$ is convergent $\lim_{n\to\infty} s_n = l$ (Finite)

also $\lim_{n\to\infty} s_{n-1} = l$

$$\therefore \quad \lim_{n\to\infty} a_n = \lim_{n\to\infty}(s_n - s_{n-1}) = l - l = 0$$

(ii) If $\lim_{n\to\infty} a_n = 0$, the series may be convergent or divergent and if $\lim_{n\to\infty} a_n \neq 0$, the series is divergent.

Illustrative Examples

Example 1:

Discuss the convergence of the series $1 + 2 + 3 + 4 + \ldots$

Solution :

$$\sum_{n=1}^{\infty} u_n = 1 + 2 + 3 + 4 \ldots$$

$$s_n = u_1 + u_2 + u_3 + \ldots + u_n$$

$$= 1 + 2 + 3 + \ldots + n$$

$$= \frac{n(n+1)}{2}$$

$$\lim_{n\to\infty} s_n = \infty$$

Hence, $\sum_{n=1}^{\infty} u_n$ is divergent series.

Example 2:

Check the convergence of the series $1 + \frac{1}{2} + \frac{1}{4} + \frac{1}{8}$

Solution :

$$s_n = 1 + \frac{1}{2} + \frac{1}{4} + \ldots \frac{1}{2^{n-1}}$$

$$= \frac{\left(1-\frac{1}{2^n}\right)}{1-\frac{1}{2}} \qquad \left[\because \text{ it is finite G.P.} \quad \therefore \left(\frac{1-r^n}{1-r}\right)\right]$$

$$= 2\left(1-\frac{1}{2^n}\right)$$

$$\lim_{n\to\infty} s_n = \lim_{n\to\infty} 2\left(1-\frac{1}{2^n}\right) = 2$$

Hence the series is a convergent series.

Example 3:

Check the convergence of the series $1-\frac{1}{5}+\frac{1}{5^2}-\frac{1}{5^3}+\ldots$

Solution :

Here

$$s_n = 1-\frac{1}{5}+\frac{1}{5^2}-\frac{1}{5^3}+\ldots+(-1)^{n-1}\frac{1}{5^{n-1}}$$

$$= \frac{1-\left(-\frac{1}{5}\right)^n}{1-\left(-\frac{1}{5}\right)} = \frac{1}{1+\frac{1}{5}} = \frac{5}{6} \text{ as } n\to\infty$$

$\therefore$ $\{s_n\}$ is a convergent sequence.

Therefore the given series is convergent.

Example 4:

Test the convergence of the series $1^2 + 3^2 + 5^2 + 7^2 + \ldots$

Solution :

For $1^2 + 3^2 + 5^2 + 7^2 + \ldots$

$$\text{the } n^{th} \text{ term } u_n = (2n-1)^2$$

$$\lim_{n\to\infty} u_n = \infty$$

$\therefore$ $\sum u_n$ is not a convergent series and hence it is a divergent series.

Example 5:

Test the convergence of the series $\frac{1}{1.3}+\frac{1}{3.5}+\frac{1}{5.7}+\ldots\ldots$

Solution :

The series can be written as

$$\frac{1}{2}\left[\left(\frac{1}{1}-\frac{1}{3}\right)+\left(\frac{1}{3}-\frac{1}{5}\right)+\left(\frac{1}{5}-\frac{1}{7}\right)+\ldots+\left(\frac{1}{2n-1}-\frac{1}{2n+1}\right)+\ldots\right]$$

$$s_n = \frac{1}{2}\left(1-\frac{1}{2n+1}\right)$$

$$s_n \to \frac{1}{2} \quad \text{as} \quad n \to \infty$$

$\therefore \quad \sum_{n=1}^{\infty} u_n$ is convergent series.

Example 6:

Test the convergence of the series $7-4-3+7-4-3+7-4-3+\ldots$

Solution :

The sequence $\{s_n\}$ of partial sum is given as

$$\begin{aligned} s_n &= 7, \text{ if } n = 3m+1, m \geq 0 \text{ a positive integer} \\ &= 0, \text{ if } n = 3m, m \geq 0 \text{ a positive integer} \\ &= 3, \text{ if } n = 3m-1, m > 0 \text{ a positive integer.} \end{aligned}$$

Hence $\{s_n\}$ is an oscillatory sequence.

Therefore $\sum_{n=1}^{\infty} u_n$ is an oscillatory series.

Exercise 1.1

1. Test the convergence of following series :

(i) $\frac{1}{1.2}+\frac{1}{2.3}+\frac{1}{3.4}+\ldots+\frac{1}{n(n+1)}+\ldots\infty.$ (ii) $1^2+2^2+3^2+\ldots n^2+\ldots$

(iii) $-1-2-3-\ldots-n\ldots$ (iv) $\sum_{n=1}^{\infty}(-1)^{n-1}$

(v) 5 – 4 – 1 + 5 – 4 – 1 + 5 – 4 – 1 + ... to ∞

2. Show that the series $\sum_{n=1}^{\infty} \frac{1}{n(n+2)}$ converges to $\frac{3}{4}$.

3. Show that the series $\sum_{n=1}^{\infty} \left(\frac{3}{4}\right)^{n-1}$ converges to 4.

4. Show that the series $\sum_{n=1}^{\infty} n(-1)^n$ oscillates infinitely.

5. Test the convergence of the following series :

(a) 1 + 3 + 5 + 7 ... ∞ (b) –1 – 8 – 27 – 64 ... ∞

Answers

1. (i) Convergent (ii) divergent to $+\infty$

(iii) diverges to $-\infty$ (iv) oscillates finitely

(v) Oscillates finitely

5. (a) Divergent

(b) Divergent

1.11 Convergence or Divergence of Geometric Series

The geometrical series $a + ar + ar^2 + ... + \infty$

(i) converges if $|r| < 1$

(ii) diverges if $r \geq 1$

(iii) Oscillates finitely if $r = -1$

(iv) Oscillates infinitely if $r < -1$.

Proof : Here

$$s_n = a + ar + ar^2 + ... ar^{n-1}$$

$$= \frac{a(1-r^n)}{1-r}, \text{ provided } r \neq 1$$

Case I

If $|r| < 1$, then $r^n \to o$ as $n \to \infty$

$$\lim_{n\to\infty} s_n = \frac{a}{1-r} \text{ which is unique and finite}$$

Thus the given series converges to $\frac{a}{1-r}$.

Case II

If $r > 1$, then $r^n \to \infty$ as $n \to \infty$

$$\lim_{n\to\infty} s_n = \lim_{n\to\infty} \frac{a(r^n - 1)}{r-1} = \infty$$

Thus the given series is divergent.

If $r = 1$, then $s_n = a + a \ldots + n$ term $= na$

$$\therefore \lim_{n\to\infty} s_n = \infty$$

Thus the series is divergent.

Case III

If $r = -1$ then

$$s_n = a - a + a - a + \ldots \text{ to } n \text{ terms}$$

$$= \begin{cases} a \text{ if } n \text{ is odd} \\ 0 \text{ if } n \text{ is even} \end{cases}$$

$\therefore$ Sequence $\{s_n\}$ oscillates finitely and hence series $\sum_{n=1}^{\infty} a_n$ oscillates finitely.

Case III

If $r < -1$

$\Rightarrow -r > 1 \quad \Rightarrow x > 1 \quad$ where $x = -r$

$\Rightarrow x^n \to \infty$ as $n \to \infty$

Now $$s_n = \frac{a(1-r^n)}{1-r} = \frac{a(1-(-x)^n)}{1+x}$$

$$= \begin{cases} \frac{a(1-x^n)}{1+x}, \text{ if } n \text{ is even} \\ \frac{a(1+x^n)}{1+x}, \text{ if } n \text{ is odd} \end{cases}$$

$$\lim_{n\to\infty} s_n = \begin{cases} -\infty, & \text{if } n \text{ is even} \\ \infty, & \text{if } n \text{ is odd} \end{cases}$$

Thus the sequence $\{s_n\}$ oscillates infinitely and hence the given series oscillates infinitely.

1.12 Series of Positive Terms

A series $\sum_{n=1}^{\infty} a_n$ for which $a_n > 0$ for all n, is called a series of positive terms.

Example of Geometric Series(convergence or divergence)

(1) $$\sum \frac{1}{2^n} = \frac{1}{2} + \frac{1}{2^2} + \frac{1}{2^3} + \ldots$$

it is geometric series with $r = \frac{1}{2}$

$$r = \frac{1}{2} < 1$$

$\therefore$ it is convergent series.

(2) $$\sum 3^n = 3 + 3^2 + 3^3 + \ldots$$

it is geometric series with $r = 3$

$$r = 3 > 1$$

$\therefore$ it is divergent.

(3) $$\frac{3}{4} - \frac{3}{4} + \frac{3}{4} - \frac{3}{4} + \ldots$$

It is geometric series with $r = -1$

$$r = -1$$

$\therefore$ the given series oscillates finitely.

***Basic Results :**

(A necessary condition for convergence). If the series $\sum_{n-1}^{\infty} a_n$ converges, then $\lim_{n\to\infty} a_n = 0$, and the converse is not always true.

1.13 Cauchy's General Principle of Convergence

A Necessary and sufficient condition for series $\sum_{n=1}^{\infty} a_n$ to converge is that to each $\varepsilon > 0$, there exists a positive integer m such that

$$|a_{m+1} + a_{m+2} \ldots a_n| < \varepsilon \text{ for all } n \geq m.$$

Illustrative Examples

Example 1:

Show that the series

$$\sum \frac{n}{n+1} = \frac{1}{2}+\frac{2}{3}+\frac{3}{4}+\ldots \text{ is not convergent.}$$

Solution:

$$a_n = \frac{n}{n+1} = \frac{1}{1+\frac{1}{n}}$$

$$\therefore \quad \lim_{n\to\infty} a_n = \lim_{n\to\infty}\frac{1}{1+\frac{1}{n}} = \frac{1}{1+0}$$

$$= 1 \neq 0$$

Hence the given series is not convergent, (by I[st] result).

Example 2 :

Test the convergence of the series $\sum \cos\left(\frac{1}{n}\right)$.

Solution :

$$\lim_{n\to\infty} a_n = \lim_{n\to\infty} \cos\left(\frac{1}{n}\right)$$

$$\Rightarrow \quad 1 \neq 0$$

Hence the series is not convergent.

Example 3:

Show that the series

$$\sum \frac{1}{n} = 1+\frac{1}{2}+\frac{1}{3}+\ldots+\frac{1}{n}+\ldots \text{ does not converge.}$$

Solution:

Suppose the given series converges. Then the sequence $<s_n>$ of partial sums of the given series is convergent. By Cauchy's principle of convergence for sequence, for $\varepsilon = \frac{1}{2}$ there exists a positive integer m such that

$$|s_n - s_m| < \frac{1}{2} \;\forall\; n \ge m$$

$$\left|\left(1+\frac{1}{2}+\ldots+\frac{1}{m}+\frac{1}{m+1}+\frac{1}{m+2}+\ldots+\frac{1}{n}\right)-\left(1+\frac{1}{2}+\ldots\frac{1}{m}\right)\right| < \frac{1}{2} \;\forall\; n \ge m$$

$$\frac{1}{m+1}+\frac{1}{m+2}+\ldots+\frac{1}{n} < \frac{1}{2} \;\forall\; n \ge m \qquad \ldots\text{(i)}$$

On taking n = 2m, we see that

$$\frac{1}{m+1}+\frac{1}{m+2}+\ldots+\frac{1}{n} = \frac{1}{m+1}+\frac{1}{m+2}+\ldots+\frac{1}{2m} > \frac{1}{2m}+\frac{1}{2m}+\ldots+\frac{1}{2m}$$

$$= m \cdot \frac{1}{2m} = \frac{1}{2}$$

$$\therefore \quad \frac{1}{m+1}+\frac{1}{m+2}+\ldots+\frac{1}{n} > \frac{1}{2} \quad (n = 2m > m) \qquad \ldots\text{(ii)}$$

Since (i) and (ii) are contradictory statement, the given series does not converge.

Positive term series

An inf nite series all of whose terms are positive is called a positive term series.

*Result :

(a) (A Fundamental test for a positive term series) A positive term series $\sum a_n$ is convergent if and only if its sequence $< s_n >$ of partial sums is bounded above.

i.e., A positive term series $\sum a_n$ converges

$\Leftrightarrow$ $s_n < k \;\forall$ n (k being some positive real number)

(b) A monotonic sequence either converge or diverges but cannot oscillate. Hence a positive term series either converges or diverges

i.e., A positive term series cannot be oscillating.

1.14 Comparison tests

Test I

If $\sum_{n=1}^{\infty} a_n$ and $\sum_{n=1}^{\infty} b_n$ are series of positive terms and $\sum_{n=1}^{\infty} b_n$ is convergent and there is a positive constant κ such that $a_n \le k\, b_n, \forall\ n > m$ then $\sum_{n=1}^{\infty} a_n$ is also convergent.

Test II

If $\sum_{n=1}^{\infty} a_n$ and $\sum_{n=1}^{\infty} b_n$ are two positive terms series and $\sum_{n=1}^{\infty} b_n$ is divergent and there is a positive constant k such that $a_n > k\, b_n, \forall\ n > m$ then $\sum_{n=1}^{\infty} a_n$ is also divergent.

Test III

If $\sum_{n=1}^{\infty} a_n$ and $\sum_{n=1}^{\infty} b_n$ are two positive terms series and κ, K are two positive numbers such that $kb_n \le a_n \le K.b_n$ for all n, then the series $\sum_{n=1}^{\infty} a_n$ and $\sum_{n=1}^{\infty} b_n$ converges or diverges together.

Test IV

If $\sum_{n=1}^{\infty} a_n$ and $\sum_{n=1}^{\infty} b_n$ are two series of positive terms such that $\lim_{n\to\infty} \frac{a_n}{b_n} = l$ (Finite and non zero), then both the series $\sum_{n=1}^{\infty} a_n$ and $\sum_{n=1}^{\infty} b_n$ converge or diverge together.

Remark:

In the above test the condition that $\lim_{n\to\infty} \frac{a_n}{b_n}$ is finite and non zero cannot be dropped.

$$\text{Consider } a_n = \frac{1}{n} \text{ and } b_n = \frac{1}{n^2}$$

$$\lim_{n\to\infty} \frac{a_n}{b_n} = \lim_{n\to\infty} n = \infty$$

Here $\sum_{n=1}^{\infty} a_n$ and $\sum_{n=1}^{\infty} b_n$ do not converge or diverge together as $\sum_{n=1}^{\infty} a_n$ is convergent and $\sum_{n=1}^{\infty} b_n$ is divergent.

Test V

If $\sum_{n=1}^{\infty} a_n$ and $\sum_{n=1}^{\infty} b_n$ are two series of positive terms such that

(a) $\lim_{n\to\infty} \frac{a_n}{b_n} = 0$ and $\sum_{n=1}^{\infty} b_n$ is convergent, then $\sum_{n=1}^{\infty} a_n$ is convergent.

(b) $\lim_{n\to\infty} \frac{a_n}{b_n} = \infty$ and $\sum_{n=1}^{\infty} b_n$ is divergent, then $\sum_{n=1}^{\infty} a_n$ is divergent.

Test VI

If $\sum_{n=1}^{\infty} a_n$ and $\sum_{n=1}^{\infty} b_n$ are two series of positive terms such that $\frac{a_n}{a_{n+1}} \geq \frac{b_n}{b_{n+1}}$ for all $n \geq m$, where m is some positive integer, then

(i) $\sum_{n=1}^{\infty} b_n$ converges $\Rightarrow \sum_{n=1}^{\infty} a_n$ converges

(ii) $\sum_{n=1}^{\infty} a_n$ diverges $\Rightarrow \sum_{n=1}^{\infty} b_n$ diverges.

1.15 Important test for Comparison

[$\frac{1}{n^p}$ Hyper harmonic series or p-series]

Theorem :

The series $\sum_{n=1}^{\infty} \frac{1}{n^p} = \frac{1}{1^p} + \frac{1}{2^p} + \frac{1}{3^p} \ldots + \frac{1}{n^p} + \ldots$

(i) converges if $p > 1$

(ii) diverges if $p \leq 1$

Proof :

Case I

if $p > 1$

$$\frac{1}{1^p} = 1$$

$$\frac{1}{2^p} + \frac{1}{3^p} < \frac{1}{2^p} + \frac{1}{2^p} = \frac{1}{2^{p-1}}$$

$$\frac{1}{4^p}+\frac{1}{5^p}+\frac{1}{6^p}+\frac{1}{7^p}<\frac{1}{4^p}+\frac{1}{4^p}+\frac{1}{4^p}+\frac{1}{4^p}=\frac{4}{4^p}=\frac{1}{4^{p-1}}=\frac{1}{(2^{p-1})^2}$$

...

...

...

$$\frac{1}{(2^n)^p}+\ldots+\frac{1}{(2^{n+1}-1)^p}<\frac{1}{(2^{p-1})^n}$$

Adding the above inequalities, we have

$$s_{2^{n+1}-1}<1+\frac{1}{2^{p-1}}+\frac{1}{(2^{p-1})^2}+\ldots\ldots+\frac{1}{(2^{p-1})^n}$$

In the above inequality the right hand side is a geometrical series with common ratio $\frac{1}{2^{p-1}}<1$

$$s_{2^{n+1}-1}<\frac{1\left[1-\frac{1}{(2^{p-1})^{n+1}}\right]}{1-\frac{1}{2^{p-1}}} \qquad \left[\because s_n=\frac{a(1-r^n)}{1-r}\right]$$

$$s_{2^{n+1}-1}<\frac{1}{1-\frac{1}{2^{p-1}}}$$

or $s_{2^{n+1}-1}<k$ where $k=\frac{1}{1-\frac{1}{2^{p-1}}}$ is fixed

But $s_n<s_{2^{n+1}-1}$ for all n

$s_n<k$ for all n

$\Rightarrow$ $<s_n>$ is bounded above sequence. But every positive terms sequence which is bounded above is convergent.

$\Rightarrow$ $<s_n>$ is convergent and hence the given series $\sum_{n=1}^{\infty}\frac{1}{n^p}$ is convergent.

Case II

If $p = 1$ in this case we have the series

$$\sum_{n=1}^{\infty}\frac{1}{n^p} = 1+\frac{1}{2}+\frac{1}{3}+\ldots+\frac{1}{n}+\ldots$$

$$1+\frac{1}{2}>\frac{1}{2}$$

$$\frac{1}{3}+\frac{1}{4}>\frac{1}{4}+\frac{1}{4} = \frac{2}{4}=\frac{1}{2}$$

$$\frac{1}{5}+\frac{1}{6}+\frac{1}{7}+\frac{1}{8}>\frac{1}{8}+\frac{1}{8}+\frac{1}{8}+\frac{1}{8}=\frac{4}{8}=\frac{1}{2}$$

we have

$$1+\frac{1}{2}>\frac{1}{2}$$

$$\frac{1}{3}+\frac{1}{4}>\frac{1}{2}$$

$$\frac{1}{5}+\frac{1}{6}+\frac{1}{7}+\frac{1}{8}>\frac{1}{2}$$

...

...

$$\frac{1}{2^{n-1}+1}+\ldots+\frac{1}{2^n}>\frac{1}{2}$$

adding the above inequality, we have

$$s_{2^n}>\frac{1}{2}+\frac{1}{2}+\frac{1}{2}+\ldots+n \text{ terms}$$

$$s_{2^n}>\frac{n}{2}$$

$\Rightarrow$ sequence $< s_{2^n} >$ is not bounded above as n can take sufficient large values.

Hence the series $\sum_{n=1}^{\infty}\frac{1}{n}$ is divergent to $+\infty$.

Case III

If $p < 1$

Here $n^p < n$ for all n or $\frac{1}{n^p}>\frac{1}{n}$ for all n.

By case II, $\sum_{n=1}^{\infty}\frac{1}{n}$ is also divergent.

$\therefore$ By comparison test $\sum_{n=1}^{\infty}\frac{1}{n^p}$ is also divergent.

This proves the result.

Note :

(a) If $\lim_{n\to\infty} a_n \neq 0$ then the positive terms series $\sum_{n=1}^{\infty} a_n$ diverges to $+\infty$

(b) If $\lim_{n\to\infty} a_n = 0$, then there is no conclusion about the behaviour of positive terms series $\sum_{n=1}^{\infty} a_n$.

For example, if $a_n = \frac{1}{n}$, then $\sum_{n=1}^{\infty} a_n$ is divergent by p - test and

if $a_n = \frac{1}{n^2}$, then $\sum_{n=1}^{\infty} a_n$ is convergent by p - test.

But in both cases, $\lim_{n\to\infty} a_n = 0$.

Illustrative Examples

Example 1 :

Test the convergence of the series

(i) $\frac{1}{1\cdot 2}+\frac{1}{2\cdot 3}+\frac{1}{3\cdot 4}+\ldots\infty$ (ii) $\frac{1}{1\cdot 3}+\frac{2}{3\cdot 5}+\frac{3}{5\cdot 7}+\ldots\infty$

(iii) $\frac{1}{\sqrt{1\cdot 2}}+\frac{1}{\sqrt{2\cdot 3}}+\frac{1}{\sqrt{3\cdot 4}}+\ldots\infty$

Solution :

(i) $\frac{1}{1\cdot 2}+\frac{1}{2\cdot 3}+\frac{1}{3\cdot 4}+\ldots\infty$

This series can be written as

$$\sum a_n = \sum\frac{1}{n(n+1)}$$

$$a_n = \frac{1}{n^2\left(1+\frac{1}{n}\right)}$$

Let $$b_n = \frac{1}{n^2}$$

$$\Rightarrow \quad \lim_{n\to\infty}\frac{a_n}{b_n} = \lim_{n\to\infty}\frac{\frac{1}{n^2(1+\frac{1}{n})}}{\frac{1}{n^2}}$$

$$= \lim_{n\to\infty}\frac{1}{1+\frac{1}{n}} = 1 \qquad \text{non zero and finite}$$

By comparison test, series $\sum a_n$ and $\sum b_n$ converges or diverges together.

But $\sum b_n = \sum \frac{1}{n^2}$ is of the form $\sum \frac{1}{n^p}$.

Here $p = 2 > 1$

$\therefore$ $\sum b_n$ is convergent.

$\Rightarrow$ $\sum a_n$ is convergent.

(ii) $\frac{1}{1.3}+\frac{2}{3.5}+\frac{3}{5.7}+\ldots\infty$

Here $$a_n = \frac{n}{(2n-1)(2n+1)} = \frac{n}{4n^2\left(1-\frac{1}{2n}\right)\left(1+\frac{1}{2n}\right)}$$

$$= \frac{1}{4n\left(1-\frac{1}{2n}\right)\left(1+\frac{1}{2n}\right)}$$

let $$b_n = \frac{1}{n}$$

$$\lim_{n\to\infty}\frac{a_n}{b_n} = \lim_{n\to\infty}\frac{1}{4}\left(\frac{1}{(1-\frac{1}{2n})(1+\frac{1}{2n})}\right) = \frac{1}{4}(\textit{finite})$$

Both series converge or diverge together.

but $\sum b_n = \sum \frac{1}{n}$ compare with $\frac{1}{n^p}$

we have $p = 1$

$\therefore$ $\sum b_n$ is divergent.

Hence given series is divergent.

(iii) $\frac{1}{\sqrt{1.2}} + \frac{1}{\sqrt{2.3}} + \frac{1}{\sqrt{3.4}} + \ldots\infty$

Here $a_n = \frac{1}{\sqrt{n(n+1)}} = \frac{1}{n\sqrt{1+\frac{1}{n}}}$

let $b_n = \frac{1}{n}$

$$\lim_{n\to\infty} \frac{a_n}{b_n} = \lim_{n\to\infty} \frac{1}{1+\frac{1}{n}} = 1$$

$\therefore$ $\sum a_n$ and $\sum b_n$ both converge or diverge together.

But $\sum b_n = \sum \frac{1}{n}$ is divergent (since $p = 1$)

Hence given series is divergent.

Example 2 :

Check the convergence of the series.

$$\frac{2}{1^p} + \frac{3}{2^p} + \frac{4}{3^p} + \ldots\infty$$

Solution :

Here $a_n = \frac{n+1}{n^p} = \frac{n\left(1+\frac{1}{n}\right)}{n^p}$

$$= \frac{\left(1+\frac{1}{n}\right)}{n^{p-1}}$$

Let $b_n = \frac{1}{n^{p-1}}$

$$\lim_{n\to\infty}\frac{a_n}{b_n} = \lim_{n\to\infty}\left(1+\frac{1}{n}\right) = 1 \text{ (finite \& non-zero)}$$

$\therefore \sum a_n$ and $\sum b_n$ converge or diverge together.

But $\sum b_n = \sum \frac{1}{n^{p-1}}$ is convergent for $p-1>1$ (i.e $p>2$) and

divergent for $p-1\le 1 \Rightarrow p\le 2$

Hence given series is convergent for $p>2$ & divergent for $p\le 2$.

Example 3:

Check the convergence of the series $\sum_{n=1}^{\infty}\frac{n+1}{n(2n-1)}$

Solution :

Here
$$a_n = \frac{n+1}{n(2n-1)} = \frac{n\left(1+\frac{1}{n}\right)}{n^2\left(2-\frac{1}{n}\right)} = \frac{1+\frac{1}{n}}{n\left(2-\frac{1}{n}\right)}$$

let
$$b_n = \frac{1}{n}$$

$$\therefore \quad \lim_{n\to\infty}\frac{a_n}{b_n} = \lim_{n\to\infty}\frac{1+\frac{1}{n}}{2-\frac{1}{n}} = \frac{1}{2} \text{ (finite \& non-zero)}$$

By comparison test $\sum a_n$ and $\sum b_n$ converges or diverges together.

But $\sum b_n = \frac{1}{n}$ is divergent $(\because p=1)$ by p-test

So given series is divergent.

Example 4 :

Test the convergence of the following series.

(i) $\sum_{n=1}^{\infty}\frac{1}{n^p(n+1)^p}$ (ii) $\frac{2^p}{1^q}+\frac{3^p}{2^q}+\frac{4^p}{3^q}+\ldots\infty$

(iii) $\sum_{n=1}^{\infty} \frac{\sqrt{n}}{n^2+1}$ (iv) $\sum_{n=1}^{\infty} \frac{\sqrt{n^2-1}}{n^3+1}$

Solution :

(i) $\sum_{n=1}^{\infty} \frac{1}{n^p (n+1)^p}$

Here $$a_n = \frac{1}{n^p (n+1)^p} = \frac{1}{n^p . n^p \left(1+\frac{1}{n}\right)^p}$$

$$= \frac{1}{n^{2p} \; 1+\frac{1}{n}^{p}}$$

Let $$b_n = \frac{1}{n^{2p}}$$

$\therefore$ $$\lim_{n\to\infty} \frac{a_n}{b_n} = \lim_{n\to\infty} \frac{\frac{1}{n^{2p}\left(1+\frac{1}{n}\right)^p}}{\frac{1}{n^{2p}}} = 1 \text{ (finite\& non-zero)}$$

By comparison test, $\sum a_n$ and $\sum b_n$ both converge or diverge together.

But $\sum b_n = \sum_n \frac{1}{n^{2p}}$ is convergent if $2p > 1 \Rightarrow p > \frac{1}{2}$

and divergent if $2p \le 1$ $\left(i.e\ p \le \frac{1}{2}\right)$

Hence given series is convergent if $p > \frac{1}{2}$ and divergent if $p \le \frac{1}{2}$.

(ii) $\frac{2^p}{1^q} + \frac{3^p}{2^q} + \frac{4^p}{3^q} + \ldots\infty$

Here $$a_n = \frac{(n+1)^p}{n^q}$$

$$= \frac{n^p \left(1+\frac{1}{n}\right)^p}{n^q} = \frac{\left(1+\frac{1}{n}\right)^p}{n^{q-p}}$$

Let $$b_n = \frac{1}{n^{q-p}}$$

$$\lim_{n\to\infty}\frac{a_n}{b_n} = \lim_{n\to\infty}\left(1+\frac{1}{n}\right)^p = 1 \text{ (finite \& non-zero)}$$

By comparison test $\sum a_n$ and $\sum b_n$ converge or diverge together.

But $$\sum b_n = \sum\frac{1}{n^{q-p}} \text{ is convergent if } q - p > 1 \Rightarrow q > p+1$$

and divergent if $q-p\le 1 \quad \Rightarrow \quad q\le p+1$

Hence given series is convergent if $q > p+1$ and divergent if $q\le p+1$.

(iii) $\displaystyle\sum_{n=1}^{\infty}\frac{\sqrt{n}}{n^2+1}$

Here $$a_n = \frac{\sqrt{n}}{n^2+1} = \frac{n^{\frac{1}{2}}}{n^2\left(1+\frac{1}{n^2}\right)} = \frac{1}{n^{\frac{3}{2}}\left(1+\frac{1}{n^2}\right)}$$

Let $$b_n = \frac{1}{n^{3/2}}$$

$$\lim_{n\to\infty}\frac{a_n}{b_n} = \lim_{n\to\infty}\frac{1}{\left(1+\frac{1}{n^2}\right)} = 1 \text{ (finite \& non-zero)}$$

By comparison test $\sum a_n$ and $\sum b_n$ both series converge or diverge together.

But $\sum b_n = \dfrac{1}{n^{\frac{3}{2}}}$ is a convergent series. $\left(\because\ p = \frac{3}{2}\right)$

Hence given series is convergent.

(iv) $\displaystyle\sum\frac{\sqrt{n^2-1}}{n^3+1}$

Here $$a_n = \frac{\sqrt{n^2-1}}{n^3+1} = \frac{n\sqrt{1-\frac{1}{n^2}}}{n^3\left(1+\frac{1}{n^3}\right)} = \frac{\sqrt{1-\frac{1}{n^2}}}{n^2\left(1+\frac{1}{n^3}\right)}$$

$$b_n = \frac{1}{n^2}$$

$$\therefore \quad \lim_{n\to\infty}\frac{a_n}{b_n} = \lim_{n\to\infty}\frac{\sqrt{1-\frac{1}{n^2}}}{1+\frac{1}{n^3}} = 1 \text{ (finite \& non-zero)}$$

$\sum a_n$ and $\sum b_n$ both converge or diverge together.

But $\sum b_n = \frac{1}{n^2}$ is convergent by p - test.

Hence $\sum a_n$ is convergent.

Example 5:

Check the convergence of the given series :

(a) $\frac{1}{5}+\frac{\sqrt{2}}{7}+\frac{\sqrt{3}}{9}+\frac{\sqrt{4}}{11}+\ldots\infty$ (b) $\frac{1}{1.2.3}+\frac{1}{2.3.4}+\frac{1}{3.4.5}+\ldots\infty$

Solution:

(a) $\frac{1}{5}+\frac{\sqrt{2}}{7}+\frac{\sqrt{3}}{9}+\frac{\sqrt{4}}{11}+\ldots.\infty$

Here $$a_n = \frac{\sqrt{n}}{2n+3} = \frac{n^{\frac{1}{2}}}{2n+3} = \frac{n^{\frac{1}{2}}}{n\left(2+\frac{3}{n}\right)}$$

$$= \frac{1}{n^{\frac{1}{2}}\left(2+\frac{3}{n}\right)}$$

Let $$b_n = \frac{1}{n^{\frac{1}{2}}}$$

$$\lim_{n\to\infty}\frac{a_n}{b_n} = \lim_{n\to\infty}\frac{1}{2+\frac{3}{n}} = \frac{1}{2} \text{ (finite \& non-zero)}$$

$\sum a_n$ and $\sum b_n$ both converge or diverge together.

But $\quad \sum b_n = \sum \frac{1}{n^{\frac{1}{2}}}$ is divergent (since $p = \frac{1}{2} < 1$)

Hence given series is divergent.

(b) $\frac{1}{1.2.3} + \frac{1}{2.3.4} + \frac{1}{3.4.5} + \ldots\infty$

Here $\quad a_n = \frac{1}{n(n+1)(n+2)} = \frac{1}{n^3\left(1+\frac{1}{n}\right)\left(1+\frac{2}{n}\right)}$

and let $\quad b_n = \frac{1}{n^3}$

now $\quad \lim_{n\to\infty} \frac{a_n}{b_n} = \lim_{n\to\infty} \frac{1}{\left(1+\frac{1}{n}\right)\left(1+\frac{2}{n}\right)} = 1$ (finite & non-zero)

$\sum a_n$ and $\sum b_n$ converge or diverge together.

But $\quad \sum b_n = \frac{1}{n^3}$ is convergent since $p = 3 > 1$, by p - test.

Hence given series is convergent.

Example 6 :

Check the convergence of the series $\sum\left(\sqrt{n^2+1}-n\right)$

Solution :

Here
$$a_n = \sum\left(\sqrt{n^2+1}-n\right)$$
$$= \sqrt{n^2+1}-n \times \frac{\sqrt{n^2+1}+n}{\sqrt{n^2+1}+n}$$
$$= \frac{1}{\sqrt{n^2+1}+n} = \frac{1}{n\left(\sqrt{1+\frac{1}{n}}+1\right)}$$

Let $\quad b_n = \frac{1}{n}$

$$\lim_{n\to\infty}\frac{a_n}{b_n} = \lim_{n\to\infty}\frac{1}{\left(\sqrt{1+\frac{1}{n^2}}+1\right)} = \frac{1}{2}$$

By comparison test $\sum a_n$ and $\sum b_n$ converge or diverge together.

But $\sum b_n = \frac{1}{n}$ is divergent (since $p = 1$ by p test)

Hence given series is divergent.

Example 7:

Check the convergence of the series :

$$\sum\left(\sqrt{n^4+1}-\sqrt{n^4-1}\right)$$

Solution :

$$a_n = \sqrt{n^4+1}-\sqrt{n^4-1}$$

$$= \sqrt{n^4+1}-\sqrt{n^4-1}\times\frac{\sqrt{n^4+1}+\sqrt{n^4-1}}{\sqrt{n^4+1}+\sqrt{n^4-1}}$$

$$= \frac{n^4+1-n^4+1}{\sqrt{n^4+1}+\sqrt{n^4-1}}$$

$$= \frac{2}{\sqrt{n^4+1}+\sqrt{n^4-1}}$$

$$= \frac{2}{n^2[\sqrt{1+\frac{1}{n^2}}+\sqrt{1-\frac{1}{n^2}}]}$$

Let $b_n = \frac{1}{n^2}$.

$$\lim_{n\to\infty}\frac{a_n}{b_n} = \lim_{n\to\infty}\frac{2}{\sqrt{1+\frac{1}{n^4}}+\sqrt{1-\frac{1}{n^4}}} = 1 \text{ (finite \& non-zero)}$$

$\Rightarrow$ $\sum a_n$ and $\sum b_n$ both converge or diverge together.

But $\sum b_n = \frac{1}{n^2}$ is convergent by *p*-series test.

Hence given series is convergent.

Example 8:

Test the convergence of the series

$$\frac{\sqrt{2}-\sqrt{1}}{1}+\frac{\sqrt{3}-\sqrt{2}}{2}+\frac{\sqrt{4}-\sqrt{3}}{3}+\ldots$$

Solution :

Here $a_n = \frac{\sqrt{n+1}-\sqrt{n}}{n}$

$$= \frac{\sqrt{n+1}-\sqrt{n}}{n}\times\frac{\sqrt{n+1}+\sqrt{n}}{\sqrt{n+1}+\sqrt{n}}$$

$$= \frac{1}{n}\left(\frac{1}{\sqrt{n+1}+\sqrt{n}}\right)=\frac{1}{n^{\frac{3}{2}}\left(\sqrt{1+\frac{1}{n}}+1\right)}$$

Let $b_n = \frac{1}{n^{\frac{3}{2}}}$

Now,

$$\lim_{n\to\infty}\frac{a_n}{b_n}=\frac{1}{\sqrt{1+\frac{1}{n}}+1}=\frac{1}{2}(finite\ \&\ non-zero)$$

By comparison test both $\sum a_n$ and $\sum b_n$ converge or diverge together.

But $\sum b_n = \frac{1}{n^{3/2}}$ is convergent as $p = \frac{3}{2} > 1$

Hence given series is convergent.

Example 9 :

Test the convergence of the series

$$\sum\{\sqrt[3]{n+1}-\sqrt[3]{n}\}$$

Solution :

Here $a_n = (n+1)^{1/3} - (n)^{1/3}$

$$= n^{\frac{1}{3}}\left(1+\frac{1}{n}\right)^{\frac{1}{3}} - n^{\frac{1}{3}}$$

$$= n^{\frac{1}{3}}\left[1+\frac{1}{3}.\frac{1}{n}+\frac{\frac{1}{3}\left(\frac{1}{3}-1\right)}{2!}.\frac{1}{n^2}+...\right]-n^{\frac{1}{3}}$$

$$= n^{\frac{1}{3}}\left[\frac{1}{3}.\frac{1}{n}-\frac{\frac{1}{3}\left(\frac{2}{3}\right)}{2!}.\frac{1}{n^2}+...\right]$$

$$= \frac{1}{n^{1-\frac{1}{3}}}\left[\frac{1}{3}-\frac{2}{9.2!}.\frac{1}{n}+...\right]$$

$$= \frac{1}{n^{\frac{2}{3}}}\left[\frac{1}{3}-\frac{2}{9.2!}.\frac{1}{n}+...\right]$$

Let $$b_n = \frac{1}{n^{\frac{2}{3}}}$$

$$\lim_{n\to\infty}\frac{a_n}{b_n} = \lim_{n\to\infty}\left[\frac{1}{3}-\frac{2}{9.2!}.\frac{1}{n}+...\right] = \frac{1}{3} \text{ (finite \& non-zero)}$$

$\sum a_n$ and $\sum b_n$ both converge or diverge together.

But $\sum b_n = \frac{1}{n^{\frac{2}{3}}}$ is divergent since $p = \frac{2}{3} < 1$

Hence given series is divergent.

Example 10:

Check the convergence of the following series :

(a) $\sum \frac{1}{\sqrt{n}} \sin\frac{1}{n}$ (b) $\sum \cot^{-1} n^2$

(c) $\sum\left(\frac{1}{n}-\log\frac{n+1}{n}\right)$ (d) $\sum_{n=1}^{\infty}\left(1-\cos\frac{\pi}{n}\right)$

Solution:

(a) Here
$$a_n = \frac{1}{\sqrt{n}}\sin\frac{1}{n}$$
$$= \frac{1}{\sqrt{n}}\left[\frac{1}{n}-\frac{1}{3!}.\left(\frac{1}{n}\right)^3+\frac{1}{5!}\left(\frac{1}{n}\right)^5 ...\right] \qquad \left[\because \sin x = x - \frac{x^3}{3!}+\frac{x^5}{5!}.....\right]$$
$$= \left[\frac{1}{n^{\frac{3}{2}}}-\frac{1}{3!n^{\frac{7}{2}}}+\frac{1}{5!}.\frac{1}{n^{\frac{11}{2}}}.....\right]$$
$$= \frac{1}{n^{\frac{3}{2}}}\left[1-\frac{1}{3!.n^2}+\frac{1}{5!n^4}...\right]$$

Let
$$b_n = \frac{1}{n^{\frac{3}{2}}}$$
$$\lim_{n\to\infty}\frac{a_n}{b_n} = 1 \neq 0 \quad \text{(a finite quantity)}$$

By comparison test both $\sum a_n$ and $\sum b_n$ converge or diverge together.

But $\sum b_n = \frac{1}{n^{\frac{3}{2}}}$ is convergent $\left(p=\frac{3}{2}>1\right)$

$\therefore$ a_n is also convergent.

(b)
$$\sum a_n = \sum \cot^{-1} n^2$$
$$a_n = \cot^{-1} n^2$$
$$= \tan^{-1}\frac{1}{n^2} \qquad \left[\because \tan^{-1}x = x-\frac{x^3}{3}+\frac{x^5}{5}-\frac{x^7}{7}+...\right]$$
$$= \frac{1}{n^2}-\frac{1}{3}\left(\frac{1}{n^2}\right)^3+\frac{1}{5}\left(\frac{1}{n^2}\right)^5-\frac{1}{7}\left(\frac{1}{n^2}\right)^7+....$$

Let $$b_n = \frac{1}{n^2}$$

$$\lim_{n\to\infty} \frac{a_n}{b_n} = 1$$

By comparison test both a_n and b_n converge or diverge together.

But $b_n = \frac{1}{n^2}$ is convergent as $p = 2 > 1$ by *p*-test

$\Rightarrow \sum a_n$ is convergent.

(c)
$$a_n = \frac{1}{n} - \log\frac{n+1}{n}$$

$$= \frac{1}{n} - \log\left(1+\frac{1}{n}\right)$$

$$= \frac{1}{n} - \left(\frac{1}{n} - \frac{1}{2n^2} + \frac{1}{3n^3} - \frac{1}{4n^4} \cdots\right)$$

$$= \frac{1}{2n^2} - \frac{1}{3n^3} + \frac{1}{4n^4} \cdots$$

$$= \frac{1}{n^2}\left[\frac{1}{2} - \frac{1}{3n} + \frac{1}{4n^2} \cdots\right]$$

Let $$b_n = \frac{1}{n^2}$$

$$\lim_{n\to\infty} \frac{a_n}{b_n} = \frac{1}{2} \text{ (finite \& non-zero)}$$

$\sum a_n$ and $\sum b_n$ both converge or diverge together.

But $\sum b_n = \frac{1}{n^2}$ is convergent as $p = 2 > 1$

so a_n is also convergent.

(d)
$$\sum a_n = \sum\left(1 - \cos\frac{\pi}{n}\right)$$

$$a_n = 1 - \cos\frac{\pi}{n}$$

$$= 1 - \left(1 - \frac{\pi^2}{\underline{|2} n^2} + \frac{1\pi^4}{\underline{|4} n^4} - \frac{1}{\underline{|6}} \cdot \frac{\pi^6}{n^6} \ldots\ldots\right)$$

$$= \frac{\pi^2}{\lfloor 2n^2} - \frac{\pi^4}{\lfloor 4n^4} + \frac{\pi^6}{\lfloor 6n^6} \ldots$$

$$= \frac{1}{n^2}\left[\frac{\pi^2}{\lfloor 2} - \frac{\pi^4}{\lfloor 4n^2} + \frac{\pi^6}{\lfloor 6n^4} \ldots\right]$$

let $b_n = \dfrac{1}{n^2}$

$$\lim_{n\to\infty}\frac{a_n}{b_n} = \lim_{n\to\infty}\left[\frac{\pi^2}{\lfloor 2} - \frac{\pi^4}{\lfloor 4n^2} + \frac{\pi^6}{\lfloor 6n^4} \ldots\right] = \frac{\pi^2}{2!}$$

$\sum a_n$ and $\sum b_n$ both converge or diverge together.

But $b_n = \dfrac{1}{n^2}$ is convergent $\because p = 2 > 1$

$\therefore$ $\sum a_n$ is a convergent series.

Example 11:

Check the convergence of the series

$$\sum \frac{x^{n-1}}{1+x^n}, \quad x > 0$$

Solution:

Let, $a_n = \dfrac{x^{n-1}}{1+x^n}$

To check the convergence, we have the following cases :

<u>Case I</u>

$0 < x < 1$

Then $x^n \to 0$ as $n \to \infty$

let $b_n = x^{n-1}$

$$\lim_{n\to\infty}\frac{a_n}{b_n} = \lim_{n\to\infty}\frac{1}{1+x^n} = \frac{1}{1+0} = 1 \qquad \text{(which is finite and non -zero)}$$

$\sum a_n$ and $\sum b_n$ both converge or diverge together.

But $\sum b_n = \sum x^{n-1}$ is a G.P series with common ratio $x < 1$

$\therefore$ $\sum b_n$ is convergent.

$\therefore \sum a_n$ is convergent.

Case II

If $x = 1$

$$\lim_{n\to\infty} a_n = \frac{1}{1+1} = \frac{1}{2} \text{ (non-zero and finite)}$$

If $\lim_{n\to\infty} a_n \neq 0$ then $\sum a_n$ do not converge.

$\Rightarrow \sum a_n$ is a divergent series.

Case III

If $x > 1$ then $0 < \frac{1}{x} < 1$ and

$\frac{1}{x_n} \to 0$ as $n \to \infty$

$$\lim_{n\to\infty} a_n = \lim_{n\to\infty} \frac{x^n}{x^n . x\left(1+\frac{1}{x^n}\right)} = \lim_{n\to\infty} \frac{1}{x\left(1+\frac{1}{x^n}\right)} = \frac{1}{x} \neq 0$$

Hence series is divergent.

Thus the given series is convergent if $x < 1$,

divergent if $x \geq 1$.

Exercise 1.2

Test the convergence or divergence of the following series :

1. $\frac{1}{2}+\frac{1}{3^2}+\frac{1}{2^3}+\frac{1}{3^4}+\ldots\infty$

2. $\frac{1}{1.4}+\frac{1}{2.5}+\frac{1}{3.6}+\ldots\infty$

3. $\frac{1}{1.2}+\frac{2}{3.4}+\frac{3}{5.6}+\ldots\infty$

4. $\frac{3}{5}+\frac{4}{5^2}+\frac{3}{5^3}+\frac{4}{5^4}+\ldots\infty$

5. $1+\frac{1}{4^{2/3}}+\frac{1}{9^{2/3}}+\frac{1}{16^{2/3}}+\ldots\infty$

6. $\frac{1}{1.2.3}+\frac{3}{2.3.4}+\frac{5}{3.4.5}+\ldots$

7. $\frac{1}{\sqrt{1}+\sqrt{2}}+\frac{1}{\sqrt{2}+\sqrt{3}}+\frac{1}{\sqrt{3}+\sqrt{4}}+\ldots$

8. $\sqrt{\frac{1}{4}}+\sqrt{\frac{2}{6}}+\sqrt{\frac{3}{8}}+\ldots+\sqrt{\frac{n}{2(n+1)}}+\ldots$

9. $1+\frac{1}{2^2}+\frac{2^2}{3^3}+\frac{3^3}{4^4}+\frac{4^4}{5^5}+\ldots$

10. $\frac{\sqrt{2}-1}{3^3-1}+\frac{\sqrt{3}-1}{4^3-1}+\frac{\sqrt{4}-1}{5^3-1}+\ldots$

11. $\frac{1}{1^p}+\frac{1}{3^p}+\frac{1}{5^p}+\ldots$

12. $\sum\frac{2n^3+5}{4n^5+1}$

13. $\sum\left(\sqrt{n^3+1}-\sqrt{n^3}\right)$

14. $\sum\frac{\sqrt[3]{3n^2+1}}{\sqrt[4]{4n^3+2n+7}}$

15. $\frac{1}{1^2}+\frac{1+2}{1^2+2^2}+\frac{1+2+3}{1^2+2^2+3^2}+\ldots$

16. $\frac{1}{1+\sqrt{2}}+\frac{2}{1+2\sqrt{3}}+\frac{3}{1+3\sqrt{4}}+\ldots$

17. $\frac{1}{a(a+b)}+\frac{1}{(a+2b)(a+3b)}+\frac{1}{(a+4b)(a+5b)}+\ldots\infty,\ a>0,\ b>0$

18. $\sum_{n=1}^{\infty}\frac{\sqrt{n+1}-\sqrt{n}}{n^p}$

19. $\sum_{n=1}^{\infty} \frac{1}{n} \sin \frac{1}{n}$

20. $\sum_{n=1}^{\infty} \frac{1}{x^n + x^{-n}}, \; x > 0$

Answers

1. Convergent
2. Convergent
3. Divergent
4. Convergent
5. Convergent
6. Convergent
7. Divergent
8. Divergent
9. Divergent
10. Convergent
11. Convergent for $p > 1$, divergent for $p \leq 1$
12. Convergent
13. Convergent
14. Divergent
15. Divergent
16. Divergent
17. Convergent
18. Convergent if $p > \frac{1}{2}$ and divergent if $p \leq \frac{1}{2}$
19. Convergent
20. Converges for $x < 1$ and $x > 1$ but diverges for $x = 1$

1.16 D'Alembert's Ratio Test

If $\sum_{n=1}^{\infty} a_n$ is a series of positive terms such that $\lim_{n \to \infty} \frac{a_n}{a_{n+1}} = l$, then

(i) For, $l > 1$, the series is convergent.

(ii) For $l < 1$, the series is divergent.

(iii) For $l = 1$, no conclusion (*i.e.*, the test fails).

Illustrative Examples

Example 1:

Discuss the convergence of the following series :

(i) $1 + \frac{2!}{2^2} + \frac{3!}{3^3} + \frac{4!}{4^4} + \ldots\infty$

(ii) $\frac{1}{1+2} + \frac{2}{1+2^2} + \frac{3}{1+2^3} + \ldots\infty$

(iii) $\frac{2}{1}+\frac{2.5.8}{1.5.9}+\frac{2.5.8.11}{1.5.9.13}+\ldots\infty$ (iv) $\frac{2!}{3}+\frac{3!}{3^2}+\frac{4!}{3^3}+\ldots\infty$

(v) $\left(\frac{1}{3}\right)^2+\left(\frac{1.2}{3.5}\right)^2+\left(\frac{1.2.3}{3.5.7}\right)^2+\left(\frac{1.2.3.4}{3.5.7.9}\right)^2\ldots\infty$

Solution :

(i) $1+\frac{2!}{2^2}+\frac{3!}{3^3}+\frac{4!}{4^4}+\ldots\infty$

Here $a_n = \frac{n!}{n^n}$, $a_{n+1} = \frac{(n+1)!}{(n+1)^{n+1}}$

$$\lim_{n\to\infty}\frac{a_n}{a_{n+1}} = \lim_{n\to\infty}\frac{n!}{n^n}\times\frac{(n+1)^{n+1}}{(n+1)!}$$

$$= \lim_{n\to\infty}\frac{n!(n+1)^n\cdot(n+1)}{n^n(n+1)n!}$$

$$= \lim_{n\to\infty}\frac{(n+1)^n}{n^n}$$

$$= \lim_{n\to\infty}\frac{n^n(1+\frac{1}{n})^n}{n^n}$$

$$= \lim_{n\to\infty}\left(1+\frac{1}{n}\right)^n = e > 1$$

$\therefore$ $\sum a_n$ is convergent.

(ii) $\frac{1}{1+2}+\frac{2}{1+2^2}+\frac{3}{1+2^3}+\ldots\infty$

Here $a_n = \frac{n}{1+2^n}$, $a_{n+1} = \frac{(n+1)}{1+2^{(n+1)}}$

$$\lim_{n\to\infty}\frac{a_n}{a_{n+1}} = \lim_{n\to\infty}\frac{n}{1+2^n}\times\frac{1+2^{(n+1)}}{(n+1)}$$

$$= \lim_{n\to\infty}\frac{n\cdot 2^{n+1}\left(1+\frac{1}{2^{n+1}}\right)}{n\cdot 2^n\left(1+\frac{1}{2^n}\right)\left(1+\frac{1}{n}\right)}$$

$$= 2 > 1$$

$\therefore \sum a_n$ is convergent.

(iii) $\frac{2}{1} + \frac{2.5.8}{1.5.9} + \frac{2.5.8.11}{1.5.9.13} + \ldots\infty$

Here
$$a_n = \frac{2.5.8.11\ \ldots(3n-1)}{1.5.9.13\ \ldots[4(n-1)+1]}$$

$$a_{n+1} = \frac{2.5.8.11\ \ldots(3n-1)(3n+2}{1.5.9.13\ \ldots[4(n-1)+1][4n+1]}$$

$$\lim_{n\to\infty}\frac{a_n}{a_{n+1}} = \lim_{n\to\infty}\frac{(4n+1)}{3n+2} = \lim_{n\to\infty}\frac{n\left(4+\frac{1}{n}\right)}{n\left(3+\frac{2}{n}\right)}$$

$$= \frac{4}{3} > 1$$

$\therefore \sum a_n$ is convergent.

(iv) $\frac{2!}{3} + \frac{3!}{3^2} + \frac{4!}{3^3} + \ldots\infty$

Here
$$a_n = \frac{(n+1)!}{3^n} \qquad a_{n+1} = \frac{(n+2)!}{3^{n+1}}$$

$$\lim_{n\to\infty}\frac{a_n}{a_{n+1}} = \lim_{n\to\infty}\frac{(n+1)!}{3^n} \times \frac{3^n \cdot 3}{(n+2)(n+1)!}$$

$$= \lim_{n\to\infty}\frac{3}{(n+2)} = 0 < 1$$

$\therefore \sum a_n$ is divergent.

(v) $\left(\frac{1}{3}\right)^2 + \left(\frac{1.2}{3.5}\right)^2 + \left(\frac{1.2.3}{3.5.7}\right)^2 + \left(\frac{1.2.3.4}{3.5.7.9}\right)^2 \ldots\infty$

Here
$$a_n = \left[\frac{1.2.3.4\ \ldots\ n}{3.5.7.9\ \ldots\ (2n+1)}\right]^2$$

$$a_{n+1} = \left[\frac{1.2.3.4\ \ldots\ n(n+1)}{3.5.7.9\ \ldots\ (2n+1)(2n+3)}\right]^2$$

$$\lim_{n\to\infty}\frac{a_n}{a_{n+1}} = \lim_{n\to\infty}\left(\frac{(2n+3)}{(n+1)}\right)^2 = \lim_{n\to\infty}\frac{n^2}{n^2}\left(\frac{\left(2+\frac{3}{n}\right)}{1+\frac{1}{n}}\right)^2$$

$$= 4 > 1$$

$\therefore\ \sum a_n$ is convergent.

Example 2:

Discuss the convergence of the following series :

(i) $\sum \frac{2^{n-1}}{3^n+1}$ (ii) $\sum \frac{1}{n!}$

(iii) $\sum \frac{x^n}{3^n \cdot n^2}$, $x > 0$ (iv) $\sum \frac{x^n}{n}$, $x > 0$

(v) $\sum \frac{n}{n^2+1}\cdot x^n$, $x > 0$ (vi) $\sum\left(\frac{n^2}{2^n}+\frac{1}{n^2}\right)$

Solution :

(i) $\sum \frac{2^{n-1}}{3^n+1}$

Here $a_n = \frac{2^{n-1}}{3^n+1}$

and $a_{n+1} = \frac{2^n}{3^{n+1}+1}$

$$\lim_{n\to\infty}\frac{a_n}{a_{n+1}} = \lim_{n\to\infty}\frac{2^{n-1}}{3^n+1}\times\frac{3^{n+1}+1}{2^n}$$

$$= \lim_{n\to\infty}\frac{2^n\cdot 3^{n+1}\left(1+\frac{1}{3^{n+1}}\right)}{2\cdot 3^n\left(1+\frac{1}{3^n}\right)2^n}$$

$$= \lim_{n\to\infty}\frac{3}{2}\left(\frac{1+\frac{1}{3^{n+1}}}{1+\frac{1}{3^n}}\right) = \frac{3}{2} > 1$$

$\therefore\ \sum a_n$ is convergent.

(ii) $\sum \frac{1}{n!}$

Here $a_n = \frac{1}{n!}, \quad a_{n+1} = \frac{1}{(n+1)!}$

$$\lim_{n\to\infty} \frac{a_n}{a_{n+1}} = \lim_{n\to\infty} \frac{(n+1)n!}{n!}$$

$$= \lim_{n\to\infty}(n+1)$$

$$= \infty > 1$$

$\therefore \sum a_n$ is convergent.

(iii) $\sum \frac{x^n}{3^n \cdot n^2}, \quad x > 0$

Here, $a_n = \frac{x^n}{3^n \cdot n^2} \quad a_{n+1} = \frac{x^{n+1}}{3^{n+1}(n+1)^2}$

$$\lim_{n\to\infty} \frac{a_n}{a_{n+1}} = \lim_{n\to\infty} \frac{x^n}{3^n \cdot n^2} \times \frac{3^{n+1}(n+1)^2}{x^{n+1}}$$

$$= \lim_{n\to\infty} \frac{x^n \cdot 3^n \cdot 3 \cdot n^2 \left(1+\frac{1}{n}\right)^2}{3^n \cdot n^2 \cdot x^n \cdot x} = \frac{3}{x}$$

$\therefore \sum a_n$ is convergent if $\frac{3}{x} > 1$ *i.e.*, $3 > x$ and divergent if $\frac{3}{x} < 1$ *i.e.*, $3 < x$.

Test fails for $\frac{3}{x} = 1 \Rightarrow 3 = x$

If $x = 3$ then $a_n = \frac{3^n}{3^n \cdot n^2} = \frac{1}{n^2}$

$\sum \frac{1}{n^2}$ is convergent by *p*-test. Here $p = 2 > 1$.

Hence given series is convergent for $3 \geq x$ and divergent for $3 < x$.

(iv) $\sum \frac{x^n}{n}, \quad x > 0$

Here $a_n = \frac{x^n}{n}, \quad a_{n+1} = \frac{x^{n+1}}{(n+1)}$

$$\lim_{n\to\infty}\frac{a_n}{a_{n+1}} = \lim_{n\to\infty}\frac{x^n}{n}\times\frac{(n+1)}{x^{n+1}}$$

$$= \lim_{n\to\infty}\frac{1}{x}\left(1+\frac{1}{n}\right)=\frac{1}{x}$$

So convergent if $\frac{1}{x}>1$ i.e., $x < 1$

divergent if $\frac{1}{x}<1$ i.e., $x > 1$

test fails if $\frac{1}{x}=1$ i.e., $x = 1$

Put $x = 1$ in given series, then we get $\sum a_n = \sum\frac{1}{n}$ which is divergent by *p*-test. so given series is convergent if $x < 1$ and divergent if $x \geq 1$.

(v) $\sum\frac{n}{n^2+1}x^n,\ x>0$

Here $a_n = \frac{nx^n}{n^2+1}$ $a_{n+1} = \frac{(n+1)\cdot x^{n+1}}{(n+1)^2+1}$

$$\lim_{n\to\infty}\frac{a_n}{a_{n+1}} = \lim_{n\to\infty}\frac{n}{n^2+1}x^n\times\frac{(n+1)^2+1}{(n+1)x^{n+1}}$$

$$= \lim_{n\to\infty}\frac{n\cdot x^n\cdot n^2\left[\left(1+\frac{1}{n}\right)^2+\frac{1}{n^2}\right]}{n\cdot n^2\left(1+\frac{1}{n^2}\right)\left(1+\frac{1}{n}\right)x^n\cdot x}=\frac{1}{x}$$

By ratio test $\sum a_n$ is convergent if $\frac{1}{x}>1$ *i.e.*, $x < 1$ and divergent if $x > 1$, test fails if $x = 1$.

Put $x = 1$ in given series then

$$a_n = \frac{n}{n^2+1} = \frac{1}{n\left(1+\frac{1}{n^2}\right)}$$

Let $b_n = \frac{1}{n}$, which is divergent

Hence series is convergent if $x < 1$ and divergent for $x \geq 1$.

(vi) $\sum\left(\frac{n^2}{2^n}+\frac{1}{n^2}\right)$

$$\sum\left(\frac{n^2}{2^n}+\frac{1}{n^2}\right) = \sum a_n + \sum b_n$$

Here $\quad a_n = \dfrac{n^2}{2^n}, \quad a_{n+1} = \dfrac{(n+1)^2}{2^{n+1}}$

$$\therefore \quad \lim_{n\to\infty}\frac{a_n}{a_{n+1}} = \lim_{n\to\infty}\frac{n^2}{2^n}\times\frac{2^{n+1}}{(n+1)^2}$$

$$= \lim_{n\to\infty}\frac{n^2\cdot 2^n\cdot 2}{2^n\cdot n^2\left(1+\frac{1}{n^2}\right)^2}$$

$$= \lim_{n\to\infty}\frac{2}{\left(1+\frac{1}{n^2}\right)} = 2 > 1$$

Hence $\sum\frac{n^2}{2n}$ is convergent.

Also $\sum b_n = \sum\frac{1}{n^2}$ is convergent by *p*-test.

Hence given series is convergent.

Example 3 :

Discuss the convergence of the series $\sum\sqrt{\dfrac{n+1}{n^3+1}}\cdot x^n, \quad x > 0$

Solution :

Here $\quad a_n = \sqrt{\dfrac{n+1}{n^3+1}}\cdot x^n$

$$a_{n+1} = \sqrt{\frac{n+2}{(n+1)^3+1}}\cdot x^{n+1}$$

$$\lim_{n\to\infty}\frac{a_n}{a_{n+1}} = \lim_{n\to\infty}\sqrt{\frac{n+1}{n^3+1}}\cdot x^n \times \sqrt{\frac{(n+1)^3+1}{(n+2)}}\cdot\frac{1}{x^{n+1}} = \frac{1}{x}$$

$\sum a_n$ is convergent if $\frac{1}{x} > 1$ *i.e.,* $x < 1$ and divergent if $\frac{1}{x} < 1$, i.e., $x > 1$.

Test fails if $\frac{1}{x} = 1 \quad \Rightarrow x = 1$

Put $x = 1$ in given series $a_n = \sqrt{\dfrac{n+1}{n^3+1}} = \dfrac{1}{n}\sqrt{\dfrac{\left(1+\frac{1}{n}\right)}{\left(1+\frac{1}{n^3}\right)}}$

Let $\sum b_n = \sum \frac{1}{n}$ which is divergent by p-test.

Hence given series is convergent, for $x < 1$ and divergent for $x \geq 1$.

Example 4: Test the convergence of the series :

$x + 2x^2 + 3x^3 + 4x^4 + \ldots \infty$

Solution :

Here

$$a_n = nx^n$$

$$a_{n+1} = (n+1)\, x^{n+1}$$

$$\lim_{n\to\infty} \frac{a_n}{a_{n+1}} = \lim_{n\to\infty} \frac{nx^n}{(n+1)\, x^{n+1}} = \frac{1}{x}$$

by D'Alembert's ratio test, this series is convergent if $\frac{1}{x} > 1$ i.e., $x < 1$ and divergent

if $\frac{1}{x} < 1$ *i.e.,* $x > 1$, test fails if $\frac{1}{x} = 1$ *i.e.,* $x = 1$

Put $x = 1$ in given series

$$\sum a_n = \sum n, \quad \lim_{n\to\infty} a_n = \lim_{n\to\infty} n = \infty$$

$\therefore$ Series is divergent for $x = 1$

Hence given series is convergent for $x < 1$, divergent for $x \geq 1$.

Example 5:

$$x + \frac{3}{5}x^2 + \frac{8}{10}x^3 + \frac{15}{17}x^4 + \ldots \frac{n^2-1}{n^2+1}x^n,$$

Check the convergence of the series.

Solution :

Let

$$a_n = \frac{n^2-1}{n^2+1}x^n, \qquad a_{n+1} = \frac{(n+1)^2-1}{(n+1)^2+1}\cdot x^{n+1}$$

$$\lim_{n\to\infty} \frac{a_n}{a_{n+1}} = \lim_{n\to\infty} \frac{n^2-1}{n^2+1}\cdot x^n \times \frac{(n+1)^2+1}{(n+1)^2-1}\cdot\frac{1}{x^{n+1}}$$

$$= \frac{1}{x}$$

$\sum a_n$ is convergent if $\frac{1}{x} > 1$, *i.e.,* $x < 1$, divergent if $\frac{1}{x} < 1$ i.e., $x > 1$, test fails if $x = 1$.

Put x = 1 in given series, we get

$$a_n = \frac{n^2-1}{n^2+1}, \lim_{n\to\infty} a_n = \lim_{n\to\infty}\frac{n^2-1}{n^2+1} = 1$$

By the result for positive term series $\sum a_n$,If $\lim_{n\to\infty} a_n \neq 0, then \sum a_n$ is definitely divergent.

$\therefore$ $\sum a_n$ is divergent for x = 1

$\sum \frac{n^2-1}{n^2+1}\cdot x^n$ is convergent for $x < 1$, divergent for $x \geq 1$.

Example 6:

Test the convergence of $\sum \frac{3^n-2}{3^n+1}\cdot x^{n-1}$, $x > 0$

Solution :

Here $$a_n = \frac{3^n-2}{3^n+1}\cdot x^{n-1}$$

and $$a_{n+1} = \frac{3^{n+1}-2}{3^{n+1}+1}.x^n$$

$$\lim_{n\to\infty}\frac{a_n}{a_{n+1}} = \lim_{n\to\infty}\frac{3^n-2}{3^n+1}x^{n-1}\times\frac{3^{n+1}+1}{3^{n+1}-2}\cdot\frac{1}{x^n} = \frac{1}{x}$$

By ratio test $\sum a_n$ is convergent if $\frac{1}{x} > 1$ *i.e.*, $x < 1$ divergent if $\frac{1}{x} < 1$ *i.e.*, $x > 1$ and test fails if $\frac{1}{x} = 1$ *i.e*, $x = 1$.

Put x = 1 in given series then $a_n = \frac{3^n-2}{3^n+1}$

$$\lim_{n\to\infty} a_n = \lim_{n\to\infty}\frac{3^n\left(1-\frac{2}{3^n}\right)}{3^n\left(1+\frac{1}{3^n}\right)} = 1$$

For the positive term series $\sum a_n$, if $\lim_{n\to\infty} a_n \neq 0$ then $\sum a_n$ is divergent.

Hence given series is convergent if $x < 1$ and divergent if $x \geq 1$.

Example 7:

Test the convergence of $\sum_{n=1}^{\infty} \frac{n \cdot x^n}{(n+1)(n+2)}$

Solution :

Here $$a_n = \frac{n \cdot x^n}{(n+1)(n+2)}$$

$$a_{n+1} = \frac{(n+1)\, x^{n+1}}{(n+2)(n+3)}$$

$$\lim_{n\to\infty} \frac{a_n}{a_{n+1}} = \lim_{n\to\infty} \frac{n \cdot x^n}{(n+1)(n+2)} \times \frac{(n+2)(n+3)}{(n+1) \cdot x^{n+1}}$$

$$= \lim_{n\to\infty} \frac{n \cdot x^n (n^2)\left(1+\frac{2}{n}\right)\left(1+\frac{3}{n}\right)}{n^2\left(1+\frac{1}{n}\right)\left(1+\frac{2}{n}\right) n\left(1+\frac{1}{n}\right) \cdot x^n \cdot x}$$

$$= \frac{1}{x}$$

By ratio test series is convergent if $\frac{1}{x} > 1$ *i.e.,* $x < 1$ and divergent if $\frac{1}{x} < 1$ *i.e.,* $x > 1$ and test fails if $\frac{1}{x} = 1$

Now put $x = 1$ in given series

$$a_n = \frac{n}{(n+1)(n+2)} = \frac{1}{n\left[1+\frac{1}{n}\right]\left[1+\frac{2}{n}\right]}$$

Let $$b_n = \frac{1}{n}$$

then $$\lim_{n\to\infty} \frac{a_n}{b_n} = 1 \therefore$$ both the series converge or diverge together.

But $\sum b_n = \sum \frac{1}{n}$ is divergent by *p*-test

Hence given series is convergent if $x < 1$, divergent if $x \geq 1$.

Example 8: Test the convergence of the series

$$1 + \frac{x^2}{2} + \frac{x^4}{4} + \frac{x^6}{6} + \ldots \infty$$

Solution :

Neglecting I[st] term

$$a_n = \frac{x^{2n}}{2n} \qquad a_{n+1} = \frac{x^{2n+2}}{2n+2}$$

$$\lim_{n\to\infty} \frac{a_n}{a_{n+1}} = \lim_{n\to\infty} \frac{x^{2n}}{2n} \times \frac{(2n+2)}{x^{2n+2}}$$

$$= \lim_{n\to\infty} \frac{1}{x^2} \cdot \frac{(2n+2)}{2n}$$

$$= \lim_{n\to\infty} \frac{1}{x^2} \cdot \frac{2n\left(1+\frac{1}{n}\right)}{2n} = \frac{1}{x^2}$$

if $\frac{1}{x^2} > 1$, $x^2 < 1$ then convergent and if $\frac{1}{x^2} < 1$ or $x^2 > 1$, then divergent and if $\frac{1}{x^2} = 1$

i.e., $x^2 = 1$ then test fail.

Put $x^2 = 1$ in given series

$$a_n = \frac{1}{2n} \text{ is divergent.}$$

Hence given series is convergent for $x^2 < 1$ and divergent for $x^2 \geq 1$.

Example 9: Test the convergence of the series

$$\frac{4}{18} + \frac{4 \cdot 12}{18 \cdot 27} + \frac{4 \cdot 12 \cdot 20}{18 \cdot 27 \cdot 36} + \ldots \infty$$

Solution :

Here $$a_n = \frac{4 \cdot 12 \cdot 20 \ldots (8n-4)}{18 \cdot 27 \cdot 36 \ldots (9n+9)}$$

$$a_{n+1} = \frac{4 \cdot 12 \cdot 20 \ldots (8n-4)(8n+4)}{18 \cdot 27 \cdot 36 \ldots (9n+9)(9n+18)}$$

$$\lim_{n\to\infty} \frac{a_n}{a_{n+1}} = \lim_{n\to\infty} \frac{9n+18}{8n+4} = \frac{9}{8} > 1$$

$\therefore$ By ratio test $\sum a_n$ is convergent.

Exercise 1.3

Discuss the convergence of the following series :

1. $1+\frac{2^p}{2!}+\frac{3^p}{3!}+\frac{4^p}{4!}+\ldots,\ (p>0)$

2. $\frac{1}{2}+\frac{1}{3}+\frac{1}{5}\ldots+\frac{1}{2^{n-1}+1}+\ldots$

3. $\frac{1^2\cdot 2^2}{1!}+\frac{2^2\cdot 3^2}{2!}+\frac{3^2\cdot 4^2}{3!}+\frac{4^2\cdot 5^2}{4!}+\ldots\infty$

4. $1+\frac{2^2}{2!}+\frac{3^2}{3!}+\frac{4^2}{4!}+\ldots\infty$

5. $\sum\frac{n^3+a}{2^n+a}$

6. $\sum\frac{n!3^n}{n^n}$

7. $\sum\frac{n!}{n^n}$

8. $\sum\frac{n^3-n+1}{n!}$

9. $\frac{1}{1\cdot 2\cdot 3}+\frac{x}{4\cdot 5\cdot 6}+\frac{x^2}{7\cdot 8\cdot 9}+\ldots\infty,\ (x>0)$

10. $1+\frac{x}{2}+\frac{x^2}{5}+\frac{x^3}{10}+\ldots+\frac{x^n}{n^2+1}+\ldots\infty$

11. $\frac{x}{1\cdot 3}+\frac{x^2}{3\cdot 5}+\frac{x^3}{5\cdot 7}+\ldots$

12. $\frac{x}{2\sqrt{3}}+\frac{x^2}{3\sqrt{4}}+\frac{x^3}{4\sqrt{5}}+\ldots\infty$

13. $\sum_{n=1}^{\infty}\frac{n^2}{3^n}$

14. $\sum_{n=1}^{\infty}\frac{x^n}{(2n)!}$

15. $\frac{x}{1+\sqrt{1}}+\frac{x^2}{2+\sqrt{2}}+\frac{x^3}{3+\sqrt{3}}+\ldots\infty$

16. $\sum_{n=1}^{\infty}\frac{x^n}{(2n-1)^2\cdot 2^n}$

17. $1+\frac{3}{2!}+\frac{5}{3!}+\frac{7}{4!}+\ldots\infty$

18. $\sum_{n=1}^{\infty}\frac{x^{n+1}}{(n+1)\sqrt{n}}$

19. $1+\frac{1^2\cdot 2^2}{1\cdot 3\cdot 5}+\frac{1^2\cdot 2^2\cdot 3^2}{1\cdot 3\cdot 5\cdot 7\cdot 9}+\ldots\infty$

20. $\sum_{n=1}^{\infty} \frac{3\cdot 6\cdot 9 \ldots 3n}{4\cdot 7\cdot 10 \ldots (3n+1)} \cdot \frac{5^n}{3n+2}$

21. $\sum \frac{\sqrt{n}}{\sqrt{n^2+1}} x^n, \ (x > 0)$

22. $\frac{1}{2\sqrt{1}} + \frac{x^2}{3\sqrt{2}} + \frac{x^4}{4\sqrt{3}} + \frac{x^6}{5\sqrt{4}} + \ldots$

23. $1 + \frac{2}{5}x + \frac{6}{9}x^2 + \frac{14}{17}x^3 + \ldots + \frac{2^{n+1}-2}{2^{n+1}+1} \cdot x^n + \ldots (x > 0)$

24. $1 + \frac{\alpha+1}{\beta+1} + \frac{(\alpha+1)(2\alpha+1)}{(\beta+1)(2\beta+1)} + \frac{(\alpha+1)(2\alpha+1)(3\alpha+1)}{(\beta+1)(2\beta+1)(3\beta+1)} + \ldots$

Answers

1. Convergent
2. Convergent
3. Convergent
4. Convergent
5. Convergent
6. Divergent
7. Convergent
8. Convergent
9. Convergent if $x \le 1$, divergent if $x > 1$
10. Convergent for $x \le 1$, divergent for $x > 1$
11. Convergent for $x \le 1$, divergent for $x > 1$
12. Convergent for $x \le 1$, divergent if $x > 1$
13. Convergent
14. Convergent
15. Convergent if $x < 1$, divergent if $x \ge 1$
16. Convergent if $x \le 2$, divergent if $x > 2$
17. Convergent
18. Convergent for $x \le 1$, divergent if $x > 1$
19. Convergent
20. Divergent
21. Convergent if $x < 1$, divergent if $x \ge 1$
22. Convergent if $x^2 \le 1$, divergent if $x^2 > 1$
23. Convergent if $x < 1$, divergent if $x \ge 1$
24. Convergent if $\beta > \alpha > 0$, divergent if $\alpha \ge \beta > 0$

1.17 Cauchy's Root Test

If $\sum a_n$ is a positive term series and $\lim_{n\to\infty} (a_n)^{1/n} = l$ then

(i) $\sum a_n$ is convergent if $l < 1$

(ii) $\sum a_n$ is divergent if $l > 1$

If $l = 1$ the test fails i.e no conclusion can be drawn about the convergence or divergence of the series.

> **Note:**
>
> **This test is useful when a_n involves expression with n^{th} power.**
>
> **If this test fails, comparison test may be used.**

Illustrative Examples

Example 1:

Check the convergence of $\sum \frac{1}{n^n}$

Solution :

Here
$$\sum a_n = \sum \frac{1}{n^n}$$
$$a_n = \frac{1}{n^n}$$
$$\lim_{n\to\infty}(a_n)^{1/n} = \lim_{n\to\infty}\left(\frac{1}{n^n}\right)^{\frac{1}{n}} = \lim_{n\to\infty}\frac{1}{n} = 0 < 1$$

By Cauchy's test, this series is convergent.

Example 2 :

Test the convergence of $\sum_{n=2}^{\infty} \frac{1}{(\log n)^n}$

Solution :

Here
$$a_n = \frac{1}{(\log n)^n}$$
$$\lim_{n\to\infty}(a_n)^{\frac{1}{n}} = \lim_{n\to\infty}\frac{1}{\log n} < 1$$

$\therefore$ series is convergent.

Example 3 :

Test the convergence of $\sum\left(\frac{n+1}{3n}\right)^n$

Solution : *Here* $a_n = \left(\frac{n+1}{3n}\right)^n$

$$\lim_{n\to\infty}(a_n)^{\frac{1}{n}} = \lim_{n\to\infty}\frac{n+1}{3n} = \lim_{n\to\infty}\frac{n\left(1+\frac{1}{n}\right)}{3n}$$

$$= \frac{1}{3} < 1$$

∴ Series is convergent.

Example 4 :

Test the convergence of $\sum\frac{(n-\log n)^n}{2^n \cdot n^n}$

Solution : Here

$$a_n = \frac{(n-\log n)^n}{2^n . n^n}$$

$$\lim_{n\to\infty}(a_n)^{\frac{1}{n}} = \lim_{n\to\infty}\frac{n-\log n}{2\cdot n}$$

$$= \lim_{n\to\infty}\frac{1-\frac{\log n}{n}}{2} = \frac{1}{2}-\lim_{n\to\infty}\frac{\log n}{2n}$$

$$= \frac{1}{2}-\frac{1}{2}\lim_{n\to\infty}\frac{\frac{1}{n}}{1} \quad \left(\because \frac{\infty}{\infty} \text{ form using L' Hospital's rule}\right)$$

$$= \frac{1}{2}-0$$

$$= \frac{1}{2} < 1$$

∴ Series is convergent.

Example 5 :

Check the convergence of $\sum\left(1+\frac{1}{\sqrt{n}}\right)^{-n^{\frac{3}{2}}}$

Solution :

Here $$a_n = \left(1+\frac{1}{\sqrt{n}}\right)^{-n^{\frac{3}{2}}}$$

$$\lim_{n\to\infty}(a_n)^{\frac{1}{n}} = \lim_{n\to\infty}\left[\frac{1}{\left(1+\frac{1}{\sqrt{n}}\right)^{n\sqrt{n}}}\right]^{\frac{1}{n}}$$

$$= \lim_{n\to\infty}\frac{1}{\left(1+\frac{1}{\sqrt{n}}\right)^{\sqrt{n}}} = \frac{1}{e}$$

Which is less than 1 $\left(\because \frac{1}{e} < 1\right)$

$\therefore$ series is convergent.

Example 6 :

Test the convergence of $\sum\left(\frac{nx}{(n+1)}\right)^n$

Solution :

Here $$a_n = \left(\frac{nx}{(n+1)}\right)^n$$

$$\lim_{n\to\infty}(a_n)^{\frac{1}{n}} = \lim_{n\to\infty}\frac{nx}{n\left(1+\frac{1}{n}\right)}$$

$$= x$$

Series is convergent if $x < 1$,

divergent if $x > 1$

and test fails if $x = 1$

Put $x = 1$ in given series

$$a_n = \left(\frac{n}{(n+1)}\right)^n = \frac{1}{\left(1+\frac{1}{n}\right)^n}$$

$$\lim_{n\to\infty} a_n = \lim_{n\to\infty}\left(\frac{1}{\left(1+\frac{1}{n}\right)^n}\right) = \frac{1}{e} \neq 0$$

($\because$ if $\lim_{n\to\infty} a_n \neq 0$ then $\sum a_n$ is not convergent)

Hence $\sum a_n$ is divergent.

Given series is convergent for $x < 1$ and divergent for $x \geq 1$

Example 7 :

Test the convergence of $\sum 5^{-n-(-1)^n}$

Solution :

Here, $\sum a_n = \sum 5^{-n-(-1)^n}$

$$\lim_{n\to\infty}(a_n)^{\frac{1}{n}} = \lim_{n\to\infty}\left[\frac{1}{5^{n+(-1)^n}}\right]^{\frac{1}{n}} = \lim_{n\to\infty}\frac{1}{5^{\frac{n}{n}+\frac{(-1)^n}{n}}}$$

$$= \lim_{n\to\infty}\frac{1}{5\cdot 5^{\frac{(-1)^n}{n}}} = \frac{1}{5\cdot 5^0}$$

$$= \frac{1}{5} < 1$$

$\therefore$ Given series is convergent.

Example 8 :

Test the convergence of $\sum \frac{(x+2)^n}{3^n \cdot n}$

Solution: Here, $a_n = \dfrac{(x+2)^n}{3^n \cdot n}$

$$(a_n)^{\frac{1}{n}} = \frac{x+2}{3 \cdot n^{\frac{1}{n}}}$$

$$\lim_{n\to\infty}(a_n)^{\frac{1}{n}} = \lim_{n\to\infty}\frac{x+2}{3 \cdot n^{\frac{1}{n}}} = \frac{x+2}{3}$$

If $\dfrac{x+2}{3} < 1$, then $\sum a_n$ is convergent i.e $x < 1$ means convergent. $\left(\because \lim\limits_{n\to\infty} n^{\frac{1}{n}} = 1\right)$

If $\dfrac{x+2}{3} > 1$ *i.e* $x > 1$ means $\sum a_n$ is divergent.

If $\dfrac{x+2}{3} = 1$ *i.e* $x = 1$ Test fails. Put $x = 1$ in given series.

$$\sum \frac{3^n}{3^n \cdot n} = \sum \frac{1}{n} \text{ which is divergent.}$$

$\therefore$ Given series is convergent for $x < 1$ and divergent for $x \geq 1$.

Example 9 :

Check the convergence of $1 + \dfrac{x}{2} + \dfrac{x^2}{3^2} + \dfrac{x^3}{4^3} + \ldots.\ (x > 0)$

Solution :

Here $a_n = \dfrac{x^{n-1}}{n^{n-1}}$ *(neglecting Ist term)*

$$(a_n)^{\frac{1}{n}} = \frac{x^{\frac{n-1}{n}}}{n^{\frac{n-1}{n}}} = \frac{x^{1-\frac{1}{n}}}{n^{1-\frac{1}{n}}}$$

$$\lim_{n\to\infty}(a_n)^{\frac{1}{n}} = \lim_{n\to\infty}\frac{x \cdot x^{\frac{-1}{n}}}{n \cdot n^{\frac{-1}{n}}}$$

$$= \lim_{n\to\infty} \frac{x}{n} = 0 < 1$$

$\therefore$ Series is convergent.

Example 10 :

Test the convergence of $\sum \frac{(1+nx)^n}{n^n}$

Solution: *Here,* $$a_n = \left(\frac{(1+nx)^n}{n^n}\right)$$

$$\lim_{n\to\infty} (a_n)^{\frac{1}{n}} = \lim_{n\to\infty} \frac{1+nx}{n} = \lim_{n\to\infty} \frac{n\left(\frac{1}{n}+x\right)}{n}$$

$$= x$$

If $x < 1 \Rightarrow \sum a_n$ is convergent

$x > 1 \Rightarrow \sum a_n$ is divergent

$x = 1$ test fails.

Put $x = 1$ in given series

$$a_n = \left(\frac{1+n}{n}\right)^n = \frac{n^n\left(\frac{1}{n}+1\right)^n}{n^n}$$

$$\lim_{n\to\infty} a_n = \lim_{n\to\infty}\left(\frac{1}{n}+1\right)^n = e > 1 \quad \therefore \text{ divergent.}$$

Given series is convergent for $x < 1$ and divergent for $x \geq 1$.

Exercise 1.4

Test the convergence or divergence of the following series.

1. $\sum\left(\frac{n}{(n+1)}\right)^{n^2}$ or $\sum\left(1+\frac{1}{n}\right)^{-n^2}$

2. $\left(\frac{2^2}{1^2}-\frac{2}{1}\right)^{-1}+\left(\frac{3^3}{2^3}-\frac{3}{2}\right)^{-2}+\left(\frac{4^4}{3^4}-\frac{4}{3}\right)^{-3}+\ldots..$

3. $\frac{1}{2}+\frac{2}{3}x+\left(\frac{3}{4}\right)^2 x^2+\left(\frac{4}{5}\right)^3 x^3+\ldots\ldots \quad (x>0)$

4. $\sum_{n=1}^{\infty}\frac{(n+1)^n \cdot x^n}{n^{n+1}}$

5. $\frac{2x}{1^2}+\frac{3^2x^2}{2^3}+\frac{4^3x^3}{3^4}+\ldots.+\frac{(n+1)^n \cdot x^n}{n^{n+1}}+\ldots..$

Answers

1. Convergent
2. Convergent
3. Convergent if $x < 1$ and divergent if $x \geq 1$.
4. Convergent if $x < 1$ and divergent if $x \geq 1$.
5. Convergent if $x < 1$ and divergent if $x \geq 1$.

1.18 Raabe's test

If $\sum_{n=1}^{\infty} a_n$ is a series of positive terms such that $\lim_{n\to\infty} n\left[\frac{a_n}{a_{n+1}}-1\right] = l$

Then the series is

(i) Convergent if $l > 1$

(ii) Divergent if $l < 1$

(iii) Test fails if $l = 1$

Remark:

(a) Raabe's test is used when D' Alembert Ratio test fails and when $\lim_{n\to\infty}\frac{a_n}{a_{n+1}}$ does not involve the number e.

(b) Logarithmic test is used when Ratio test fails and $\lim_{n\to\infty}\frac{a_n}{a_{n+1}}$ contains e.

(c) Raabe's test is inconclusive when $\lim_{n\to\infty} n\left[\frac{a_n}{a_{n+1}}-1\right]=1$

1.19 Logarithmic Test

If the series $\sum_{n=1}^{\infty} a_n$ is positive term series such that $\lim_{n\to\infty} n \log \dfrac{a_n}{a_{n+1}} = l$ then the series is

(i) convergent if $l > 1$

(ii) divergent if $l < 1$

(iii) test fails if $l = 1$

1.20 Gauss Test

If $\sum_{n=1}^{\infty} a_n$ is a series of positive terms such that $\dfrac{a_n}{a_{n+1}}$ can be expanded in the form

$$\frac{a_n}{a_{n+1}} = 1+\frac{\lambda}{n}+0\left(\frac{1}{n^2}\right)$$

Then $\sum a_n$ converges if $\lambda > 1$ and diverges if $\lambda \le 1$.

Note:

(a) For application of Gauss test, expand $\dfrac{a_n}{a_{n+1}}$ in power of $\dfrac{1}{n}$ as $\dfrac{a_n}{a_{n+1}} = 1+\dfrac{\lambda}{n}+0\left(\dfrac{1}{n^2}\right)$, where $0\left(\dfrac{1}{n^2}\right)$ stands for the terms of higher power of $\dfrac{1}{n}$.

(b) The test never fails as we know that the series diverges for $\lambda = 1$. Moreover the test is applied after the failure of Ratio test and when it is possible to expand $\dfrac{a_n}{a_{n+1}}$ in power of $\dfrac{1}{n}$ by binomial theorem or by any other method.

Illustrative Examples

Example 1:

Discuss the convergence of the following series

(a) $1+\frac{2^2}{3^2}+\frac{2^2\cdot 4^2}{3^2\cdot 5^2}+\frac{2^2\cdot 4^2\cdot 6^2}{3^2\cdot 5^2\cdot 7^2}+\ldots\infty$

(b) $1+\frac{3}{7}x+\frac{3\cdot 6}{7\cdot 10}x^2+\frac{3\cdot 6\cdot 9}{7\cdot 10\cdot 13}x^3+\ldots\infty$

(c) $1+\frac{2}{1}\cdot\frac{1}{2}+\frac{2\cdot 4}{1\cdot 3}\cdot\frac{1}{3}+\frac{2\cdot 4\cdot 6}{1\cdot 3\cdot 5}\cdot\frac{1}{4}+\ldots\infty$

(d) $x+\frac{2^2x^2}{2!}+\frac{3^3x^3}{3!}+\frac{4^4x^4}{4!}+\frac{5^5x^5}{5!}+\ldots\infty$

(e) $1+\frac{1}{2}\cdot\frac{x^2}{4}+\frac{1\cdot 3\cdot 5}{2\cdot 4\cdot 6}\cdot\frac{x^4}{8}+\frac{1\cdot 3\cdot 5\cdot 7\cdot 9}{2\cdot 4\cdot 6\cdot 8\cdot 10}\cdot\frac{x^6}{12}+\ldots\infty$

Solution :

(a) We have , $1+\frac{2^2}{3^2}+\frac{2^2\cdot 4^2}{3^2\cdot 5^2}+\frac{2^2\cdot 4^2\cdot 6^2}{3^2\cdot 5^2\cdot 7^2}+\ldots\infty$

Neglecting 1st term

$$a_n = \frac{2^2\cdot 4^2\cdot 6^2\ldots(2n)^2}{3^2\cdot 5^2\cdot 7^2\ldots(2n+1)^2}$$

$$a_{n+1} = \frac{2^2\cdot 4^2\cdot 6^2\ldots(2n)^2(2n+2)^2}{3^2\cdot 5^2\cdot 7^2\ldots(2n+1)^2(2n+3)^2}$$

$$\lim_{n\to\infty}\frac{a_n}{a_{n+1}} = \lim_{n\to\infty}\frac{(2n+3)^2}{(2n+2)^2}=1$$

$\therefore$ D' Alembert's ratio test fail, Now we apply Raabe's test.

$$\frac{a_n}{a_{n+1}}-1 = \frac{(2n+3)^2}{(2n+2)^2}-1$$

$$= \frac{4n^2+9+12n-4n^2-4-8n}{(2n+2)^2}=\frac{4n+5}{(2n+2)^2}$$

$$\lim_{n\to\infty} n\left[\frac{a_n}{a_{n+1}}-1\right] = \lim_{n\to\infty}\frac{4+\frac{5}{n}}{\left(2+\frac{2}{2n}\right)^2}=1$$

$\therefore$ Raabe's test fails.

Now by Gauss test

$$\frac{a_n}{a_{n+1}} = \frac{(2n+3)^2}{(2n+2)^2}$$

$$= \left(1+\frac{3}{2n}\right)^2\left(1+\frac{2}{2n}\right)^{-2}$$

$$= \left(1+\frac{3}{n}+0\left(\frac{1}{n^2}\right)\right)\left(1-\frac{2}{n}+0\left(\frac{1}{n^2}\right)\right)$$

$$= 1+\frac{1}{n}+0\left(\frac{1}{n^2}\right)$$

Comparing with $1+\frac{\lambda}{n}+0\left(\frac{1}{n^2}\right) \quad \Rightarrow \lambda = 1$

$\therefore$ by Gauss test series is divergent.

(b) $1+\frac{3}{7}x+\frac{3\cdot 6}{7\cdot 10}x^2+\frac{3\cdot 6\cdot 9}{7\cdot 10\cdot 13}x^3+....\infty$

Neglecting 1st term

$$a_n = \frac{3\cdot 6\cdot 9....3n}{7\cdot 10\cdot 13....(3n+4)}x^n$$

$$a_{n+1} = \frac{3\cdot 6\cdot 9....(3n)(3n+3)}{7\cdot 10\cdot 13....(3n+4)(3n+7)}x^{n+1}$$

$$\lim_{n\to\infty}\frac{a_n}{a_{n+1}} = \lim_{n\to\infty}\frac{3n+7}{3n+3}\cdot\frac{1}{x}$$

$$= \frac{1}{x}$$

Series is convergent if $\frac{1}{x}>1 \quad$ *i.e* $x < 1$

divergent if $\frac{1}{x} < 1$ i.e $x > 1$

and test fails if $x = 1$

Now apply Raabe's test, put $x = 1$, we get

$$\frac{a_n}{a_{n+1}} - 1 = \frac{3n+7}{3n+3} - 1 = \frac{3n+7-3n-3}{3n+3}$$

$$\lim_{n\to\infty} n\left[\frac{a_n}{a_{n+1}} - 1\right] = \lim_{n\to\infty} \frac{4n}{3n+3} = \frac{4}{3} > 1 \quad \therefore \sum a_n \text{ convergent.}$$

Hence given series is convergent for $x \le 1$

and divergent for $x > 1$

(c) $1 + \frac{2}{1}\cdot\frac{1}{2} + \frac{2\cdot 4}{1\cdot 3}\cdot\frac{1}{3} + \frac{2\cdot 4\cdot 6}{1\cdot 3\cdot 5}\cdot\frac{1}{4} + \ldots\infty$

Neglecting first term

$$a_n = \frac{2\cdot 4\cdot 6 \ldots\ldots (2n)}{1\cdot 3\cdot 5 \ldots (2n-1)}\cdot\frac{1}{(n+1)}$$

$$a_{n+1} = \frac{2\cdot 4\cdot 6 \ldots\ldots (2n)(2n+2)}{1\cdot 3\cdot 5 \ldots (2n-1)(2n+1)} \times \frac{1}{(n+2)}$$

$$\lim_{n\to\infty} \frac{a_n}{a_{n+1}} = \lim_{n\to\infty} \frac{2n+1}{2n+2}\cdot\frac{n+2}{(n+1)} = 1$$

This shows D'Alembert's ratio test fail,

Now applying Raabe's test.

$$\left(\frac{a_n}{a_{n+1}} - 1\right) = \frac{n}{2n^2 + 4n + 2}$$

$$\lim_{n\to\infty} n\left[\frac{a_n}{a_{n+1}} - 1\right] = \frac{1}{2} < 1$$

By Raabe's test $\sum a_n$ is divergent.

(d) $x + \frac{2^2 x^2}{2!} + \frac{3^3 x^3}{3!} + \frac{4^4 x^4}{4!} + \frac{5^5 x^5}{5!} + \ldots\infty$

Here $$a_n = \sum \frac{n^n x^n}{n!}$$

$$a_{n+1} = \frac{(n+1)^{n+1} x^{n+1}}{(n+1)!}$$

$$\frac{a_n}{a_{n+1}} = \frac{n^n x^n}{n!} \times \frac{(n+1)!}{(n+1)^{n+1} x^{n+1}} = \frac{n^n}{(n+1)^n} \cdot \frac{1}{x}$$

$$\lim_{n\to\infty} \frac{a_n}{a_{n+1}} = \lim_{n\to\infty} \frac{1}{x} \left[\frac{1}{\left(1+\frac{1}{n}\right)^n} \right] = \frac{1}{ex}$$

series is convergent for $\frac{1}{ex} > 1$ *i.e* $ex < 1$

i.e $x < \frac{1}{e}$

divergent for $\frac{1}{ex} < 1$ *i.e* $x > \frac{1}{e}$

test fails for $x = \frac{1}{e}$

Since $\frac{a_n}{a_{n+1}}$ contains e, apply Logrithmic test.

Put $x = \frac{1}{e}$ in given series.

$$\frac{a_n}{a_{n+1}} = \frac{e}{\left(1+\frac{1}{n}\right)^n}$$

$$\log \frac{a_n}{a_{n+1}} = \log e - n \log \left(1+\frac{1}{n}\right)$$

$$= 1 - n\left(\frac{1}{n} - \frac{1}{2n^2} + \frac{1}{3n^3} \ldots\right)$$

$$\lim_{n\to\infty}\log\frac{a_n}{a_{n+1}} = \lim_{n\to\infty}\left(\frac{1}{2n}-\frac{1}{3n^2}+...\right)$$

$$\lim_{n\to\infty} n\log\frac{a_n}{a_{n+1}} = \lim_{n\to\infty} n[\frac{1}{2n}-\frac{1}{3n^2}+.........]$$

$$= \frac{1}{2} < 1$$

$\therefore \sum a_n$ is divergent for $x = \frac{1}{e}$

Hence given series is convergent for $x < \frac{1}{e}$ and divergent for $x \geq \frac{1}{e}$

(e) $1+\frac{1}{2}\frac{x^2}{4}+\frac{1\cdot3\cdot5}{2\cdot4\cdot6}\frac{x^4}{8}+\frac{1\cdot3\cdot5\cdot7\cdot9}{2\cdot4\cdot6\cdot8\cdot10}\frac{x^6}{12}+......$

Neglecting I[st] term

$$a_n = \frac{1\cdot3\cdot5......(4n-3)x^{2n}}{2\cdot4\cdot6.....(4n-2)\cdot4n}$$

$$a_{n+1} = \frac{1\cdot3\cdot5......(4n-3)(4n-1)(4n+1)x^{2n+2}}{2\cdot4\cdot6.....(4n-2)(4n)(4n+2)(4n+4)}$$

$$\frac{a_n}{a_{n+1}} = \frac{x^{2n}}{4n}\times\frac{4n(4n+2)}{(4n-1)(4n+1)}\frac{(4n+4)}{x^{2n+2}}$$

$$= \frac{1}{X^2}\frac{(4n+2)(4n+4)}{(4n-1)(4n+1)} = \frac{1}{x^2}\frac{\left(4+\frac{2}{n}\right)\left(4+\frac{4}{n}\right)}{\left(4-\frac{1}{n}\right)\left(4+\frac{1}{n}\right)}$$

$$\lim_{n\to\infty}\frac{a_n}{a_{n+1}} = \frac{1}{x^2}$$

If $\frac{1}{x^2} > 1$ or $x^2 < 1$, the series is convergent.

If $\frac{1}{x^2} < 1$ or $x^2 > 1$, the series is divergent.

If $\frac{1}{x^2} = 1$, the test fails

Put $\frac{1}{x^2} = 1$ in given series and apply Raabe's test

$$\frac{a_n}{a_{n+1}} = \frac{(4n+2)(4n+4)}{(4n-1)(4n+1)}$$

$$\frac{a_n}{a_{n+1}} - 1 = \frac{16n^2+24n+8}{(4n-1)(4n+1)} - 1 = \frac{24n+9}{(4n-1)(4n+1)}$$

$$\lim_{n\to\infty} n\left[\frac{a_n}{a_{n+1}} - 1\right] = \lim_{n\to\infty} n\left[\frac{24n+9}{(4n-1)(4n+1)}\right] = \frac{24}{16} = \frac{3}{2} > 1$$

$\therefore$ the series is convergent

$\Rightarrow$ so the series is convergent for $x^2 \le 1$ and divergent for $x^2 > 1$

Example 2: Check the convergence of

$$1^p + \left(\frac{1}{2}\right)^p + \left(\frac{1\cdot3}{2\cdot4}\right)^p + \left(\frac{1\cdot3\cdot5}{2\cdot4\cdot6}\right)^p + \ldots.$$

Solution: Neglecting 1st term

$$a_n = \left[\frac{1\cdot3\cdot5....(2n-1)}{2\cdot4\cdot6....(2n)}\right]^p$$

$$a_{n+1} = \left[\frac{1\cdot3\cdot5....(2n-1)(2n+1)}{2\cdot4\cdot6....(2n)(2n+2)}\right]^p$$

$$\lim_{n\to\infty} \frac{a_n}{a_{n+1}} = \lim_{n\to\infty} \left(\frac{2n+2}{2n+1}\right)^p = 1$$

$\therefore$ Ratio test fails

Now apply Gauss test

$$\frac{a_n}{a_{n+1}} = \left(1+\frac{1}{n}\right)^p \left(1+\frac{1}{2n}\right)^{-p} = \left[1+\frac{p}{n}+0\left(\frac{1}{n^2}\right)\right]\left[1-\frac{p}{2n}+0\left(\frac{1}{n^2}\right)\right]$$

$$= \left[1+\left(p-\frac{p}{2}\right)\frac{1}{n}+0\left(\frac{1}{n^2}\right)\right]$$

$$= 1+\frac{p}{2}\cdot\frac{1}{n}+0\left(\frac{1}{n^2}\right)$$

Here $\lambda = \frac{p}{2}$

By Gauss test $\sum a_n$ convergent if $\frac{p}{2}>1$ i.e $p>2$

and divergent for $p \le 2$.

Example 3:

Check the convergence of the series

$$\frac{a}{b}+\frac{a(a+1)}{b(b+1)}+\frac{a(a+1)(a+2)}{b(b+1(b+2))}+\ldots\infty$$

Solution :

Neglecting I[st] term

$$a_n = \frac{a(1+a)\ldots(n+a)}{b(1+b)\ldots(n+b)}$$

$$a_{n+1} = \frac{a(1+a)\ldots(n+a)(n+1+a)}{b(1+b)\ldots(n+b)(n+1+b)}$$

$$\lim_{n\to\infty}\frac{a_n}{a_{n+1}} = \frac{n+1+b}{n+1+a} = \lim_{n\to\infty}\frac{n\left(1+\frac{1+b}{n}\right)}{n\left(1+\frac{1+a}{n}\right)}=1$$

D' Alembert ratio test fail.

Now apply Raabe's test

$$n\left[\frac{a_n}{a_{n+1}}-1\right] = n\left[\frac{n+1+b}{n+1+a}-1\right]$$

$$= \frac{n(n+1+b-n-1-a)}{n+1+a}=\frac{n(b-a)}{n\left(1+\frac{1}{n}+\frac{a}{n}\right)}$$

$$\lim_{n\to\infty} n\left[\frac{a_n}{a_{n+1}} - 1\right] = \lim_{n\to\infty} n\frac{b-a}{1+\frac{1}{n}+\frac{a}{n}} = b-a$$

$\sum a_n$ is convergent if $b - a > 1$ *i.e.,* $b > 1 + a$ and divergent if $b - a < 1$ *i.e.,* $b < a + 1$

and test fails when $b - a = 1$ *i.e.,* $b = 1 + a$

then the given series by putting $b = 1 + a$ becomes,

$$a_n = \frac{a(1+a)\ldots(n+a)}{(1+a)(2+a)\ldots(n+a)(n+1+a)}$$

$$= \frac{a}{n\left(1+\frac{1}{n}+\frac{a}{n}\right)}$$

Let $b_n = \frac{1}{n}$ $\qquad \lim_{n\to\infty}\frac{a_n}{b_n} = a$ (which is finite & non -zero)

$\therefore$ By comparison test both series converge or diverge together

As $b_n = \frac{1}{n}$ and by p-test $\sum b_n$ is divergent.

$\therefore$ $\sum a_n$ is convergent when $b > 1 + a$ and divergent for $b \leq a + 1$

Example 4 :

Test the convergence of $\sum \frac{n!}{x(x+1)(x+2)\ldots(x+n-1)}$

Solution :

Here $a_n = \frac{n!}{x(x+1)(x+2)\ldots(x+n-1)}$

$$a_{n+1} = \frac{(n+1)!}{x(x+1)(x+2)\ldots(x+n)}$$

$$\lim_{n\to\infty}\frac{a_n}{a_{n+1}} = \lim_{n\to\infty}\frac{x+n}{(n+1)} = \lim_{n\to\infty}\frac{n\left(1+\frac{x}{n}\right)}{n\left(1+\frac{1}{n}\right)} = 1$$

∴ D'Alembert ratio test fails

Applying Gauss test

$$\frac{a_n}{a_{n+1}} = \left(1+\frac{x}{n}\right)\left(1+\frac{1}{n}\right)^{-1}$$

$$= \left(1+\frac{x}{n}\right)\left[1-\frac{1}{n}+0\left(\frac{1}{n^2}\right)\right]$$

$$= \left[1+\frac{x-1}{n}+0\left(\frac{1}{n^2}\right)\right]$$

Here $\lambda = x - 1$

By Gauss test series is convergent when $x - 1 > 1$ *i.e.*, $x > 2$ and divergent when $x - 1 \leq 1$ *i.e.*, $x \leq 2$.

Example 5 :

Check the convergence of the series $\frac{x}{1\cdot 2}+\frac{x^2}{3\cdot 4}+\frac{x^3}{5\cdot 6}+\frac{x^4}{7\cdot 8}+\ldots\infty$ $(x > 0)$

Solution :

Here $a_n = \frac{x^n}{(2n-1)(2n)}$

$$a_{n+1} = \frac{x^{n+1}}{(2n+1)(2n+2)}$$

$$\lim_{n\to\infty}\frac{a_n}{a_{n+1}} = \lim_{n\to\infty}\frac{(2n+1)(2n+2)}{2n(2n-1)}\cdot\frac{1}{x}=\frac{1}{x}$$

By D'Alembert ratio test series is convergent for $x < 1$ and divergent for $x > 1$

test fails if $x = 1$

Applying Raabe's test, put $x = 1$

$$\lim_{n\to\infty} n\left[\frac{a_n}{a_{n+1}}-1\right] = \lim_{n\to\infty} n\left[\frac{4n^2+6n+2}{4n^2-2n}-1\right]$$

$$= \lim_{n\to\infty} n\left[\frac{4n^2+6n+2-4n^2+2n}{4n^2-2n}\right]$$

$$= \lim_{n\to\infty} n\left[\frac{8n+2}{4n^2-2n}\right]$$

$$= \lim_{n\to\infty} \frac{8n^2 + 2n}{4n^2 - 2n} = 2 > 1$$

$\therefore$ Series is convergent.

Thus the series is convergent for $x \leq 1$ and divergent for $x > 1$.

Example 6: Check the convergence of the given series:

$$1 + \frac{a}{1!} + \frac{a(a+1)}{2!} + \frac{a(a+1)(a+2)}{3!} + \ldots$$

Solution :

Neglecting Ist term

$$a_n = \frac{a(a+1)(a+2)\ldots(a+n-1)}{n!}$$

$$a_{n+1} = \frac{a(a+1)(a+2)\ldots(a+n-1)(a+n)}{(n+1)!}$$

$$\lim_{n\to\infty} \frac{a_n}{a_{n+1}} = \lim_{n\to\infty} \frac{n+1}{a+n} = \lim_{n\to\infty} \frac{1+\frac{1}{n}}{1+\frac{a}{n}}$$

$$= 1 \qquad \therefore \text{ Ratio test fails.}$$

Now applying Raabe's test

$$\lim_{n\to\infty} n\left[\frac{a_n}{a_{n+1}} - 1\right] = \lim_{n\to\infty} n\left[\frac{n+1}{a+n} - 1\right]$$

$$= \lim_{n\to\infty} \frac{n(1-a)}{a+n} = \lim_{n\to\infty} \frac{1-a}{\frac{a}{n}+1}$$

$$= 1 - a$$

By Raabe's test, series is convergent if $1 - a > 1$ *i.e.*, $a < 0$, divergent if $1 - a < 1$ *i.e.*, $a > 0$.

If $1 - a = 1$ *i.e.*, $a = 0$ test fails

In this case series become $1 + 0 + 0 \ldots$

This is convergent.

Hence $\sum a_n$ convergent if $a \leq 0$ and divergent for $a > 0$.

Example 7 :

Test the convergence of the series $\sum \frac{1\cdot 3\cdot 5\ldots(2n-1)}{2\cdot 4\cdot 6\ldots(2n)}\cdot x^{2n}$, $x>0$

Solution :

Here, $$a_n = \frac{1\cdot 3\cdot 5\ldots(2n-1)}{2\cdot 4\cdot 6\ldots(2n)}\cdot x^{2n}$$

$$a_{n+1} = \frac{1\cdot 3\cdot 5\ldots(2n-1)(2n+1)}{2\cdot 4\cdot 6\ldots(2n)(2n+2)}\cdot\frac{x^{2n+2}}{2^{2n+2}}$$

$$\lim_{n\to\infty}\frac{a_n}{a_{n+1}} = \lim_{n\to\infty}\frac{2n+2}{2n+1}\frac{(2n+2)}{(2n)}\cdot\frac{1}{x^2} = \frac{1}{x^2}$$

By D'Alembert ratio test $\sum a_n$ is convergent if $\frac{1}{x^2}>1$ *i.e.,* $x^2<1$

and divergent if $\frac{1}{x^2}<1$ i.e., $x^2>1$, if $1=x^2$ test fails.

Put $x^2 = 1$ in given series and apply Raabe's test

$$\lim_{n\to\infty} n\left[\frac{a_n}{a_{n+1}}-1\right] = \lim_{n\to\infty} n\left[\frac{2n+2}{2n+1}\cdot\frac{2n+2}{2n}-1\right]$$

$$= \lim_{n\to\infty}\frac{n(6n+4)}{2n(2n+1)} = \frac{3}{2}>1 \quad \therefore \text{ Convergent by Raabe's test.}$$

Hence given series is convergent if $x^2 \le 1$ and divergent for $x^2 > 1$.

Example 8 :

Check the convergence of $\sum \frac{(n!)^2}{(2n)!}x^{2n}$, $x>0$

Solution :

Here $$a_n = \frac{(n!)^2}{(2n)!}x^{2n}$$

$$a_{n+1} = \frac{[(n+1)!]^2}{(2n+2)!}x^{2n+2}$$

we have, $$\frac{a_n}{a_{n+1}} = \left(\frac{n!}{(n+1)!}\right)^2\cdot\frac{(2n+2)!}{(2n)!}\,\frac{x^{2n}}{x^{2n+2}}$$

$$= \frac{(2n+2)(2n+1)}{(n+1)^2} \cdot \frac{1}{x^2} = \frac{2(2n+1)}{(n+1)} \cdot \frac{1}{x^2}$$

$$\lim_{n\to\infty} \frac{a_n}{a_{n+1}} = \lim_{n\to\infty} \frac{2\left(2+\frac{1}{n}\right)}{\left(1+\frac{1}{n}\right)} \cdot \frac{1}{x^2} = \frac{4}{x^2}$$

By ratio test, the series is convergent for $\frac{4}{x^2} > 1$ *i.e.,* $x^2 < 4$ and divergent for $\frac{4}{x^2} < 1$ i.e., $x^2 > 4$, test fail for $x^2 = 4$.

Put $x^2 = 4$ and apply Raabe's test

$$\lim_{n\to\infty} n\left[\frac{a_n}{a_{n+1}} - 1\right] = \lim_{n\to\infty} n\left[\frac{1}{4} \cdot \frac{2(2n+1)}{n+1} - 1\right]$$

$$= \lim_{n\to\infty} \frac{-n}{2(n+1)} = -\frac{1}{2} < 1$$

By Raabe's test series is divergent.

Hence given series is convergent for $x^2 < 4$ and divergent for $x^2 \geq 4$.

Example 9 :

Check the convergence of $1+\frac{(1!)^2}{2!}x^2 + \frac{(2!)^2}{4!}x^4 + \frac{(3!)^2}{6!}x^6 + \ldots \infty \; (x > 0)$

Solution :

Here $$a_n = \frac{(n!)^2}{(2n)!} x^{2n}$$

$$a_{n+1} = \frac{[(n+1)!]^2}{(2n+2)!} x^{2n+2}$$

$$\lim_{n\to\infty} \frac{a_n}{a_{n+1}} = \lim_{n\to\infty} \frac{n!\,n!}{(2n)!} \times \frac{(2n+2)(2n+1)2n!}{(n+1)(n+1)n!\,n!} \cdot \frac{1}{x^2}$$

$$= \lim_{n\to\infty} \frac{2(2n+1)}{(n+1)} \cdot \frac{1}{x^2}$$

$$= \frac{4}{x^2}$$

By Ratio test Σa_n is convergent for $\frac{4}{x^2} > 1$ *i.e.,* $x^2 < 4$ and divergent for $\frac{4}{x^2} < 1$ *i.e.,* $x^2 > 4$ test, fail if $x^2 = 4$.

Put $x^2 = 4$ and apply Raabe's test

$$\lim_{n\to\infty} n\left[\frac{a_n}{a_{n+1}} - 1\right] = \lim_{n\to\infty} n\left[\frac{4n+2-4n-4}{4(n+1)}\right]$$

$$= \lim_{n\to\infty} \frac{-2n}{4(n+1)} = \lim_{n\to\infty} \frac{n}{n}\left[\frac{-2}{4\left(1+\frac{1}{n}\right)}\right]$$

$$= -\frac{1}{2} < 1$$

By Raabe's test series is divergent.

Hence given series is convergent for $x^2 < 4$ and divergent for $x^2 \geq 4$.

Example 10 :

Check the convergence of $\frac{x^2}{2\log 2} + \frac{x^3}{3\log 3} + \frac{x^4}{4\log 4} + \ldots \infty$

Solution :

Here $$a_n = \frac{x^{n+1}}{(n+1)\log(n+1)}$$

$$a_{n+1} = \frac{x^{n+2}}{(n+2)\log(n+2)}$$

$$\frac{a_n}{a_{n+1}} = \frac{x^{n+1}}{(n+1)\log(n+1)} \times \frac{(n+2)\log(n+2)}{x^{n+2}}$$

$$= \frac{(n+2)\log(n+2)}{(n+1)\log(n+1)} \cdot \frac{1}{x} = \frac{\left(1+\frac{2}{n}\right)\log n\left(1+\frac{2}{n}\right)}{\left(1+\frac{1}{n}\right)\log n\left(1+\frac{1}{n}\right)} \cdot \frac{1}{x}$$

$$= \frac{\left(1+\frac{2}{n}\right)[\log n + \log\left(1+\frac{2}{n}\right)]}{\left(1+\frac{1}{n}\right)\left[\log n + \log\left(1+\frac{1}{n}\right)\right]} \cdot \frac{1}{x}$$

$$= \frac{1}{x} \times \frac{\left(1+\frac{2}{n}\right)\left[\log n + \frac{2}{n} - \frac{1}{2}\left(\frac{2}{n}\right)^2 + \frac{1}{3}\left(\frac{2}{n}\right)^3 + \ldots\right]}{\left(1+\frac{1}{n}\right)\left[\log n + \frac{1}{n} - \frac{1}{2}\left(\frac{1}{n}\right)^2 + \frac{1}{3}\left(\frac{1}{n}\right)^3 + \ldots\right]}$$

$$\lim_{n\to\infty} \frac{a_n}{a_{n+1}} = \lim_{n\to\infty} \frac{1}{x} \times \frac{\left(1+\frac{2}{n}\right)\left[\log n + \frac{2}{n} - \frac{1}{2}\left(\frac{2}{n}\right)^2 \ldots\right]}{\left(1+\frac{1}{n}\right)\left[\log n + \frac{1}{n} - \frac{1}{2}\left(\frac{1}{n}\right)^2 \ldots\right]}$$

$$= \frac{1}{x}$$

By ratio test,the series is convergent for $x < 1$ and divergent for $x > 1$ and test fail for $x = 1$. Put $x = 1$ in given series, we get

$$\frac{1}{2\log 2} + \frac{1}{3\log 3} + \frac{1}{4\log 4} + \ldots$$

$$a_n = \frac{1}{(n+1)\log(n+1)}$$

$$\lim_{n\to\infty} a_n = 0$$

Now by applying Cauchy's Integral test $\sum a_n$ is divergent.

Hence given series is convergent for $x < 1$ and divergent for $x \geq 1$.

Example 11 :

Check the convergence of $1 + \frac{a}{b}x + \frac{a(a+1)}{b(b+1)}x^2 + \frac{a(a+1)(a+2)}{b(b+1)(b+2)}x^3 + \ldots$

Solution :

Neglecting 1st term

$$a_n = \frac{a(a+1)(a+2)\ldots(a+n-1)}{b(b+1)(b+2)\ldots(b+n-1)}x^n \quad (x \geq 0)$$

$$a_{n+1} = \frac{a(a+1)(a+2)\ldots(a+n)}{b(b+1)(b+2)\ldots(b+n)} x^{n+1}$$

$$\lim_{n\to\infty} \frac{a_n}{a_{n+1}} = \lim_{n\to\infty} \frac{b+n}{a+n} \cdot \frac{1}{x}$$

$$= \frac{1}{x}$$

By ratio test $\sum a_n$ convergent if $\frac{1}{x} > 1$ *i.e.*, $x < 1$ and divergent for $\frac{1}{x} < 1$ *i.e.*, $x > 1$ and for $x = 1$ test fails.

Now apply Raabe's test for $x = 1$

$$\lim_{n\to\infty} n\left[\frac{a_n}{a_{n+1}} - 1\right] = \lim_{n\to\infty} \frac{(b-a)n}{a+n}$$

$$= \lim_{n\to\infty} \frac{b-a}{\frac{a}{n} + \left(1 + \frac{1}{n}\right)} = (b-a)$$

By Raabe's test $\sum a_n$ is convergent if $b - a > 1$ and divergent if $b - a < 1$ *i.e.*, $b < 1 + a$

For $b - a = 1$, Apply Gauss test

$$\frac{a_n}{a_{n+1}} = \frac{b+n}{a+n} = \frac{a+1+n}{a+n}$$

$$= \frac{a+n+1}{a+n} = \frac{n\left[1 + \frac{a+1}{n}\right]}{n\left[1 + \frac{a}{n}\right]}$$

$$= \left[1 + \left(\frac{a+1}{n}\right)\right]\left[1 - \left(\frac{a}{n}\right) + \left(\frac{a}{n}\right)^2 \ldots\right]$$

$$= 1 + \left(\frac{a+1}{n}\right) - \frac{a}{n} + 0\left(\frac{1}{n^2}\right)$$

$$= 1 + \frac{1}{n} + 0\left(\frac{1}{n^2}\right)$$

By Gauss test $\sum a_n$ is divergent if $x = 1$ and $b - a = 1$.

Hence given series $\sum a_n$ is convergent when $x < 1$ and divergent when $x > 1$.

For $x = 1$ if $b - a > 1$ it is convergent and it divergent for $b - a \leq 1$.

Exercise 1.5

Discuss the convergence and divergence of the following series :

1. $\frac{1^2}{4^2}+\frac{1^2\cdot 5^2}{4^2\cdot 8^2}+\frac{1^2\cdot 5^2\cdot 9^2}{4^2\cdot 8^2\cdot 12^2}+....\infty$

2. $\frac{1^2}{2^2}+\frac{1^2\cdot 3^2}{2^2\cdot 4^2}x+\frac{1^2\cdot 3^2\cdot 5^2}{2^2\cdot 4^2\cdot 6^2}x^2+...\infty$

3. $1+\frac{2x}{2!}+\frac{3^2x^2}{3!}+\frac{4^3x^3}{4!}+...\infty$

4. $\frac{a}{b}+\frac{a(a+d)x}{b(b+d)}+\frac{a(a+d)(a+2d)x^2}{b(b+d)(b+2d)}+...\infty$

5. $1+\frac{1}{2}x+\frac{1\cdot 3}{2\cdot 4}x^2+\frac{1\cdot 3\cdot 5}{2\cdot 4\cdot 6}x^3+...\infty$

6. $\frac{x}{1\cdot 2}+\frac{x^2}{2\cdot 3}+\frac{x^3}{3\cdot 4}+\frac{x^4}{4\cdot 5}+....\infty$

7. $\sum\frac{4\cdot 7\cdot 10....(3n+1)x^n}{1\cdot 2\cdot 3....n}$

8. $\frac{1}{2}x+x^2+\frac{9x^3}{8}+x^4+\frac{25x^5}{32}+...\infty$

9. $\frac{1}{2}+\frac{1\cdot 3}{2\cdot 4}+\frac{1\cdot 3\cdot 5}{2\cdot 4\cdot 6}+.....$

10. $\frac{1^2}{2^2}+\frac{1^2.3^2}{2^2.4^2}+\frac{1^2.3^2.5^2}{2^2.4^2.6^2}+......$

11. $\frac{x}{1}+\frac{1}{2}\cdot\frac{x^3}{3}+\frac{1\cdot 3}{2\cdot 4}\cdot\frac{x^5}{5}+\frac{1\cdot 3\cdot 5}{2\cdot 4\cdot 6}\cdot\frac{x^7}{7}+....$

12. $1+\frac{x}{2}+\frac{2!}{3^2}x^2+\frac{3!}{4^3}x^3+\frac{4!}{5^4}x^4+.....$

13. $\frac{a+x}{1!}+\frac{(a+2x)^2}{2!}+\frac{(a+3x)^3}{3!}+....\infty$

14. $x^2(\log 2)^q+x^3(\log 3)^q+x^4(\log 4)^q+......\infty$

15. $1+\frac{\alpha\beta}{1\cdot\gamma}x+\frac{\alpha(\alpha+1)\beta(\beta+1)}{1\cdot 2\cdot\gamma(\gamma+1)}x^2+\frac{\alpha(\alpha+1)(\alpha+2)\beta(\beta+1)(\beta+2)}{1\cdot 2\cdot 3\cdot\gamma(\gamma+1)(\gamma+2)}x^3+....$

16. $1+\frac{2}{3}x+\frac{2\cdot 3}{3\cdot 5}x^2+\frac{2\cdot 3\cdot 4}{3\cdot 5\cdot 7}x^3+...\infty$

17. $1+\frac{2^2}{3^2}x+\frac{2^2\cdot 4^2}{3^2\cdot 5^2}x^2+\frac{2^2\cdot 4^2\cdot 6^2}{3^2\cdot 5^2\cdot 7^2}x^3+....\infty$

18. $\sum_{n=1}^{\infty}\frac{(2n)!}{(n!)^2}x^n$

19. $\sum\frac{n^2(n+1)^2}{n!}$

20. $\frac{1}{(\log 2)^k}+\frac{1}{(\log 3)^k}+\frac{1}{(\log 4)^k}+....\infty$

Answers

1. Convergent
2. Convergent for $x < 1$, divergent for $x \geq 1$.
3. Convergent for $x \leq \frac{1}{e}$, divergent for $x > \frac{1}{e}$.
4. Convergent for $x < 1$ or $x = 1$ and $b > a + d$, divergent for $x > 1$ or $x = 1$ and $b \leq a + d$.
5. Convergent for $x < 1$, divergent for $x \geq 1$.
6. Convergent for $x \leq 1$, divergent for $x > 1$.
7. Convergent for $x < \frac{1}{3}$, divergent for $x \geq \frac{1}{3}$.
8. Convergent for $x < 2$, divergent for $x \geq 2$.
9. Divergent
10. Divergent.
11. Convergent if $x^2 \leq 1$ and divergent if $x^2 > 1$.
12. Convergent if $x < e$ and divergent if $x \geq e$.
13. Converges for $x < \frac{1}{e}$ and divergent for $x \geq \frac{1}{e}$.
14. Convergent for $x < 1$ and divergent for $x \geq 1$.
15. Converges if $x < 1$ and diverges if $x > 1$. If $x = 1$ then the series converges if $\gamma > \alpha + \beta$ and diverges if $\gamma \leq \alpha + \beta$.
16. Convergent for $x < 2$, divergent for $x \geq 2$.
17. Convergent for $x < 1$, divergent for $x \geq 1$.
18. Convergent for $x < \frac{1}{4}$, divergent for $x \geq \frac{1}{4}$.
19. Convergent.
20. Divergent.

1.21 Cauchy's Integral Test

If f(x) is a non negative , monotonically decreasing and integrable function of $x \geq 1$ such that $f(n) = a_n$ $\forall\, n \in N$, then the series $\sum a_n$ and the integral $\int_1^{\infty} f(x)$ converges or diverges together.

> **Note:**
>
> **If** $x \geq \kappa$ **then** $\sum a_n$ **and** $\int_{\kappa}^{\infty} f(x)dx$ **converges or diverges together.**

Illustrative Examples

Example 1:

Using integral test , discuss the convergence of the following series

(i) $\sum \frac{1}{2n+3}$ (ii) $\sum \frac{1}{n(n+1)}$

(iii) $\sum \frac{1}{\sqrt{n}}$ (iii) $\sum \frac{1}{(n+1)^2}$

(v) $\sum \frac{2n^3}{n^4+3}$ (vi) $\sum_{n=2}^{\infty} \frac{1}{n\sqrt{n^2-1}}$

Solution :

(i) *Let* $a_n = \frac{1}{2n+3} = f(n)$

$f(x) = \frac{1}{2x+3}$ For $x \geq 1, f(x)$ is +ve and monotonic decreasing,

$\therefore$ Cauchy integral test is applicable.

$$= \int_1^{\infty} f(x)dx = \int_1^{\infty} \frac{1}{2x+3}dx$$

$$= \frac{1}{2}\left[\log(2x+3)\right]_1^{\infty} = \frac{1}{2}\left[\log\infty - \log 5\right]$$

$= \infty$ $\therefore$ divergent.

(ii) $\sum \frac{1}{n((n+1))}$

Let $a_n = \frac{1}{n(n+1)} = f(n)$

$f(x) = \frac{1}{x(x+1)}$ for $x \geq 1$, $f(x)$ is +ve and monotonic decreasing.

$\therefore$ by cauchy's integral test.

$$= \int_1^{\infty} \frac{1}{x(x+1)} dx = \int_1^{\infty} \left(\frac{1}{x} - \frac{1}{x+1}\right) dx$$

$$= [\log x - \log(1+x)]_1^{\infty} = \left[\log\left(\frac{x}{1+x}\right)\right]_1^{\infty}$$

$$= \left[\log 1 - \log\left(1+\frac{1}{x}\right)\right]_1^{\infty} = \left[\log\left(\frac{1}{1+\frac{1}{x}}\right)\right]_1^{\infty}$$

$$= \log 1 - \log 1 + \log 2$$

$= \log 2$ (Finite) $\Rightarrow \int_1^{\infty} f(x) dx$ converges and

$\therefore$ given series is convergent.

(iii) $\sum \frac{1}{\sqrt{n}}$

Let $a_n = \frac{1}{\sqrt{n}} = f(n)$

$f(x) = \frac{1}{\sqrt{x}} = \frac{1}{(x)^{\frac{1}{2}}}$ for $x \geq 1$, $f(x)$ is (+)ve and monotonic decreasing.

$\therefore$ by cauchy's integral test.

$$\int_1^{\infty} f(x) dx = \int_1^{\infty} x^{\frac{-1}{2}} dx = \left[\frac{x^{\frac{1}{2}}}{\frac{1}{2}}\right]_1^{\infty}$$

$$= 2[\infty - 1] = \infty \quad \Rightarrow \int_1^{\infty} f(x)\,dx \text{ diverges and}$$

$\therefore$ given series is divergent.

(iv) $\sum \frac{1}{(n+1)^2}$

Let $a_n = \sum \frac{1}{(n+1)^2} = f(n)$

$f(x) = \frac{1}{(x+1)^2}$ for $x \geq 1$, $f(x)$ is $(+)$ve and monotonic decreasing.

$\therefore$ by cauchy's integral test.

$$\int_1^{\infty} f(x)\,dx = \int_1^{\infty} \frac{1}{(x+1)^2}\,dx$$

$$\int_1^{\infty} \frac{dx}{(x+1)^2} = \left[-\frac{1}{(x+1)}\right]_1^{\infty} = -\left[\frac{1}{\infty+1} - \frac{1}{2}\right]$$

$$= 0 + \frac{1}{2} = \frac{1}{2} \quad \text{(Finite)}$$

$\Rightarrow \int_1^{\infty} f(x)\,dx$ converges and hence by integral test $\sum a_n$ is also converges.

(v) $\sum \frac{2n^3}{n^4+3}$

Solution :

Let $a_n = \sum \frac{2n^3}{n^4+3} = f(n)$

$\therefore f(x) = \frac{2x^3}{x^4+3}$

For $x \geq 1$, $f(x)$ is $(+)$ve and monotonic decreasing.

$\therefore$ Cauchy's integral test is applicable.

$$\therefore \int_1^{\infty} f(x)\,dx = \int_1^{\infty} \frac{2x^3}{x^4+3}\,dx$$

$$= \frac{2}{4}\left[\log\left(x^4+3\right)\right]_1^{\infty} = \infty$$

$\Rightarrow \int_1^{\infty} f(x)dx$ diverges and hence by Cauchy's integral test $\sum a_n$ also diverges.

(vi) $\sum_{n=2}^{\infty} \frac{1}{n\sqrt{n^2-1}}$

Solution :

Let $a_n = \frac{1}{n\sqrt{n^2-1}} = f(n)$

$$f(x) = \frac{1}{x\sqrt{x^2-1}}$$

For $x \geq 2$, $f(x)$ is (+)ve and monotonic decreasing.

∴ Cauchy's integral test is applicable.

$$\int_2^{\infty} f(x)dx = \int_2^{\infty} \frac{1}{x\sqrt{x^2-1}} = \left[\sec^{-1} x\right]_2^{\infty}$$

$$= \left[\cos^{-1}\frac{1}{x}\right]_2^{\infty} = \left[\cos^{-1} 0 - \cos^{-1}\left(\frac{1}{2}\right)\right]$$

$$= \frac{\pi}{2} - \frac{\pi}{3} = \frac{\pi}{6} \quad \text{(Finite)}$$

$\int_2^{\infty} f(x)dx$ is convergent and hence $\sum a_n$ is also convergent.

Example 2 :

Discuss the convergence of the following series using integral test:

(i) $\sum n \cdot e^{-n^2}$ (ii) $\sum_{n=3}^{\infty} \frac{1}{n\log n(\log\log n)^p}$, $p > 0$

(iii) $\sum_{n=2}^{\infty} \frac{1}{n(\log n)^2}$ (iv) $\sum_{n=1}^{\infty} \frac{1}{(n+1)\log(n+1)}$

Solution :

(i) Let $a_n = n \cdot e^{-n^2} = f(n)$

$$f(x) = x \cdot e^{-x^2}$$

For $x \geq 1$, $f(x)$ is (+)ve and monotonic decreasing.

$\therefore$ Cauchy's integral test is applicable.

$$\int_1^{\infty} f(x)dx = \int_1^{\infty} x \cdot e^{-x^2} dx$$

Put $x^2 = t$

$2xdx = dt$

$xdx = \frac{dt}{2}$

$$= \frac{1}{2}\int_1^{\infty} e^{-t} dt = \frac{1}{2}\left[\frac{e^{-t}}{-1}\right]_1^{\infty}$$

$$= -\frac{1}{2}\left[e^{-\infty} - \frac{1}{e}\right] = \frac{1}{2e} \quad \text{(Finite)}$$

$\therefore \int_1^{\infty} f(x)dx$ is convergent, hence $\sum a_n$ is also convergent.

(ii) Let $f(n) = a_n = \dfrac{1}{n\log n(\log\log n)^p}, \quad p \geq 0$

$$\therefore \quad f(x) = \frac{1}{x\log x(\log\log x)^p}$$

For $x \geq 3$, $p > 0$, $f(x)$ is (+)ve and monotonic decreasing

$\therefore$ Cauchy's integral test is applicable.

Case I :

When $p > 1$ $I_n = \int_3^n \dfrac{1}{x\log x(\log\log x)^p} \cdot dx$ as $n \to \infty$

let $\log(\log x) = t$

$$\frac{1}{\log x} \cdot \frac{1}{x} dx = dt$$

So $I_n = \int_3^n \dfrac{1}{t^p} dt$

$$= \left[\frac{t^{-p+1}}{-p+1}\right]_3^n = \left[\frac{(\log(\log x))^{-p+1}}{-p+1}\right]_3^{\infty}$$

$$= \frac{1}{-p+1}\left[\lim_{n\to\infty}(\log\log n)^{-p+1} - (\log\log 3)^{-p+1}\right]$$

$$= \frac{1}{1-p}\left[(\log\log\infty)^{-p+1} - (\log\log 3)^{-p+1}\right]$$

$$= \frac{1}{1-p}\left[(\infty)^{-ve} - \frac{1}{(\log\log 3)^{p-1}}\right] = \frac{-1}{1-p}\left[\frac{1}{(\log\log 3)^{p-1}}\right]$$

= Finite

$\therefore$ convergent for $p > 1$

Case II :

For $p < 1$

$$I_n = \frac{1}{1-p}\left[(\infty)^{+ve} - \log(\log 3)^{1-p}\right] = \infty$$,divergent for $p < 1$.

Case III :

$$p = 1$$

$$I_n = \lim_{n\to\infty}\int_3^n \frac{1}{x\log x(\log\log x)}dx$$ Let $\log\log x = t$

$$= \int \frac{1}{t}dt = \log t$$ $\frac{1}{\log x}\cdot\frac{1}{x}dx = dt$

$$= \lim_{n\to\infty}\left[\log\log\log 3\right]_3^n$$

$$= \lim_{n\to\infty}\left[\log\log\log n - \log\log\log 3\right]$$

$= \infty$ divergent.

$\therefore$ Series is convergent for $p > 1$ and divergent for $0 < p \le 1$

(iii) $\sum_{n=2}^{\infty}\frac{1}{n(\log n)^2}$

Solution :

$$a_n = \frac{1}{n(\log n)^2} = f(n)$$

$$f(x) = \frac{1}{x(\log x)^2}$$

For $x \geq 2, f(x)$ is $(+)$ve and monotonic decreasing.

$\therefore$ Cauchy's integral test is applicable.

$$\int_2^\infty f(x)dx = \int_2^\infty \frac{1}{x(\log x)^2}dx$$

$$= \left[-\frac{1}{\log x}\right]_2^\infty$$

$$= -\left[\frac{1}{\log\infty} - \frac{1}{\log 2}\right] = \frac{1}{\log 2} = \text{convergent.}$$

$\int_2^\infty f(x)dx$ is convergent and hence $\sum a_n$ is also convergent.

(iv) $\sum_{n=1}^{\infty} \frac{1}{(n+1)\log(n+1)}$

$$a_n = \frac{1}{(n+1)\log(n+1)} = f(n)$$

$$f(x) = \frac{1}{(x+1)\log(x+1)}$$

For $x \geq 1, f(x)$ is $(+)$ve and monotonic decreasing.

$\therefore$ Cauchy's integral test is applicable.

$$\int_1^\infty \frac{1}{(x+1)\log(x+1)}dx = \left[\log(\log(x+1))\right]_1^\infty$$

$$= \log\log(\infty+1) - \log\log 2$$

$$= \infty \quad \therefore \text{ divergent.}$$

$\int_1^\infty f(x)dx$ is divergent and hence $\sum a_n$ also divergent.

Exercise – 1.6

Using integral test, test the convergence or divergence of the following series :

1. $\sum \frac{n}{(n^2+1)^2}$
2. $\sum \frac{1}{n^2+1}$
3. $\sum_{n=1}^{\infty} \frac{1}{n^p}$
4. $\sum_{n=2}^{\infty} \frac{1}{n(\log n)^p}$, $(p > 0)$
5. $\sum_{n=1}^{\infty} \frac{8\tan^{-1} n}{1+n^2}$

Answer

1. Convergent
2. Convergent
3. Converges if $p > 1$ and diverges if $0 < p \leq 1$.
4. Converges if $p > 1$ and diverges if $0 < p \leq 1$.
5. Convergent

1.22 Alternating Series

A series of the form $a_1 - a_2 + a_3 - a_4 + \ldots$ where $a_n > 0$ for all $n \varepsilon N$ is called an alternating series, and is denoted by $\sum_{n=1}^{\infty} (-1)^{n-1} a_n$.

For Example :

1. $1 - \frac{1}{\sqrt{2}} + \frac{1}{\sqrt{3}} - \frac{1}{\sqrt{4}} + \frac{1}{\sqrt{5}} - \frac{1}{\sqrt{6}} + \ldots = \sum_{n=1}^{\infty} (-1)^{n-1} \frac{1}{\sqrt{n}}$
2. $1 - \frac{1}{3} + \frac{1}{5} - \frac{1}{7} + \ldots = \sum_{n=1}^{\infty} (-1)^{n-1} \frac{1}{2n-1}$

1.23 Leibnitz's Test for Convergence of Alternating Series

The alternating series $\sum_{n=1}^{\infty}(-1)^{n-1}a_n = a_1 - a_2 + a_3 - a_4 + \ldots$ ($a_n > 0$ for all n) is convergent

(i) $a_{n+1} < a_n$ for all n

(ii) $\lim_{n\to\infty} a_n = 0$

Note :

(a) If $\lim_{n\to\infty} a_n \neq 0$ then the series $\sum_{n=1}^{\infty}(-1)^{n-1}a_n$ is oscillating.

(b) The alternating series will not be convergent if any one of the two conditions is not satisfied.

Illustrative Examples

Discuss the convergence of the following series.

1. $1 - \frac{1}{\sqrt{2}} + \frac{1}{\sqrt{3}} - \frac{1}{\sqrt{4}} + \ldots \infty$

2. $\frac{1}{\log 2} - \frac{1}{\log 3} + \frac{1}{\log 4} - \frac{1}{\log 5} + \ldots \infty$

3. $1 - \frac{1}{2!} + \frac{1}{4!} - \frac{1}{6!} + \ldots \infty$

4. $\frac{5}{2} - \frac{7}{4} + \frac{9}{6} - \frac{11}{8} + \ldots \infty$

5. $\frac{1}{\sqrt{2}+1} - \frac{1}{\sqrt{3}+1} + \frac{1}{\sqrt{4}+1} - \frac{1}{\sqrt{5}+1} + \ldots \infty$

Example 1 :

$$1 - \frac{1}{\sqrt{2}} + \frac{1}{\sqrt{3}} - \frac{1}{\sqrt{4}} + \ldots \infty$$

Solution :

Here $\quad \sum(-1)^{n-1} \cdot a_n = \sum(-1)^{n-1} \cdot \frac{1}{\sqrt{n}}$

$$a_n = \frac{1}{\sqrt{n}}$$

$$a_{n+1} = \frac{1}{\sqrt{n+1}}$$

$$a_n - a_{n+1} = \frac{1}{\sqrt{n}} - \frac{1}{\sqrt{n+1}}$$

$$\frac{\sqrt{n+1}-\sqrt{n}}{\sqrt{n}\sqrt{n+1}} > 0$$

or $$\sqrt{n} < \sqrt{n+1}$$

$$\frac{1}{\sqrt{n+1}} < \frac{1}{\sqrt{n}}$$

This shows $a_{n+1} < a_n \ \forall n$

$$\lim_{n\to\infty} a_n = \lim_{n\to\infty}\frac{1}{\sqrt{n}} = 0$$ Both conditions of Leibnitz's test are satisfied

Hence the given series is convergent.

Example 2 :

$$\frac{1}{\log 2} - \frac{1}{\log 3} + \frac{1}{\log 4} - \frac{1}{\log 5} + \ldots \infty$$

Solution :

Here $$\sum (-1)^{n-1} a_n = \sum (-1)^{n-1} \frac{1}{\log n}$$

$$a_n = \frac{1}{\log(n+1)}$$

$$a_{n+1} = \frac{1}{\log(n+2)}$$

$$a_n - a_{n+1} = \frac{1}{\log(n+1)} - \frac{1}{\log(n+2)}$$

$$= \frac{\log(n+2) - \log(n+1)}{\log(n+1)\log(n+2)}$$

since $\log(n+2) > \log(n+1)$

this shows $a_n - a_{n+1} > 0$

$a_{n+1} < a_n$ for all n

and $\lim_{n\to\infty} a_n = \lim_{n\to\infty} \frac{1}{\log(n+1)} = \frac{1}{\infty} = 0$

Hence given series is convergent by Leibnitz test.

Example 3 :

$$1 - \frac{1}{2!} + \frac{1}{4!} - \frac{1}{6!} + \ldots \infty$$

Solution :

Here $\sum (-1)^{n-1} \cdot \frac{1}{2(n-1)!} = \sum (-1)^{n-1} a_n$

$$a_n = \frac{1}{2(n-1)!}$$

$$a_{n+1} = \frac{1}{2n!}$$

$$2n! > 2\,(n-1)!$$

$$\frac{1}{2n!} < \frac{1}{2(n-1)!}$$

$a_{n+1} < a_n$ for all n.

and $\lim_{n\to\infty} a_n = \lim_{n\to\infty} \frac{1}{2(n-1)!} = \frac{1}{\infty} = 0$

Both conditions of Leibnitz's test are satisfied.

Hence the given series is convergent.

Example 4 :

$$\frac{5}{2} - \frac{7}{4} + \frac{9}{6} - \frac{11}{8} + \ldots \infty$$

Solution :

Let $$\sum(-1)^{n-1}a_n = \sum(-1)^{n-1}\frac{2n+3}{2n}$$

Here $$a_n = \frac{2n+3}{2n}$$

$$\lim_{n\to\infty} a_n = \lim_{n\to\infty}\frac{2n+3}{2n}$$

$$= \lim_{n\to\infty}\left(1+\frac{3}{2n}\right) = 1 \text{ not equal to } 0$$

Hence the given series is oscillating.

Example 5

Check the convergence of

$$\frac{1}{\sqrt{2}+1} - \frac{1}{\sqrt{3}+1} + \frac{1}{\sqrt{4}+1} - \frac{1}{\sqrt{5}+1} + \ldots\ldots$$

Solution: Let $\sum \; (-1)^{n-1}\frac{1}{\sqrt{n+1}+1} = \sum(-1)^{n-1}a_n$

$$a_n = \frac{1}{\sqrt{n+1}+1}$$

$$a_{n+1} = \frac{1}{\sqrt{n+2}+1}$$

$$n+2 > n+1 \;\forall\; n$$

$$\sqrt{n+2} > \sqrt{n+1} \text{ for all } n$$

$$\sqrt{n+2}+1 > \sqrt{n+1}+1 \text{ for all } n$$

$$\frac{1}{\sqrt{n+2}+1} < \frac{1}{\sqrt{n+1}+1} \text{ for all } n$$

$$a_{n+1} < a_n \text{ for all } n$$

also $$\lim_{n\to\infty} a_n = \lim_{n\to\infty} \frac{1}{\sqrt{n+1}+1} = 0$$

Hence the given series is convergent.

Example 6

Discuss the convergence of the following series:

$$\frac{1}{a} - \frac{1}{a+b} + \frac{1}{a+2b} - \frac{1}{a+3b} + \ldots.\infty \qquad (a > 0, b > 0)$$

Solution :

Let $$\sum (-1)^{n-1} a_n = \sum (-1)^{n-1} \cdot \frac{1}{a+(n-1)b}$$

$$a_n = \frac{1}{a+(n-1)b}$$

$$a_{n+1} = \frac{1}{a+nb}$$

$$nb > (n-1)\, b \text{ for all } n.$$

$$a + nb > a + (n-1)\, b \text{ for all } n.$$

$$\frac{1}{a+nb} < \frac{1}{a+(n-1)b} \text{ for all } n.$$

$\Rightarrow$ $$a_{n+1} < a_n \text{ for all } n.$$

Also $$\lim_{n\to\infty} a_n = \lim_{n\to\infty} \frac{1}{a+(n-1)b} = \frac{1}{\infty} = 0$$

$\Rightarrow$ by Leibnitz test, given series is convergent.

Example 7

Check the convergence of

$$\frac{x}{1+x} - \frac{x^2}{1+x^2} + \frac{x^3}{1+x^3} - \frac{x^4}{1+x^4} + \ldots.\infty \quad (0 < x < 1)$$

Solution :

Let $$\sum (-1)^{n-1} a_n = \sum (-1)^{n-1} \cdot \frac{x^n}{1+x^n} \qquad (0 < x < 1)$$

$$a_n = \frac{x^n}{1+x^n}$$

$x^{n+1} < x^n$ for all n, $0 < x < 1$

$x^{n+1} - x^n < 0$

$x^{n+1} < x^n$

$x^{n+1} + 1 < x^n + 1$

$$a_n - a_{n+1} = \frac{x^n}{1+x^n} - \frac{x^{n+1}}{1+x^{n+1}}$$

$$= \frac{x^n\left(1+x^{n+1}\right) - x^{n+1}\left(1+x^n\right)}{\left(1+x^n\right)\left(1+x^{n+1}\right)}$$

$$= \frac{x^n \cdot x^n \cdot x^{n+1} - x^{n+1} - x^{n+1} - x^{n+1} \cdot x^n}{\left(1+x^n\right)\left(1+x^{n+1}\right)}$$

$$> 0$$

$$a_n - a_{n+1} > 0$$

$$\Rightarrow \quad a_{n+1} < a_n \quad \forall\, n$$

and $$\lim_{n\to\infty} a_n = \lim_{n\to\infty}\frac{x^n}{1+x^n} = \lim_{n\to\infty}\frac{1}{\left(1+\frac{1}{x^n}\right)}$$

$$= \frac{1}{\infty} = 0$$

both conditions of Leibnitz test are satisfy.

$\therefore$ given series is convergent.

Example 8

Check the convergence of the given series:

$$\sum_{n=1}^{\infty} \frac{(-1)^{n-1}}{3n-2}$$

Solution :

$$a_n = \frac{1}{3n-2} \qquad a_{n+1} = \frac{1}{3n+1}$$

$$n+1 > n; \ \forall\, n$$

$$3n+3 > 3n \;\forall\; n$$
$$3n+1 > 3n-2 \;\forall\; n$$
$$\frac{1}{3n+1} < \frac{1}{3n-2} \;\forall\; n$$

$\Rightarrow$ $$a_{n+1} < a_n \text{ for all } n$$

and $$\lim_{n\to\infty} a_n = \lim_{n\to\infty}\frac{1}{3n-2} = 0$$

Hence by Leibnitz test given series is convergent.

Example 9

Test the convergence of the given series

$$\sum_{n=1}^{\infty}\frac{n(-1)^{n-1}}{5^n}$$

Solution :

Let $$a_n = \frac{n}{5^n}$$

$$a_{n+1} = \frac{(n+1)}{5^{n+1}}$$

$$a_n - a_{n+1} = \frac{n}{5^n} - \frac{(n+1)}{5^{n+1}} = \frac{5n-n-1}{5^{n+1}}$$

$$= \frac{4n-1}{5^{n+1}} > 0 \;\forall\; n \geq 1$$

$$a_n > a_{n+1}$$

$\Rightarrow$ $$a_{n+1} < a_n \text{ for all } n$$

and $$\lim_{n\to\infty} a_n = \lim_{n\to\infty}\frac{n}{5^n} = \frac{\infty}{\infty} \text{ form}$$

Using L'Hospital rule.

$$= \lim_{n\to\infty}\frac{1}{5^n \log 5} = \frac{1}{\infty} = 0$$

both conditions of Leibnitz test are satisfy.

$\therefore$ given series is convergent.

Example 10

Check the convergence of the given series.

$$\sum_{n=1}^{\infty} \frac{\cos n\pi}{n^2+1}$$

Solution :

Given series is

$$\sum_{n=1}^{\infty} \frac{\cos n\pi}{n^2+1} = \sum (-1)^n \frac{1}{n^2+1}$$

$$= \sum (-1)^n \cdot a_n$$

$$a_n = \frac{1}{n^2+1}, \quad a_{n+1} = \frac{1}{(n+1)^2+1}$$

$$n+1 > n$$

$$(n+1)^2 > n^2$$

$$(n+1)^2 + 1 > n^2 + 1$$

$$\frac{1}{(n+1)^2+1} < \frac{1}{n^2+1}$$

$$a_{n+1} < a_n \text{ for all } n.$$

Also $\lim\limits_{n \to \infty} a_n = \lim\limits_{n \to \infty} \frac{1}{n^2+1} = \frac{1}{\infty} = 0$

$\therefore$ by Leibnitz test given series is convergent.

Exercise 1.7

Discuss the convergence or divergence of the following series.

1. $1 - \frac{1}{2} + \frac{1}{3} - \frac{1}{4} + \ldots.$

2. $2 - \frac{3}{2} + \frac{4}{3} - \frac{5}{4} + \ldots$

3. $\frac{1}{2^3} - \frac{1}{3^3}(1+2) + \frac{1}{4^3}(1+2+3) - \frac{1}{5^3}(1+2+3+4) + \ldots..$

4. $\sum_{n=1}^{\infty} \frac{(-1)^{n-1} \cdot n}{2n-1}$

5. $\frac{\log 2}{2^2} - \frac{\log 3}{3^2} + \frac{\log 4}{4^2} \ldots.\infty$

6. $1-\frac{1}{2^2}+\frac{1}{3^2}-\frac{1}{4^2}+\ldots\infty$

7. $\frac{1}{6}-\frac{2}{11}+\frac{3}{16}-\frac{4}{21}+\ldots\infty$

8. $\frac{1}{1\cdot 2}-\frac{1}{3\cdot 4}+\frac{1}{5\cdot 6}-\frac{1}{7\cdot 8}+\ldots\infty$

9. $\sum_{n=2}^{\infty}\frac{(-1)^{n-1}\cdot x^n}{n(n-1)}$, $0 < x < 1$

10. $\sum_{n=1}^{\infty}(-1)^{n-1}\cdot\frac{n}{n^2+1}$

11. $\log\left(\frac{1}{2}\right)-\log\left(\frac{2}{3}\right)+\log\left(\frac{3}{4}\right)-\log\left(\frac{4}{5}\right)+\ldots\infty$

Answers

1. Convergent
2. Not convergent
3. Convergent
4. Not convergent
5. Convergent
6. Convergent
7. $\sum_{n=1}^{\infty}\frac{n(-1)^{n-1}}{5n-1}$, not convergent
8. Convergent
9. Convergent
10. Convergent
11. Convergent

1.24 Absolute Convergence of a series :

A series $\sum_{n=1}^{\infty}a_n$ is said to be absolutely convergent if the series $\sum_{n=1}^{\infty}|a_n|$ is convergent.

OR

If a convergent series whose terms are not all positive , remains convergent when all its terms are made positive , then it is called an absolutely convergent series.

Illustrations:

1. The series $\sum a_n = 1-\frac{1}{2^2}+\frac{1}{3^2}-\frac{1}{4^2}+\ldots$ is absolutely convergent , since

$$\sum |a_n| = 1+\frac{1}{2^2}+\frac{1}{3^2}+\frac{1}{4^2}+...$$ is a convergent series by p – test.

2. The series $\sum a_n = 1-\frac{1}{2}+\frac{1}{2^2}-\frac{1}{2^3}+....$ is absolutely convergent , since

$\sum |a_n| = 1+\frac{1}{2}+\frac{1}{2^2}+\frac{1}{2^3}+....$ being a geometric series with common ratio $\frac{1}{2}<1$ is convergent.

1.25 Conditional convergence:

A series $\sum_{n=1}^{\infty} a_n$ is said to be conditionally convergent , if

(i) $\sum_{n=1}^{\infty} a_n$ is convergent , and

(ii) $\sum_{n=1}^{\infty} a_n$ is not absolutely convergent.

(i.e) A series is said to be conditionally convergent if it is convergent but does not converge absolutely.

For Example :

The series $$\sum_{n=1}^{\infty} a_n = 1-\frac{1}{2}+\frac{1}{3}-\frac{1}{4}+\frac{1}{5}....$$ is conditionally convergent series , since

$$\sum_{n=1}^{\infty} a_n = \sum_{n=1}^{\infty} (-1)^{n-1} \cdot \frac{1}{n}$$ is convergent by Leibnitz 's test

where

$$\sum_{n=1}^{\infty} |a_n| = \sum_{n=1}^{\infty} \frac{1}{n}$$ is divergent by p – test.

Hence $\sum_{n=1}^{\infty} a_n$ is a conditionally convergent series.

An important observation :

Every absolutely convergent series is convergent , but converse need not be true.

For Example :

$$\sum \frac{(-1)^{n-1}}{n} = 1-\frac{1}{2}+\frac{1}{3}-\frac{1}{4}+...$$

$$= 1 - \frac{1}{2} + \frac{1}{3} - \frac{1}{4}$$,it is convergent.

But the series $\sum \left| \frac{(-1)^{n-1}}{n} \right| = \sum \frac{1}{n}$ is divergent.

Note :

(a) The divergence of $\sum |a_n|$ does not imply the divergence of $\sum a_n$.

For example, $\sum \left| \frac{(-1)^{n-1}}{n} \right| = \sum \frac{1}{n}$ is divergent,

where as $\sum \frac{(-1)^{n-1}}{n}$ is convergent.

(b) Since $\sum_{n=1}^{\infty} |a_n|$ is a series of positive terms,

$\therefore$ all the tests used for testing the convergence of positive term series are also applicable for testing the absolute convergence of the alternating series.
However these tests cannot give any information about the convergence of the alternating series.

Illustrative Examples

Test the convergence and absolute convergence of the following series:

1. $\frac{1}{5} - \frac{1}{7} + \frac{1}{9} - \frac{1}{11} + \ldots.$

2. $1 - \frac{1}{2^p} + \frac{1}{3^p} - \frac{1}{4^p} + \ldots \; (p > 0)$

3. $\sum_{n=1}^{\infty} (-1)^{n-1} \sin \frac{1}{n}$

4. $\sum_{n=1}^{\infty} \frac{(-1)^{n-1} \cdot n}{5n - 7}$

5. $1 - \frac{1}{4 \cdot 3} + \frac{1}{4^2 \cdot 5} - \frac{1}{4^3 \cdot 7} + \ldots$

Example 1

1. $\frac{1}{5}-\frac{1}{7}+\frac{1}{9}-\frac{1}{11}+\dots\infty$

Solution:

Given series is alternating series and each term is numerically less than the preceeding term

$$a_n = \sum(-1)^{n-1}\cdot\frac{1}{2n+3}$$

$$\lim_{n\to\infty} a_n = \lim_{n\to\infty}\frac{1}{2n+3} = \frac{1}{\infty} = 0$$

By Leibnitz test given series is convergent.

$$\sum|a_n| = \frac{1}{5}+\frac{1}{7}+\frac{1}{9}+\frac{1}{11}+\dots$$

$$a_n = \frac{1}{2n+3} = f(n)$$

$$f(x) = \frac{1}{2x+3}$$

For $x \geq 1$, $f(x)$ is (+) ve and monotonic decreasing

$\therefore$ Cauchy integral test is applicable

$$\int_1^\infty f(x)\,dx = \int_1^\infty \frac{1}{2x+3}dx$$

$$= \frac{1}{2}\left[\log(2x+3)\right]_1^\infty$$

$$= \frac{1}{2}[\log\infty - \log 5] = \infty$$

By integral test $\sum|a_n|$ divergent.

Hence given series is conditionally convergent.

Example 2 $\quad 1-\frac{1}{2^p}+\frac{1}{3^p}-\frac{1}{4^p}+\dots\ (p>0)$

Solution :

Here $\quad \sum_{n=1}^{\infty} a_n(-1)^{n-1} = \sum_{n=1}^{\infty}(-1)^{n-1}\cdot\frac{1}{n^p}$

$$a_n = \frac{1}{n^p}$$

$$\lim_{n\to\infty} a_n = \lim_{n\to\infty}\frac{1}{n^p} = \frac{1}{\infty} = 0$$

Also $(n+1)^p > n^p$ for all n

$$\frac{1}{(n+1)^p} < \frac{1}{n^p} \text{ for all } n$$

$$\Rightarrow a_{n+1} < a_n \quad \forall\, n$$

Thus by Leibnitz's test $\sum_{n=1}^{\infty} a_n$ is convergent.

Now

$$|a_n| = \frac{1}{n^p}$$

$$\sum_{n=1}^{\infty}|a_n| = \sum_{n=1}^{\infty}\frac{1}{n^p} \text{ is convergent if } p > 1 \text{ and divergent if } p \le 1.$$

∴ Hence given series is absolutely convergent if $p > 1$ and conditionally convergent if $0 < p < 1$.

Example 3 $\sum_{n=1}^{\infty}(-1)^{n-1}\sin\frac{1}{n}$

Solution: Here $\sum(-1)^{n-1}a_n = \sum(-1)^{n-1}\sin\frac{1}{n}$

$$|a_n| = \sin\frac{1}{n}$$

$$= \frac{1}{n} - \frac{1}{n^3 3!} + \frac{1}{n^5 5!}\ldots\ldots$$

$$= \frac{1}{n}\left[1 - \frac{1}{n^2 3!} + \frac{1}{n^4 5!}\ldots\right]$$

Let $b_n = \frac{1}{n}$

$b_n = \frac{1}{n}$ is divergent so $\sum a_n$ is also divergent.

Hence given series is conditionally convergent.

Example 4. $\sum_{n=1}^{\infty} \frac{(-1)^{n-1} \cdot n}{5n-7}$

Solution:

$$\sum_{n=1}^{\infty} (-1)^{n-1} a_n = \sum_{n=1}^{\infty} (-1)^{n-1} \frac{n}{5n-7}$$

$$a_n = \frac{n}{5n-7}$$

$$a_{n+1} = \frac{n+1}{5n-2}$$

$$a_n - a_{n+1} = \frac{n}{5n-7} - \frac{n+1}{5n-2} = \frac{5n^2 - 2n - 5n^2 + 7}{(5n-7)(5n-2)}$$

$$= \frac{7-2n}{(5n-7)(5n-2)} > 0 \quad \text{for all } n \text{ except } n = 1$$

$$\Rightarrow \quad a_{n+1} < a_n$$

$$\lim_{n\to\infty} a_n = \frac{5}{5n\left(1 - \frac{7}{5n}\right)} = 0$$

By Leibnitz test given series is convergent

$|a_n|$ is also convergent.

Hence given series is absolutely convergent.

Example 5 $1 - \frac{1}{4 \cdot 3} + \frac{1}{4^2 \cdot 5} - \frac{1}{4^3 \cdot 7} + \ldots$

Solution :

$$\sum (-1)^{n-1} \frac{1}{4^{n-1} \cdot (2n-1)} = \sum (-1)^{n-1} a_n$$

$$|a_n| = \frac{1}{4^{n-1}(2n-1)}$$

$$|a_{n+1}| = \frac{1}{4^n (2n+1)}$$

$$\left|\frac{a_n}{a_{n+1}}\right| = \frac{1}{4^{n-1}(2n-1)} \times 4^n \cdot (2n+1)$$

$$= 4\frac{2n\left(1+\frac{1}{2n}\right)}{2n\left(1-\frac{1}{2n}\right)}$$

$$\lim_{n\to\infty}\left|\frac{a_n}{a_{n+1}}\right| = 4 > 1$$

by D'Alembert test given series is absolutely convergent.

Example 6

Test the convergence of the series:

$$\sum_{n=1}^{\infty}(-1)^{n-1} \cdot \frac{n+2}{2^n+5}$$

Solution :

$$|a_n| = \frac{n+2}{2^n+5}$$

$$|a_{n+1}| = \frac{n+3}{2^{n+1}+5}$$

$$\left|\frac{a_n}{a_{n+1}}\right| = \frac{n+2}{2^n+5} \times \frac{2^{n+1}+5}{n+3}$$

$$= \frac{n\left(1+\frac{2}{n}\right)2^n\left(2+\frac{5}{2n}\right)}{2^n\left(1+\frac{5}{2n}\right)n\left(1+\frac{3}{n}\right)}$$

$$\lim_{n\to\infty}\left|\frac{a_n}{a_{n+1}}\right| = \frac{\left(1+\frac{2}{n}\right)\left(2+\frac{5}{2n}\right)}{\left(1+\frac{5}{2n}\right)\left(1+\frac{3}{n}\right)}$$

$$= 2 > 1$$

$\therefore$ convergent

Hence given series is absolutely convergent.

Example 7

Test whether the series $1+\frac{1}{2^2}-\frac{1}{3^2}-\frac{1}{4^2}+\frac{1}{5^2}+\frac{1}{6^2}-\frac{1}{7^2}-\frac{1}{8^2}+.......$ is convergent or not ?

Solution :

$$\sum|a_n| = 1+\frac{1}{2^2}+\frac{1}{3^2}+\frac{1}{4^2}+...$$

$$= \sum\frac{1}{n^2}$$

By p – test $\qquad p = 2 > 1$

Thus given series is absolutely convergent and hence convergent.

Example 8

Prove that the following series converge absolutely :

$$\sum_{n=1}^{\infty}\frac{(-1)^{n-1}\sin n\alpha}{n^2}, \quad \alpha \text{ real}$$

Solution:

$$\sum_{n=1}^{\infty} a_n = \sum_{n=1}^{\infty}\frac{(-1)^{n-1}\sin n\alpha}{n^2},$$

$$a_n = \sum\frac{\sin n\alpha(-1)^{n-1}}{n^2}$$

$$|a_n| = \sum\left|\frac{(-1)^{n-1}\sin n\alpha}{n^2}\right|$$

$$= \sum\frac{1}{n^2} \qquad \because |\sin n\alpha| \le 1$$

$$|a_n| = \frac{1}{n^2}$$

Hence $|a_n|$ is convergent.

Hence given series is absolutely convergent.

Example 9

Check the convergence of the series

$$\sum_{n=1}^{\infty} \frac{(-1)^{n-1} \cos^2 n\alpha}{n\sqrt{n}}, \quad \alpha \text{ real}$$

Solution :

$$\sum_{n=1}^{\infty} \frac{(-1)^{n-1} \cos^2 n\alpha}{n\sqrt{n}} = \sum_{n=1}^{\infty} a_n$$

$$\sum_{n=1}^{\infty} |a_n| = \sum_{n=1}^{\infty} \frac{(-1)^{n-1} \cos^2 n\alpha}{n\sqrt{n}}$$

$$= \sum_{n=1}^{\infty} \frac{1}{n\sqrt{n}} \qquad \because |\cos^2 n\alpha| = 1$$

$$= \sum_{n=1}^{\infty} \frac{1}{n^{3/2}}$$

Thus by p test $\quad p = \frac{3}{2} > 1$, convergent.

Hence, $\sum_{n=1}^{\infty} \frac{(-1)^{n-1} \cos^2 n\alpha}{n\sqrt{n}}$ is absolutely convergent.

Example 10

Prove that the following series is convergent :

$$\sum_{n=1}^{\infty} (-1)^{n-1} \frac{n^3}{(n+1)!}$$

Solution :

$$\sum_{n=1}^{\infty} (-1)^{n-1} \frac{n^3}{(n+1)!} = \sum a_n$$

$$|a_n| = \frac{n^3}{(n+1)!}$$

$$|a_{n+1}| = \frac{(n+1)^3}{(n+2)!}$$

$$\left|\frac{a_n}{a_{n+1}}\right| = \frac{n^3}{(n+1)!} \times \frac{(n+2)!}{(n+1)^3}$$

$$= \frac{n^3}{(n+1)!} \times \frac{(n+2)(n+1)!}{n^3\left(1+\frac{1}{n}\right)^3} = \frac{(n+2)}{\left(1+\frac{1}{n}\right)^3}$$

$$\lim_{n\to\infty} \left|\frac{a_n}{a_{n+1}}\right| = \infty$$

$\therefore$ $|a_n|$ is convergent by D' Alembert test.

Thus given series is absolutely convergent.

Example 10

For what values of x are the following series convergent :

$$\frac{1}{1-x} + \frac{1}{2(1-x)^2} + \frac{1}{3(1-x)^3} + \ldots\ldots$$

Solution :

Let $$a_n = \frac{1}{n(1-x)^n}$$

and

$$a_{n+1} = \frac{1}{(n+1)(1-x)^{n+1}}$$

$$\left|\frac{a_{n+1}}{a_n}\right| = \frac{n}{n+1}\left|\frac{(1-x)^n}{(1-x)^{n+1}}\right|$$

$$\lim_{n\to\infty} \left|\frac{a_{n+1}}{a_n}\right| = \frac{1}{|1-x|}$$

By ratio test $\sum_{n=1}^{\infty} |a_n|$ is divergent if $\frac{1}{|1-x|} < 1$

i.e $$|(1-x)| > 1$$

or

$(1-x) > 1$ or $(1-x) < -1$

Hence for $x < 0$ or $x > 2$ series is absolutely convergent, hence convergent.

Also for $x = 2$ the series $-1+\frac{1}{2}-\frac{1}{3}+\frac{1}{4}$.... which is convergent series.

$$a_n = \frac{1}{n}$$

$$a_{n+1} = \frac{1}{n+1}$$

$$a_n > a_{n+1}$$

$$\Rightarrow \quad a_{n+1} < a_n$$

Also $$\lim_{n\to\infty} a_n = 0$$

Hence $\sum a_n$ is convergent by Leibnitz test.

Exercise 1.8

Test the convergence and absolute convergence of the series (1 – 10)

1. $1-\frac{1}{3!}+\frac{1}{5!}-\frac{1}{7!}+\ldots$

2. $\frac{1}{\sqrt{2}+1}-\frac{1}{\sqrt{3}+1}+\frac{1}{\sqrt{4}+1}-\frac{1}{\sqrt{5}+1}+\ldots$

3. $\sum_{n=1}^{\infty}\frac{(-1)^{n-1}\cdot n}{n^2+1}$

4. $\sum_{n=1}^{\infty}(-1)^{n-1}\cdot\frac{2^n}{n!}$

5. $1-\frac{1}{2\sqrt{2}}+\frac{1}{3\sqrt{3}}-\frac{1}{4\sqrt{4}}+\ldots\ldots$

6. $1-\frac{1}{2^3}(1+2)+\frac{1}{3^3}(1+2+3)-\frac{1}{4^4}(1+2+3+4)+\ldots$

7. $\sum_{n=2}^{\infty}\frac{(-1)^n}{n(\log n)^2}$

8. $\sum_{n=1}^{\infty}\frac{(-1)^{n-1}\cdot n}{5^n}$

9. $\sum_{n=1}^{\infty}(-1)^{n-1}\left[\frac{1}{n^2}+\frac{1}{(n+1)^2}\right]$

10. $\frac{1}{2(\log 2)^p}-\frac{1}{3(\log 3)^p}+\frac{1}{4(\log 4)^p}\ldots\infty \quad (p>0)$

11. Prove that the series

$\frac{\sin x}{1^3}-\frac{\sin 2x}{2^3}+\frac{\sin 3x}{3^3}\ldots$ converges absolutely.

12. Discuss the convergence of the series

$1+\frac{x}{1!}+\frac{x^2}{2!}+\ldots$ for all value of x.

13. Discuss the convergence and absolute convergence of the series

$x-\frac{x^3}{3}+\frac{x^5}{5}\ldots$, x being real.

14. Test the convergence of the series

$\sum_{n=1}^{\infty}\frac{(-1)^n(x+1)^n}{2^n\cdot n^2}$.

15. For what values of x are following series convergent.

(a) $x-\frac{x^2}{2}+\frac{x^3}{3}-\frac{x^4}{4}+\frac{x^5}{5}\ldots$

(b) $1-\frac{x^2}{2}+\frac{x^4}{4}-\frac{x^6}{6}+\ldots$

16. Discuss the convergence of the series

$\sum_{n=1}^{\infty}(-1)^n\frac{\tan^{-1}n}{1+n^2}$.

Answers

1. Absolutely convergent 2. Conditionally convergent
3. Conditionally convergent
4. Absolutely convergent
5. Absolutely convergent
6. Conditionally convergent
7. Absolutely convergent
8. Absolutely convergent
9. Absolutely convergent
10. Absolutely convergent if $p > 1$ and Conditionally convergent if $0 < p \leq 1$.
11. Absolutely convergent for all x.
12. Convergent if $-1 \leq x \leq 1$ and absolutely Convergent if $-1 < x < 1$.
13. Given series is convergent if $-3 \leq x \leq 1$.
14. (a) $-1 < x \leq 1$

    (b) $-1 \leq x \leq 1$
15. Convergent.

CHAPTER 2

Matrices

CONTENTS

Introduction, Elementary Transformation rank, Inverse by Gauss-Jordan Method, Normal form, Consistency of System of Linear Equations, Linear, Dependence and Independence of vectors, Eigen values, Eigen vectors, Cayley Hamilton Theorem, etc.

INTRODUCTION

The study of matrices originated from the idea of various types of linear problems. It has a special relationship with system of linear equations which occurs in many engineering processes. It provides an impo tant list in the study and development of linear algebra. Also the matrices occur in presentation of linear system models in applied engineering and control systems.

2.1 DEFINITION

Let us consider a set of simultaneous equations

$$2x + 3y + 3z + 2t = 0$$

$$3x + 2y + 5z + 3t = 0$$

$$4x + 5y + 6z + 7t = 0$$

$$2x + 3y + 4z + 5t = 0$$

Now, we write the co-efficients of x, y, z, t of the above equations and enclose them with in brackets and then we get,

$$A = \begin{bmatrix} 2 & 3 & 3 & 2 \\ 3 & 2 & 5 & 3 \\ 4 & 5 & 6 & 7 \\ 2 & 3 & 4 & 5 \end{bmatrix}$$

The above system of numbers, arranged in a rectangular array in rows and columns and bounded by the brackets, is called a matrix.

It has got 4 rows and 4 columns and in all $4 \times 4 = 16$ elements. It is termed as 4×4 matrix. In the double subscripts of an element, the first subscript determines the row and the second subscript determines the column in which the element lies, a_{ij} lies in the ith row and jth column.

2.2 VARIOUS TYPES OF MATRICES

1. **Row Matrix:** A matrix which has only one row and any number of columns, is called row matrix.

   e.g. : $[1 \quad 2 \quad 3 \quad 4]$

2. **Column Matrix:** A matrix which has only one column and any number of rows is called column matrix.

   e.g. : $\begin{bmatrix} 1 \\ 2 \\ 3 \\ 4 \end{bmatrix}$

3. **Null Matrix or Zero Matrix:** A matrix in which all the elements are zero is called a zero matrix or null matrix

   e.g. $\begin{bmatrix} 0 & 0 & 0 \\ 0 & 0 & 0 \end{bmatrix}$

4. **Square Matrix:** A matrix in which the number of rows is equal to the number of columns, is called a square matrix.

   e.g. $\begin{bmatrix} 2 & 5 \\ 1 & 4 \end{bmatrix}$

5. **Diagonal Matrix:** A square matrix is called a diagonal matrix, if all its non-diagonal elements are zero.

   e.g. $\begin{bmatrix} 1 & 0 & 0 \\ 0 & 2 & 0 \\ 0 & 0 & 3 \end{bmatrix}$

6. **Unit or Identity Matrix:** A square matrix is called a unit matrix, if all the diagonal elements are unity and non-diagonal elements are zero.

e.g. $\begin{bmatrix} 1 & 0 & 0 \\ 0 & 1 & 0 \\ 0 & 0 & 1 \end{bmatrix}$

7. **Singular and Non-Singular Matrices:** A square matrix 'A' is called a singular matrix if $|A| = 0$, i.e., if determinant formed by elements of 'A' is zero. If $|A| \neq 0$, then matrix is known as non-singular matrix.

8. **Symmetric Matrix:** A square matrix $A = [a_j]$ is said to be symmetric if $a_{ij} = a_{ji}$ for all i and j or $A' = A$

e.g. $\begin{bmatrix} a & h & g \\ h & b & f \\ g & f & c \end{bmatrix}, \begin{bmatrix} 1 & 2 & 3 \\ 2 & 4 & 5 \\ 3 & 5 & 6 \end{bmatrix}$

9. **Skew-Symmetric Matrix:** A square matrix is called skew symmetric matrix if

(1) $a_{ij} = -a_{ji}$ for all i and j or $A' = -A$

(2) All diagonal elements are zero.

e.g. $\begin{bmatrix} 0 & -h & -g \\ h & 0 & -f \\ g & f & 0 \end{bmatrix}$

10. **Triangular Matrix:** A square matrix all of whose elements below the leading diagonal are zero, is called an upper triangular matrix.

A square matrix, all of whose elements above the leading diagonal are zero, is called a lower triangular matrix.

e.g. $\begin{bmatrix} 1 & 2 & 3 \\ 0 & 4 & 5 \\ 0 & 0 & 6 \end{bmatrix}$ *upper–triangular matrix* $\begin{bmatrix} 1 & 0 & 0 \\ 2 & 3 & 0 \\ 5 & 6 & 7 \end{bmatrix}$ *lower–triangular matrix*

11. **Equal Matrix:** Two matrix A and B are said to be equal if and only if first matrix is of same order of second matrix i.e., each element a_{ij} of 'A' is equal to the corresponding element b_{ij} of 'B'

12. **Transpose of a Matrix:** If in a given matrix 'A', we interchange the rows and the corresponding columns, the new matrix obtained is called the transpose of the matrix 'A' and is denoted by A' or A^T.

e.g. $A = \begin{bmatrix} 1 & 2 & 3 \\ 4 & 5 & 6 \\ 7 & 8 & 9 \end{bmatrix}, A' = \begin{bmatrix} 1 & 4 & 7 \\ 2 & 5 & 8 \\ 3 & 6 & 9 \end{bmatrix}$

13. **Orthogonal Matrix:** A square matrix 'A' is called an orthogonal matrix if the product of the matrix 'A' and the transpose matrix A' is an identity matrix.

e.g. $A.A' = I$ where I is an identity matrix

If $|A| = 1$, matrix A is proper.

14. Conjugate of a Matrix:

Let $$A = \begin{bmatrix} 1+i & 2-3i & 4 \\ 7+2i & -i & 3-2i \end{bmatrix}$$

Conjugate of matrix A is denoted by $\bar{A}$.

$$\therefore \quad \bar{A} = \begin{bmatrix} 1-i & 2+3i & 4 \\ 7-2i & i & 3+2i \end{bmatrix}$$

15. Matrix A^θ : Transpose of the conjugate of a matrix A is denoted by A^θ.

Let $$A = \begin{bmatrix} 1+i & 2-3i & 4 \\ 7+2i & -i & 3-2i \end{bmatrix}$$

$$\bar{A} = \begin{bmatrix} 1-i & 2+3i & 4 \\ 7-2i & i & 3+2i \end{bmatrix}$$

$$(\bar{A})' = \begin{bmatrix} 1-i & 7-2i \\ 2+3i & i \\ 4 & 3+2i \end{bmatrix}$$

or $$A^\theta = \begin{bmatrix} 1-i & 7-2i \\ 2+3i & i \\ 4 & 3+2i \end{bmatrix}.$$

16. Unitary Matrix: A square matrix A is said to be unitary if $A^\theta A = I$

e.g. $$A = \begin{bmatrix} \frac{1+i}{2} & \frac{-1+i}{2} \\ \frac{1+i}{2} & \frac{1-i}{2} \end{bmatrix}, \quad A^\theta = \begin{bmatrix} \frac{1-i}{2} & \frac{1-i}{2} \\ \frac{-1-i}{2} & \frac{1+i}{2} \end{bmatrix}$$

$$A.A^\theta = I$$

17. Hermition Matrix: A square matrix $A = [aij]$ is called Hermition matrix, if every $(ij)^{th}$ element of 'A' is equal to the conjugate complex $(ji)^{th}$ element of A.

In other words $\quad a_{ij} = \bar{a}_{ji}$

e.g. $$\begin{bmatrix} 1 & 2+3 & 3+i \\ 2-3i & 2 & 1-2i \\ 3-i & 1+2i & 5 \end{bmatrix}$$

Necessary and sufficient condition for a matrix A to be Hermition is that $A = A^{\theta}$ i.e., conjugate transpose of A.

or $$A = (\bar{A})'.$$

18. Skew-Hermition Matrix: A square matrix $A = [aij]$ will be called a skew-Hermition matrix if every $(ij)^{th}$ element of A is equal to negative conjugate complex of $(ji)^{th}$ element of A.

In other words, $a_{ij} = -\bar{a}_{ji}$

All the elements in the principal diagonal will be of the form.

$$a_{ii} = -\bar{a}_{ii} \text{ or } a_{ii} + a_{ii} = 0$$

if $a_{ii} = a + ib$ then $\bar{a}_{ii} = a - ib$

$(a + ib) + (a - ib) = 0$ or $2a = 0$ or $a = 0$

so a_{ii} is pure imaginary or $a_{ii} = 0$

Hence all the diagonal elements of a skew Hermition matrix are either zero or pure imaginary.

e.g. $$\begin{bmatrix} i & 2-3i & 4+5i \\ -(2+3i) & 0 & 2i \\ -(4-5i) & 2i & -3i \end{bmatrix}$$

The necessary and sufficient condition for a matrix A to be skew - Hermition is that,

$$A^{\theta} = -A$$

$$(\bar{A})' = -A$$

19. Idempotent Matrix: A matrix, such that $A^2 = A$ is called Idempotent matrix.

e.g. If $$A = \begin{bmatrix} 2 & -2 & -4 \\ -1 & 3 & 4 \\ 1 & -2 & -3 \end{bmatrix} \text{ then } A^2 = \begin{bmatrix} 2 & -2 & -4 \\ -1 & 3 & 4 \\ 1 & -2 & -3 \end{bmatrix}\begin{bmatrix} 2 & -2 & -4 \\ -1 & 3 & 4 \\ 1 & -2 & -3 \end{bmatrix}$$

$$= A$$

20. Periodic Matrix: A matrix A will be called a periodic matrix, if

$$A^{k+1} = A$$

where 'k' is a +ve integer. If k is the least +ve integer, for which $A^{k+1} = A$, then k is said to be the period of A. If we choose $k = 1$, we get $A^2 = A$ and we call it to be an idempotent matrix.

21. Nilpotent Matrix: A matrix will be called a nilpotent matrix, if $A^k = 0$ (null matrix), where k is a +ve integer; if however k is the least +ve integer for which $A^k = 0$, then k is the index of the nilpotent matrix.

e.g. $$A = \begin{bmatrix} ab & b^2 \\ -a^2 & -ab \end{bmatrix}, A^2 = \begin{bmatrix} ab & b^2 \\ -a^2 & -ab \end{bmatrix}\begin{bmatrix} ab & b^2 \\ -a^2 & -ab \end{bmatrix} = \begin{bmatrix} 0 & 0 \\ 0 & 0 \end{bmatrix} = 0$$

'A' is a nilpotent matrix whose index is 2.

22. Involutory Matrix: A matrix 'A' will be called an involutory matrix if $A^2 = I$ (unit matrix). Since $I^2 = I$ (always)

$\therefore$ unit matrix is involutory.

Find the value of x, y, z, and a which satisfy the matrix equation.

$$\begin{bmatrix} x+3 & 2y+x \\ z-1 & 4a-6 \end{bmatrix} = \begin{bmatrix} 0 & -7 \\ 3 & 2a \end{bmatrix}$$

As the given matrices are equal, so their corresponding elements are equal.

$x + 3 = 0$ or $x = -3$

$2y + x = -7$

$2y - 3 = -7 \Rightarrow y = -2$

$z - 1 = 3 \Rightarrow z = 4$

$4a - 6 = 2a \Rightarrow a = 3$

$x = -3, y = -2, z = 4, a = 3$ **Ans.**

2.3 ADDITION OF MATRICES

If A and B be two matrices of the same order, then their sum, $A + B$ is defined as the matrix, each element of which is the sum of the corresponding elements of A and B.

Thus if $$A = \begin{bmatrix} 4 & 2 & 5 \\ 1 & 3 & -6 \end{bmatrix}, B = \begin{bmatrix} 1 & 0 & 2 \\ 3 & 1 & 4 \end{bmatrix}$$

then $$A + B = \begin{bmatrix} 4+1 & 2+0 & 5+2 \\ 1+3 & 3+1 & -6+4 \end{bmatrix} = \begin{bmatrix} 5 & 2 & 7 \\ 4 & 4 & -2 \end{bmatrix}$$

If $A = [a_{ij}],\ B = [b_{ij}]$ then

$$A + B = [a_{ij} + b_{ij}]$$

Theorem 1: Show that any square matrix can be expressed as the sum of two matrices one symmetric and other anti symmetric.

Let 'A' be a given square matrix

Then $$A = \frac{1}{2}(A + A') + \frac{1}{2}(A - A')$$

Now $(A + A')' = A' + A = A + A'$

$\therefore$ $A + A'$ is a symmetric matrix.

Also $(A - A')' = A' - A = -(A - A')$

$A - A'$ or $\frac{1}{2}(A - A')$ is a anti- symmetric matrix.

$$A = \frac{1}{2}(A + A') + \frac{1}{2}(A - A')$$

= symmetric matrix + anti-symmetric matrix

Theorem 2: Write matrix '*A*' given below as the sum of a symmetric and a skew- symmetric matrix.

$$A = \begin{bmatrix} 1 & 2 & 4 \\ -2 & 5 & 3 \\ -1 & 6 & 3 \end{bmatrix}$$

We have,

$$A = \begin{bmatrix} 1 & 2 & 4 \\ -2 & 5 & 3 \\ -1 & 6 & 3 \end{bmatrix}$$

on transposing, we get

$$\therefore \quad A' = \begin{bmatrix} 1 & -2 & -1 \\ 2 & 5 & 6 \\ 4 & 3 & 3 \end{bmatrix}$$

on adding A and A', we get

$$A + A' = \begin{bmatrix} 1 & 2 & 4 \\ -2 & 5 & 3 \\ -1 & 6 & 3 \end{bmatrix} + \begin{bmatrix} +1 & -2 & -1 \\ 2 & 5 & 6 \\ 4 & 3 & 3 \end{bmatrix} = \begin{bmatrix} 2 & 0 & 3 \\ 0 & 10 & 9 \\ 3 & 9 & 6 \end{bmatrix} \quad \text{...(1)}$$

on subtracting A' from A, weget

$$A - A' = \begin{bmatrix} 1 & 2 & 4 \\ -2 & 5 & 3 \\ -1 & 6 & 3 \end{bmatrix} - \begin{bmatrix} 1 & -2 & -1 \\ 2 & 5 & 6 \\ 4 & 3 & 3 \end{bmatrix} = \begin{bmatrix} 0 & 4 & 5 \\ -4 & 0 & -3 \\ -5 & 3 & 0 \end{bmatrix} \quad \text{...(2)}$$

on adding (1) and (2), we have

$$2A = \begin{bmatrix} 2 & 0 & 3 \\ 0 & 10 & 9 \\ 3 & 9 & 6 \end{bmatrix} + \begin{bmatrix} 0 & 4 & 5 \\ -4 & 0 & -3 \\ -5 & 3 & 0 \end{bmatrix}$$

$$A = \begin{bmatrix} 1 & 0 & 3/2 \\ 0 & 5 & 9/2 \\ 3/2 & 9/2 & 3 \end{bmatrix} + \begin{bmatrix} 0 & 2 & 5/2 \\ -2 & 0 & -3/2 \\ -5/2 & 3/2 & 0 \end{bmatrix}$$

= symmetric matrix + skew-symmetric matrix

2.4 PROPERTIES OF MATRIX ADDITION

Only matrix of same order can be added or subtracted.

(a) ***Commutative law*** $A + B = B + A$

(b) ***Associative law*** $A + (B + C) = (A + B) + C$

2.5 SUBTRACTION OF MATRICES

The difference of two matrices is a matrix, each element of which is obtained by subtracting the elements of the second matrix from the corresponding element of the first.

$$A - B = [a_{ij} - b_{ij}]$$

Thus,
$$\begin{bmatrix} 8 & 6 & 4 \\ 1 & 2 & 0 \end{bmatrix} - \begin{bmatrix} 3 & 5 & 1 \\ 7 & 6 & 4 \end{bmatrix} = \begin{bmatrix} 5 & 1 & 3 \\ -6 & -4 & -4 \end{bmatrix}$$

2.6 SCALAR MULTIPLE OF A MATRIX

If a matrix is multiplied by a scalar quantity k, then each element is multiplied by k i.e.,

$$A = \begin{bmatrix} 2 & 3 & 4 \\ 5 & 7 & 8 \\ 6 & 2 & 3 \end{bmatrix}$$

$$3A = \begin{bmatrix} 6 & 9 & 12 \\ 15 & 21 & 24 \\ 18 & 6 & 9 \end{bmatrix}$$

Illustrative Examples

Example 1 If $A = \begin{bmatrix} -1 & 7 & 1 \\ 2 & 3 & 4 \\ 5 & 0 & 5 \end{bmatrix}$, Then Represent it as $A = B + C$, where B is symmetric and C is skew-symmetric.

Solution: We have,

$$A = \begin{bmatrix} -1 & 7 & 1 \\ 2 & 3 & 4 \\ 5 & 0 & 5 \end{bmatrix}$$

on transposing, we get $A' = \begin{bmatrix} -1 & 2 & 5 \\ 7 & 3 & 0 \\ 1 & 4 & 5 \end{bmatrix}$

an adding A and A', we get

$$A + A' = \begin{bmatrix} -2 & 9 & 6 \\ 9 & 6 & 4 \\ 6 & 4 & 10 \end{bmatrix} \quad ...(1)$$

on subtracting A' from A, weget

$$A - A' = \begin{bmatrix} 0 & 5 & -4 \\ -5 & 0 & 4 \\ 4 & -4 & 0 \end{bmatrix} \quad ...(2)$$

on adding (1) and (2), we have

$$2A = \begin{bmatrix} -2 & 14 & 2 \\ 4 & 6 & 8 \\ 10 & 0 & 10 \end{bmatrix}$$

$$= \begin{bmatrix} -2 & 9 & 6 \\ 9 & 6 & 4 \\ 6 & 4 & 10 \end{bmatrix} + \begin{bmatrix} 0 & 5 & -4 \\ -5 & 0 & 4 \\ 4 & -4 & 0 \end{bmatrix}$$

$$A = \begin{bmatrix} -1 & 9/2 & 3 \\ 9/2 & 3 & 2 \\ 3 & 2 & 5 \end{bmatrix} + \begin{bmatrix} 0 & 5/2 & -2 \\ -5/2 & 0 & 2 \\ 2 & -2 & 0 \end{bmatrix}$$

symmetric + Skew-symmetric

Example 2 Express $\begin{bmatrix} 1 & 2 & 0 \\ 3 & 7 & 1 \\ 5 & 9 & 3 \end{bmatrix}$ as a sum of symmetric and skew- symmetric matrix.

Solution: Let,

$$A = \begin{bmatrix} 1 & 2 & 0 \\ 3 & 7 & 1 \\ 5 & 9 & 3 \end{bmatrix}$$

On transposing, we have

$$A' = \begin{bmatrix} 1 & 3 & 5 \\ 2 & 7 & 9 \\ 0 & 1 & 3 \end{bmatrix}$$

Adding A + A',

$$A + A' = \begin{bmatrix} 2 & 5 & 5 \\ 5 & 14 & 10 \\ 5 & 10 & 6 \end{bmatrix} \quad ...(1)$$

On subtracting, A' from A we have

$$A - A' = \begin{bmatrix} 0 & -1 & -5 \\ 1 & 0 & -8 \\ 5 & 8 & 0 \end{bmatrix} \quad ...(2)$$

Adding (1) + (2) $\Rightarrow$ we get

$$2A = \begin{bmatrix} 2 & 5 & 5 \\ 5 & 14 & 10 \\ 5 & 10 & 6 \end{bmatrix} + \begin{bmatrix} 0 & -1 & -5 \\ 1 & 0 & -8 \\ 5 & 8 & 0 \end{bmatrix}$$

$$A = \begin{bmatrix} 1 & 5/2 & 5/2 \\ 5/2 & 7 & 5 \\ 5/2 & 5 & 3 \end{bmatrix} + \begin{bmatrix} 0 & -1/2 & -5/2 \\ 1/2 & 0 & -4 \\ 5/2 & 4 & 0 \end{bmatrix}$$

= Symmetric + skew-symmetric

Example 3. If $A = \begin{bmatrix} 0 & 2 & 0 \\ 1 & 0 & 3 \\ 1 & 1 & 2 \end{bmatrix}$, and $B = \begin{bmatrix} 1 & 2 & 1 \\ 2 & 1 & 0 \\ 0 & 0 & 3 \end{bmatrix}$, then Find $2A + 3B$

Solution:

$$2A + 3B = \begin{bmatrix} 0 & 4 & 0 \\ 2 & 0 & 6 \\ 2 & 2 & 4 \end{bmatrix} + \begin{bmatrix} 3 & 6 & 3 \\ 6 & 3 & 0 \\ 0 & 0 & 9 \end{bmatrix}$$

$$= \begin{bmatrix} 3 & 10 & 3 \\ 8 & 3 & 6 \\ 2 & 2 & 13 \end{bmatrix}$$

Example 4. If 'A' is a skew-symmetric matrix of odd order, then By taking an example prove that the determinant of 'A' is always '0'

Solution: Let (1) $A = \begin{bmatrix} 0 & 2 & 3 \\ -2 & 0 & 4 \\ -3 & -4 & 0 \end{bmatrix}$, '$A$' is odd order matrix, then

$$|A| = 0\,(0 + 16) - 2\,(0 + 12) + 3\,(8 + 0)$$
$$= 0 - 24 + 24 = 0$$

(2) If $A = \begin{bmatrix} 0 & 1 \\ -1 & 0 \end{bmatrix}$ i.e., 'A' is of even order matrix, then

$|A| = 0 + 1 = 1$ i.e., a real number.

2.7 MULTIPLICATION

The product of two matrices A and B is only possible if the number of columns of 'A' is equal to the number of rows in B.

Let $A = [a_{ij}]$ be an mxn matrix and

$B = [b_{ij}]$ be an nxp matrix.

Then, the product AB of these matrices is an mxp matrix $C = [c_{ij}]$ where

$$c_{ij} = a_{i1}\, b_{1j} + a_{i2}\, b_{2j} + \ldots\ldots\ldots\ldots\ldots\ldots + a_{in} b_{nj}$$

2.8 TO PROVE (*AB*)' = *B'A'*, WE HAVE A THEOREM

Theorem 3 : If A and B are two matrices conformal for product AB, then show that $(AB)' = B'A'$

Let $A = (a_{ij})$ be an mxn matrix,

$B = (b_{ij})$ be an nxp matrix

then $AB = mxp$ matrix

then $(AB)' = pxm$ matrix

further B' is an pxn matrix

A' is an nxm matrix

then $B'A'$ is an pxm matrix

Thus (AB)' and B'A' is of same order

Now $(j, i)^{th}$ element of (AB)'

$= (i, j)^{th}$ element of B'A'

$$\sum_{k=1}^{n} a_{ik}\, b_{kj} \qquad \ldots(1)$$

Also the jth row of B' is b_{1j}, b_{2j} .... b_{nj}

and i^{th} column of A' is $a_{i1}, a_{i2}, \ldots\ldots\ldots\ldots a_{in}$

$\therefore$ $(j, i)^{th}$ element of $B'A' = \sum_{k=1}^{n} b_{jk}\, a_{ik}$

As the matrices (AB)' and B'A' are of same order and their corresponding elements are equal, we have (AB)' = B'A'

2.9 PROPERTIES OF MATRIX MULTIPLICATION

(1) Multiplication of Matrices is not commutative

$$AB \neq BA$$

(2) Matrix multiplication is associative, if confirmability is assured.

$$A(BC) = (AB)C$$

(3) Matrix multiplication is distributive with respect to addition

$$A(B + C) = AB + AC$$

(4) Multiplication of matrix A by unit matrix is a matrix itself

$$AI = IA = A$$

(5) Multiplicative inverse of a matrix exists

$$\text{if } |A| \neq 0$$

$$A.A^{-1} = A^{-1}.A = I$$

Example 5 If $A = \begin{bmatrix} 0 & 1 & 2 \\ 1 & 2 & 3 \\ 2 & 3 & 4 \end{bmatrix}$ and $B = \begin{bmatrix} 1 & -2 \\ -1 & 0 \\ 2 & -1 \end{bmatrix}$

Then, obtain the product AB and explain why BA is not defined?

Solution: Since the number of columns of A (3×3) = number of rows of B (3×2) therefore the product AB is defined.

$$AB = \begin{bmatrix} 0 & 1 & 2 \\ 1 & 2 & 3 \\ 2 & 3 & 4 \end{bmatrix}_{3 \times 3} \begin{bmatrix} 1 & -2 \\ -1 & 0 \\ 2 & -1 \end{bmatrix}_{3 \times 2}$$

$$= \begin{bmatrix} 0-1+4 & 0+0-2 \\ 1-2+6 & -2+0-3 \\ 2-3+8 & -4+0-4 \end{bmatrix} = \begin{bmatrix} 3 & -2 \\ 5 & -5 \\ 7 & -8 \end{bmatrix}_{3 \times 2}$$

Now, for BA,

the number of columns of B $(3 \times 2) \neq$ number of rows of A(3×3).

therefore the product BA is not defined.

Example 6 If $A = \begin{bmatrix} 1 & 2 & 2 \\ 2 & 1 & 2 \\ 2 & 2 & 1 \end{bmatrix}$, Showthat $A^2 - 4A - 5I = 0$, where I,O are unit and null matrix of order '3' respectively. Use this result to find A^{-1}.

Solution: We have $A = \begin{bmatrix} 1 & 2 & 2 \\ 2 & 1 & 2 \\ 2 & 2 & 1 \end{bmatrix}$, then $A^2 = \begin{bmatrix} 1 & 2 & 2 \\ 2 & 1 & 2 \\ 2 & 2 & 1 \end{bmatrix}\begin{bmatrix} 1 & 2 & 2 \\ 2 & 1 & 2 \\ 2 & 2 & 1 \end{bmatrix}$

$$= \begin{bmatrix} 1+4+4 & 2+2+4 & 2+4+2 \\ 2+2+4 & 4+1+4 & 4+2+2 \\ 2+4+2 & 4+2+2 & 4+4+1 \end{bmatrix}$$

$$= \begin{bmatrix} 9 & 8 & 8 \\ 8 & 9 & 8 \\ 8 & 8 & 9 \end{bmatrix}$$

$$A^2 - 4A - 5I = \begin{bmatrix} 9 & 8 & 8 \\ 8 & 9 & 8 \\ 8 & 8 & 9 \end{bmatrix} - \begin{bmatrix} 4 & 8 & 8 \\ 8 & 4 & 8 \\ 8 & 8 & 4 \end{bmatrix} - \begin{bmatrix} 5 & 0 & 0 \\ 0 & 5 & 0 \\ 0 & 0 & 5 \end{bmatrix}$$

$$= \begin{bmatrix} 9-4-5 & 8-8-0 & 8-8-0 \\ 8-8-0 & 9-4-5 & 8-8-0 \\ 8-8-0 & 8-8-0 & 9-4-5 \end{bmatrix} = \begin{bmatrix} 0 & 0 & 0 \\ 0 & 0 & 0 \\ 0 & 0 & 0 \end{bmatrix}$$

Now to find A^{-1},

$5I = A^2 - 4A$

multiplying the above equation on both sides by A^{-1},

$5A^{-1} = A - 4I$

$$A^{-1} = \frac{1}{5}\left[\begin{bmatrix} 1 & 2 & 2 \\ 2 & 1 & 2 \\ 2 & 2 & 1 \end{bmatrix} - \begin{bmatrix} 4 & 0 & 0 \\ 0 & 4 & 0 \\ 0 & 0 & 4 \end{bmatrix}\right]$$

$$= \frac{1}{5}\begin{bmatrix} -3 & 2 & 2 \\ 2 & -3 & 2 \\ 2 & 2 & -3 \end{bmatrix}$$

Example 7 Show that, by means of an example that in matrices AB = 0 does not necessarily mean that either A = 0 or B = 0, where '0' stands for null matrix.

Solution: Let $A = \begin{bmatrix} 1 & -1 \\ 1 & -1 \end{bmatrix}$, $B = \begin{bmatrix} -1 & -1 \\ -1 & -1 \end{bmatrix}$

$$AB = \begin{bmatrix} -1+1 & -1+1 \\ -1+1 & -1+1 \end{bmatrix} = \begin{bmatrix} 0 & 0 \\ 0 & 0 \end{bmatrix}$$

where neither A = 0 nor B = 0

but AB = 0

Example 8 Prove that If AB = AC, it is not necessarily true that B = C i.e., like ordinary algebra, the equal matrices in the identity cannot be cancelled.

Solution: Let $A = \begin{bmatrix} 1 & -1 \\ 1 & -1 \end{bmatrix}$, $B = \begin{bmatrix} -1 & -1 \\ -1 & -1 \end{bmatrix}$, $C = \begin{bmatrix} -2 & -2 \\ -2 & -2 \end{bmatrix}$

then $AB = \begin{bmatrix} 0 & 0 \\ 0 & 0 \end{bmatrix}$ and

$AC = \begin{bmatrix} 0 & 0 \\ 0 & 0 \end{bmatrix}$ but $B \neq C$

Example 9 Represent each of the transformations $x_1 = 3y_1 + 2y_2$, $y_1 = z_1 + 2z_2$ and $x_2 = -y_1 + 4y_2$, $y_2 = 3z_1$

by the use of matrices and find the composite transformation which expresses x_1, x_2 in terms of z_1, z_2.

Solution: Given equation in the matrix form is,

$$\begin{bmatrix} x_1 \\ x_2 \end{bmatrix} = \begin{bmatrix} 3 & 2 \\ -1 & 4 \end{bmatrix} \begin{bmatrix} y_1 \\ y_2 \end{bmatrix} \qquad \ldots(1)$$

$$\begin{bmatrix} y_1 \\ y_2 \end{bmatrix} = \begin{bmatrix} 1 & 2 \\ 3 & 0 \end{bmatrix} \begin{bmatrix} z_1 \\ z_2 \end{bmatrix} \qquad \ldots(2)$$

putting the value of y_1, y_2 from (2) in (1)

$$\begin{bmatrix} x_1 \\ x_2 \end{bmatrix} = \begin{bmatrix} 3 & 2 \\ -1 & 4 \end{bmatrix} \begin{bmatrix} 1 & 2 \\ 3 & 0 \end{bmatrix} \begin{bmatrix} z_1 \\ z_2 \end{bmatrix}$$

$$= \begin{bmatrix} 3+6 & 6+0 \\ -1+12 & -2+0 \end{bmatrix} \begin{bmatrix} z_1 \\ z_2 \end{bmatrix} = \begin{bmatrix} 9z_1 + 6z_2 \\ 11z_1 - 2z_2 \end{bmatrix}$$

$$x_1 = 9z_1 + 6z_2$$

$$\Rightarrow \quad x_2 = 11z_1 - 2z_2$$

Example 10 If A, B and C are matrices such that the products AC and BC obey the commutative law, then

Prove that

$$(AB + BA)\,C = C(AB + BA)$$

Solution: Let (AB + BA) C = (AB)C + (BA)C (distributive law)

$= A(BC) + B(AC)$ (Associative law)

$= A(CB) + B(CA)$ $\because$ BC, AC obey the commutative law

$= (AC)B + (BC)A$ Associative law

$= (CA)B + (CB)A$ CA, CB obey commutative law

$= C(AB) + C(BA)$ Associative law

$= C[AB + BA]$ distributive law

Example 11. Prove that the product of two matrices

$$A = \begin{bmatrix} \cos^2\theta & \cos\theta\sin\theta \\ \cos\theta\sin\theta & \sin^2\theta \end{bmatrix} \textit{ and } B = \begin{bmatrix} \cos^2\phi & \cos\phi\sin\phi \\ \cos\phi\sin\phi & \sin^2\phi \end{bmatrix}$$

is zero when θ and ϕ differ by an odd multiple of π/2.

Solution: We have

$$AB = \begin{bmatrix} \cos^2\theta\cos^2\phi + \cos\theta\sin\theta\cos\phi\sin\phi & \cos^2\theta\cos\phi\sin\phi + \sin^2\phi\sin\theta\cos\theta \\ \cos\theta\sin\theta\cos^2\phi + \sin^2\theta\cos\phi\sin\phi & \cos\theta\sin\theta\cos\phi\sin\phi + \sin^2\theta\sin^2\phi \end{bmatrix}$$

$$= \begin{bmatrix} \cos\theta\cos\phi\,(\cos\theta\cos\phi + \sin\theta\sin\phi) & \cos\theta\sin\phi\,(\cos\theta\cos\phi + \sin\theta\sin\phi) \\ \sin\theta\cos\phi\,(\cos\theta\cos\phi + \sin\theta\sin\phi) & \sin\theta\sin\phi\,(\cos\theta\cos\phi + \sin\theta\sin\phi) \end{bmatrix}$$

or

$$= \begin{bmatrix} \cos\theta\cos\phi\cos(\theta-\phi) & \cos\theta\sin\phi\cos(\theta-\phi) \\ \sin\theta\cos\phi\cos(\theta-\phi) & \sin\theta\sin\phi\cos(\theta-\phi) \end{bmatrix} \quad - \quad (A)$$

Given $\theta - \phi = (2n + 1)\,\pi/2$

and $\cos(\theta - \phi) = \cos(2n + 1)\,\pi/2 = 0$

$\therefore$ from equation A, the product becomes $\begin{bmatrix} 0 & 0 \\ 0 & 0 \end{bmatrix} = 0$

Example 12 Verify that $A = \frac{1}{3}\begin{bmatrix} 1 & 2 & 2 \\ 2 & 1 & -2 \\ -2 & 2 & -1 \end{bmatrix}$ is orthogonal.

Solution: Here $A = \frac{1}{3}\begin{bmatrix} 1 & 2 & 2 \\ 2 & 1 & -2 \\ -2 & 2 & -1 \end{bmatrix}$

$$\therefore \quad A' = \frac{1}{3}\begin{bmatrix} 1 & 2 & -2 \\ 2 & 1 & 2 \\ 2 & -2 & -1 \end{bmatrix}$$

$$\text{and } AA' = \frac{1}{9}\begin{bmatrix} 1 & 2 & 2 \\ 2 & 1 & -2 \\ -2 & 2 & -1 \end{bmatrix}\begin{bmatrix} 1 & 2 & -2 \\ 2 & 1 & 2 \\ 2 & -2 & -1 \end{bmatrix}$$

$$= \frac{1}{9}\begin{bmatrix} 9 & 0 & 0 \\ 0 & 9 & 0 \\ 0 & 0 & 9 \end{bmatrix} = \begin{bmatrix} 1 & 0 & 0 \\ 0 & 1 & 0 \\ 0 & 0 & 1 \end{bmatrix} = I$$

Hence 'A' is an orthogonal matrix.

Example 13 Determine the values of α, β,γ when $\begin{bmatrix} 0 & 2\beta & \gamma \\ \alpha & \beta & -\gamma \\ \alpha & -\beta & \gamma \end{bmatrix}$ is orthogonal.

Solution: Let

$$A = \begin{bmatrix} 0 & 2\beta & \gamma \\ \alpha & \beta & -\gamma \\ \alpha & -\beta & \gamma \end{bmatrix}$$

on transposing A, we have

$$A' = \begin{bmatrix} 0 & \alpha & \alpha \\ 2\beta & \beta & -\beta \\ \gamma & -\gamma & \gamma \end{bmatrix}$$

If A is orthogonal, then AA' = I

$$\therefore \quad AA' = \begin{bmatrix} 0 & 2\beta & \gamma \\ \alpha & \beta & -\gamma \\ \alpha & -\beta & \gamma \end{bmatrix}\begin{bmatrix} 0 & \alpha & \alpha \\ 2\beta & \beta & -\beta \\ \gamma & -\gamma & \gamma \end{bmatrix} = \begin{bmatrix} 1 & 0 & 0 \\ 0 & 1 & 0 \\ 0 & 0 & 1 \end{bmatrix}$$

$$= \begin{bmatrix} 0+4\beta^2+\gamma^2 & 0+2\beta^2-\gamma^2 & 0-2\beta^2+\gamma^2 \\ 0+2\beta^2-\gamma^2 & \alpha^2+\beta^2+\gamma^2 & \alpha^2-\beta^2-\gamma^2 \\ 0-2\beta^2+\gamma^2 & \alpha^2-\beta^2-\gamma^2 & \alpha^2+\beta^2+\gamma^2 \end{bmatrix} = \begin{bmatrix} 1 & 0 & 0 \\ 0 & 1 & 0 \\ 0 & 0 & 1 \end{bmatrix}$$

$\Rightarrow \quad 4\beta^2 + \gamma^2 = 1 \qquad \ldots(1)$

$-2\beta^2 + \gamma^2 = 0 \quad \Rightarrow \quad \beta^2 = \frac{\gamma^2}{2}$

$(1) \quad \Rightarrow \quad 4.\frac{\gamma^2}{2} + \gamma^2 = 1$

$$2\gamma^2 + \gamma^2 = 1 \Rightarrow 3\gamma^2 = 1 \Rightarrow \boxed{\gamma = \pm\frac{1}{\sqrt{3}}}$$

Also $$\beta^2 = \frac{1}{6} \Rightarrow \boxed{\beta = \pm\frac{1}{\sqrt{6}}}$$

Also, we have $$\alpha^2 = 1 - \beta^2 - \gamma^2$$

$$= 1 - \frac{1}{6} - \frac{1}{3}$$

$$\alpha^2 = \frac{6-1-2}{6} = \frac{1}{2}$$

$$\Rightarrow \quad \boxed{\alpha = \pm\frac{1}{\sqrt{2}}}$$

Example 14. Write the following transformation in matrix form.

$$x_1 = \frac{\sqrt{3}}{2}y_1 + \frac{1}{2}y_2 \quad \text{....(1)}; \quad x_2 = \frac{-1}{2}y_1 + \frac{\sqrt{3}}{2}y_2 \quad \text{...(2)}$$

Hence, find the transformation in matrix form which expresses y_1, y_2 in terms fo x_1, x_2.

Solution: Let
$$\begin{bmatrix} x_1 \\ x_2 \end{bmatrix} = \begin{bmatrix} \frac{\sqrt{3}}{2} & \frac{1}{2} \\ -\frac{1}{2} & \frac{\sqrt{3}}{2} \end{bmatrix} \begin{bmatrix} y_1 \\ y_2 \end{bmatrix}$$

Multiply (1) by $\frac{1}{2}$ and (2) by $\frac{\sqrt{3}}{2}$ and adding, we get

$$+\frac{1}{2}x_1 = \frac{\sqrt{3}}{4}y_1 + \frac{1}{4}y_2$$

$$\underset{(+)}{}\frac{\sqrt{3}}{2}x_2 = \underset{(+)}{}-\frac{\sqrt{3}}{4}y_1 + \underset{(+)}{}\frac{3}{4}y_2$$

$$+\frac{1}{2}x_1 + \frac{\sqrt{3}}{2}x_2 = \frac{3}{4}y_2 + \frac{1}{4}y_2$$

$$\Rightarrow \quad \boxed{y_2 = \frac{\sqrt{3}}{2}x_2 + \frac{1}{2}x_1} \quad \text{...(3)}$$

put the value of y_2 from equation (3) to equation (1), we get

$$x_1 = \frac{\sqrt{3}}{2} y_1 + \frac{\sqrt{3}}{4} x_2 + \frac{1}{4} x_1$$

or
$$x_1 - \frac{1}{4} x_1 - \frac{\sqrt{3}}{4} x_2 = \frac{\sqrt{3}}{2} y_1$$

$$\frac{3}{4} x_1 - \frac{\sqrt{3}}{4} x_2 = \frac{\sqrt{3}}{2} y_1$$

$$\Rightarrow \quad \boxed{\frac{\sqrt{3}x_1}{2} - \frac{x_2}{2} = y_1} \qquad \text{...(4)}$$

By (3) and (4), we have

$$\begin{bmatrix} y_1 \\ y_2 \end{bmatrix} = \begin{bmatrix} \sqrt{3/2} & -1/2 \\ +1/2 & \sqrt{3/2} \end{bmatrix} \begin{bmatrix} x_1 \\ x_2 \end{bmatrix}$$

$$\Rightarrow \quad \boxed{\begin{aligned} y_1 &= \frac{\sqrt{3}}{2} x_1 - \frac{1}{2} x_2 \\ y_2 &= \tfrac{1}{2} x_1 + \frac{\sqrt{3}}{2} x_2 \end{aligned}}$$

which is the required Answers.

Example 15 If $A = \begin{bmatrix} 0 & -\tan \alpha/2 \\ \tan \alpha/2 & 0 \end{bmatrix}$ and I is a unit matrix, showthat $I + A = (I - A) \begin{bmatrix} \cos\alpha & -\sin\alpha \\ \sin\alpha & \cos\alpha \end{bmatrix}$

Hint: Let $I = \begin{bmatrix} 1 & 0 \\ 0 & 1 \end{bmatrix}$, $A = \begin{bmatrix} 0 & -\tan\alpha/2 \\ \tan\alpha/2 & 0 \end{bmatrix}$, then we have

$$I + A = \begin{bmatrix} 1 & -\tan\alpha/2 \\ \tan\alpha/2 & 1 \end{bmatrix}$$

and
$$I - A = \begin{bmatrix} 1 & \tan\alpha/2 \\ -\tan\alpha/2 & 1 \end{bmatrix}$$

$$(I - A) \begin{bmatrix} \cos\alpha & -\sin\alpha \\ \sin\alpha & \cos\alpha \end{bmatrix} = \begin{bmatrix} 1 & \tan\alpha/2 \\ -\tan\alpha/2 & 1 \end{bmatrix} \begin{bmatrix} \cos\alpha & -\sin\alpha \\ \sin\alpha & \cos\alpha \end{bmatrix}$$

After multiplying the above matrices, we will get the answer.

Exercise 2.1

1. If $\begin{bmatrix} 4 & 1 & 2 \\ 0 & 5 & 3 \end{bmatrix}\begin{bmatrix} 3 & 4 & 5 \\ -1 & 0 & -2 \\ 3 & 4 & 7 \end{bmatrix} = \begin{bmatrix} 8x+3y & 6z & 3z \\ 4 & 12 & 26x-5y \end{bmatrix}$, find the value of x, y, z.

2. Given $A = \begin{bmatrix} 1 & 2 \\ 3 & 4 \end{bmatrix}$, $B = \begin{bmatrix} 2 & 1 \\ 4 & 2 \end{bmatrix}$, $C = \begin{bmatrix} 5 & 1 \\ 7 & 4 \end{bmatrix}$, verify that A(B + C) = AB + AC and (A+B) C = AC + BC.

3. Show that the matrix

$$A = \begin{bmatrix} 1 & -3 & -4 \\ -1 & 3 & 4 \\ 1 & -3 & -4 \end{bmatrix}$$ is nilpotent of index 2.

4. If $A = \begin{bmatrix} 1 & 0 & 5 & 3 \\ -2 & 1 & 6 & 1 \\ 3 & 2 & 7 & 1 \\ 4 & -4 & -2 & 0 \end{bmatrix}$ Find A + A' and A – A'.

5. If $A = \begin{bmatrix} 3 & 1 & -1 \\ 0 & 1 & 2 \end{bmatrix}$, Find AA' and A'A

Answers

1. x = 1, y = 3, z = 4

4. $A + A' = \begin{bmatrix} 2 & -2 & 8 & 7 \\ -2 & 2 & 8 & -3 \\ 8 & 8 & 14 & -1 \\ 7 & -3 & -1 & 0 \end{bmatrix}$ and

$$A - A' = \begin{bmatrix} 0 & 2 & 2 & -1 \\ -2 & 0 & 4 & 5 \\ -2 & -4 & 0 & 3 \\ 1 & -5 & -3 & 0 \end{bmatrix}$$

5. $AA' = \begin{bmatrix} 11 & -1 \\ -1 & 5 \end{bmatrix}$ and $A'A = \begin{bmatrix} 9 & 3 & -3 \\ 3 & 2 & 1 \\ -3 & 1 & 5 \end{bmatrix}$

Note: Let A and B be any two matrices such that AB = 0 and A is non-singular. then B is singular.

2.10 ADJOINT OF A SQUARE MATRIX

Let the determinant of a square matrix A be $|A|$

If $A = \begin{bmatrix} a_1 & a_2 & a_3 \\ b_1 & b_2 & b_3 \\ c_1 & c_2 & c_3 \end{bmatrix}$, then $|A| = \begin{vmatrix} a_1 & a_2 & a_3 \\ b_1 & b_2 & b_3 \\ c_1 & c_2 & c_3 \end{vmatrix}$

The matrix formed by the co-factors of the elements in $|A|$ is,

$$\begin{bmatrix} A_1 & A_2 & A_3 \\ B_1 & B_2 & B_3 \\ C_1 & C_2 & C_3 \end{bmatrix}$$

where $A_1 = \begin{vmatrix} b_2 & b_3 \\ c_2 & c_3 \end{vmatrix} = (b_2 c_3 - c_2 b_3)$

$$A_2 = -\begin{vmatrix} b_1 & b_3 \\ c_1 & c_3 \end{vmatrix} = -(b_1 c_3 - b_3 c_1)$$

$$A_3 = +\begin{vmatrix} b_1 & b_2 \\ c_1 & c_2 \end{vmatrix} = (b_1 c_2 - c_1 b_2)$$

$$B_1 = -\begin{vmatrix} a_2 & a_3 \\ c_2 & c_3 \end{vmatrix} = -(a_2 c_3 - c_2 a_3)$$

$$B_2 = \begin{vmatrix} a_1 & a_3 \\ c_1 & c_3 \end{vmatrix} = a_1 c_3 - c_1 a_3$$

$$B_3 = -\begin{vmatrix} a_1 & a_2 \\ c_1 & c_2 \end{vmatrix} = -(a_1 c_2 - a_2 c_1)$$

$$C_1 = \begin{vmatrix} a_2 & a_3 \\ b_2 & b_3 \end{vmatrix} = a_2 b_3 - a_3 b_2$$

$$C_2 = -\begin{vmatrix} a_1 & a_3 \\ b_1 & b_3 \end{vmatrix} = -(a_1 b_3 - b_1 a_3)$$

$$C_3 = \begin{vmatrix} a_1 & a_2 \\ b_1 & b_2 \end{vmatrix} = a_1 b_2 - a_2 b_1$$

Then the transpose of the matrix of co-factors

$$\begin{bmatrix} A_1 & B_1 & C_1 \\ A_2 & B_2 & C_2 \\ A_3 & B_3 & C_3 \end{bmatrix}$$

is called the adjoint of the matrix A and is written as adj. A.

2.11 PROPERTY OF ADJOINT

The product of a matrix 'A' and its adjoint is equal to unit matrix multiplied by the determinant A.

If 'A' be a square matrix, then (adj. A). A = A. (adj. A) = $|A|.I$

2.12 INVERSE OF A MATRIX

If A and B be two square matrices of the same order, such that

AB = BA = I

then 'B' is called the inverse of A i.e., B = A^{-1} and 'A' is the inverse of B.

Condition for a square matrix 'A' to possess an inverse is that matrix 'A' is non-singular i.e. $|A| \neq 0$

To find the inverse of a matrix 'A' with the help of its adjoint, we have

$$A^{-1} = \frac{adj.\ A}{|A|}$$

Illustrative Examples

Example 1 If $A = \begin{bmatrix} 3 & -3 & 4 \\ 2 & -3 & 4 \\ 0 & -1 & 1 \end{bmatrix}$, find A^{-1}

Solution: We have $|A| = 3(-3+4) + 3(2-0) + 4(-2+0)$

$= 3 + 6 - 8 = 1 \neq 0$

$\Rightarrow$ inverse of 'A' exists, then we have

$a_{11} = -3 + 4 = 1$ $\quad a_{21} = -1$ $\quad a_{31} = 0$

$a_{12} = -2$ $\quad a_{22} = 3$ $\quad a_{32} = -4$

$a_{13} = -2$ $\quad a_{23} = 3$ $\quad a_{33} = -3$

and Adj. A = $\begin{bmatrix} 1 & -1 & 0 \\ -2 & 3 & -4 \\ -2 & 3 & -3 \end{bmatrix}$

$\Rightarrow$ $A^{-1} = \frac{adj.A}{|A|} = \begin{bmatrix} 1 & -1 & 0 \\ -2 & 3 & -4 \\ -2 & 3 & -3 \end{bmatrix}$ Hence the result.

Example 2 If $A=\frac{1}{9}\begin{bmatrix}-8 & 1 & 4\\ 4 & 4 & 7\\ 1 & -8 & 4\end{bmatrix}$, Prove that $A^{-1} = A'$, A' being the transpose of A.

Solution: We have,

$$A=\frac{1}{9}\begin{bmatrix}-8 & 1 & 4\\ 4 & 4 & 7\\ 1 & -8 & 4\end{bmatrix}, \text{ and } A'=\frac{1}{9}\begin{bmatrix}-8 & 4 & 1\\ 1 & 4 & -8\\ 4 & 7 & 4\end{bmatrix}$$

then,

$$AA'=\frac{1}{81}\begin{bmatrix}-8 & 1 & 4\\ 4 & 4 & 7\\ 1 & -8 & 4\end{bmatrix}\begin{bmatrix}-8 & 4 & 1\\ 1 & 4 & -8\\ 4 & 7 & 4\end{bmatrix}$$

$$=\frac{1}{81}\begin{bmatrix}64+1+16 & -32+4+28 & -8-8+16\\ -32+4+28 & 16+16+49 & 4-32+28\\ -8-8+16 & 4-32+28 & 1+64+16\end{bmatrix}$$

$$=\frac{1}{81}\begin{bmatrix}81 & 0 & 0\\ 0 & 81 & 0\\ 0 & 0 & 81\end{bmatrix}=\begin{bmatrix}1 & 0 & 0\\ 0 & 1 & 0\\ 0 & 0 & 1\end{bmatrix}$$

$$\Rightarrow \quad AA'=I \Rightarrow \boxed{A'=A^{-1}}$$

Example 3 If A and B are non-singular matrices of the same order, then $(AB)^{-1} = B^{-1}.A^{-1}$

Hence prove that $(A^{-1})^m = (A^m)^{-1}$ for any positive integer 'm'.

Solution: We know that

$$(AB).(B^{-1}.A^{-1}) = [(A\,B)(B^{-1}A^{-1})] = [A\,(BB^{-1})\,A^{-1}] = (AI\,A^{-1}) = I$$

Also,

$$(B^{-1}.A^{-1})(AB) = [(B^{-1}A^{-1})(AB)] = [B^{-1}(A^{-1}A)\,B] = (B^{-1}\,I\,B) = I$$

By definition of the inverse of a matrix $B^{-1}A^{-1}$ is the inverse of AB.

or $(AB)^{-1} = B^{-1}A^{-1}$.

$$\begin{aligned}\Rightarrow \quad (A^m)^{-1} &= [AA^{m-1}]^{-1} = (A^{m-1})^{-1}A^{-1}\\ &= [A\,A^{m-2}]^{-1}A^{-1}\\ &= [(A^{m-2})^{-1}A^{-1}]\,A^{-1}\\ &= [A^{m-2}]^{-1}[A^{-1}]^2\\ &= [A\,A^{m-3}]^{-1}[A^{-1}]^2\\ &= [A^{m-3}]^{-1}[A^{-1}]^3\\ &= A^{-1}(A^{-1})^{m-3} = (A^{-1})^m.\end{aligned}$$

Hence the result.

Theorem 4: P.T. the inverse of matrix in unique

Proof : Let B and C be two inverse matrices of a given matrix say A then AB = I = BA, and AC = I = CA

C.(AB) = (AC). B (Associative law)

$\Rightarrow$ I . C = B . I ($\because$ AB = AC = I

or B = C

Hence the invese of a matrix 'A' is unique.

Example 4 Find the adjoint and inverse of

$$A = \begin{bmatrix} 1 & 2 & 5 \\ 3 & 1 & 4 \\ 1 & 1 & 2 \end{bmatrix}$$

Solution: For the given A, we have

$|A| = 1(2-4) - 2(6-4) + 5(3-1)$

$= -2 - 4 + 10 = 4 \neq 0$

$a_{11} = -2 \quad a_{21} = 1 \quad a_{31} = 3$

$a_{12} = -2 \quad a_{22} = -3 \quad a_{32} = 11$

$a_{13} = 2 \quad a_{23} = 1 \quad a_{33} = -5$

$$adj.\ A = \frac{1}{4}\begin{bmatrix} -2 & 1 & 3 \\ -2 & -3 & 11 \\ 2 & 1 & -5 \end{bmatrix}$$ and Inverse can be find by using any method.

Example 5 Let I be a matrix of order 'n' and adj. (2I) = 2^k.I, then prove that k equals n–1

Solution: Let, $I = \begin{bmatrix} 1 & 0 & 0 \\ 0 & 1 & 0 \\ 0 & 0 & 1 \end{bmatrix}$, then

$a_{11} = 4, \quad a_{21} = 0, \quad a_{31} = 0$

$a_{12} = 0, \quad a_{22} = 4, \quad a_{32} = 0$

$a_{13} = 0, \quad a_{23} = 0, \quad a_{33} = 4$

and $2I = \begin{bmatrix} 2 & 0 & 0 \\ 0 & 2 & 0 \\ 0 & 0 & 2 \end{bmatrix}$

$\Rightarrow \quad |2I| = 2(4) = 8$

$$\Rightarrow \quad adj.(2I) = \begin{bmatrix} 4 & 0 & 0 \\ 0 & 4 & 0 \\ 0 & 0 & 4 \end{bmatrix} = 4\begin{bmatrix} 1 & 0 & 0 \\ 0 & 1 & 0 \\ 0 & 0 & 1 \end{bmatrix} = 2^2.I$$

Note If I is a unit matrix of order 3 then adj (2I) = 2^2 . I and adj (I) = I itself

Exercise - 2.2

1. Find the adjoint of matrix $\begin{bmatrix} 3 & 5 & 7 \\ 2 & -3 & 1 \\ 1 & 1 & 2 \end{bmatrix}$ and verify the theorem A(adj A) = (adj A) A = $|A|I_3$.

2. Find the adjoint of matrix $\begin{bmatrix} 2 & 1 & 3 \\ 3 & 1 & 2 \\ 1 & 2 & 3 \end{bmatrix}$ and show that A(adj A) = (adj A) A = $|A|\ I_3$

3. Calculate the inverse of the following matrix, where ever exists:-

(a) $\begin{bmatrix} 1 & 2 & 1 \\ 3 & 1 & 2 \\ 0 & 1 & 2 \end{bmatrix}$

(b) $\begin{bmatrix} 0 & 0 & 1 \\ 0 & 1 & 0 \\ 1 & 0 & 0 \end{bmatrix}$

(c) $\begin{bmatrix} 9 & 5 & 6 \\ 7 & -1 & 8 \\ 3 & 4 & 2 \end{bmatrix}$

4. Find inverse of the matrix A, where A = $\begin{bmatrix} 1 & 1 & 3 \\ 1 & 3 & -3 \\ -2 & -4 & -4 \end{bmatrix}$

5. Let $D = \begin{bmatrix} d_1 & 0 & 0 & 0 \\ 0 & d_2 & 0 & 0 \\ 0 & 0 & d_3 & 0 \\ 0 & 0 & 0 & d_4 \end{bmatrix}$ where none of d_1, d_2, d_3, d_4 is zero. Find D^{-1}.

Answers

1. $\begin{bmatrix} -7 & -3 & 26 \\ -3 & -1 & 11 \\ 5 & 2 & -19 \end{bmatrix}$

2. $\begin{bmatrix} -1 & 3 & -1 \\ -7 & 3 & +5 \\ 5 & -3 & -1 \end{bmatrix}$

3. (a) $\begin{bmatrix} 0 & \frac{1}{3} & \frac{-1}{3} \\ \frac{2}{3} & \frac{-2}{9} & \frac{-1}{9} \\ \frac{-1}{3} & \frac{1}{9} & \frac{5}{9} \end{bmatrix}$ (b) $\begin{bmatrix} 0 & 0 & 1 \\ 0 & 1 & 0 \\ 1 & 0 & 0 \end{bmatrix}$ (c) $\begin{bmatrix} \frac{17}{35} & \frac{-1}{5} & \frac{-23}{35} \\ -1/7 & 0 & 3/7 \\ -31/70 & \frac{3}{10} & \frac{22}{35} \end{bmatrix}$

4. $\begin{bmatrix} 3 & 1 & \frac{3}{2} \\ \frac{-5}{4} & \frac{-1}{4} & \frac{-3}{4} \\ \frac{-1}{4} & -\frac{1}{4} & \frac{-1}{4} \end{bmatrix}$

5. $\begin{bmatrix} \frac{1}{d_1} & 0 & 0 & 0 \\ 0 & \frac{1}{d_2} & 0 & 0 \\ 0 & 0 & \frac{1}{d_3} & 0 \\ 0 & 0 & 0 & \frac{1}{d_4} \end{bmatrix}$

2.13 SOLUTION OF SIMULTANEOUS SYSTEM OF LINEAR EQUATIONS

Let the equations be

$$a_1x + a_2y + a_3z = d_1$$
$$b_1x + b_2y + b_3z = d_2$$
$$c_1x + c_2y + c_3z = d_3$$

we write the above eqn. in the matrix form,

$$\begin{bmatrix} a_1x + a_2y + a_3z \\ b_1x + b_2y + b_3z \\ c_1x + c_2y + c_3z \end{bmatrix} = \begin{bmatrix} d_1 \\ d_2 \\ d_3 \end{bmatrix}$$

or $$\begin{bmatrix} a_1 & a_2 & a_3 \\ b_1 & b_2 & b_3 \\ c_1 & c_2 & c_3 \end{bmatrix} \begin{bmatrix} x \\ y \\ z \end{bmatrix} = \begin{bmatrix} d_1 \\ d_2 \\ d_3 \end{bmatrix}$$

$$AX = B \qquad ...(1)$$

where $A = \begin{bmatrix} a_1 & a_2 & a_3 \\ b_1 & b_2 & b_3 \\ c_1 & c_2 & c_3 \end{bmatrix}$, $X = \begin{bmatrix} x \\ y \\ z \end{bmatrix}$, $B = \begin{bmatrix} d_1 \\ d_2 \\ d_3 \end{bmatrix}$

Multiplying (1) by A^{-1}, we have

$$X = A^{-1} B,$$

Illustrative Examples

Example 1 Solve with the help of matrices, the simultaneous equation.

$$x + y + z = 3, \; x + 2y + 3z = 4, \qquad x + 4y + 9z = 6$$

Solution: the given eqn. in matrix form is,

$$\begin{bmatrix} 1 & 1 & 1 \\ 1 & 2 & 3 \\ 1 & 4 & 9 \end{bmatrix} \begin{bmatrix} x \\ y \\ z \end{bmatrix} = \begin{bmatrix} 3 \\ 4 \\ 6 \end{bmatrix}$$

$$AX = B$$

where, $A = \begin{bmatrix} 1 & 1 & 1 \\ 1 & 2 & 3 \\ 1 & 4 & 9 \end{bmatrix}$, $X = \begin{bmatrix} x \\ y \\ z \end{bmatrix}$ $B = \begin{bmatrix} 3 \\ 4 \\ 6 \end{bmatrix}$

and $|A| = 1\,(18 - 12) - 1(9 - 3) + 1\,(4 - 2)$

$$= 6 - 6 + 2 = 2 \neq 0$$

Also we have
$$\begin{matrix} a_{11} = 6, & a_{21} = -5, & a_{31} = 1 \\ a_{12} = -6, & a_{22} = 8, & a_{32} = -2 \\ a_{13} = 2, & a_{23} = -3, & a_{33} = 1 \end{matrix}$$

$$\Rightarrow \quad A^{-1} = \frac{1}{2}\begin{bmatrix} 6 & -5 & 1 \\ -6 & 8 & -2 \\ 2 & -3 & 1 \end{bmatrix}$$

Now, $X = A^{-1}B$

$$= \frac{1}{2}\begin{bmatrix} 6 & -5 & 1 \\ -6 & 8 & -2 \\ 2 & -3 & 1 \end{bmatrix} \begin{bmatrix} 3 \\ 4 \\ 6 \end{bmatrix}$$

$$\begin{bmatrix} x \\ y \\ z \end{bmatrix} = \frac{1}{2}\begin{bmatrix} 18 - 20 + 6 \\ -18 + 32 - 12 \\ 6 - 12 + 6 \end{bmatrix}$$

$$\Rightarrow \qquad \begin{bmatrix} x \\ y \\ z \end{bmatrix} = \frac{1}{2}\begin{bmatrix} 4 \\ 2 \\ 0 \end{bmatrix} = \begin{bmatrix} 2 \\ 1 \\ 0 \end{bmatrix}.$$

$\boxed{x = 2, y = 1, z = 0}$ Answer

Example 2 By the matrix inversion, solve the system,

$$\begin{bmatrix} 1 & 1 & 1 \\ 2 & 5 & 7 \\ 2 & 1 & -1 \end{bmatrix}\begin{bmatrix} x & u \\ y & v \\ z & w \end{bmatrix} = \begin{bmatrix} 9 & 2 \\ 52 & 15 \\ 0 & -1 \end{bmatrix}$$

Solution: Let $A = \begin{bmatrix} 1 & 1 & 1 \\ 2 & 5 & 7 \\ 2 & 1 & -1 \end{bmatrix}$, $X = \begin{bmatrix} x & u \\ y & v \\ z & w \end{bmatrix}$ and $B = \begin{bmatrix} 9 & 2 \\ 52 & 15 \\ 0 & -1 \end{bmatrix}$

$$\begin{aligned} |A| &= 1(-5-7) - 1(-2-14) + 1(2-10) \\ &= -12 + 16 - 8 \\ &= -4 \neq 0 \end{aligned}$$

$$\begin{array}{lll} a_{11} = -12, & a_{21} = 2, & a_{31} = +2 \\ a_{12} = 16, & a_{22} = -3, & a_{32} = -5 \\ a_{13} = -8, & a_{23} = 1, & a_{33} = +3 \end{array}$$

$$\Rightarrow \qquad A^{-1} = \frac{-1}{4}\begin{bmatrix} -12 & 2 & 2 \\ 16 & -3 & -5 \\ -8 & 1 & +3 \end{bmatrix}$$

Now, $X = A^{-1}B$

$$\begin{bmatrix} x & u \\ y & v \\ z & w \end{bmatrix} = \frac{-1}{4}\begin{bmatrix} -12 & 2 & 2 \\ 16 & -3 & -5 \\ -8 & 1 & 3 \end{bmatrix}\begin{bmatrix} 9 & 2 \\ 52 & 15 \\ 0 & -1 \end{bmatrix}$$

$$= \frac{-1}{4}\begin{bmatrix} -108+104+0 & -24+30-2 \\ 144-156+0 & 32-45+5 \\ -72+52+0 & -16+15-3 \end{bmatrix}$$

$$= \frac{-1}{4}\begin{bmatrix} -4 & 4 \\ -12 & -8 \\ -20 & -4 \end{bmatrix}$$

$$= \begin{bmatrix} 1 & -1 \\ 3 & 2 \\ 5 & 1 \end{bmatrix}$$

$$\Rightarrow \boxed{\begin{matrix} x = 1, \ y = 3, \ z = 5 \\ u = -1, \ v = 2, \ w = 1 \end{matrix}}$$

Example 3. Represent each of the transformations

$x_1 = 3y_1 + 2y_2,\ x_2 = -y_1 + 4y_2$ and

$y_1 = z_1 + 2z_2,\ y_2 = -3z_1$

by the use of matrices, find the composite transformation which expresses x_1, x_2 in terms of z_1, z_2.

Solution: Let $\begin{bmatrix} x_1 \\ x_2 \end{bmatrix} = \begin{bmatrix} 3 & 2 \\ -1 & 4 \end{bmatrix} \begin{bmatrix} y_1 \\ y_2 \end{bmatrix}$ ...(1)

and $\begin{bmatrix} y_1 \\ y_2 \end{bmatrix} = \begin{bmatrix} 1 & 2 \\ -3 & 0 \end{bmatrix} \begin{bmatrix} z_1 \\ z_2 \end{bmatrix}$ ...(2)

Putting the value of $\begin{bmatrix} y_1 \\ y_2 \end{bmatrix}$ from (2) in (1)

$$\begin{bmatrix} x_1 \\ x_2 \end{bmatrix} = \begin{bmatrix} 3 & 2 \\ -1 & 4 \end{bmatrix} \begin{bmatrix} 1 & 2 \\ -3 & 0 \end{bmatrix} \begin{bmatrix} z_1 \\ z_2 \end{bmatrix}$$

$$= \begin{bmatrix} 3-6 & 6+0 \\ -1-12 & -2+0 \end{bmatrix} \begin{bmatrix} z_1 \\ z_2 \end{bmatrix}$$

$$= \begin{bmatrix} -3 & 6 \\ -13 & -2 \end{bmatrix} \begin{bmatrix} z_1 \\ z_2 \end{bmatrix}$$

$$\Rightarrow \boxed{\begin{matrix} x_1 = -3z_1 + 6z_2 \\ x_2 = -13z_1 - 2z_2 \end{matrix}}$$

2.14 ELEMENTARY TRANSFORMATIONS

Any one of the following operations on a matrix is called an elementary transformation.

(1) Interchanging any two rows (or columns). This transformation is indicated by R_{ij}, if the i_{th} and j_{th} rows are interchanged.

(2) Multiplication of the elements of any row (or column) by a non-zero scalar quantity k. (k R i)

(3) Addition of constant multiplication of the elements of any row to the corresponding elements of any other row. ($R_i + k R_j$)

Note: Two matrices are said to be equivalent if one is obtained from the other by elementary transformations. The symbol ~ is used for equivalence.

Example 1. Transform $\begin{bmatrix} 1 & 3 & 3 \\ 2 & 4 & 10 \\ 3 & 8 & 4 \end{bmatrix}$ into a unit matrix.

Solution: Let $A = \begin{bmatrix} 1 & 3 & 3 \\ 2 & 4 & 10 \\ 3 & 8 & 4 \end{bmatrix}$, then applying elementary operations,

$R_2 \to R_2 - 2R_1$

$R_3 \to R_3 - 3R_1$

we get $A = \begin{bmatrix} 1 & 3 & 3 \\ 0 & -2 & 4 \\ 0 & -1 & -5 \end{bmatrix}$

$R_1 \to R_1 + (R_2 + R_3)$

$R_2 \to R_2 + R_1$

$R_3 \to R_3 - \frac{1}{2}R_2$

$A = \begin{bmatrix} 1 & 0 & 2 \\ 0 & 1 & 7 \\ 0 & 0 & -7 \end{bmatrix}$

$R_2 \to R_2 + R_3, \qquad R_3 \to -\frac{R_3}{7}$

$A = \begin{bmatrix} 1 & 0 & 2 \\ 0 & 1 & 0 \\ 0 & 0 & 1 \end{bmatrix}$

$R_1 \to R_1 - 2R_3$

$A = \begin{bmatrix} 1 & 0 & 0 \\ 0 & 1 & 0 \\ 0 & 0 & 1 \end{bmatrix}$

By applying elementary transformations we transformed the above matrix into the unit matrix.

2.15 ELEMENTARY MATRICES

A matrix obtained from a unit matrix by applying any of the elementory transformation is called a elementary matrix.

for e.g., Let $A = \begin{bmatrix} 1 & 0 & 0 \\ 0 & 1 & 0 \\ 0 & 0 & 1 \end{bmatrix}$

apply $R_2 \to R_2 + 3R_3$

then $A = \begin{bmatrix} 1 & 0 & 0 \\ 0 & 1 & 3 \\ 0 & 0 & 1 \end{bmatrix} \to$ elementary matrix

Theorem 5:

Every elementary row transformation of a matrix can be affected by pre-multiplication with the corresponding elementory matrix.

Proof: Consider the matrix $A = \begin{bmatrix} 1 & 2 & 3 \\ 4 & 5 & 6 \\ 2 & 3 & 4 \end{bmatrix}$

apply $R_1 \to R_1 + 2R_3$, we get the matrix

$$B = \begin{bmatrix} 5 & 8 & 11 \\ 4 & 5 & 6 \\ 2 & 3 & 4 \end{bmatrix} \quad ...(1)$$

Now, we shall show that pre-multiplication of 'A' by corresponding elementary matrix $R_1 + 2R_3$ will give us 'B'.

Now, if $I = \begin{bmatrix} 1 & 0 & 0 \\ 0 & 1 & 0 \\ 0 & 0 & 1 \end{bmatrix}$, then

elementary matrix $= \begin{bmatrix} 1 & 0 & 2 \\ 0 & 1 & 0 \\ 0 & 0 & 1 \end{bmatrix}$

$(R_1 + 2R_3)$

elementary matrix X A $= \begin{bmatrix} 1 & 0 & 2 \\ 0 & 1 & 0 \\ 0 & 0 & 1 \end{bmatrix} \times \begin{bmatrix} 1 & 2 & 3 \\ 4 & 5 & 6 \\ 2 & 3 & 4 \end{bmatrix}$

$$= \begin{bmatrix} 1+4 & 2+6 & 3+8 \\ 4 & 5 & 6 \\ 2 & 3 & 4 \end{bmatrix} = \begin{bmatrix} 5 & 8 & 11 \\ 4 & 5 & 6 \\ 2 & 3 & 4 \end{bmatrix} = B \quad \text{... (2)}$$

Thus, from (1) and (2) we can show that matrix 'B' is same.

we can show that every elementary column transformation of a matrix can be affected by post-multiplication with the corresponding elementary matrix.

2.16 GAUSS- JORDON METHOD

(To compute the inverse of a matrix from elementary matrices)

If 'A' is reduced to 'I' by elementary transformations, then

$PA = I, \qquad P = P_n P_{n-1} \ldots\ldots\ldots\ldots\ldots P_2 P_1$

= elementary matrices

$P = A^{-1}$

Working Rule: Write A = IA. perform elementary transformations on 'A' on the left side, on 'I' on the right side, so that 'A' is reduced to I, and I of the right hand side is reduced to P getting I = PA

Then 'P' is the inverse of 'A'

2.17 THE INVERSE OF A SYMMETRIC MATRIX

The elementary transformations are to be transformed so that the property of being symmetric is preserved. This requires that the transformations occur in pairs, a row transformation must be followed immediately by the same column transformation.

Illustrative Examples

Example 1 Use Gauss- Jordon reduction method (elementory matrix method) to compute the inverse of the matrix $\begin{bmatrix} 3 & -3 & 4 \\ 2 & -3 & 4 \\ 0 & -1 & 1 \end{bmatrix}$ by applying elementory row transformation.

Solution: Let $A = \begin{bmatrix} 3 & -3 & 4 \\ 2 & -3 & 4 \\ 0 & -1 & 1 \end{bmatrix}$

We have to apply those Elementary row transformation which will reduce A = IA to I = PA, then matrix P will be the the inverse of matrix A.

$$\begin{bmatrix} 3 & -3 & 4 \\ 2 & -3 & 4 \\ 0 & -1 & 1 \end{bmatrix} = \begin{bmatrix} 1 & 0 & 0 \\ 0 & 1 & 0 \\ 0 & 0 & 1 \end{bmatrix} A$$

$R_1 \to R_1 - R_2$

$$\begin{bmatrix} 1 & 0 & 0 \\ 2 & -3 & 4 \\ 0 & -1 & 1 \end{bmatrix} = \begin{bmatrix} 1 & -1 & 0 \\ 0 & 1 & 0 \\ 0 & 0 & 1 \end{bmatrix} A$$

$R_2 \to R_2 - 2R_1$

$$\begin{bmatrix} 1 & 0 & 0 \\ 0 & -3 & 4 \\ 0 & -1 & 1 \end{bmatrix} = \begin{bmatrix} 1 & -1 & 0 \\ -2 & 3 & 0 \\ 0 & 0 & 1 \end{bmatrix} A$$

$R_3 \to R_3 - \frac{1}{3} R_2$

$R_2 \to -\frac{1}{3} R_2$

$$\begin{bmatrix} 1 & 0 & 0 \\ 0 & 1 & -4/3 \\ 0 & 0 & \frac{-1}{3} \end{bmatrix} = \begin{bmatrix} 1 & -1 & 0 \\ 2/3 & -1 & 0 \\ 2/3 & -1 & 1 \end{bmatrix} A$$

$R_3 \to R_3 - R_2$

$$\begin{bmatrix} 1 & 0 & 0 \\ 0 & 1 & -4/3 \\ 0 & -1 & 1 \end{bmatrix} = \begin{bmatrix} 1 & -1 & 0 \\ 2/3 & -1 & 0 \\ 0 & 0 & 0 \end{bmatrix} A$$

$R_3 \to R_3 + R_2$

$$\begin{bmatrix} 1 & 0 & 0 \\ 0 & 1 & -4/3 \\ 0 & 0 & -1/3 \end{bmatrix} = \begin{bmatrix} 1 & -1 & 0 \\ 2/3 & -1 & 0 \\ 2/3 & -1 & 1 \end{bmatrix} A.$$

$R_2 \to R_2 - R_3$

$$\begin{bmatrix} 1 & 0 & 0 \\ 0 & 1 & -1 \\ 0 & 0 & -1/3 \end{bmatrix} = \begin{bmatrix} 1 & -1 & 0 \\ 0 & 0 & -1 \\ 2/3 & -1 & 1 \end{bmatrix} A$$

$R_3 \to -3R_3$

$$\begin{bmatrix} 1 & 0 & 0 \\ 0 & 1 & -1 \\ 0 & 0 & 1 \end{bmatrix} = \begin{bmatrix} 1 & -1 & 0 \\ 0 & 0 & -1 \\ -2 & 3 & -3 \end{bmatrix} A$$

$R_2 \rightarrow R_2 + R_3$

$$\begin{bmatrix} 1 & 0 & 0 \\ 0 & 1 & 0 \\ 0 & 0 & 1 \end{bmatrix} = \begin{bmatrix} 1 & -1 & 0 \\ -2 & 3 & -4 \\ -2 & 3 & -3 \end{bmatrix} A$$

$\Rightarrow \quad I = PA$

$\therefore \quad P = A^{-1}$

Thus, $A^{-1} = \begin{bmatrix} 1 & -1 & 0 \\ -2 & 3 & -4 \\ -2 & 3 & -3 \end{bmatrix}$

Example 2. Find the inverse of the matrix by Gauss Jordon method

$$A = \begin{bmatrix} 1 & -1 & 1 \\ 4 & 1 & 0 \\ 8 & 1 & 1 \end{bmatrix}$$

Solution: Let $A = I\,A$

$$\begin{bmatrix} 1 & -1 & 1 \\ 4 & 1 & 0 \\ 8 & 1 & 1 \end{bmatrix} = \begin{bmatrix} 1 & 0 & 0 \\ 0 & 1 & 0 \\ 0 & 0 & 1 \end{bmatrix} A$$

$R_2 \rightarrow R_2 - 4\,R_1$

$R_3 \rightarrow R_3 - 8\,R_1$

$$\begin{bmatrix} 1 & -1 & 1 \\ 0 & 5 & -4 \\ 0 & 9 & -7 \end{bmatrix} = \begin{bmatrix} 1 & 0 & 0 \\ -4 & 1 & 0 \\ -8 & 0 & 1 \end{bmatrix} A$$

$R_3 \rightarrow R_3 + 9\,R_1$

$R_2 \rightarrow \frac{1}{5}\,R_2$

$$\begin{bmatrix} 1 & -1 & 1 \\ 0 & 1 & -4/5 \\ 9 & 0 & 2 \end{bmatrix} = \begin{bmatrix} 1 & 0 & 0 \\ -4/5 & 1/5 & 0 \\ 1 & 0 & 1 \end{bmatrix} A$$

$R_1 \to R_1 + R_2$

$R_3 \to \frac{1}{2} R_3$

$$\begin{bmatrix} 1 & 0 & 1/5 \\ 0 & 1 & -4/5 \\ 9/2 & 0 & 1 \end{bmatrix} = \begin{bmatrix} 1/5 & 1/5 & 0 \\ -4/5 & 1/5 & 0 \\ 1/2 & 0 & 1/2 \end{bmatrix} A$$

$R_1 \to R_1 - \frac{1}{5} R_3$

$$\begin{bmatrix} \frac{1}{10} & 0 & 0 \\ 0 & 1 & -4/5 \\ 9/2 & 0 & 1 \end{bmatrix} = \begin{bmatrix} \frac{1}{10} & \frac{1}{5} & \frac{-1}{10} \\ -4/5 & 1/5 & 0 \\ 1/2 & 0 & 1/2 \end{bmatrix} A$$

$R_1 \to 10 R_1$

$$\begin{bmatrix} 1 & 0 & 0 \\ 0 & 1 & -4/5 \\ 9/2 & 0 & 1 \end{bmatrix} = \begin{bmatrix} 1 & 2 & -1 \\ -4/5 & 1/5 & 0 \\ 1/2 & 0 & 1/2 \end{bmatrix} A$$

$R_3 \to R_3 - \frac{9}{2} R_1$

$$\begin{bmatrix} 1 & 0 & 0 \\ 0 & 1 & -4/5 \\ 0 & 0 & 1 \end{bmatrix} = \begin{bmatrix} 1 & 2 & -1 \\ -4/5 & 1/5 & 0 \\ -4 & -9 & 5 \end{bmatrix} A$$

$R_2 \to R_2 + \frac{4}{5} R_3$

$$\begin{bmatrix} 1 & 0 & 0 \\ 0 & 1 & 0 \\ 0 & 0 & 1 \end{bmatrix} = \begin{bmatrix} 1 & 2 & -1 \\ -4 & -7 & +4 \\ -4 & -9 & 5 \end{bmatrix} A$$

Thus, $$A^{-1} = \begin{bmatrix} 1 & 2 & -1 \\ -4 & -7 & 4 \\ -4 & -9 & 5 \end{bmatrix}$$

Example 3 Find inverse by Gauss Jordon Method of the given matrix $A = \begin{bmatrix} 2 & 3 & 1 \\ 1 & 2 & 3 \\ 3 & 1 & 2 \end{bmatrix}$

Solution: Let $$\begin{bmatrix} 2 & 3 & 1 \\ 1 & 2 & 3 \\ 3 & 1 & 2 \end{bmatrix} = \begin{bmatrix} 1 & 0 & 0 \\ 0 & 1 & 0 \\ 0 & 0 & 1 \end{bmatrix} A$$

$R_1 \leftrightarrow R_2$

$$\begin{bmatrix} 1 & 2 & 3 \\ 2 & 3 & 1 \\ 3 & 1 & 2 \end{bmatrix} = \begin{bmatrix} 0 & 1 & 0 \\ 1 & 0 & 0 \\ 0 & 0 & 1 \end{bmatrix} A$$

$R_2 \to R_2 - 2R_1$
$R_3 \to R_3 - 3R_1$

$$\begin{bmatrix} 1 & 2 & 3 \\ 0 & -1 & -5 \\ 0 & -5 & -7 \end{bmatrix} = \begin{bmatrix} 0 & 1 & 0 \\ 1 & -2 & 0 \\ 0 & -3 & 1 \end{bmatrix} A$$

$R_2 \to -R_2$
$R_3 \to R_3 - 5R_2$

$$\begin{bmatrix} 1 & 2 & 3 \\ 0 & 1 & 5 \\ 0 & 0 & 18 \end{bmatrix} = \begin{bmatrix} 0 & 1 & 0 \\ -1 & 2 & 0 \\ -5 & 7 & 1 \end{bmatrix} A$$

$R_3 \to \frac{1}{18} R_3$

$$\begin{bmatrix} 1 & 2 & 3 \\ 0 & 1 & 5 \\ 0 & 0 & 1 \end{bmatrix} = \begin{bmatrix} 0 & 1 & 0 \\ -1 & 2 & 0 \\ -5/18 & 7/18 & 1/18 \end{bmatrix} A$$

$R_1 \to R_1 - 2R_2$

$$\begin{bmatrix} 1 & 0 & -7 \\ 0 & 1 & 5 \\ 0 & 0 & 1 \end{bmatrix} = \begin{bmatrix} 2 & -3 & 0 \\ -1 & 2 & 0 \\ -5/18 & 7/18 & 1/18 \end{bmatrix} A$$

$R_1 \to R_1 + 7R_3$
$R_2 \to R_2 - 5R_3$

$$\begin{bmatrix} 1 & 0 & 0 \\ 0 & 1 & 0 \\ 0 & 0 & 1 \end{bmatrix} = \begin{bmatrix} 1/18 & -5/18 & 7/18 \\ 7/18 & 1/18 & -5/18 \\ -5/18 & 7/18 & 1/18 \end{bmatrix} A$$

$$\begin{bmatrix} 1 & 0 & 0 \\ 0 & 1 & 0 \\ 0 & 0 & 1 \end{bmatrix} = \frac{1}{18}\begin{bmatrix} 1 & -5 & 7 \\ 7 & 1 & -5 \\ -5 & 7 & 1 \end{bmatrix} A$$

Thus, $$A^{-1} = \frac{1}{18}\begin{bmatrix} 1 & -5 & 7 \\ 7 & 1 & -5 \\ -5 & 7 & 1 \end{bmatrix}$$

Example 4 If $A = \begin{bmatrix} 1 & 1 & 2 \\ 1 & 2 & 3 \\ 2 & 3 & 1 \end{bmatrix}$, find A^{-1}

Solution: Firstly, By Gauss-Jordan Method

Write A as $\boxed{A = IA}$.

we have $$\begin{bmatrix} 1 & 1 & 2 \\ 1 & 2 & 3 \\ 2 & 3 & 1 \end{bmatrix} = \begin{bmatrix} 1 & 0 & 0 \\ 0 & 1 & 0 \\ 0 & 0 & 1 \end{bmatrix} A$$

$$\downarrow$$

Make it an identity Matrix because $I = A^{-1}A$ then first term on right hand side is inverse of given matrix.

By Row Elementary operation

$R_2 \to R_2 - R_1$, $R_3 \to R_3 - 2R_1$

$$\Rightarrow \begin{bmatrix} 1 & 1 & 2 \\ 0 & 1 & 1 \\ 0 & 1 & -3 \end{bmatrix} = \begin{bmatrix} 1 & 0 & 0 \\ -1 & 1 & 0 \\ -2 & 0 & 1 \end{bmatrix} A$$

$R_1 \to R_1 - R_2$, $R_3 \to R_3 - R_2$

$$\Rightarrow \begin{bmatrix} 1 & 0 & 1 \\ 0 & 1 & 1 \\ 0 & 0 & -4 \end{bmatrix} = \begin{bmatrix} 2 & -1 & 0 \\ -1 & 1 & 0 \\ -1 & -1 & 1 \end{bmatrix} A$$

$$R_3 \to -\frac{1}{4}R_3$$

$$\Rightarrow \quad \begin{bmatrix} 1 & 0 & 1 \\ 0 & 1 & 1 \\ 0 & 0 & 1 \end{bmatrix} = \begin{bmatrix} 2 & -1 & 0 \\ -1 & 1 & 0 \\ 1/4 & 1/4 & -1/4 \end{bmatrix} A$$

$$R_1 \to R_1 - R_3, \quad R_2 \to R_2 - R_3$$

$$\Rightarrow \quad \begin{bmatrix} 1 & 0 & 0 \\ 0 & 1 & 0 \\ 0 & 0 & 1 \end{bmatrix} = \begin{bmatrix} 7/4 & -5/4 & 1/4 \\ -5/4 & 3/4 & 1/4 \\ 1/4 & 1/4 & -1/4 \end{bmatrix} A$$

Compare with $I = A^{-1} A$

So,
$$\boxed{A^{-1} = \frac{1}{4}\begin{bmatrix} 7 & -5 & 1 \\ -5 & 3 & 1 \\ 1 & 1 & -1 \end{bmatrix}} \quad \text{Ans.}$$

Example 5 If $A = \begin{bmatrix} 4 & 3 & 1 \\ 2 & 1 & -2 \\ 1 & 2 & 1 \end{bmatrix}, A^{-1} = ?$

Solution: Let $A = IA$

$$\begin{bmatrix} 4 & 3 & 1 \\ 2 & 1 & -2 \\ 1 & 2 & 1 \end{bmatrix} = \begin{bmatrix} 1 & 0 & 0 \\ 0 & 1 & 0 \\ 0 & 0 & 1 \end{bmatrix} A$$

$$R_1 \leftrightarrow R_3$$

$$\begin{bmatrix} 1 & 2 & 1 \\ 2 & 1 & -2 \\ 4 & 3 & 1 \end{bmatrix} = \begin{bmatrix} 0 & 0 & 1 \\ 0 & 1 & 0 \\ 1 & 0 & 0 \end{bmatrix} A$$

$$R_2 \to R_2 - 2R_1, \quad R_3 \to R_3 - 4R_1$$

$$\begin{bmatrix} 1 & 2 & 1 \\ 0 & -3 & -4 \\ 0 & -5 & -3 \end{bmatrix} = \begin{bmatrix} 0 & 0 & 1 \\ 0 & 1 & -2 \\ 1 & 0 & -4 \end{bmatrix} A$$

$$R_2 \to -\frac{R_2}{3}$$

$$\begin{bmatrix} 1 & 2 & 1 \\ 0 & 1 & 4/3 \\ 0 & -5 & -3 \end{bmatrix} = \begin{bmatrix} 0 & 0 & 1 \\ 0 & -1/3 & 2/3 \\ 1 & 0 & -4 \end{bmatrix} A$$

$R_1 \to R_1 - 2R_2, \quad R_3 \to R_3 + 5R_2$

$$\begin{bmatrix} 1 & 0 & -5/3 \\ 0 & 1 & 4/3 \\ 0 & 0 & 11/3 \end{bmatrix} = \begin{bmatrix} 0 & 2/3 & -1/3 \\ 0 & -1/3 & 2/3 \\ 1 & -5/3 & -2/3 \end{bmatrix} A$$

$R_3 \to \frac{3}{11} R_3$

$$\begin{bmatrix} 1 & 0 & -5/3 \\ 0 & 1 & 4/3 \\ 0 & 0 & 1 \end{bmatrix} = \begin{bmatrix} 0 & 2/3 & -1/3 \\ 0 & -1/3 & 2/3 \\ 3/11 & -5/11 & -2/11 \end{bmatrix} A$$

$R_1 \to R_1 + \frac{5}{3} R_3, \quad R_2 \to R_2 - \frac{4}{3} R_3$

$$\begin{bmatrix} 1 & 0 & 0 \\ 0 & 1 & 0 \\ 0 & 0 & 1 \end{bmatrix} = \begin{bmatrix} 5/11 & -1/11 & -7/11 \\ -4/11 & 3/11 & 10/11 \\ 3/11 & -5/11 & -2/11 \end{bmatrix} A$$

Compare with $\boxed{I = A^{-1} A}$

So, $$\boxed{A^{-1} = \frac{1}{11} \begin{bmatrix} 5 & -1 & -7 \\ -4 & 3 & 10 \\ 3 & -5 & -2 \end{bmatrix} \text{ Ans.}}$$

Example 6 If $A = \begin{bmatrix} 2 & -1 & 1 \\ -1 & 2 & -1 \\ 1 & -1 & 2 \end{bmatrix}$, Find A^{-1} = ?

Solution: Let A = IA

$$\begin{bmatrix} 2 & -1 & 1 \\ -1 & 2 & -1 \\ 1 & -1 & 2 \end{bmatrix} = \begin{bmatrix} 1 & 0 & 0 \\ 0 & 1 & 0 \\ 0 & 0 & 1 \end{bmatrix} A$$

$R_1 \leftrightarrow R_3$

$$\begin{bmatrix} 1 & -1 & 2 \\ -1 & 2 & -1 \\ 2 & -1 & 1 \end{bmatrix} = \begin{bmatrix} 0 & 0 & 1 \\ 0 & 1 & 0 \\ 1 & 0 & 0 \end{bmatrix} A$$

$R_2 \to R_2 + R_1, \quad R_3 \to R_3 - 2R_1$

$$\begin{bmatrix} 1 & -1 & 2 \\ 0 & 1 & 1 \\ 0 & 1 & -3 \end{bmatrix} = \begin{bmatrix} 0 & 0 & 1 \\ 0 & 1 & 1 \\ 1 & 0 & -2 \end{bmatrix} A$$

$R_1 \to R_1 + R_2, \quad R_3 \to R_3 - R_2$

$$\begin{bmatrix} 1 & 0 & 3 \\ 0 & 1 & 1 \\ 0 & 0 & -4 \end{bmatrix} = \begin{bmatrix} 0 & 1 & 2 \\ 0 & 1 & 1 \\ 1 & -1 & -3 \end{bmatrix} A$$

$R_3 \to R_3 / -4$

$$\begin{bmatrix} 1 & 0 & 3 \\ 0 & 1 & 1 \\ 0 & 0 & 1 \end{bmatrix} = \begin{bmatrix} 0 & 1 & 2 \\ 0 & 1 & 1 \\ -1/4 & 1/4 & 3/4 \end{bmatrix} A$$

$R_1 \to R_1 - 3R_3, \quad R_2 \to R_2 - R_3$

$$\begin{bmatrix} 1 & 0 & 0 \\ 0 & 1 & 0 \\ 0 & 0 & 1 \end{bmatrix} = \begin{bmatrix} 3/4 & 1/4 & -1/4 \\ 1/4 & 3/4 & 1/4 \\ -1/4 & 1/4 & 3/4 \end{bmatrix} A$$

Compare with $\boxed{I = A^{-1} A}$

So, $\boxed{A^{-1} = \frac{1}{4}\begin{bmatrix} 3 & 1 & -1 \\ 1 & 3 & 1 \\ -1 & 1 & 3 \end{bmatrix}}$ Ans.

Example 7 If $A = \begin{bmatrix} 2 & 1 & 1 \\ 0 & 1 & 0 \\ 1 & 1 & 2 \end{bmatrix}$, Find $A^{-1} = ?$

Solution: Let $A = IA$

$$\begin{bmatrix} 2 & 1 & 1 \\ 0 & 1 & 0 \\ 1 & 1 & 2 \end{bmatrix} = \begin{bmatrix} 1 & 0 & 0 \\ 0 & 1 & 0 \\ 0 & 0 & 1 \end{bmatrix} A$$

$R_1 \leftrightarrow R_3$

$$\begin{bmatrix} 1 & 1 & 2 \\ 0 & 1 & 0 \\ 2 & 1 & 1 \end{bmatrix} = \begin{bmatrix} 0 & 0 & 1 \\ 0 & 1 & 0 \\ 1 & 0 & 0 \end{bmatrix} A$$

$R_3 \to R_3 - 2\,R_1$

$$\begin{bmatrix} 1 & 1 & 2 \\ 0 & 1 & 0 \\ 0 & -1 & -3 \end{bmatrix} = \begin{bmatrix} 0 & 0 & 1 \\ 0 & 1 & 0 \\ 1 & 0 & -2 \end{bmatrix} A$$

$R_1 \to R_1 - R_2,\quad R_3 \to R_3 + R_2$

$$\begin{bmatrix} 1 & 0 & 2 \\ 0 & 1 & 0 \\ 0 & 0 & -3 \end{bmatrix} = \begin{bmatrix} 0 & -1 & 1 \\ 0 & 1 & 0 \\ 1 & 1 & -2 \end{bmatrix} A$$

$R_3 \to R_3 / -3$

$$\begin{bmatrix} 1 & 0 & 2 \\ 0 & 1 & 0 \\ 0 & 0 & 1 \end{bmatrix} = \begin{bmatrix} 0 & -1 & 1 \\ 0 & +1 & 0 \\ -1/3 & -1/3 & 2/3 \end{bmatrix} A$$

$R_1 \to R_1 - 2\,R_3$

$$\begin{bmatrix} 1 & 0 & 0 \\ 0 & 1 & 0 \\ 0 & 0 & 1 \end{bmatrix} = \begin{bmatrix} 2/3 & -1/3 & -1/3 \\ 0 & 1 & 0 \\ -1/3 & -1/3 & 2/3 \end{bmatrix} A$$

Compare it with $\boxed{I = A^{-1} A}$

So, $$\boxed{A^{-1} = \frac{1}{3}\begin{bmatrix} 2 & -1 & -1 \\ 0 & 3 & 0 \\ -1 & -1 & 2 \end{bmatrix}}$$ Ans.

Example 8 If $A = \begin{bmatrix} 1 & 1 & 1 & 1 \\ -2 & 1 & -1 & -2 \\ -4 & -2 & -3 & 1 \\ 1 & 1 & 1 & 0 \end{bmatrix}$, Find $A^{-1} = ?$

Solution: Let A = IA

$$\begin{bmatrix} 1 & 1 & 1 & 1 \\ -2 & 1 & -1 & -2 \\ -4 & -2 & -3 & 1 \\ 1 & 1 & 1 & 0 \end{bmatrix} = \begin{bmatrix} 1 & 0 & 0 & 0 \\ 0 & 1 & 0 & 0 \\ 0 & 0 & 1 & 0 \\ 0 & 0 & 0 & 1 \end{bmatrix} A$$

$R_2 \to R_2 + 2R_1, \quad R_3 \to R_3 + 4R_1, R_4 \to R_4 - R_1$

$$\begin{bmatrix} 1 & 1 & 1 & 1 \\ 0 & 3 & 1 & 0 \\ 0 & 2 & 1 & 5 \\ 0 & 0 & 0 & -1 \end{bmatrix} = \begin{bmatrix} 1 & 0 & 0 & 0 \\ 2 & 1 & 0 & 0 \\ 4 & 0 & 1 & 0 \\ -1 & 0 & 0 & 1 \end{bmatrix} A$$

$R_2 \to R_2/3$

$$\begin{bmatrix} 1 & 1 & 1 & 1 \\ 0 & 1 & 1/3 & 0 \\ 0 & 2 & 1 & 5 \\ 0 & 0 & 0 & -1 \end{bmatrix} = \begin{bmatrix} 1 & 0 & 0 & 0 \\ 2/3 & 1/3 & 0 & 0 \\ 4 & 0 & 1 & 0 \\ -1 & 0 & 0 & 1 \end{bmatrix} A$$

$R_1 \to R_1 - R_2, \quad R_3 \to R_3 - 2R_2$

$$\begin{bmatrix} 1 & 0 & 2/3 & 1 \\ 0 & 1 & 1/3 & 0 \\ 0 & 0 & 1/3 & 5 \\ 0 & 0 & 0 & -1 \end{bmatrix} = \begin{bmatrix} 1/3 & -1/3 & 0 & 0 \\ 2/3 & 1/3 & 0 & 0 \\ 8/3 & -2/3 & 1 & 0 \\ -1 & 0 & 0 & 1 \end{bmatrix} A$$

$R_3 \to 3R_3$

$$\begin{bmatrix} 1 & 0 & 2/3 & 1 \\ 0 & 1 & 1/3 & 0 \\ 0 & 0 & 1 & 15 \\ 0 & 0 & 0 & -1 \end{bmatrix} = \begin{bmatrix} 1/3 & -1/3 & 0 & 0 \\ 2/3 & 1/3 & 0 & 0 \\ 8 & -2 & 3 & 0 \\ -1 & 0 & 0 & 1 \end{bmatrix} A$$

$R_1 \to R_1 - \frac{2}{3}R_3, \quad R_2 \to R_2 - \frac{1}{3}R_3$

$$\begin{bmatrix} 1 & 0 & 0 & -9 \\ 0 & 1 & 0 & -5 \\ 0 & 0 & 1 & 15 \\ 0 & 0 & 0 & -1 \end{bmatrix} = \begin{bmatrix} -5 & 1 & -2 & 0 \\ -2 & 1 & -1 & 0 \\ 8 & -2 & 3 & 0 \\ -1 & 0 & 0 & 1 \end{bmatrix} A$$

$R_4 \to (-1)R_4$

$$\begin{bmatrix} 1 & 0 & 0 & -9 \\ 0 & 1 & 0 & -5 \\ 0 & 0 & 1 & 15 \\ 0 & 0 & 0 & 1 \end{bmatrix} = \begin{bmatrix} -5 & 1 & -2 & 0 \\ -2 & 1 & -1 & 0 \\ 8 & -2 & 3 & 0 \\ 1 & 0 & 0 & -1 \end{bmatrix} A$$

$R_1 \to R_1 + 9R_4, \quad R_2 \to R_2 + 5R_4, R_3 \to R_3 - 15R_4$

$$\begin{bmatrix} 1 & 0 & 0 & 0 \\ 0 & 1 & 0 & 0 \\ 0 & 0 & 1 & 0 \\ 0 & 0 & 0 & 1 \end{bmatrix} = \begin{bmatrix} 4 & 1 & -2 & -9 \\ 3 & 1 & -1 & -5 \\ -7 & -2 & 3 & 15 \\ 1 & 0 & 0 & -1 \end{bmatrix} A$$

Compare it with $I = A^{-1} A$

So, $$A^{-1} = \begin{bmatrix} 4 & 1 & -2 & -9 \\ 3 & +1 & -1 & -5 \\ -7 & -2 & 3 & 15 \\ 1 & 0 & 0 & -1 \end{bmatrix}$$ Ans.

Example 9 If $A = \begin{bmatrix} 0 & 1 & 2 \\ 1 & 2 & 3 \\ 3 & 1 & 1 \end{bmatrix}$, Find A^{-1} = ?

Solution: Let $A = IA$

$$\begin{bmatrix} 0 & 1 & 2 \\ 1 & 2 & 3 \\ 3 & 1 & 1 \end{bmatrix} = \begin{bmatrix} 1 & 0 & 0 \\ 0 & 1 & 0 \\ 0 & 0 & 1 \end{bmatrix} A$$

$R_1 \leftrightarrow R_2$

$$\begin{bmatrix} 1 & 2 & 3 \\ 0 & 1 & 2 \\ 3 & 1 & 1 \end{bmatrix} = \begin{bmatrix} 0 & 1 & 0 \\ 1 & 0 & 0 \\ 0 & 0 & 1 \end{bmatrix} A$$

$R_3 \to R_3 - 3R_1$

$$\begin{bmatrix} 1 & 2 & 3 \\ 0 & 1 & 2 \\ 0 & -5 & -8 \end{bmatrix} = \begin{bmatrix} 0 & 1 & 0 \\ 1 & 0 & 0 \\ 0 & -3 & 1 \end{bmatrix} A$$

$R_1 \to R_1 - 2R_2,\ R_3 \to R_3 + 5R_2$

$$\begin{bmatrix} 1 & 0 & -1 \\ 0 & 1 & 2 \\ 0 & 0 & 1 \end{bmatrix} = \begin{bmatrix} -2 & 1 & 0 \\ 1 & 0 & 0 \\ 5 & -3 & 1 \end{bmatrix} A$$

$R_2 \to -\dfrac{R_3}{2}$

$$\begin{bmatrix} 1 & 0 & -1 \\ 0 & 1 & 2 \\ 0 & 0 & 1 \end{bmatrix} = \begin{bmatrix} -2 & 1 & 0 \\ 1 & 0 & 0 \\ 5/2 & -3/2 & 1/2 \end{bmatrix} A$$

$R_1 \to R_1 + R_3,\ R_2 \to R_2 - 2R_3$

$$\begin{bmatrix} 1 & 0 & 0 \\ 0 & 1 & 0 \\ 0 & 0 & 1 \end{bmatrix} = \begin{bmatrix} 1/2 & -1/2 & 1/2 \\ -4 & 3 & -1 \\ 5/2 & -3/2 & 1/2 \end{bmatrix} A$$

Compare it with $\boxed{I = A^{-1} A}$

So,
$$\boxed{A^{-1} = -\frac{1}{2}\begin{bmatrix} -1 & 1 & -1 \\ 8 & -6 & 2 \\ -5 & 3 & -1 \end{bmatrix}}$$ Ans.

Example 10 If $A = \begin{bmatrix} 2 & -6 & -2 & -3 \\ 5 & -13 & -4 & -7 \\ -1 & 4 & 1 & 2 \\ 0 & 1 & 0 & 1 \end{bmatrix}$, Find $A^{-1} = ?$

Solution: Let A = IA

$$\begin{bmatrix} 2 & -6 & -2 & -3 \\ 5 & -13 & -4 & -7 \\ -1 & 4 & 1 & 2 \\ 0 & 1 & 0 & 1 \end{bmatrix} = \begin{bmatrix} 1 & 0 & 0 & 0 \\ 0 & 1 & 0 & 0 \\ 0 & 0 & 1 & 0 \\ 0 & 0 & 0 & 1 \end{bmatrix} A$$

$R_1 \leftrightarrow (-1)R_3$

$$\begin{bmatrix} 1 & -4 & -1 & -2 \\ 5 & -13 & -4 & -7 \\ -2 & +6 & +2 & +3 \\ 0 & 1 & 0 & 1 \end{bmatrix} = \begin{bmatrix} 0 & 0 & -1 & 0 \\ 0 & 1 & 0 & 0 \\ -1 & 0 & 0 & 0 \\ 0 & 0 & 0 & 1 \end{bmatrix} A$$

$R_2 \to R_2 - 5R_1,\quad R_3 \to R_3 + 2R_1$

$$\begin{bmatrix} 1 & -4 & -1 & -2 \\ 0 & 7 & 1 & 3 \\ 0 & -2 & 0 & -1 \\ 0 & 1 & 0 & 1 \end{bmatrix} = \begin{bmatrix} 0 & 0 & -1 & 0 \\ 0 & 1 & 5 & 0 \\ -1 & 0 & -2 & 0 \\ 0 & 0 & 0 & 1 \end{bmatrix} A$$

$R_2 \leftrightarrow R_4$

$$\begin{bmatrix} 1 & -4 & -1 & -2 \\ 0 & 1 & 0 & 1 \\ 0 & -2 & 0 & -1 \\ 0 & 7 & 1 & 3 \end{bmatrix} = \begin{bmatrix} 0 & 0 & -1 & 0 \\ 0 & 0 & 0 & 1 \\ -1 & 0 & -2 & 0 \\ 0 & 1 & 5 & 0 \end{bmatrix} A$$

$R_1 \to R_1 + 4R_2,\ \ R_3 \to R_3 + 2R_2,\ \ R_4 \to R_4 - 7R_2$

$$\begin{bmatrix} 1 & 0 & -1 & 2 \\ 0 & 1 & 0 & 1 \\ 0 & 0 & 0 & +1 \\ 0 & 0 & 1 & -4 \end{bmatrix} = \begin{bmatrix} 0 & 0 & -1 & 4 \\ 0 & 0 & 0 & 1 \\ -1 & 0 & -2 & +2 \\ 0 & 1 & 5 & -7 \end{bmatrix} A$$

$R_3 \leftrightarrow R_4$

$$\begin{bmatrix} 1 & 0 & -1 & 2 \\ 0 & 1 & 0 & 1 \\ 0 & 0 & 1 & -4 \\ 0 & 0 & 0 & +1 \end{bmatrix} = \begin{bmatrix} 0 & 0 & -1 & 4 \\ 0 & 0 & 0 & 1 \\ 0 & 1 & 5 & -7 \\ -1 & 0 & -2 & +2 \end{bmatrix} A$$

$R_1 \to R_1 + R_3$

$$\begin{bmatrix} 1 & 0 & 0 & -2 \\ 0 & 1 & 0 & 1 \\ 0 & 0 & 1 & -4 \\ 0 & 0 & 0 & +1 \end{bmatrix} = \begin{bmatrix} 1 & 1 & 4 & -3 \\ 0 & 0 & 0 & 1 \\ 1 & 1 & 5 & -7 \\ -1 & 0 & -2 & +2 \end{bmatrix} A$$

$R_1 \to R_1 + 2R_4,\ \ R_2 \to R_2 - R_4,\ \ R_3 \to R_3 + 4R_4$

$$\begin{bmatrix} 1 & 0 & 0 & 0 \\ 0 & 1 & 0 & 0 \\ 0 & 0 & 1 & 0 \\ 0 & 0 & 0 & 1 \end{bmatrix} = \begin{bmatrix} -1 & 1 & 0 & 1 \\ 1 & 0 & 2 & -1 \\ -3 & 1 & -3 & 1 \\ -1 & 0 & -2 & 2 \end{bmatrix} A$$

Compare it with $\boxed{I = A^{-1} A}$

So. $A^{-1} = \begin{bmatrix} -1 & 1 & 0 & 1 \\ 1 & 0 & 2 & -1 \\ -3 & 1 & -3 & 1 \\ -1 & 0 & -2 & 2 \end{bmatrix}$ Ans.

Exercise - 2.3

Solve the following system of equations using matrix method (or by matrix inversion)

1. $3x + y + 2z = 3$
   $2x - 3y - z = -3$
   $x + 2y + z = 4$
2. $2x + 3y + 5z = 15$
   $3x + 5y + 2z = 12$
   $5x + 2y + 3z = 13$
3. $x + z = 7$
   $2x + y = 7$
   $3x + 2y + z = 17$
4. $x + y + z + 6 = 0$
   $x + 2y + 3z + 14 = 0$
   $x + 4y + 9z + 36 = 0$
5. $2x - 3y + z = 9$
   $x + y + z = 6$
   $x - y + z = 2$
6. $x + 4y + 6z = -25$
   $6x + y + 4z = -17$
   $4x + 6y + z = -24$
7. $x + y + z = \lambda$
   $ax + by + cz = \mu$
   $a^2x + b^2y + c^2z = \gamma$

Find inverse of the following matrices using Gauss Jordan Method:-

8. $\begin{bmatrix} 1 & 1 & 1 \\ 1 & 2 & -3 \\ 2 & -1 & 3 \end{bmatrix}$

9. $\begin{bmatrix} 2 & -2 & 1 \\ 1 & 2 & 2 \\ 2 & 1 & -2 \end{bmatrix}$

10. $\begin{bmatrix} 0 & 1 & 2 \\ 1 & 2 & 3 \\ 3 & 1 & 1 \end{bmatrix}$

11. $\begin{bmatrix} 2 & 1 & -1 \\ 0 & 2 & 1 \\ 5 & 2 & -3 \end{bmatrix}$

12. $\begin{bmatrix} 1 & 1 & 1 & 1 \\ 1 & 2 & 3 & -4 \\ 2 & 3 & 5 & -5 \\ 3 & -4 & -5 & 8 \end{bmatrix}$

Answers

1. $x = 1, \quad y = 2, \quad z = -1$
2. $x = 1, \quad y = 1, \quad z = 2$
3. $x = 2, \quad y = 3, \quad z = 5$
4. $x = -1, y = -2, \quad z = -3$
5. $x = 11, \quad y = 2, \quad z = -7$
6. $x = -1, \quad y = -3, \quad z = -2$
7. $x = \dfrac{\lambda bc - \mu(b+c) + \gamma}{(a-b)(a-c)}, y = \dfrac{\lambda ca - \mu(c+a) + \gamma}{(b-a)(b-c)}, \; z = \dfrac{\lambda ab - \mu(a+b) + \gamma}{(c-b)(c-a)}$
8. $A^{-1} = \dfrac{1}{11}\begin{bmatrix} -3 & 4 & 5 \\ 9 & -1 & -4 \\ 5 & -3 & -1 \end{bmatrix}$
9. $\dfrac{1}{9}\begin{bmatrix} 2 & 1 & 2 \\ -2 & 2 & 1 \\ 1 & 2 & -2 \end{bmatrix}$
10. $\begin{bmatrix} 1/2 & -1/2 & 1/2 \\ -4 & 3 & -1 \\ 5/2 & -3/2 & 1/2 \end{bmatrix}$
11. $\begin{bmatrix} 8 & -1 & -3 \\ -5 & 1 & 2 \\ 10 & -1 & -4 \end{bmatrix}$
12. $\dfrac{1}{18}\begin{bmatrix} 2 & 16 & -6 & 4 \\ 22 & 41 & -30 & -1 \\ -10 & -44 & 30 & -2 \\ 4 & -13 & 6 & -1 \end{bmatrix}$

2.18 RANK OF A MATRIX

The rank of a matrix is said to be r if

(a) It has atleast one non-zero minor of order r.

(b) Every minor of A of order higher than r is zero.

(i) Non-zero row is that row in which at least one element is not zero.

(ii) The rank of the product matrix AB of two matrices A and B is less than the rank of either the matrices A or B.

(iii) Corresponding to every matrix A of rank r, there exist non-singular matrices P and Q such that

$$PAQ = \begin{bmatrix} Ir & 0 \\ 0 & 0 \end{bmatrix}$$

Note: Non zero row is that row which does not contain all the elements zero.

Illustrative Examples

Example 1 Find the rank of the matrix.

$$A = \begin{bmatrix} 1 & 3 & 5 & 7 \\ -3 & 2 & 4 & 5 \end{bmatrix}$$

Solution: Here, we have

$$A = \begin{bmatrix} 1 & 3 & 5 & 7 \\ -3 & 2 & 4 & 5 \end{bmatrix}$$

Apply $R_2 \to R_2 + 3R_1$

$$A = \begin{bmatrix} 1 & 3 & 5 & 7 \\ 0 & 11 & 19 & 26 \end{bmatrix}$$

The number of non-zero row is 2, therefore $\boxed{Rank\,(A) = 2}$

Example 2 Find the rank of the matrix

$$A = \begin{bmatrix} -1 & 2 & -2 \\ 1 & 2 & 1 \\ -1 & -1 & 0 \end{bmatrix}$$

Solution: We have,

$$A = \begin{bmatrix} -1 & 2 & -2 \\ 1 & 2 & 1 \\ -1 & -1 & 0 \end{bmatrix}.$$

$R_2 \to R_2 + R_1, \quad R_3 \to R_3 - R_1$

$$A = \begin{bmatrix} -1 & 2 & -2 \\ 0 & 4 & -1 \\ 0 & -3 & 2 \end{bmatrix}$$

$R_3 \to R_3 + R_2$ $\qquad\qquad$ $R_3 \to 4R_3 - R_2$

$$A = \begin{bmatrix} -1 & 2 & -2 \\ 0 & 4 & -1 \\ 0 & 1 & 1 \end{bmatrix} = \begin{bmatrix} -1 & 2 & -2 \\ 0 & 4 & -1 \\ 0 & 0 & 5 \end{bmatrix}$$

The number of non-zero row is 3, therefore $\boxed{Rank(A) = 3}$

Example 3 Find the rank of the matrix

$$A = \begin{bmatrix} 1 & 2 & 3 & -4 \\ -2 & 3 & 7 & -1 \\ 1 & 9 & 16 & -13 \end{bmatrix}$$

Solution: Here we have,

$$A = \begin{bmatrix} 1 & 2 & 3 & -4 \\ -2 & 3 & 7 & -1 \\ 1 & 9 & 16 & -13 \end{bmatrix}$$

$R_2 \to R_2 + 2R_1$, $R_3 \to R_3 - R_1$

$$A = \begin{bmatrix} 1 & 2 & 3 & -4 \\ 0 & 7 & 13 & -9 \\ 0 & 7 & 13 & -9 \end{bmatrix}$$

$R_3 \to R_3 - R_2$

$$A = \begin{bmatrix} 1 & 2 & 3 & -4 \\ 0 & 7 & 13 & -9 \\ 0 & 0 & 0 & 0 \end{bmatrix}$$

The number of non-zero row is 2, therefore $\boxed{Rank\,(A) = 2}$

Example 4 For which value of 'b' the rank of the matrix

$$A = \begin{bmatrix} 1 & 5 & 4 \\ 0 & 3 & 2 \\ b & 13 & 10 \end{bmatrix} \text{ is 2.}$$

Solution: Here, we have

$$A = \begin{bmatrix} 1 & 5 & 4 \\ 0 & 3 & 2 \\ b & 13 & 10 \end{bmatrix}$$

$R_1 \to 2R_1$

$$A = \begin{bmatrix} 2 & 10 & 8 \\ 0 & 3 & 2 \\ b & 13 & 10 \end{bmatrix}$$

$R_3 \to R_3 - R_1$ $\qquad$ $R_3 \to R_3 - R_2$

$$A = \begin{bmatrix} 2 & 10 & 8 \\ 0 & 3 & 2 \\ b-2 & 3 & 2 \end{bmatrix} = \begin{bmatrix} 2 & 10 & 8 \\ 0 & 3 & 2 \\ b-2 & 0 & 0 \end{bmatrix}$$

If rank of A is 2, then b–2 must be zero

i.e., $\quad b - 2 = 0 \quad \Rightarrow \quad \boxed{b = 2}$

Example 5 Find the rank of the matrix

$$A = \begin{bmatrix} 2 & 3 & 4 & 5 \\ 3 & 4 & 5 & 6 \\ 4 & 5 & 6 & 7 \\ 9 & 10 & 11 & 12 \end{bmatrix}$$

Solution: Here we have,

$$A = \begin{bmatrix} 2 & 3 & 4 & 5 \\ 3 & 4 & 5 & 6 \\ 4 & 5 & 6 & 7 \\ 9 & 10 & 11 & 12 \end{bmatrix}$$

$$R_2 \to R_2 - \frac{3}{2}R_1,\ R_3 \to R_3 - 2R_1,\ R_4 \to R_4 - \frac{9}{2}R_1$$

$$A = \begin{bmatrix} 2 & 3 & 4 & 5 \\ 0 & \frac{-1}{2} & -1 & \frac{-3}{2} \\ 0 & -1 & -2 & -3 \\ 0 & -7/2 & -7 & \frac{-21}{2} \end{bmatrix}$$

$R_3 \to R_3 - 2R_1, \quad R_4 \to R_4 - 7R_2$

$$A = \begin{bmatrix} 2 & 3 & 4 & 5 \\ 0 & \frac{-1}{2} & -1 & \frac{-3}{2} \\ 0 & 0 & 0 & 0 \\ 0 & 0 & 0 & 0 \end{bmatrix}$$

Here, the number of non-zero row is 2 therefore $\boxed{Rank\,(A) = 2}$

Example 6 Find the rank of the matrix

$$A = \begin{bmatrix} 1 & 3 & 4 & 2 \\ 2 & -1 & 3 & 2 \\ 3 & -5 & 2 & 2 \\ 6 & -3 & 8 & 6 \end{bmatrix}$$

Solution: We have,

$$A = \begin{bmatrix} 1 & 3 & 4 & 2 \\ 2 & -1 & 3 & 2 \\ 3 & -5 & 2 & 2 \\ 6 & -3 & 8 & 6 \end{bmatrix}$$

$R_2 \to R_2 - 2R_1, \quad R_3 \to R_3 - 3R_1, \quad R_4 \to R_4 - 6R_1$

$$A = \begin{bmatrix} 1 & 3 & 4 & 2 \\ 0 & -7 & -5 & -2 \\ 0 & -14 & -10 & -4 \\ 0 & -21 & -16 & -6 \end{bmatrix}$$

$R_3 \to R_3 - 2R_2, \quad R_4 \to R_4 - 3R_2, \; R_3 \leftrightarrow R_4$

$$A = \begin{bmatrix} 1 & 3 & 4 & 2 \\ 0 & -7 & -5 & -2 \\ 0 & 0 & 0 & 0 \\ 0 & 0 & -1 & 0 \end{bmatrix} = \begin{bmatrix} 1 & 3 & 4 & 2 \\ 0 & -7 & -5 & -2 \\ 0 & 0 & -1 & 0 \\ 0 & 0 & 0 & 0 \end{bmatrix}$$

Here, the number of non-zero row is 3, therefore $\boxed{Rank\,(A) = 3}$

Example 7 Find the rank of the matrix

$$A = \begin{bmatrix} 1 & 2 & -1 & 4 \\ 2 & 4 & 3 & 4 \\ 1 & 2 & 3 & 4 \\ -1 & -2 & 6 & -7 \end{bmatrix}$$

Solution: We have,

$$A = \begin{bmatrix} 1 & 2 & -1 & 4 \\ 2 & 4 & 3 & 4 \\ 1 & 2 & 3 & 4 \\ -1 & -2 & 6 & -7 \end{bmatrix}$$

$R_2 \to R_2 - 2R_1, \quad R_3 \to R_3 - R_1, \quad R_4 \to R_4 + R_1$

$$A = \begin{bmatrix} 1 & 2 & -1 & 4 \\ 0 & 0 & 5 & -4 \\ 0 & 0 & 4 & 0 \\ 0 & 0 & 5 & -3 \end{bmatrix}$$

$R_3 \to \frac{1}{4} R_3, \; R_4 \to R_4 - 5R_3$

$$A = \begin{bmatrix} 1 & 2 & -1 & 4 \\ 0 & 0 & 5 & -4 \\ 0 & 0 & 1 & 0 \\ 0 & 0 & 5 & -3 \end{bmatrix} = \begin{bmatrix} 1 & 2 & -1 & 4 \\ 0 & 0 & 5 & -4 \\ 0 & 0 & 1 & 0 \\ 0 & 0 & 0 & -3 \end{bmatrix}$$

Here, the number of non-zero row is 4, therefore $\boxed{Rank\,(A) = 4}$

2.19 RANK OF MATRIX BY TRIANGULAR FORM

Rank = Number of non-zero rows in upper triangular matrix.

Example 8 Find the rank of the matrix

$$A = \begin{bmatrix} -1 & 2 & 3 & -2 \\ 2 & -5 & 1 & 2 \\ 3 & -8 & 5 & 2 \\ 5 & -12 & -1 & 6 \end{bmatrix}$$

Solution: Here we have

$$A = \begin{bmatrix} -1 & 2 & 3 & -2 \\ 2 & -5 & 1 & 2 \\ 3 & -8 & 5 & 2 \\ 5 & -12 & -1 & 6 \end{bmatrix}$$

$$R_1 \to -R_1 = \begin{bmatrix} 1 & -2 & -3 & 2 \\ 2 & -5 & 1 & 2 \\ 3 & -8 & 5 & 2 \\ 5 & -12 & -1 & 6 \end{bmatrix}$$

$R_2 \to R_2 - 2R_1, \quad R_3 \to R_3 - 3R_1, \quad R_4 \to R_4 - 5R_1$

$$A = \begin{bmatrix} 1 & -2 & -3 & 2 \\ 0 & -1 & 7 & -2 \\ 0 & -2 & 14 & -4 \\ 0 & -2 & 14 & -4 \end{bmatrix}$$

$R_3 \to R_3 - 2R_2, \quad R_4 \to R_4 - 2R_2$

$$A = \begin{bmatrix} 1 & -2 & -3 & 2 \\ 0 & -1 & 7 & -2 \\ 0 & 0 & 0 & 0 \\ 0 & 0 & 0 & 0 \end{bmatrix}$$

Here the 4th and 3rd order minors are zero. But a minor of second order

$$\begin{vmatrix} -3 & 2 \\ 7 & -2 \end{vmatrix} = +6 - 14 = -8 \neq 0$$

Rank = number of non-zero rows = 2.

Example 9 Find the rank of matrix

$$A = \begin{bmatrix} 2 & 3 & -2 & 4 \\ 3 & -2 & 1 & 2 \\ 3 & 2 & 3 & 4 \\ -2 & 4 & 0 & 5 \end{bmatrix}$$

Solution: Here we have

$$A = \begin{bmatrix} 2 & 3 & -2 & 4 \\ 3 & -2 & 1 & 2 \\ 3 & 2 & 3 & 4 \\ -2 & 4 & 0 & 5 \end{bmatrix}$$

$R_1 \to \frac{1}{2} R_1$

$$A = \begin{bmatrix} 1 & \frac{3}{2} & -1 & 2 \\ 3 & -2 & 1 & 2 \\ 3 & 2 & 3 & 4 \\ -2 & 4 & 0 & 5 \end{bmatrix}$$

$R_2 \to R_2 - 3R_1, \quad R_3 \to R_3 - 3R_1, \quad R_4 \to R_4 + 2R_1$

$$A = \begin{bmatrix} 1 & \frac{3}{2} & -1 & 2 \\ 0 & \frac{-13}{2} & 4 & -4 \\ 0 & \frac{-5}{2} & 6 & -2 \\ 0 & 7 & -2 & 9 \end{bmatrix}$$

$R_4 \to R_4 + \frac{14}{5} R_3$

$$A = \begin{bmatrix} 1 & \frac{3}{2} & -1 & 2 \\ 0 & \frac{-13}{2} & 4 & -4 \\ 0 & \frac{-5}{2} & 6 & -2 \\ 0 & 0 & \frac{74}{5} & \frac{17}{5} \end{bmatrix}$$

Rank = number of non-zero rows = 4

$$\boxed{Rank\,(A) = 4}$$

Example 10 Find the rank of the matrix

$$A = \begin{bmatrix} 2 & 3 & -1 & 1 \\ 1 & -1 & -2 & -4 \\ 3 & 1 & 3 & -2 \\ 6 & 3 & 0 & -7 \end{bmatrix}$$

Solution: Here we have

$$A = \begin{bmatrix} 2 & 3 & -1 & 1 \\ 1 & -1 & -2 & -4 \\ 3 & 1 & 3 & -2 \\ 6 & 3 & 0 & -7 \end{bmatrix}$$

$R_1 \leftrightarrow R_2$

$$A = \begin{bmatrix} 1 & -1 & -2 & -4 \\ 2 & 3 & -1 & 1 \\ 3 & 1 & 3 & -2 \\ 6 & 3 & 0 & -7 \end{bmatrix}$$

$R_2 \to R_2 - 2R_1,\quad R_3 \to R_3 - 3R_1,\quad R_4 \to R_4 - 6R_1$

$$A = \begin{bmatrix} 1 & -1 & -2 & -4 \\ 0 & 5 & 3 & 9 \\ 0 & 4 & 9 & 10 \\ 0 & 9 & 12 & 17 \end{bmatrix}$$

$R_3 \to R_3 - \frac{4}{5}R_2,\quad R_4 \to R_4 - \frac{9}{5}R_2$

$$A = \begin{bmatrix} 1 & -1 & -2 & -4 \\ 0 & 5 & 3 & 9 \\ 0 & 0 & 33/5 & 14/5 \\ 0 & 0 & 33/5 & 4/5 \end{bmatrix}$$

$R_4 \to R_4 - R_3$

$$A = \begin{bmatrix} 1 & -1 & -2 & -4 \\ 0 & 5 & 3 & 9 \\ 0 & 0 & 33/5 & 14/5 \\ 0 & 0 & 0 & -2 \end{bmatrix}$$

Rank = number of non zero rows = 4

$$\boxed{Rank\,(A) = 4}$$

Example 11 Find the rank of the matrix

$$A = \begin{bmatrix} 3 & -4 & -1 & 2 \\ 1 & 7 & 3 & 1 \\ 5 & -2 & 5 & 4 \\ 9 & -3 & 7 & 7 \end{bmatrix}$$

Solution: Here we have $R_1 \leftrightarrow R_2$

$$A = \begin{bmatrix} 3 & -4 & -1 & 2 \\ 1 & 7 & 3 & 1 \\ 5 & -2 & 5 & 4 \\ 9 & -3 & 7 & 7 \end{bmatrix} = \begin{bmatrix} 1 & 7 & 3 & 1 \\ 3 & -4 & -1 & 2 \\ 5 & -2 & 5 & 4 \\ 9 & -3 & 7 & 7 \end{bmatrix}$$

$R_2 \to R_2 - 3R_1, \quad R_3 \to R_3 - 5R_1, \quad R_4 \to R_4 - 9R_1$

$$A = \begin{bmatrix} 1 & 7 & 3 & 1 \\ 0 & -25 & -10 & -1 \\ 0 & -37 & -10 & -1 \\ 0 & -66 & -20 & -2 \end{bmatrix}$$

$R_3 \to R_3 - \frac{37}{25} R_2, \; R_4 \to R_4 - \frac{66}{25} R_2$

$$A = \begin{bmatrix} 1 & 7 & 3 & 1 \\ 0 & -25 & -10 & -1 \\ 0 & 0 & 24/5 & 12/25 \\ 0 & 0 & 32/5 & 16/25 \end{bmatrix}$$

$R_4 \to R_4 - \frac{4}{3} R_3$

$$A = \begin{bmatrix} 1 & 7 & 3 & 1 \\ 0 & -25 & -10 & -1 \\ 0 & 0 & 24/5 & 12/5 \\ 0 & 0 & 0 & 0 \end{bmatrix}$$

Rank = number of non zero rows = 3

$$\boxed{Rank\,(A) = 3}$$

Example 12 Find the rank of the matrix

$$A = \begin{bmatrix} 3 & -1 & 2 \\ -6 & 2 & 4 \\ -3 & 1 & 2 \end{bmatrix}$$

Solution: Here we have

$$A = \begin{bmatrix} 3 & -1 & 2 \\ -6 & 2 & 4 \\ -3 & 1 & 2 \end{bmatrix}$$

$R_2 \to R_2 + 2R_1, \quad R_3 \to R_3 + R_1$

$$A = \begin{bmatrix} 3 & -1 & 2 \\ 0 & 0 & 8 \\ 0 & 0 & 4 \end{bmatrix}$$

$R_3 \to 2R_3 - R_2$

$$A=\begin{bmatrix}3 & -1 & 2\\ 0 & 0 & 8\\ 0 & 0 & 0\end{bmatrix}$$

$R_2 \to \frac{1}{8} R_2$

$$A=\begin{bmatrix}3 & -1 & 2\\ 0 & 0 & 1\\ 0 & 0 & 0\end{bmatrix}$$

Rank = Number of non zero rows = 2

$$\boxed{Rank\,(A)=2}$$

Example 13 Find the rank of the matrix

$$A=\begin{bmatrix}1 & 2 & 1 & 0\\ -2 & 4 & 3 & 0\\ 1 & 0 & 2 & -8\end{bmatrix}$$

Solution: Here we have

$$A=\begin{bmatrix}1 & 2 & 1 & 0\\ -2 & 4 & 3 & 0\\ 1 & 0 & 2 & -8\end{bmatrix}$$

$R_2 \to R_2 + 2R_1, \quad R_3 \to R_3 - R_1$

$$A=\begin{bmatrix}1 & 2 & 1 & 0\\ 0 & 8 & 5 & 0\\ 0 & -2 & 1 & -8\end{bmatrix}$$

$R_3 \to 4R_3 + R_2$

$$A=\begin{bmatrix}1 & 2 & 1 & 0\\ 0 & 8 & 5 & 0\\ 0 & 0 & 9 & -32\end{bmatrix}$$

Rank = Number of non zero rows = 3

$$\boxed{Rank\,(A)=3}$$

Example 14 Find the rank of the matrix

$$A=\begin{bmatrix}2 & 4 & 3 & -2\\ -3 & -2 & -1 & 4\\ 6 & -1 & 7 & 2\end{bmatrix}$$

Solution: Here we have

$$A = \begin{bmatrix} 2 & 4 & 3 & -2 \\ -3 & -2 & -1 & 4 \\ 6 & -1 & 7 & 2 \end{bmatrix}$$

$R_1 \to \frac{1}{2} R_1$

$$A = \begin{bmatrix} 1 & 2 & 3/2 & -1 \\ -3 & -2 & -1 & 4 \\ 6 & -1 & 7 & 2 \end{bmatrix}$$

$R_2 \to R_2 + 3\,R_1, \quad R_3 \to R_3 - 6\,R_1$

$$A = \begin{bmatrix} 1 & 2 & 3/2 & -1 \\ 0 & 4 & 7/2 & -1 \\ 0 & -13 & -2 & 8 \end{bmatrix}$$

$R_3 \to R_3 + \frac{13}{4} R_2$

$$A = \begin{bmatrix} 1 & 2 & 3/2 & -1 \\ 0 & 4 & 7/2 & -1 \\ 0 & 0 & 75/8 & 19/4 \end{bmatrix}$$

Rank = Number of non zero rows = 3

$$\boxed{Rank\,(A) = 3}$$

Example 15 Find the rank of the matrix

$$A = \begin{bmatrix} 3 & 4 & 1 & 1 \\ 2 & 4 & 3 & 6 \\ -1 & -2 & 6 & 4 \\ 1 & -1 & 2 & -3 \end{bmatrix}$$

Solution: Here we have

$$A = \begin{bmatrix} 3 & 4 & 1 & 1 \\ 2 & 4 & 3 & 6 \\ -1 & -2 & 6 & 4 \\ 1 & -1 & 2 & -3 \end{bmatrix}$$

$R_2 \leftrightarrow R_4$

$$A = \begin{bmatrix} 1 & -1 & 2 & -3 \\ 2 & 4 & 3 & 6 \\ -1 & -2 & 6 & 4 \\ 3 & 4 & 1 & 1 \end{bmatrix}$$

$R_2 \to R_2 - 2R_1, \quad R_3 \to R_3 + R_1, \quad R_4 \to R_4 - 3R_1$

$$A = \begin{bmatrix} 1 & -1 & 2 & -3 \\ 0 & 6 & -1 & 12 \\ 0 & -3 & 8 & 1 \\ 0 & 7 & -5 & 10 \end{bmatrix}$$

$R_3 \to 2R_3 + R_2, \quad R_4 \to R_4 - \frac{7}{6} R_2$

$$A = \begin{bmatrix} 1 & -1 & 2 & -3 \\ 0 & 6 & -1 & 12 \\ 0 & 0 & 15 & 14 \\ 0 & 0 & -23/6 & -4 \end{bmatrix}$$

Rank = Number of non zero rows = 4

$$\boxed{Rank\,(A) = 4}$$

Example 16 Find the rank of the matrix

$$A = \begin{bmatrix} 1 & 4 & 3 & -2 & 1 \\ -2 & -3 & -1 & 4 & 3 \\ -1 & 6 & 7 & 2 & 9 \\ -3 & 3 & 6 & 6 & 12 \end{bmatrix}$$

Solution: Here we have

$$A = \begin{bmatrix} 1 & 4 & 3 & -2 & 1 \\ -2 & -3 & -1 & 4 & 3 \\ -1 & 6 & 7 & 2 & 9 \\ -3 & 3 & 6 & 6 & 12 \end{bmatrix}$$

$R_2 \to R_2 + 2R_1, \quad R_3 \to R_3 + R_1, \quad R_4 \to R_4 + 3R_1$

$$A = \begin{bmatrix} 1 & 4 & 3 & -2 & 1 \\ 0 & 5 & 5 & 0 & 5 \\ 0 & 10 & 10 & 0 & 10 \\ 0 & 15 & 15 & 0 & 15 \end{bmatrix}$$

$R_3 \to R_3 - 2R_2, \quad R_4 \to R_4 - 3R_2$

$$A = \begin{bmatrix} 1 & 4 & 3 & -2 & 1 \\ 0 & 5 & 5 & 0 & 5 \\ 0 & 0 & 0 & 0 & 0 \\ 0 & 0 & 0 & 0 & 0 \end{bmatrix}$$

$R_2 \to \frac{1}{5} R_2$

$$A = \begin{bmatrix} 1 & 4 & 3 & -2 & 1 \\ 0 & 1 & 1 & 0 & 1 \\ 0 & 0 & 0 & 0 & 0 \\ 0 & 0 & 0 & 0 & 0 \end{bmatrix}$$

Rank = Number of non zero rows = 2

$$\boxed{Rank\,(A) = 2}$$

Example 17 Find the rank of the matrix

$$A = \begin{bmatrix} 1 & 2 & 3 & 0 \\ 2 & 4 & 3 & 2 \\ 3 & 2 & 1 & 3 \\ 6 & 8 & 7 & 5 \end{bmatrix}$$

Solution: Here we have

$$A = \begin{bmatrix} 1 & 2 & 3 & 0 \\ 2 & 4 & 3 & 2 \\ 3 & 2 & 1 & 3 \\ 6 & 8 & 7 & 5 \end{bmatrix}$$

$R_2 \to R_2 - 2R_1, \quad R_3 \to R_3 - 3R_1, \quad R_4 \to R_4 - 6R_1$

$$A = \begin{bmatrix} 1 & 2 & 3 & 0 \\ 0 & 0 & -3 & 2 \\ 0 & -4 & -8 & 3 \\ 0 & -4 & -11 & 5 \end{bmatrix}$$

$R_4 \to R_4 - R_3$

$$A = \begin{bmatrix} 1 & 2 & 3 & 0 \\ 0 & 0 & -3 & 2 \\ 0 & -4 & -8 & 3 \\ 0 & 0 & -3 & 2 \end{bmatrix}$$

$R_4 \to R_4 - R_2$

$$A = \begin{bmatrix} 1 & 2 & 3 & 0 \\ 0 & 0 & -3 & 2 \\ 0 & -4 & -8 & 3 \\ 0 & 0 & 0 & 0 \end{bmatrix}$$

Rank = Number of non zero rows = 3

$$\boxed{Rank\,(A) = 3}$$

Example 18 Find the rank of the matrix

$$A = \begin{bmatrix} 2 & -4 & 3 & 1 & 0 \\ 1 & -2 & 1 & -4 & 2 \\ 0 & 1 & -1 & 3 & 1 \\ 4 & -7 & 4 & -4 & 5 \end{bmatrix}$$

Solution: Here we have $A = \begin{bmatrix} 2 & -4 & 3 & 1 & 0 \\ 1 & -2 & 1 & -4 & 2 \\ 0 & 1 & -1 & 3 & 1 \\ 4 & -7 & 4 & -4 & 5 \end{bmatrix}$

$R_1 \leftrightarrow R_2$

$$A = \begin{bmatrix} 1 & -2 & 1 & -4 & 2 \\ 2 & -4 & 3 & 1 & 0 \\ 0 & 1 & -1 & 3 & 1 \\ 4 & -7 & 4 & -4 & 5 \end{bmatrix}$$

$R_2 \to R_2 - 2\,R_1, \quad R_4 \to R_4 - 4\,R_1$

$$A = \begin{bmatrix} 1 & -2 & 1 & -4 & 2 \\ 0 & 0 & 1 & 9 & -4 \\ 0 & 1 & -1 & 3 & 1 \\ 0 & 1 & 0 & 12 & -3 \end{bmatrix}$$

$R_4 \to R_4 - R_3$

$$A = \begin{bmatrix} 1 & -2 & 1 & -4 & 2 \\ 0 & 0 & 1 & 9 & -4 \\ 0 & 1 & -1 & 3 & 1 \\ 0 & 0 & 1 & 9 & -4 \end{bmatrix}$$

$R_4 \to R_4 - R_2$

$$A = \begin{bmatrix} 1 & -2 & 1 & -4 & 2 \\ 0 & 0 & 1 & 9 & -4 \\ 0 & 1 & -1 & 3 & 1 \\ 0 & 0 & 0 & 0 & 0 \end{bmatrix}$$

Rank = Number of non zero rows = 3

$$\boxed{Rank\,(A) = 3}$$

Example 19 Find the rank of the matrix

$$A = \begin{bmatrix} 1 & 2 & 5 & 6 \\ 2 & 10 & 10 & 12 \\ -1 & -2 & -5 & -6 \\ 0 & 0 & 0 & 0 \end{bmatrix}$$

Solution: Here we have

$$A = \begin{bmatrix} 1 & 2 & 5 & 6 \\ 2 & 10 & 10 & 12 \\ -1 & -2 & -5 & -6 \\ 0 & 0 & 0 & 0 \end{bmatrix}$$

$R_2 \to R_2 - 2\,R_1, \quad R_3 \to R_3 + R_1$

$$A = \begin{bmatrix} 1 & 2 & 5 & 6 \\ 0 & 6 & 0 & 0 \\ 0 & 0 & 0 & 0 \\ 0 & 0 & 0 & 0 \end{bmatrix}$$

$R_2 \to \frac{1}{6} R_2$

$$A = \begin{bmatrix} 1 & 2 & 5 & 6 \\ 0 & 1 & 0 & 0 \\ 0 & 0 & 0 & 0 \\ 0 & 0 & 0 & 0 \end{bmatrix}$$

Rank = Number of non zero rows = 2

$$\boxed{Rank\,(A) = 2}$$

Example 20 Find the rank of the matrix

$$A = \begin{bmatrix} 0 & 1 & -3 & -1 \\ 1 & 0 & 1 & 1 \\ 3 & 1 & 0 & 2 \\ 1 & 1 & -2 & 0 \end{bmatrix}$$

Solution: Here we have

$$A = \begin{bmatrix} 0 & 1 & -3 & -1 \\ 1 & 0 & 1 & 1 \\ 3 & 1 & 0 & 2 \\ 1 & 1 & -2 & 0 \end{bmatrix}$$

$R_1 \leftrightarrow R_2$

$$A = \begin{bmatrix} 1 & 0 & 1 & 1 \\ 0 & 1 & -3 & -1 \\ 3 & 1 & 0 & 2 \\ 1 & 1 & -2 & 0 \end{bmatrix}$$

$R_3 \to R_3 - 3R_1, \quad R_4 \to R_4 - R_1$

$$A = \begin{bmatrix} 1 & 0 & 1 & 1 \\ 0 & 1 & -3 & -1 \\ 0 & 1 & -3 & -1 \\ 0 & 1 & -3 & -1 \end{bmatrix}$$

$R_3 \to R_3 - R_2, \quad R_4 \to R_4 - R_3$

$$A = \begin{bmatrix} 1 & 0 & 1 & 1 \\ 0 & 1 & -3 & -1 \\ 0 & 0 & 0 & 0 \\ 0 & 0 & 0 & 0 \end{bmatrix}$$

Rank = Number of non zero rows = 2

$$\boxed{Rank\,(A) = 2}$$

2.20 NORMAL FORM OF A MATRIX (CANONICAL FORM)

Every non- zero matrix of order mxn can be reduced by means of elementary row and column operation into equivalent matrix of any of the following forms;

(i) $\begin{bmatrix} I_r & : & 0 \\ \cdots & \cdots & \cdots \\ 0 & : & 0 \end{bmatrix}$ (ii) $\begin{bmatrix} I_r \\ \cdots \\ 0 \end{bmatrix}$

(iii) $[I_r \;:\; 0]$ (iv) $[I_r]$

Where I_r is the identity matrix of order r and 0 repersent zero matix of any order which is called its normal form or canonical form. The number r so obtained is called the rank of A and we write $\rho(A) = r$.

The form $\begin{bmatrix} I_r & 0 \\ 0 & 0 \end{bmatrix}$ is called the first canonical form of A. Since both row and column transformation may be used here, the element 1 of the first row obtained can be moved in first column. Then both the first row and first column can be cleared of other non-zero elements. Similarly, the element 1 of the second row can be brought into the second column and so on

Illustrative Examples

Example 1 Find the rank of the matrix $A = \begin{bmatrix} 1 & 2 & -1 & 3 \\ -2 & -4 & 4 & -7 \\ 1 & 2 & 1 & 2 \end{bmatrix}$ by reducing it to normal form.

Solution: Let

$$A = \begin{bmatrix} 1 & 2 & -1 & 3 \\ -2 & -4 & 4 & -7 \\ 1 & 2 & 1 & 2 \end{bmatrix}$$

$C_2 \to C_2 - 2C_1, \quad C_3 \to C_3 + C_1, \quad C_4 \to C_4 - 3C_1$

$$A \sim \begin{bmatrix} 1 & 0 & 0 & 0 \\ -2 & 0 & 2 & -1 \\ 1 & 0 & 2 & -1 \end{bmatrix}$$

$C_3 \to C_3 + 2C_4, \quad C_4 \to (-1)C_4$

$$A \sim \begin{bmatrix} 1 & 0 & 0 & 0 \\ -2 & 0 & 0 & 1 \\ 1 & 0 & 0 & 1 \end{bmatrix}$$

$R_2 \to R_2 + 2R_1, \quad R_3 \to R_3 - R_1$

$$A \sim \begin{bmatrix} 1 & 0 & 0 & 0 \\ 0 & 0 & 0 & 1 \\ 0 & 0 & 0 & 1 \end{bmatrix}$$

$R_3 \to R_3 - R_2$

$$A \sim \begin{bmatrix} 1 & 0 & 0 & 0 \\ 0 & 0 & 0 & 1 \\ 0 & 0 & 0 & 0 \end{bmatrix}$$

$C_2 \leftrightarrow C_4$

$$A \sim \begin{bmatrix} 1 & 0 & 0 & 0 \\ 0 & 1 & 0 & 0 \\ 0 & 0 & 0 & 0 \end{bmatrix} \sim \begin{bmatrix} I_2 & : & 0 \\ 0 & : & 0 \end{bmatrix}$$

Which is required normal form.

Rank of Matrix (A) = 2

Example 2 Find the rank of the matrix $A = \begin{bmatrix} 9 & 0 & 2 & 3 \\ 0 & 1 & 5 & 6 \\ 4 & 5 & 3 & 0 \end{bmatrix}$ by reducing it to normal form.

Solution: We have $A = \begin{bmatrix} 9 & 0 & 2 & 3 \\ 0 & 1 & 5 & 6 \\ 4 & 5 & 3 & 0 \end{bmatrix}$

$R_1 \to R_1 (1/9)$

$$A \sim \begin{bmatrix} 1 & 0 & 2/9 & 3/9 \\ 0 & 1 & 5 & 6 \\ 4 & 5 & 3 & 0 \end{bmatrix}$$

$R_3 \to R_3 + R_1(-4)$

$$A \sim \begin{bmatrix} 1 & 0 & 2/9 & 3/9 \\ 0 & 1 & 5 & 6 \\ 0 & 5 & 19/9 & -12/9 \end{bmatrix}$$

$C_3 \to C_3 + C_1\left(\frac{-2}{9}\right)$

$$A \sim \begin{bmatrix} 1 & 0 & 0 & 3/9 \\ 0 & 1 & 5 & 6 \\ 0 & 5 & 19/9 & -12/9 \end{bmatrix}$$

$C_4 \to C_4 + \left(\frac{-3}{9}\right) C_1$

$$A \sim \begin{bmatrix} 1 & 0 & 0 & 0 \\ 0 & 1 & 5 & 6 \\ 0 & 5 & 19/9 & -12/9 \end{bmatrix}$$

$R_3 \to R_3 - R_2 (5)$ $\quad A \sim \begin{bmatrix} 1 & 0 & 0 & 0 \\ 0 & 1 & 5 & 6 \\ 0 & 0 & \frac{-206}{9} & \frac{-282}{9} \end{bmatrix}$

$C_3 \to C_3 + C_2(-5)$
$C_4 \to C_4 + C_2(-6)$
$$A \sim \begin{bmatrix} 1 & 0 & 0 & 0 \\ 0 & 1 & 0 & 0 \\ 0 & 0 & \frac{-206}{9} & \frac{-282}{9} \end{bmatrix}$$

$R_3 \to R_3\left(\frac{-9}{206}\right)$
$$A \sim \begin{bmatrix} 1 & 0 & 0 & 0 \\ 0 & 1 & 0 & 0 \\ 0 & 0 & 1 & \frac{141}{103} \end{bmatrix}$$

$C_4 \to C_4 + C_3\left(\frac{-141}{103}\right)$
$$A \sim \begin{bmatrix} 1 & 0 & 0 & 0 \\ 0 & 1 & 0 & 0 \\ 0 & 0 & 1 & 0 \end{bmatrix} = [I_3, 0]$$

Which is required normal form.
Rank of matrix (A) = 3

Example 3 Reduce the matrix $A = \begin{bmatrix} 1 & -1 & 2 & -3 \\ 4 & 1 & 0 & 2 \\ 0 & 3 & 0 & 4 \\ 0 & 1 & 0 & 2 \end{bmatrix}$ to normal form $\begin{bmatrix} I_r & 0 \\ 0 & 0 \end{bmatrix}$.

Solution: We have
$$A = \begin{bmatrix} 1 & -1 & 2 & -3 \\ 4 & 1 & 0 & 2 \\ 0 & 3 & 0 & 4 \\ 0 & 1 & 0 & 2 \end{bmatrix}$$

$C_2 \to C_2 + C_1, \quad C_3 \to C_3 - 2C_1, \quad C_4 \to C_4 + 3C_1$
$$A \sim \begin{bmatrix} 1 & 0 & 0 & 0 \\ 4 & 5 & -8 & 14 \\ 0 & 3 & 0 & 4 \\ 0 & 1 & 0 & 2 \end{bmatrix}$$

$R_2 \to R_2 - 4R_1$
$$A \sim \begin{bmatrix} 1 & 0 & 0 & 0 \\ 0 & 5 & -8 & 14 \\ 0 & 3 & 0 & 4 \\ 0 & 1 & 0 & 2 \end{bmatrix}$$

$R_4 \leftrightarrow R_2$ $\quad A \sim \begin{bmatrix} 1 & 0 & 0 & 0 \\ 0 & 1 & 0 & 2 \\ 0 & 3 & 0 & 4 \\ 0 & 5 & -8 & 14 \end{bmatrix}$

$C_4 \to C_4 - 2\,C_2$ $\quad A \sim \begin{bmatrix} 1 & 0 & 0 & 0 \\ 0 & 1 & 0 & 0 \\ 0 & 3 & 0 & -2 \\ 0 & 5 & -8 & 4 \end{bmatrix}$

$R_3 \to R_3 - 3R_2$
$R_4 \to R_5 - 5R_2$ $\quad A \sim \begin{bmatrix} 1 & 0 & 0 & 0 \\ 0 & 1 & 0 & 0 \\ 0 & 0 & 0 & -2 \\ 0 & 0 & -8 & 4 \end{bmatrix}$

$C_3 \leftrightarrow C_4$ $\quad A \sim \begin{bmatrix} 1 & 0 & 0 & 0 \\ 0 & 1 & 0 & 0 \\ 0 & 0 & -2 & 0 \\ 0 & 0 & 4 & -8 \end{bmatrix}$

$C_3 \to \dfrac{-1}{2} C_3$
$C_4 \to \dfrac{-1}{8} C_4$ $\quad A \sim \begin{bmatrix} 1 & 0 & 0 & 0 \\ 0 & 1 & 0 & 0 \\ 0 & 0 & 1 & 0 \\ 0 & 0 & -2 & 1 \end{bmatrix}$

$R_4 \to R_4 + 2R_3$ $\quad A \sim \begin{bmatrix} 1 & 0 & 0 & 0 \\ 0 & 1 & 0 & 0 \\ 0 & 0 & 1 & 0 \\ 0 & 0 & 0 & 1 \end{bmatrix} = I_4$

Hence $\rho(A) = 4$

Example 4 Find the rank of the following matrix by reducing it to normal form $A = \begin{bmatrix} 1 & 2 & -1 & 3 \\ 4 & 1 & 2 & 1 \\ 3 & -1 & 1 & 2 \\ 1 & 2 & 0 & 1 \end{bmatrix}$

Solution: We have $\quad A = \begin{bmatrix} 1 & 2 & -1 & 3 \\ 4 & 1 & 2 & 1 \\ 3 & -1 & 1 & 2 \\ 1 & 2 & 0 & 1 \end{bmatrix}$

$R_2 \to R_2 - 4R_1$
$R_3 \to R_3 \to 3R_1$
$R_4 \to R_4 - R_1$
$$A \sim \begin{bmatrix} 1 & 2 & -1 & 3 \\ 0 & -7 & 6 & -11 \\ 0 & -7 & 4 & -7 \\ 0 & 0 & 1 & -2 \end{bmatrix}$$

$C_2 \to C_2 - 2C_1$
$C_3 \to C_3 \to C_1$
$C_4 \to C_4 - 3C_1$
$$A \sim \begin{bmatrix} 1 & 0 & 0 & 0 \\ 0 & -7 & 6 & -11 \\ 0 & -7 & 4 & -7 \\ 0 & 0 & 1 & -2 \end{bmatrix}$$

$R_3 \to R_3 - R_2$
$$A \sim \begin{bmatrix} 1 & 0 & 0 & 0 \\ 0 & -7 & 6 & -11 \\ 0 & 0 & -2 & 4 \\ 0 & 0 & 1 & -2 \end{bmatrix}$$

$C_3 \to C_3 + \frac{6}{7}C_2$
$C_4 \to C_4 - \frac{11}{7}C_2$
$$A \sim \begin{bmatrix} 1 & 0 & 0 & 0 \\ 0 & -7 & 0 & 0 \\ 0 & 0 & -2 & 4 \\ 0 & 0 & 1 & -2 \end{bmatrix}$$

$R_4 \to R_4 + \frac{1}{2}R_3$
$$A \sim \begin{bmatrix} 1 & 0 & 0 & 0 \\ 0 & -7 & 0 & 0 \\ 0 & 0 & -2 & 4 \\ 0 & 0 & 0 & 0 \end{bmatrix}$$

$C_4 \to C_4 + 2C_3$
$$A \sim \begin{bmatrix} 1 & 0 & 0 & 0 \\ 0 & -7 & 0 & 0 \\ 0 & 0 & -2 & 0 \\ 0 & 0 & 0 & 0 \end{bmatrix}$$

$R_2 \to \frac{-1}{7}R_2$
$R_3 \to \frac{-1}{2}R_3$
$$A \sim \begin{bmatrix} 1 & 0 & 0 & 0 \\ 0 & 1 & 0 & 0 \\ 0 & 0 & 1 & 0 \\ 0 & 0 & 0 & 0 \end{bmatrix} = \begin{bmatrix} I_3 & 0 \\ 0 & 0 \end{bmatrix}$$

Rank of A = 3

Example 5 Reduce the matrix to normal form and find its rank if $A = \begin{bmatrix} 2 & 3 & 4 & 5 \\ 3 & 4 & 5 & 6 \\ 4 & 5 & 6 & 7 \\ 9 & 10 & 11 & 12 \end{bmatrix}$

Solution: We have $A = \begin{bmatrix} 2 & 3 & 4 & 5 \\ 3 & 4 & 5 & 6 \\ 4 & 5 & 6 & 7 \\ 9 & 10 & 11 & 12 \end{bmatrix}$

$R_2 \to R_2 - \frac{3}{2}R_1$
$R_3 \to R_3 - 2R_1$
$R_4 \to R_4 - \frac{9}{2}R_1$

$$A \sim \begin{bmatrix} 2 & 3 & 4 & 5 \\ 0 & -1/2 & -1 & -3/2 \\ 0 & -1 & -2 & -3 \\ 0 & -7/2 & -7 & -21/2 \end{bmatrix}$$

$C_2 \to C_2 - \frac{3}{2}C_1$
$C_3 \to C_3 - 2C_1$
$C_4 \to C_4 - \frac{5}{2}C_1$

$$A \sim \begin{bmatrix} 2 & 0 & 0 & 0 \\ 0 & -1/2 & -1 & -3/2 \\ 0 & -1 & -2 & -3 \\ 0 & -7/2 & -7 & -21/2 \end{bmatrix}$$

$C_1 \to \frac{1}{2}C_1$
$C_2 \to -2C_2$
$C_3 \to -C_3$
$C_4 \to \frac{-2}{7}C_4$

$$A \sim \begin{bmatrix} 1 & 0 & 0 & 0 \\ 0 & 1 & 2 & 3 \\ 0 & 1 & 2 & 3 \\ 0 & 1 & 2 & 3 \end{bmatrix}$$

$R_3 \to R_3 - R_2$
$R_4 \to R_4 - R_2$

$$A \sim \begin{bmatrix} 1 & 0 & 0 & 0 \\ 0 & 1 & 2 & 3 \\ 0 & 0 & 0 & 0 \\ 0 & 0 & 0 & 0 \end{bmatrix}$$

$C_3 \to C_3 - 2C_2$
$C_4 \to C_4 - 3C_2$

$$A \sim \begin{bmatrix} 1 & 0 & 0 & 0 \\ 0 & 1 & 0 & 0 \\ 0 & 0 & 0 & 0 \\ 0 & 0 & 0 & 0 \end{bmatrix} = \begin{bmatrix} I_2 & 0 \\ 0 & 0 \end{bmatrix}$$

Hence its rank = 2

Example 6 Reduce the matrix A to its normal form, when $A = \begin{bmatrix} 1 & 2 & -1 & 4 \\ 2 & 4 & 3 & 4 \\ 1 & 2 & 3 & 4 \\ -1 & -2 & 6 & -7 \end{bmatrix}$

Hence find the rank of A.

Solution: The given matrix is $A = \begin{bmatrix} 1 & 2 & -1 & 4 \\ 2 & 4 & 3 & 4 \\ 1 & 2 & 3 & 4 \\ -1 & -2 & 6 & -7 \end{bmatrix}$

$R_2 \to R_2 - 2R_1$
$R_3 \to R_3 - R_1$
$R_4 \to R_4 + R_1$
$A \sim \begin{bmatrix} 1 & 2 & -1 & 4 \\ 0 & 0 & 5 & -4 \\ 0 & 0 & 4 & 0 \\ 0 & 0 & 5 & -3 \end{bmatrix}$

$C_2 \to C_2 - 2C_1$
$C_3 \to C_3 + C_1$
$C_4 \to C_4 - 4C_1$
$A \sim \begin{bmatrix} 1 & 0 & 0 & 0 \\ 0 & 0 & 5 & -4 \\ 0 & 0 & 4 & 0 \\ 0 & 0 & 5 & -3 \end{bmatrix}$

$C_3 \leftrightarrow C_2$
$A \sim \begin{bmatrix} 1 & 0 & 0 & 0 \\ 0 & 5 & 0 & -4 \\ 0 & 4 & 0 & 0 \\ 0 & 5 & 0 & -3 \end{bmatrix}$

$R_3 \to R_3 - \frac{4}{5}R_2$
$R_4 \to R_4 - R_2$
$A \sim \begin{bmatrix} 1 & 0 & 0 & 0 \\ 0 & 5 & 0 & -4 \\ 0 & 0 & 0 & 16/5 \\ 0 & 0 & 0 & 1 \end{bmatrix}$

$C_4 \leftrightarrow C_3$
$A \sim \begin{bmatrix} 1 & 0 & 0 & 0 \\ 0 & 5 & -4 & 0 \\ 0 & 0 & 16/5 & 0 \\ 0 & 0 & 1 & 0 \end{bmatrix}$

$R_2 \to R_2 + \frac{5}{4}R_3$
$R_4 \to R_4 - \frac{5}{16}R_3$
$A \sim \begin{bmatrix} 1 & 0 & 0 & 0 \\ 0 & 5 & 0 & 0 \\ 0 & 0 & 16/5 & 0 \\ 0 & 0 & 0 & 0 \end{bmatrix}$

$R_2 \to 1/5\, R_2$
$R_3 \to 5/16\, R_3$
$A \sim \begin{bmatrix} 1 & 0 & 0 & 0 \\ 0 & 1 & 0 & 0 \\ 0 & 0 & 1 & 0 \\ 0 & 0 & 0 & 0 \end{bmatrix} = \begin{bmatrix} I_3 & 0 \\ 0 & 0 \end{bmatrix}$

Which is required normal form

Hence the rank of given matrix is = 3

2.21 TO CALCULATE P AND Q WHERE PAQ = $\begin{bmatrix} I_r & 0 \\ 0 & 0 \end{bmatrix}$

If A is a matrix of order mxn then write $A = I_m A I_n$ where I_m and I_n are mth and nth order unit matrices respectively. The elementry row operation on A can be effected by premultiplication with corresponding elementary matrix i.e. application of the same to I_m or to the matrix obtained from I_m in subsequent steps. Similarly, application of an elementary column operation to A is equivalent to application of the same to I_n or the matrix obtained from I_n in subsequent steps. In the end when we get $\begin{bmatrix} I_r & 0 \\ 0 & 0 \end{bmatrix}$ in place of A on left hand side, we have P in place of I_m and Q in place of I_n on the right hand side.

Example 7 Find the non-singular matrices P and Q such that PAQ is in normal form where $A = \begin{bmatrix} 2 & 2 & -6 \\ -1 & 2 & 2 \end{bmatrix}$

Solution: We have $[A]_{2\times 3} = I_2 \cdot A \cdot I_3$

$$\begin{bmatrix} 2 & 2 & -6 \\ -1 & 2 & 2 \end{bmatrix} = \begin{bmatrix} 1 & 0 \\ 0 & 1 \end{bmatrix} A \begin{bmatrix} 1 & 0 & 0 \\ 0 & 1 & 0 \\ 0 & 0 & 1 \end{bmatrix}$$

$R_1 \to \frac{1}{2} R_1$

$$\begin{bmatrix} 1 & 1 & -3 \\ -1 & 2 & 2 \end{bmatrix} = \begin{bmatrix} 1/2 & 0 \\ 0 & 1 \end{bmatrix} A \begin{bmatrix} 1 & 0 & 0 \\ 0 & 1 & 0 \\ 0 & 0 & 1 \end{bmatrix}$$

$R_2 \to R_2 + R_1$

$$\begin{bmatrix} 1 & 1 & -3 \\ 0 & 3 & -1 \end{bmatrix} = \begin{bmatrix} 1/2 & 0 \\ 1/2 & 1 \end{bmatrix} A \begin{bmatrix} 1 & 0 & 0 \\ 0 & 1 & 0 \\ 0 & 0 & 1 \end{bmatrix}$$

$C_2 \to C_2 - C_1, \quad C_3 \to C_3 + 3C_1$

$$\begin{bmatrix} 1 & 0 & 0 \\ 0 & 3 & -1 \end{bmatrix} = \begin{bmatrix} 1/2 & 0 \\ 1/2 & 1 \end{bmatrix} A \begin{bmatrix} 1 & -1 & 3 \\ 0 & 1 & 0 \\ 0 & 0 & 1 \end{bmatrix}$$

$C_3 \to \frac{1}{3} C_2$

$$\begin{bmatrix} 1 & 0 & 0 \\ 0 & 1 & -1 \end{bmatrix} = \begin{bmatrix} 1/2 & 0 \\ 1/2 & 1 \end{bmatrix} A \begin{bmatrix} 1 & -1/3 & 3 \\ 0 & 1/3 & 0 \\ 0 & 0 & 1 \end{bmatrix}$$

$C_3 \to C_3 + C_2$

$$\begin{bmatrix} 1 & 0 & 0 \\ 0 & 1 & 0 \end{bmatrix} = \begin{bmatrix} 1/2 & 0 \\ 1/2 & 1 \end{bmatrix} A \begin{bmatrix} 1 & -1/3 & 8/3 \\ 0 & 1/3 & 1/3 \\ 0 & 0 & 1 \end{bmatrix}$$

$$[I_2\,,\,0] = P\,A\,Q$$

Hence $P = \begin{bmatrix} 1/2 & 0 \\ 1/2 & 1 \end{bmatrix}$ and $Q = \begin{bmatrix} 1 & -1/3 & 8/3 \\ 0 & 1/3 & 1/3 \\ 0 & 0 & 1 \end{bmatrix}$

Example 8 If $A = \begin{bmatrix} 3 & -3 & 4 \\ 2 & -3 & 4 \\ 0 & -1 & 1 \end{bmatrix}$. Find two non singular matrices P & Q such that PAQ = I

Solution: $[A]_{3\times 3} = I_3\,A\,I_3$

$$\begin{bmatrix} 3 & -3 & 4 \\ 2 & -3 & 4 \\ 0 & -1 & 1 \end{bmatrix} = \begin{bmatrix} 1 & 0 & 0 \\ 0 & 1 & 0 \\ 0 & 0 & 1 \end{bmatrix} A \begin{bmatrix} 1 & 0 & 0 \\ 0 & 1 & 0 \\ 0 & 0 & 1 \end{bmatrix}$$

$R_1 \to R_1 - R_2$

$$\begin{bmatrix} 1 & 0 & 0 \\ 2 & -3 & 4 \\ 0 & -1 & 1 \end{bmatrix} = \begin{bmatrix} 1 & -1 & 0 \\ +0 & 1 & 0 \\ 0 & 0 & 1 \end{bmatrix} A \begin{bmatrix} 1 & 0 & 0 \\ 0 & 1 & 0 \\ 0 & 0 & 1 \end{bmatrix}$$

$R_2 \to R_2 - 2\,R_1$

$$\begin{bmatrix} 1 & 0 & 0 \\ 0 & -3 & 4 \\ 0 & -1 & 1 \end{bmatrix} = \begin{bmatrix} 1 & -1 & 0 \\ -2 & 3 & 0 \\ 0 & 0 & 1 \end{bmatrix} A \begin{bmatrix} 1 & 0 & 0 \\ 0 & 1 & 0 \\ 0 & 0 & 1 \end{bmatrix}$$

$C_2 \to -\,C_2$

$$\begin{bmatrix} 1 & 0 & 0 \\ 0 & 3 & 4 \\ 0 & 1 & 1 \end{bmatrix} = \begin{bmatrix} 1 & -1 & 0 \\ -2 & 3 & 0 \\ 0 & 0 & 1 \end{bmatrix} A \begin{bmatrix} 1 & 0 & 0 \\ 0 & -1 & 0 \\ 0 & 0 & 1 \end{bmatrix}$$

$R_2 \leftrightarrow R_3$

$$\begin{bmatrix} 1 & 0 & 0 \\ 0 & 1 & 1 \\ 0 & 3 & 4 \end{bmatrix} = \begin{bmatrix} 1 & -1 & 0 \\ 0 & 0 & 1 \\ -2 & 3 & 0 \end{bmatrix} A \begin{bmatrix} 1 & 0 & 0 \\ 0 & -1 & 0 \\ 0 & 0 & 1 \end{bmatrix}$$

$R_3 \to R_3 - 3\,R_2$

$$\begin{bmatrix} 1 & 0 & 0 \\ 0 & 1 & 1 \\ 0 & 0 & 1 \end{bmatrix} = \begin{bmatrix} 1 & -1 & 0 \\ 0 & 0 & 1 \\ -2 & 3 & -3 \end{bmatrix} A \begin{bmatrix} 1 & 0 & 0 \\ 0 & -1 & 0 \\ 0 & 0 & 1 \end{bmatrix}$$

$C_3 \to C_3 - C_2$

$$\begin{bmatrix} 1 & 0 & 0 \\ 0 & 1 & 0 \\ 0 & 0 & 1 \end{bmatrix} = \begin{bmatrix} 1 & -1 & 0 \\ 0 & 0 & 1 \\ -2 & 3 & -3 \end{bmatrix} A \begin{bmatrix} 1 & 0 & 0 \\ 0 & -1 & 1 \\ 0 & 0 & 1 \end{bmatrix}$$

$I_3 = PAQ$

i.e. $P = \begin{bmatrix} 1 & -1 & 0 \\ 0 & 0 & 1 \\ -2 & 3 & -3 \end{bmatrix}$ $\quad Q = \begin{bmatrix} 1 & 0 & 0 \\ 0 & -1 & 1 \\ 0 & 0 & 1 \end{bmatrix}$

Example 9 Find the non singular matrices P and Q such that PAQ is in the normal form where

$$A = \begin{bmatrix} 3 & 1 & 2 & 1 \\ 1 & 4 & 6 & 1 \\ 2 & -3 & 1 & -2 \end{bmatrix}$$

Solution: $A = I_3 A I_4$

$$\begin{bmatrix} 3 & 1 & 2 & 1 \\ 1 & 4 & 6 & 1 \\ 2 & -3 & 1 & -2 \end{bmatrix} = \begin{bmatrix} 1 & 0 & 0 \\ 0 & 1 & 0 \\ 0 & 0 & 1 \end{bmatrix} A \begin{bmatrix} 1 & 0 & 0 & 0 \\ 0 & 1 & 0 & 0 \\ 0 & 0 & 1 & 0 \\ 0 & 0 & 0 & 1 \end{bmatrix}$$

$R_1 \leftrightarrow R_2$

$$\begin{bmatrix} 1 & 4 & 6 & 1 \\ 3 & 1 & 2 & 1 \\ 2 & -3 & 1 & -2 \end{bmatrix} = \begin{bmatrix} 0 & 1 & 0 \\ 1 & 0 & 0 \\ 0 & 0 & 1 \end{bmatrix} A \begin{bmatrix} 1 & 0 & 0 & 0 \\ 0 & 1 & 0 & 0 \\ 0 & 0 & 1 & 0 \\ 0 & 0 & 0 & 1 \end{bmatrix}$$

$R_2 \to R_2 - 3R_1$
$R_3 \to R_3 - 2R_1$

$$\begin{bmatrix} 1 & 4 & 6 & 1 \\ 0 & -11 & -16 & -2 \\ 0 & -11 & -11 & -4 \end{bmatrix} = \begin{bmatrix} 0 & 1 & 0 \\ 1 & -3 & 0 \\ 0 & -2 & 1 \end{bmatrix} A \begin{bmatrix} 1 & 0 & 0 & 0 \\ 0 & 1 & 0 & 0 \\ 0 & 0 & 1 & 0 \\ 0 & 0 & 0 & 1 \end{bmatrix}$$

$C_2 \to C_2 - 4C_1$
$C_3 \to C_3 - 6C_1$
$C_4 \to C_4 - C_1$

$$\begin{bmatrix} 1 & 0 & 0 & 0 \\ 0 & -11 & -16 & -2 \\ 0 & -11 & -11 & -4 \end{bmatrix} = \begin{bmatrix} 0 & 1 & 0 \\ 1 & -3 & 0 \\ 0 & -2 & 1 \end{bmatrix} A \begin{bmatrix} 1 & -4 & -6 & -1 \\ 0 & 1 & 0 & 0 \\ 0 & 0 & 1 & 0 \\ 0 & 0 & 0 & 1 \end{bmatrix}$$

$R_2 \leftrightarrow R_3$

$$\begin{bmatrix} 1 & 0 & 0 & 0 \\ 0 & -11 & -11 & -4 \\ 0 & -11 & -16 & -2 \end{bmatrix} = \begin{bmatrix} 0 & 1 & 0 \\ 0 & -2 & 1 \\ 1 & -3 & 0 \end{bmatrix} A \begin{bmatrix} 1 & -4 & -6 & -1 \\ 0 & 1 & 0 & 0 \\ 0 & 0 & 1 & 0 \\ 0 & 0 & 0 & 1 \end{bmatrix}$$

$R_2 \to R_2\,(-1/11)$

$$\begin{bmatrix} 1 & 0 & 0 & 0 \\ 0 & 1 & 1 & 4/11 \\ 0 & -11 & -16 & -2 \end{bmatrix} = \begin{bmatrix} 0 & 1 & 1 \\ 0 & 2/11 & -1/11 \\ 1 & -3 & 0 \end{bmatrix} A \begin{bmatrix} 1 & -4 & -6 & -1 \\ 0 & 1 & 0 & 0 \\ 0 & 0 & 1 & 0 \\ 0 & 0 & 0 & 1 \end{bmatrix}$$

$R_3 \to R_3 + 11\,R_2$

$$\begin{bmatrix} 1 & 0 & 0 & 0 \\ 0 & 1 & 1 & 4/11 \\ 0 & 0 & -5 & 2 \end{bmatrix} = \begin{bmatrix} 0 & 1 & 0 \\ 0 & 2/11 & -1/11 \\ 1 & -1 & -1 \end{bmatrix} A \begin{bmatrix} 1 & -4 & -6 & -1 \\ 0 & 1 & 0 & 0 \\ 0 & 0 & 1 & 0 \\ 0 & 0 & 0 & 1 \end{bmatrix}$$

$C_3 \to C_3 - C_2,\ C_4 \to C_4 - \frac{4}{11} C_2$

$$\begin{bmatrix} 1 & 0 & 0 & 0 \\ 0 & 1 & 0 & 0 \\ 0 & 0 & -5 & 2 \end{bmatrix} = \begin{bmatrix} 0 & 1 & 0 \\ 0 & 2/11 & -1/11 \\ 1 & -1 & -1 \end{bmatrix} A \begin{bmatrix} 1 & -4 & -2 & 5/11 \\ 0 & 1 & -1 & -4/11 \\ 0 & 0 & 1 & 0 \\ 0 & 0 & 0 & 1 \end{bmatrix}$$

$R_3 \to \frac{-1}{5} R_3$

$$\begin{bmatrix} 1 & 0 & 0 & 0 \\ 0 & 1 & 0 & 0 \\ 0 & 0 & 1 & -2/5 \end{bmatrix} = \begin{bmatrix} 0 & 1 & 0 \\ 0 & 2/11 & -1/11 \\ -1/5 & 1/5 & 1/5 \end{bmatrix} A \begin{bmatrix} 1 & -4 & -2 & 5/11 \\ 0 & 1 & -1 & -4/1 \\ 0 & 0 & 1 & 0 \\ 0 & 0 & 0 & 1 \end{bmatrix}$$

$$C_4 \to C_4 + C_3 \left(\frac{2}{5}\right) \begin{bmatrix} 1 & 0 & 0 & 0 \\ 0 & 1 & 0 & 0 \\ 0 & 0 & 1 & 0 \end{bmatrix} = \begin{bmatrix} 0 & 1 & 0 \\ 0 & 2/11 & -1/11 \\ -1/5 & 1/5 & 1/5 \end{bmatrix} A \begin{bmatrix} 1 & -4 & -2 & -19/55 \\ 0 & +1 & -1 & -42/55 \\ 0 & 0 & 1 & 2/5 \\ 0 & 0 & 0 & 1 \end{bmatrix}$$

$$[I_3,\ o] = PAQ$$

Hence $P = \begin{bmatrix} 0 & 1 & 0 \\ 0 & 2/11 & -1/11 \\ -1/5 & 1/5 & 1/5 \end{bmatrix}$ $Q = \begin{bmatrix} 1 & -4 & -2 & -19/55 \\ 0 & 1 & -1 & -42/55 \\ 0 & 0 & 1 & 2/5 \\ 0 & 0 & 0 & 1 \end{bmatrix}$

Verification

$$PAQ = \begin{bmatrix} 0 & 1 & 0 \\ 0 & 2/11 & -1/11 \\ -1/5 & 1/5 & 1/5 \end{bmatrix} \begin{bmatrix} 3 & 1 & 2 & 1 \\ 1 & 4 & 6 & 1 \\ 2 & -3 & 1 & -2 \end{bmatrix} \begin{bmatrix} 1 & -4 & -2 & -19/55 \\ 0 & 1 & -1 & -42/55 \\ 0 & 0 & 1 & 2/5 \\ 0 & 0 & 0 & 1 \end{bmatrix}$$

$$= \begin{bmatrix} 1 & 4 & 6 & 1 \\ 0 & 1 & 1 & 4/11 \\ 0 & 0 & 1 & -2/5 \end{bmatrix} \begin{bmatrix} 1 & -4 & -2 & -19/55 \\ 0 & 1 & -1 & -42/55 \\ 0 & 0 & 1 & 2/5 \\ 0 & 0 & 0 & 1 \end{bmatrix}$$

$$= \begin{bmatrix} 1 & 0 & 0 & 0 \\ 0 & 1 & 0 & 0 \\ 0 & 0 & 1 & 0 \end{bmatrix}$$

Exercise - 2.4

Question: Find the rank of following matrices:

1. $\begin{bmatrix} 2 & -1 & 1 \\ 1 & 0 & 2 \\ 3 & -1 & 3 \end{bmatrix}$

2. $\begin{bmatrix} 1 & -2 & 3 & 4 \\ 5 & 4 & 1 & 6 \\ 2 & 3 & -1 & 1 \end{bmatrix}$

3. $\begin{bmatrix} 1 & 2 & -1 & 3 \\ 2 & 4 & -4 & 7 \\ -1 & -2 & -1 & -2 \end{bmatrix}$

4. $\begin{bmatrix} 1 & 1 & -1 & 1 \\ 1 & -1 & 2 & -1 \\ 3 & 1 & 0 & 1 \end{bmatrix}$

5. $\begin{bmatrix} 1 & 2 & 3 & 4 \\ -2 & 0 & 5 & 7 \end{bmatrix}$

6. $\begin{bmatrix} 1 & 2 & 3 \\ 1 & 4 & 2 \\ 2 & 6 & 5 \end{bmatrix}$

7. $\begin{bmatrix} 2 & 3 & 4 & -1 \\ 5 & 2 & 0 & -1 \\ -4 & 5 & 12 & -1 \end{bmatrix}$

Question: Find the rank of the following matrices after reducing them to normal form:

8. $\begin{bmatrix} 1 & 1 & 2 \\ 1 & 2 & 3 \\ 0 & -1 & -1 \end{bmatrix}$

9. $\begin{bmatrix} 1 & -1 & 2 & -1 \\ 4 & 2 & -1 & 2 \\ 2 & 2 & -2 & 0 \end{bmatrix}$

10. $\begin{bmatrix} 2 & 3 & -1 & -1 \\ 1 & -1 & -2 & -4 \\ 3 & 1 & 3 & -2 \\ 6 & 3 & 0 & -7 \end{bmatrix}$ 11. $\begin{bmatrix} 8 & 1 & 3 & 6 \\ 0 & 3 & 2 & 2 \\ -8 & -1 & -3 & 4 \end{bmatrix}$

Question: Find two non singular matrices P and Q such PAQ is in normal form for the matrix A, where A is

12. $\begin{bmatrix} 1 & 1 & 1 \\ 1 & -1 & -1 \\ 3 & 1 & 1 \end{bmatrix}$ 13. $\begin{bmatrix} 3 & 2 & -1 & 5 \\ 5 & 1 & 4 & -2 \\ 1 & -4 & 11 & -19 \end{bmatrix}$

14. $\begin{bmatrix} 1 & 1 & 2 \\ 1 & 2 & 3 \\ 0 & -1 & -1 \end{bmatrix}$

Answers

1.	2	2.	2	3.	2	4.	2
5.	2	6.	2	7.	2	8.	2
9.	3	10.	3	11.	3		

12. $P = \begin{bmatrix} 1/2 & 1/2 & 0 \\ 1/2 & -1/2 & 0 \\ -2 & -1 & 1 \end{bmatrix}$ $Q = \begin{bmatrix} 1 & 0 & 0 \\ 0 & 1 & -1 \\ 0 & 0 & 1 \end{bmatrix}$ and $PAQ = \begin{bmatrix} I_2 & 0 \\ 0 & 0 \end{bmatrix}$

13. $P = \begin{bmatrix} 0 & 0 & 1 \\ 1/2 & 0 & -3/2 \\ -1/2 & 1/3 & -1/6 \end{bmatrix}$ $Q = \begin{bmatrix} 1 & 4/7 & 9/119 & 9/217 \\ 0 & 1/7 & -1/7 & -1/7 \\ 0 & 0 & -1/17 & 0 \\ 0 & 0 & 0 & 1/31 \end{bmatrix}$

14. $PAQ = \begin{bmatrix} I_2 & 0 \\ 0 & 0 \end{bmatrix}$ where $P = \begin{bmatrix} 1 & 0 & 0 \\ -1 & 1 & 0 \\ -1 & 1 & 1 \end{bmatrix}$ and $Q = \begin{bmatrix} 1 & -1 & -1 \\ 0 & 1 & -1 \\ 0 & 0 & 1 \end{bmatrix}$

2.22 LINEAR SYSTEM OF EQUATIONS

A linear system of n equations in n unknowns x_1, x_2 .... x_n is a set of equations of the form

$$a_{11}\, x_1 + \ldots\ldots\ldots\ldots\ldots + a_{1n}\, x_n = b_1$$
$$a_{21}\, x_1 + \ldots\ldots\ldots\ldots\ldots + a_{2n}\, x_n = b_2$$
$$\ldots\ldots\ldots\ldots\ldots\ldots\ldots\ldots\ldots\ldots\ldots\ldots$$

$a_{n1} x_1 + \dots\dots\dots\dots\dots\dots + a_{nn} x_n = b_n$ ...(1)

where aij ($1 \leq i$, $j \leq n$) are the known coefficients, b_i ($1 \leq i \leq n$) are given numbers.

Note: The system is called homogeneous if all the bi's are zero.

Otherwise it is non-Homogeneous.

In the matrix natation, the system (1) can be writen as

$$\boxed{AX = B}$$

where $A = \begin{bmatrix} a_{11} & a_{12} & \dots\dots\dots & a_{1n} \\ a_{21} & a_{22} & \dots\dots\dots & a_{2n} \\ \dots & \dots & \dots & \dots \\ a_{n1} & a_{n2} & \dots\dots\dots & a_{nn} \end{bmatrix}$, $X = \begin{bmatrix} x_1 \\ \vdots \\ \vdots \\ x_n \end{bmatrix}$, $B = \begin{bmatrix} b_1 \\ \vdots \\ \vdots \\ b_n \end{bmatrix}$

The augmented matrix of the system (1) is

$$C = [A : B] = \begin{bmatrix} a_{11} & a_{12} & \dots\dots\dots & a_{1n} & b_1 \\ a_{21} & a_{22} & \dots\dots\dots & a_{2n} & b_2 \\ \dots & \dots & \dots & \dots & \dots \\ a_{n1} & a_{n2} & \dots\dots\dots & a_{nn} & b_n \end{bmatrix}$$

A solution of (1) is the set of number x_1 .... x_n that satisfy all the n equations.

2.23 SOLUTION OF LINEAR SYSTEM OF EQUATIONS

The matrix of the cofficients of x, y, z is reduced into echlon form by elementary row transformation. At the end of the row transformation the value of z is calculated from the last equation and value of y and x are calculated by backward substitution.

Illustrative Examples

Example 1 Solve the following equations

$$x - y + 2z = 3$$
$$x + 2y + 3z = 5$$
$$3x - 4y - 5z = -13$$

Solution: In the matrix form, the equations are written in the following form

$$\begin{bmatrix} 1 & -1 & 2 \\ 1 & 2 & 3 \\ 3 & -4 & -5 \end{bmatrix} \begin{bmatrix} x \\ y \\ z \end{bmatrix} = \begin{bmatrix} 3 \\ 5 \\ -13 \end{bmatrix}$$

$R_2 \to R_2 - R_1$

$R_3 \to R_3 - 3R_1$

$$\begin{bmatrix} 1 & -1 & 2 \\ 0 & 3 & 1 \\ 0 & -1 & -11 \end{bmatrix}\begin{bmatrix} x \\ y \\ z \end{bmatrix} = \begin{bmatrix} 3 \\ 2 \\ -22 \end{bmatrix}$$

$$R_3 \to R_3 + \frac{1}{3}R_2$$

$$\begin{bmatrix} 1 & -1 & 2 \\ 0 & 3 & 1 \\ 0 & 0 & \frac{-32}{3} \end{bmatrix}\begin{bmatrix} x \\ y \\ z \end{bmatrix} = \begin{bmatrix} 3 \\ 2 \\ \frac{-64}{3} \end{bmatrix}$$

$$x - y + 2z = 3 \quad \text{....... (1)}$$

$$3y + z = 2 \quad \text{....... (2)}$$

$$\frac{+32}{3}z = \frac{+64}{3}$$

$$\boxed{z = 2}$$

Putting the value of z in (2), we get

$$3y + 2 = 2$$

$$\boxed{y = 0}$$

Putting the vaue of y, z in (1), we get

$$x - 0 + 2\,(2) = 3$$

$$x + 4 = 3$$

$$\boxed{x = -1}$$

Hence $\boxed{x = -1,\ y = 0,\ z = 2}$

2.24 TYPES OF LINEAR EQUATIONS

1. Non-Homogeneous
2. Homogeneous

(i) Consistent:

A system of equations is said to be consistent, if they have one or more solution i.e.

$x + 2y = 4$	$x + 2y = 4$
$3x + 2y = 2$	$3x + 6y = 12$
Unique solution	Infinite solution.

(ii) Inconsistent:

If a system of equation has no solution, it is said to be inconsistent.

$$x + 2y = 4$$
$$3x + 6y = 5$$

2.25 CONSISTENCY OF A SYSTEM OF LINEAR EQUATIONS

$$a_{11}x_1 + a_{12}x_2 + \dots\dots + a_{1n}x_n = b_1$$
$$a_{21}x_1 + a_{22}x_2 + \dots\dots + a_{2n}x_n = b_2$$
$$\dots\dots\dots\dots\dots\dots$$
$$a_{m1}x_1 + a_{m2}x_2 + \dots\dots + a_{mn}x_n = b_m$$

In matrix notation, these system of equations are written as:

$$AX = B$$

where
$$A = \begin{bmatrix} a_{11} & a_{12} & \dots\dots & a_{1n} \\ a_{21} & a_{22} & \dots\dots & a_{2n} \\ \dots & \dots & \dots & \dots \\ a_{m1} & a_{m2} & \dots\dots & a_{mn} \end{bmatrix}, \quad X = \begin{bmatrix} x_1 \\ x_2 \\ \vdots \\ x_n \end{bmatrix}, \quad B = \begin{bmatrix} b_1 \\ b_2 \\ \vdots \\ b_m \end{bmatrix}$$

Augmented matrix is

$$C = [A : B] = \begin{bmatrix} a_{11} & a_{12} & \dots\dots & a_{1n} & b_1 \\ a_{21} & a_{22} & \dots\dots & a_{2n} & b_2 \\ \dots & \dots & \dots & \dots & \dots \\ a_{m1} & a_{m2} & \dots\dots & a_{mn} & b_m \end{bmatrix}$$

(a) **Consistent equation:**

If Rank A = Rank [A : B]

(i) Unique solution: Rank A = Rank [A : B] = n (no. of unknowns)

(ii) Infinite solution : Rank A = Rank [A : B] $<$ n (no. of unknowns)

(b) **Inconsistent equation:**

If Rank A $\neq$ Rank C

Thus we can say that the system of linear equations are either non-Homogenous or Homogeneous, it depends on bi, then system of non-Homogeneous equations are either consistent or Inconsistent according to which system has unique solution or Infinitely many solution.

IN BRIEF

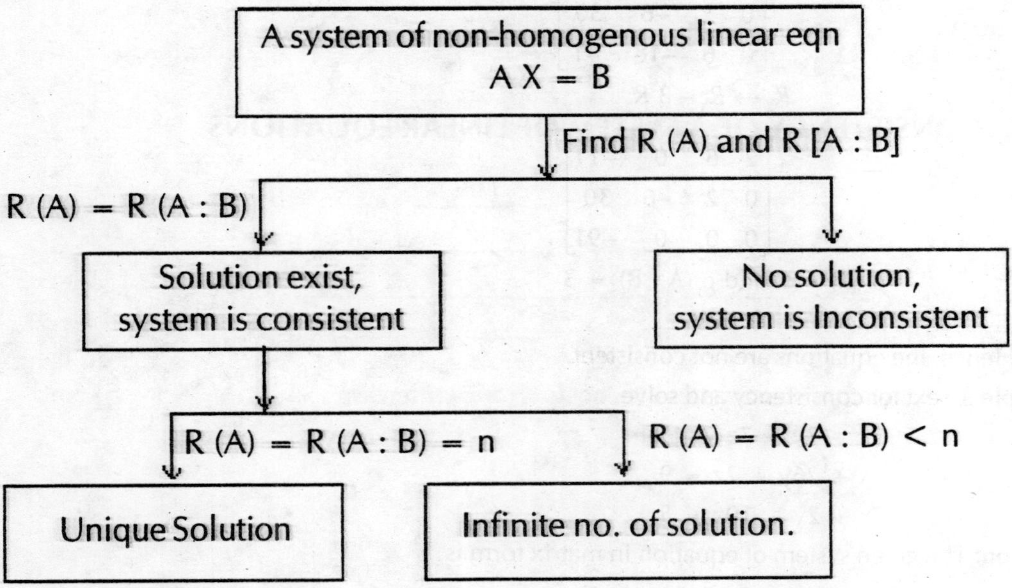

- n = no. of unknowns.

Example 2 Show that the equations

$$2x + 6y = -11$$
$$6x + 20y - 6z = -3$$
$$6y - 18z = -1$$

are not consistent.

Solution: The given system of equations in matrix form is:

$$\begin{bmatrix} 2 & 6 & 0 \\ 6 & 20 & -6 \\ 0 & 6 & -18 \end{bmatrix} \begin{bmatrix} x \\ y \\ z \end{bmatrix} = \begin{bmatrix} -11 \\ -3 \\ -1 \end{bmatrix}$$

$$A X = B$$

where $A = \begin{bmatrix} 2 & 6 & 0 \\ 6 & 20 & -6 \\ 0 & 6 & -18 \end{bmatrix}$ $X = \begin{bmatrix} x \\ y \\ z \end{bmatrix}$ $B = \begin{bmatrix} -11 \\ -3 \\ -1 \end{bmatrix}$

Augmented matrix [A : B] is

$$= \begin{bmatrix} 2 & 6 & 0 & -11 \\ 6 & 20 & -6 & -3 \\ 0 & 6 & -18 & -1 \end{bmatrix}$$

$$R_2 \to R_2 - 3\,R_1$$

$$\begin{bmatrix} 2 & 6 & 0 & -11 \\ 0 & 2 & -6 & 30 \\ 0 & 6 & -18 & -1 \end{bmatrix}$$

$R_3 \to R_3 - 3R_2$

$$\begin{bmatrix} 2 & 6 & 0 & -11 \\ 0 & 2 & -6 & 30 \\ 0 & 0 & 0 & -91 \end{bmatrix}$$

$\rho(A) = 2$ and $\rho(A : B) = 3$

$\therefore \quad \rho(A) \neq \rho(A : B)$

Hence, the equations are not consistent.

Example 3 Text for consistency and solve

$$5x + 3y + 7z = 4$$
$$3x + 26y + 2z = 9$$
$$7x + 2y + 10z = 5$$

Solution: The given system of equation in matrix form is

$$\begin{bmatrix} 5 & 3 & 7 \\ 3 & 26 & 2 \\ 7 & 2 & 10 \end{bmatrix} \begin{bmatrix} x \\ y \\ z \end{bmatrix} = \begin{bmatrix} 4 \\ 9 \\ 5 \end{bmatrix}$$

$$AX = B$$

where $A = \begin{bmatrix} 5 & 3 & 7 \\ 3 & 26 & 2 \\ 7 & 2 & 10 \end{bmatrix}, \quad X = \begin{bmatrix} x \\ y \\ z \end{bmatrix}, \quad B = \begin{bmatrix} 4 \\ 9 \\ 5 \end{bmatrix}$

Augmented matrix (A : B) is

$$\begin{bmatrix} 5 & 3 & 7 & : & 4 \\ 3 & 26 & 2 & : & 9 \\ 7 & 2 & 10 & : & 5 \end{bmatrix}$$

$$R_1 \to R_1 \left(\frac{1}{5}\right)$$

$$\begin{bmatrix} 1 & \frac{3}{5} & \frac{7}{5} & : & \frac{4}{5} \\ 3 & 26 & 2 & : & 9 \\ 7 & 2 & 10 & : & 5 \end{bmatrix}$$

$R_2 \to R_2 - 3R_1$

$R_3 \to R_3 - 7R_1$

$$\begin{bmatrix} 1 & \frac{3}{5} & \frac{7}{5} & : & \frac{4}{5} \\ 0 & \frac{121}{5} & \frac{-11}{5} & : & \frac{33}{5} \\ 0 & \frac{-11}{5} & \frac{1}{5} & : & \frac{-3}{5} \end{bmatrix} \quad R_3 \to R_3 + \frac{1}{11} R_2$$

$$\begin{bmatrix} 1 & \frac{3}{5} & \frac{7}{5} & : & \frac{4}{5} \\ 0 & \frac{121}{5} & \frac{-11}{5} & : & \frac{33}{5} \\ 0 & 0 & 0 & : & 0 \end{bmatrix}$$

$\rho(A) = 2$ and $\rho(A : B) = 2$

$\therefore \quad \rho(A) = \rho(A : B)$

Hence, the given system of equations are consistent.

But $\rho(A) = \rho(A : B) < n$ (no. of unknowns)

So, its solutions are infinite.

$$x + \frac{3}{5}y + \frac{7}{5}z = \frac{4}{5} \qquad \ldots(1)$$

$$\frac{121}{5}y - \frac{11}{5}z = \frac{33}{5} \Rightarrow 11y - z = 3$$

Let $\quad z = k \quad$ then $\quad 11y = 3 + k$

$$\boxed{y = \frac{3}{11} + \frac{k}{11}}$$

Put value of y and z in (1)

$$x + \frac{3}{5}\left(\frac{3}{11} + \frac{k}{11}\right) + \frac{7}{5}k = \frac{4}{5}$$

$$x + \frac{9}{55} + \frac{3k}{55} + \frac{7k}{5} = \frac{4}{5}$$

$$x = \frac{4}{5} - \frac{9}{55} - \left(\frac{3k}{55} + \frac{7k}{5}\right)$$

$$\boxed{x = \frac{7}{11} - \frac{16}{11}k}$$

Example 4 Test for consistency of the following system of equations.

$$x_1 + 2x_2 + 3x_3 + 4x_4 = 5$$
$$6x_1 + 7x_2 + 8x_3 + 9x_4 = 10$$
$$11x_1 + 12x_2 + 13x_3 + 14x_4 = 15$$
$$16x_1 + 17x_2 + 18x_3 + 19x_4 = 20$$
$$21x_1 + 22x_2 + 23x_3 + 24x_4 = 25$$

Solution: The given system of equations in matrix form is

$$\begin{bmatrix} 1 & 2 & 3 & 4 \\ 6 & 7 & 8 & 9 \\ 11 & 12 & 13 & 14 \\ 16 & 17 & 18 & 19 \\ 21 & 22 & 23 & 24 \end{bmatrix} \begin{bmatrix} x_1 \\ x_2 \\ x_3 \\ x_4 \end{bmatrix} = \begin{bmatrix} 5 \\ 10 \\ 15 \\ 20 \\ 25 \end{bmatrix}$$

$$A\ X = B$$

The augmented matrix (A : B) is

$$\begin{bmatrix} 1 & 2 & 3 & 4 & 5 \\ 6 & 7 & 8 & 9 & 10 \\ 11 & 12 & 13 & 14 & 15 \\ 16 & 17 & 18 & 19 & 20 \\ 21 & 22 & 23 & 24 & 25 \end{bmatrix}$$

$R_2 \to R_2 - 6R_1$
$R_3 \to R_3 - 11R_1$
$R_4 \to R_4 - 16R_1$
$R_5 \to R_5 - 21R_1$

$$\begin{bmatrix} 1 & 2 & 3 & 4 & 5 \\ 0 & -5 & -10 & -15 & -20 \\ 0 & -10 & -20 & -30 & -40 \\ 0 & -15 & -30 & -45 & -60 \\ 0 & -20 & -40 & -60 & -80 \end{bmatrix} \Rightarrow (-1)^4 \begin{bmatrix} 1 & 2 & 3 & 4 & 5 \\ 0 & 5 & 10 & 15 & 20 \\ 0 & 10 & 20 & 30 & 40 \\ 0 & 15 & 30 & 45 & 60 \\ 0 & 20 & 40 & 60 & 80 \end{bmatrix}$$

$R_3 \to R_3 - 2R_2$
$R_4 \to R_4 - 3R_2$
$R_5 \to R_5 - 4R_2$

$$\begin{bmatrix} 1 & 2 & 3 & 4 & 5 \\ 0 & 5 & 10 & 15 & 20 \\ 0 & 0 & 0 & 0 & 0 \\ 0 & 0 & 0 & 0 & 0 \\ 0 & 0 & 0 & 0 & 0 \end{bmatrix}$$

$\rho(A) = 2$ and $\rho(A : B) = 2$

$\therefore \quad \rho(A) = \rho(A : B)$

Hence the given system of equations are consistent.

But $\rho(A) = \rho(A : B) < n$

So, its solution is infinite.

Example 5 For what value of k ,the system

$$x + y + z = 1$$
$$2x + y + 4z = k$$
$$4x + y + 10z = k^2 \text{ has a solution.}$$

Solution: The given system of equations in matrix form are

$$\begin{bmatrix} 1 & 1 & 1 \\ 2 & 1 & 4 \\ 4 & 1 & 10 \end{bmatrix} \begin{bmatrix} x \\ y \\ z \end{bmatrix} = \begin{bmatrix} 1 \\ k \\ k^2 \end{bmatrix}$$

$$A\ X = B$$

Augmented matrix is (A : B)

$$\begin{bmatrix} 1 & 1 & 1 & 1 \\ 2 & 1 & 4 & k \\ 4 & 1 & 10 & k^2 \end{bmatrix}$$

$R_2 \to R_2 - 2R_1$

$R_3 \to R_3 - 4R_1$

$$\begin{bmatrix} 1 & 1 & 1 & 1 \\ 0 & -1 & 2 & k-2 \\ 0 & -3 & 6 & k^2-4 \end{bmatrix}$$

$R_3 \to R_3 - 3R_2$

$$\begin{bmatrix} 1 & 1 & 1 & 1 \\ 0 & -1 & 2 & k-2 \\ 0 & 0 & 0 & k^2-3k+2 \end{bmatrix}$$

If the given system has solution, then

$$\rho(A) = \rho(A : B)$$

and $\rho(A : B) = 2$ if $k^2 - 3k + 2 = 0$

$$k^2 - 2k - k + 2 = 0$$
$$(k-2)(k-1) = 0$$
$$\boxed{k = 2,\ k = 1}$$

Case I : when $k = 1$, we have

$x + y + z = 1$ ...(1)

$-y + 2z = 1 - 2 = -1$ ...(2)

Let $z = \lambda$

Putting value of $z = \lambda$ in (2) $y = 2\lambda + 1$

Putting the value of y and z in (1)

$x + (2\lambda + 1) + \lambda = 0$

$x + 3\lambda + 1 = 0$

$\boxed{x = -3\lambda - 1}$

Case II: when $k = 2$

$x + y + z = 1$ ...(3)

$-y + 2z = 2 - 2 = 0$ ...(4)

Let $\boxed{z = c}$

Putting the value of z in (4)

$-y + 2c = 0$

$\boxed{y = 2c}$

Putting the value of y and z in (3)

$x + 2c + c = 1$

$\boxed{x = 1 - 3c}$

Example 6 Investigate the values of λ and μ so that the equations:

$$2x + 3y + 5z = 9$$
$$7x + 3y - 2z = 8$$
$$2x + 3y + \lambda z = \mu$$

have (i) no solution

(ii) a unique solution

(iii) an infinite no. of solution.

Solution: The given system of equations in matrix form is

$$\begin{bmatrix} 2 & 3 & 5 \\ 7 & 3 & -2 \\ 2 & 3 & \lambda \end{bmatrix} \begin{bmatrix} x \\ y \\ z \end{bmatrix} = \begin{bmatrix} 9 \\ 8 \\ \mu \end{bmatrix}$$

Augmented matrix is (A : B)

$$\begin{bmatrix} 2 & 3 & 5 & 9 \\ 7 & 3 & -2 & 8 \\ 2 & 3 & \lambda & \mu \end{bmatrix}$$

$R_2 \to R_2 - \frac{7}{2} R_1$

$R_3 \to R_3 - R_1$

$$\begin{bmatrix} 2 & 3 & 5 & 9 \\ 0 & \frac{-15}{2} & \frac{-39}{2} & \frac{-47}{2} \\ 0 & 0 & \lambda - 5 & \mu - 9 \end{bmatrix}$$

(i) No Solution: Rank A ≠ Rank (A : B)

$\lambda - 5 = 0 \qquad \mu - 9 \neq 0$

$\boxed{\lambda = 5} \qquad \boxed{\mu \neq 9}$

(ii) An infinite no. of solutions: Rank A = Rank (A : B) < n

$\lambda - 5 = 0 \qquad \mu - 9 = 0$

$\boxed{\lambda = 5} \qquad \boxed{\mu = 9}$

(iii) A unique solution Rank A = Rank (A : B) = n

$\lambda - 5 \neq 0$

$\lambda \neq 5$, μ is arbitrary

Example 7 Determine for what value of λ and μ the following equation have

$$x + y + z = 6$$
$$x + 2y + 3z = 10$$
$$x + 2y + \lambda z = \mu$$

(i) no solution

(ii) a unique solution

(iii) infinite no. of solution.

Solution: The given system of equation in matrix form is

$$\begin{bmatrix} 1 & 1 & 1 \\ 1 & 2 & 3 \\ 1 & 2 & \lambda \end{bmatrix} \begin{bmatrix} x \\ y \\ z \end{bmatrix} = \begin{bmatrix} 6 \\ 10 \\ \mu \end{bmatrix}$$

$$AX = B$$

The augmented matrix is

$$\begin{bmatrix} 1 & 1 & 1 & 6 \\ 1 & 2 & 3 & 10 \\ 1 & 2 & \lambda & \mu \end{bmatrix}$$

$R_2 \to R_2 - R_1$

$R_3 \to R_3 - R_1$

$$\begin{bmatrix} 1 & 1 & 1 & 6 \\ 0 & 1 & 2 & 4 \\ 0 & 1 & \lambda-1 & \mu-6 \end{bmatrix}$$

$R_3 \to R_3 - R_2$

$$\begin{bmatrix} 1 & 1 & 1 & 6 \\ 0 & 1 & 2 & 4 \\ 0 & 0 & \lambda-3 & \mu-10 \end{bmatrix}$$

(i) No solution: $R(A) \neq R(A : B)$

$\lambda - 3 = 0 \qquad \mu - 10 \neq 0$

$\boxed{\lambda = 3} \qquad \boxed{\mu \neq 10}$

(ii) A unique Solution: $R(A) = R(A : B) = n$

$\lambda - 3 \neq 0$

$\boxed{\lambda \neq 3}$, μ is arbitrary.

(iii) Infinite Solution: $R(A) = R(A : B) < n$

$\lambda - 3 = 0 \qquad \mu - 10 = 0$

$\boxed{\lambda = 3} \qquad \boxed{\mu = 10}$

Example 8 Show that the equations

$$-2x + y + z = a$$
$$x - 2y + z = b$$
$$x + y - 2z = c$$

have no solution unless $a + b + c = 0$. In which case they have infinitely many solution? Find these solutions when $a = 1$, $b = 1$ and $c = -2$.

Solution: The given system of equation in matrix form is

$$\begin{bmatrix} -2 & 1 & 1 \\ 1 & -2 & 1 \\ 1 & 1 & -2 \end{bmatrix} \begin{bmatrix} x \\ y \\ z \end{bmatrix} = \begin{bmatrix} a \\ b \\ c \end{bmatrix}$$

$AX = B$

The augmented matrix is (A : B)

$$\begin{bmatrix} -2 & 1 & 1 & a \\ 1 & -2 & 1 & b \\ 1 & 1 & -2 & c \end{bmatrix}$$

$R_1 \leftrightarrow R_2$

$$\begin{bmatrix} 1 & -2 & 1 & b \\ -2 & 1 & 1 & a \\ 1 & 1 & -2 & c \end{bmatrix}$$

$R_2 \to R_2 + 2R_1$

$R_3 \to R_3 - R_1$

$$\begin{bmatrix} 1 & -2 & 1 & b \\ 0 & -3 & 3 & a+2b \\ 0 & 3 & -3 & c-b \end{bmatrix}$$

$R_3 \to R_3 + R_2$

$$\begin{bmatrix} 1 & -2 & 1 & b \\ 0 & -3 & 3 & a+2b \\ 0 & 0 & 0 & a+b+c \end{bmatrix}$$

Case I : If $a+b+c \neq 0$

$\rho(A) = 2$ and $\rho(A : B) = 3$

$\therefore$ $\rho(A) \neq \rho(A : B)$

Hence, the system being inconsistent, have no solution.

Case II: If $a+b+c = 0$

$\rho(A) = 2, \quad \rho(A : B) = 2$

$\therefore$ $\rho(A) = \rho(A : B)$

Hence, the given system of equations are consistent.

But $\rho(A) = \rho(A : B) < n$

So, its have infinite no. of solutions.

Case III : On putting $a = 1, \quad b = 1, \quad c = -2$

$$\begin{bmatrix} 1 & -2 & 1 & 1 \\ 0 & -3 & 3 & 3 \\ 0 & 0 & 0 & 0 \end{bmatrix}$$

$x - 2y + z = 1 \quad \text{...(1)}$

$-3y + 3z = 3 \quad \text{...(2)}$

$-y + z = 1$

Put $z = k$

$\therefore$ $\boxed{y = k - 1}$

put the value of y and z in (1)

$x - 2(k-1) + k = 1$

$x - 2k + 2 + k = 1$

$x - k + 2 = 1$

$\boxed{x = k - 1}$

$\therefore$ $\boxed{x = k - 1, \quad y = k - 1, \quad z = k}$

2.26 HOMOGENEOUS EQUATIONS

For a system of homogeneous linear equations $AX = 0$

(i) $X = 0$ is always a solution. This solution in which each unknown has the value zero is called the null solution or the Trivial solution. Thus a Homogeneous system is always consistent.

- A system of Homogeneous linear equations has either the trivial solution or infinite no. of solutions.

(ii) If $\rho(A)$ = no. of unknowns the system has only the trivial solution.

(iii) If $\rho(A) <$ no. of unknowns, the system has infinite no of non-trivial solutions

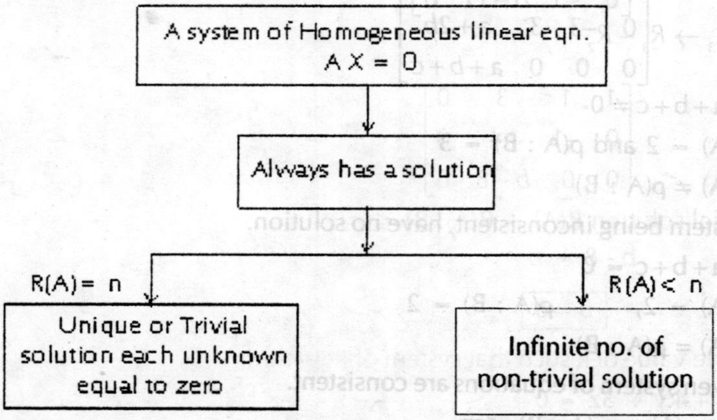

Example 9 Determine the value of b such that the system of Homogeneous equations

$$2x + y + 2z = 0$$
$$x + y + 3z = 0$$
$$4x + 3y + bz = 0$$

has (i) Trivial solution

(ii) Non-trivial solution. Find Non-trivial solution using matrix method.

Solution: (i) ***For Trivial solution:*** We know that $x = 0, y = 0, z = 0$, so b can have any value.

(ii) ***For Non-Trivial Solution:*** The given system of equation in matrix form is

$$\begin{bmatrix} 2 & 1 & 2 \\ 1 & 1 & 3 \\ 4 & 1 & b \end{bmatrix} \begin{bmatrix} x \\ y \\ z \end{bmatrix} = \begin{bmatrix} 0 \\ 0 \\ 0 \end{bmatrix}$$

$$AX = B$$

Augmented matrix is (A : B)

$$\begin{bmatrix} 2 & 1 & 2 & 0 \\ 1 & 1 & 3 & 0 \\ 4 & 1 & b & 0 \end{bmatrix}$$

$R_1 \leftrightarrow R_2$

$$\begin{bmatrix} 1 & 1 & 3 & 0 \\ 2 & 1 & 2 & 0 \\ 4 & 1 & b & 0 \end{bmatrix}$$

$R_2 \to R_2 - 2R_1$

$R_3 \to R_3 - 4R_1$

$$\begin{bmatrix} 1 & 1 & 3 & 0 \\ 0 & -1 & -4 & 0 \\ 0 & -1 & b-12 & 0 \end{bmatrix}$$

$R_3 \to R_3 - R_2$

$$\begin{bmatrix} 1 & 1 & 3 & 0 \\ 0 & -1 & -4 & 0 \\ 0 & 0 & b-8 & 0 \end{bmatrix}$$

For Non-trivial solution R(A) = R(A : B) < n

$$b - 8 = 0$$

$$\boxed{b = 8}$$

Example 10 Find the value of k such that system of equations

$$x + ky + 3z = 0$$
$$4x + 3y + kz = 0$$
$$2x + y + 2z = 0$$

has non trivial solution.

Solution: The given system of equation in matrix form is

$$\begin{bmatrix} 1 & k & 3 \\ 4 & 3 & k \\ 2 & 1 & 2 \end{bmatrix} \begin{bmatrix} x \\ y \\ z \end{bmatrix} = \begin{bmatrix} 0 \\ 0 \\ 0 \end{bmatrix}$$

$$AX = 0$$

The augmented matrix is (A : B)

$$\begin{bmatrix} 1 & k & 3 & 0 \\ 4 & 3 & k & 0 \\ 2 & 1 & 2 & 0 \end{bmatrix}$$

$R_1 \leftrightarrow R_3$

$$\begin{bmatrix} 2 & 1 & 2 & 0 \\ 4 & 3 & k & 0 \\ 1 & k & 3 & 0 \end{bmatrix}$$

$R_2 \to R_2 - 2R_1$

$$R_3 \to R_3 - \frac{1}{2} R_1$$

$$\begin{bmatrix} 2 & 1 & 2 & 0 \\ 0 & 1 & k-4 & 0 \\ 0 & k-\frac{1}{2} & 2 & 0 \end{bmatrix}$$

$$R_3 \to R_3 - (k - \frac{1}{2}) R_2$$

$$\begin{bmatrix} 2 & 1 & 2 & 0 \\ 0 & 1 & k-4 & 0 \\ 0 & 0 & 2-\left(k-\frac{1}{2}\right)(k-4) & 0 \end{bmatrix}$$

For Non-trivial solution R(A) = R(A : B) < n

$$2 - \left(K - \frac{1}{2}\right)(k-4) = 0$$

$$2 - k^2 + 4k + \frac{k}{2} - 2 = 0$$

$$-k^2 + \frac{9}{2} k = 0$$

$$k\left(-k + \frac{9}{2}\right) = 0$$

$$\boxed{k = 0,\ k = \frac{9}{2}}$$

Exercise - 2.5

Check the consistency of the following system of equations. Also find the solution set if consistent.

1. $x + 2y - z = 3$

   $3x - y + 2z = 1$

   $2x - 2y + 3z = 2$

   $x - y + z = -1$

2. $x + 3y - z = 4$

   $2x + y + z = 7$

   $2x - 4y + 4z = 6$

   $3x + 4y = 1$

3. $x_1 + 2x_2 - x_3 = 6$
$3x_1 - x_2 - 2x_3 = 3$
$4x_1 + 3x_2 + x_3 = 9$

4. $x - 4y - 3z = -16$
$2x + 7y + 12z = 48$
$4x - y + 6z = 16$
$5x - 5y + 3z = 0$

5. Discuss the consistency of the equation
$x + 2y + 3z + 4t = 0$
$2x + 3y + 4z - 1 = 0$
$3x + 4y + t = 2$
$4x + z + 2t = 3$ Find the solution set if consistent.

6. For what value of λ will the equations
$3x - y + \lambda z = 1$
$2x + y + z = 2$
$x + 2y - \lambda z = -1$ Fail to have a unique solution.
Will the equations have any solution for this value of λ?

7. Use the test of rank to show that the following system of equations is inconsistent:
$2x - y + z = 4$
$3x - y + z = 6$
$4x - y + 2z = 7$
$-x + y - z = 9$

8. Show that the following equations are consistent and solve them:
$x + 2y - 5z = -9$
$3x - y + 2z = 5$
$2x + 3y - z = 3$
$4x - 5y + z = -3$

9. Solve the system of equations:
$\lambda x + 2y - 2z = 1$
$4x + 2\lambda y - z = 2$
$6x + 6y + \lambda z = 3$ considering specially the case when $\lambda = 2$

10. For what value of a and b the equation
$x + y + 5z = 0$
$x + 2y + 3az = b$
$x + 3y + az = 1$ have

(i) No solution (ii) unique solution

(iii) Infinitely many solutions.

Solve the following system of equations:

11. $x - y + z = 0$
$-3x + y - 4z = 0$
$7x - 3y - 9z = 0$
$4x - 2y + 5z = 0$

12. $x + 3y - 2z = 0$
$2x - y + 4z = 0$
$x - 11y + 14z = 0$

13. $2w + 3x - y - z = 0$
$4w - 6x - 2y + 2z = 0$
$-6w + 12x + 3y - 4z = 0$

14. $3x + 4y - z - 6w = 0$
$2x + 3y + 2z - 3w = 0$
$2x + y - 14z - 9w = 0$
$x + 3y + 13z + 3w = 0$

15. Find the value of k such that the following system of equations has a non-trivial solution:
$(3k - 8)x + 3y + 3z = 0$
$3x + (3k - 8)y + 3z = 0$
$3x + 3y + (3k - 8)z = 0$

16. Show that the only real value of λ for which the equations
$x + 2y + 3z = \lambda x$
$3x + y + 2z = \lambda y$
$2x + 3y + z = \lambda z$ have a non-zero solution is 6.

Answers

2. *No solution*

8. $\frac{1}{2}, \frac{3}{2}, \frac{5}{2}$

9. $x = \frac{1}{2} - k,\ y = k,\ z = 0$

10. (i) $a = 1,\ b \neq \frac{1}{2}$ (ii) $a \neq 1,\ b \in R$ (iii) $a = 1,\ b = \frac{1}{2}$

11. $x = y = z = 0$

12. $x = \frac{-10}{7}k,\quad y = \frac{8}{7}k,\quad z = k$

13. $x = \frac{1}{3}k_1,\quad y = k_2,\quad z = k_1,\quad w = \frac{1}{2}k_2$

14. $x = 11k_1 + 6k_2,\quad y = -k_1 + 3k_2,\quad z = k_1,\quad w = k_2$

15. Non trivial solution if $k = \frac{11}{3}$ or $k = \frac{2}{3}$

2.27 Linear Dependence and Independence of Vectors

Vectors (matrices) $x_1, x_2, \ldots\ldots\ldots\ldots x_n$ are said to be dependent if

(1) All the vectors (row or column matrices) are of same order.

(2) 'n' scalars $\lambda_1, \lambda_2, \ldots\ldots\ldots\ldots \lambda_n$ (not all zero) exist such that

$$\lambda_1 x_1 + \lambda_2 x_2 + \ldots\ldots\ldots\ldots + \lambda_n x_n = 0$$

otherwise they are linearly independent.

Illustrative Examples

Example 1 Examine the following vectors for linear dependence and find the relation if exists.

$$x_1 = (1, 2, 4),\ x_2 = (2, -1, 3),\ x_3 = (0, 1, 2),\ x_4 = (-3, 7, 2)$$

Solution: Consider the matrix equation

$$\lambda_1 x_1 + \lambda_2 x_2 + \lambda_3 x_3 + \lambda_4 x_4 = 0 \qquad \ldots(1)$$

$$\Rightarrow \quad \lambda_1 (1, 2, 4) + \lambda_2 (2, -1, 3) + \lambda_3 (0, 1, 2) + \lambda_4 (-3, 7, 2) = 0$$

$$\lambda_1 + 2\lambda_2 + 0\lambda_3 - 3\lambda_4 = 0$$

$$2\lambda_1 - \lambda_2 + \lambda_3 + 7\lambda_4 = 0$$

$$4\lambda_1 + 3\lambda_2 + 2\lambda_3 + 2\lambda_4 = 0$$

This is the homogeneous system.

$$\begin{bmatrix} 1 & 2 & 0 & -3 \\ 2 & -1 & 1 & 7 \\ 4 & 3 & 2 & 2 \end{bmatrix} \begin{bmatrix} \lambda_1 \\ \lambda_2 \\ \lambda_3 \\ \lambda_4 \end{bmatrix} = \begin{bmatrix} 0 \\ 0 \\ 0 \end{bmatrix}$$

or $\quad A\lambda = 0$

$R_2 \to R_2 - 2R_1, \quad R_3 \to R_3 - 4R_1$

$$\begin{bmatrix} 1 & 2 & 0 & -3 \\ 0 & -5 & 1 & 13 \\ 0 & -5 & 2 & 14 \end{bmatrix} \begin{bmatrix} \lambda_1 \\ \lambda_2 \\ \lambda_3 \\ \lambda_4 \end{bmatrix} = \begin{bmatrix} 0 \\ 0 \\ 0 \end{bmatrix}$$

$R_3 \to R_3 - R_2$

$$\begin{bmatrix} 1 & 2 & 0 & -3 \\ 0 & -5 & 1 & 13 \\ 0 & 0 & 1 & 1 \end{bmatrix} \begin{bmatrix} \lambda_1 \\ \lambda_2 \\ \lambda_3 \\ \lambda_4 \end{bmatrix} = \begin{bmatrix} 0 \\ 0 \\ 0 \end{bmatrix}$$

$$\begin{aligned} \lambda_1 + 2\lambda_2 - 3\lambda_4 &= 0 \\ -5\lambda_2 + \lambda_3 + 13\lambda_4 &= 0 \\ \lambda_3 + \lambda_4 &= 0 \end{aligned} \qquad (2)$$

Let $\lambda_4 = t, \quad \lambda_3 = -t$

$-5\lambda_2 - t + 13t = 0$

then, from (2),

$\Rightarrow \qquad \lambda_2 = \dfrac{12t}{5}$

Again, we have $\qquad \lambda_1 + \dfrac{24t}{5} - 3t = 0$

or $\qquad \lambda_1 = \dfrac{-9t}{5}$

Hence the given vectors are linearly dependent.
substituting the values of 'λ' in (1)
we get,

$$\frac{-9t}{5}x_1 + \frac{12t}{5}x_2 - t x_3 + t x_4 = 0$$

or $\qquad -\dfrac{9x_1}{5} + \dfrac{12x_2}{5} - x_3 + x_4 = 0$

or $\qquad 9x_1 - 12x_2 + 5x_3 - 5x_4 = 0 \qquad$ Ans.

Example 2 Is the system of vectors

$x_1 = (2, 2, 1)^T, x_2 = (1, 3, 1)^T, x_3 = (1, 2, 2)^T$

linearly dependent

Solution: Here $\qquad x_1 = \begin{bmatrix} 2 \\ 2 \\ 1 \end{bmatrix}, \ x_2 = \begin{bmatrix} 1 \\ 3 \\ 1 \end{bmatrix}, \ x_3 = \begin{bmatrix} 1 \\ 2 \\ 2 \end{bmatrix}$

Consider the matrix equation

$\lambda_1 x_1 + \lambda_2 x_2 + \lambda_3 x_3 = 0 \qquad$ ...(1)

$$\lambda_1 \begin{bmatrix} 2 \\ 2 \\ 1 \end{bmatrix} + \lambda_2 \begin{bmatrix} 1 \\ 3 \\ 1 \end{bmatrix} + \lambda_3 \begin{bmatrix} 1 \\ 2 \\ 2 \end{bmatrix} = 0$$

$$2\lambda_1 + \lambda_2 + \lambda_3 = 0$$
$$2\lambda_1 + 3\lambda_2 + 2\lambda_3 = 0$$
$$\lambda_1 + \lambda_2 + 2\lambda_3 = 0$$

which is a homogeneous equation

$$\begin{bmatrix} 2 & 1 & 1 \\ 2 & 3 & 2 \\ 1 & 1 & 2 \end{bmatrix} \begin{bmatrix} \lambda_1 \\ \lambda_2 \\ \lambda_3 \end{bmatrix} = \begin{bmatrix} 0 \\ 0 \\ 0 \end{bmatrix}$$

$R_1 \leftrightarrow R_3$

$$\begin{bmatrix} 1 & 1 & 2 \\ 2 & 3 & 2 \\ 2 & 1 & 1 \end{bmatrix} \begin{bmatrix} \lambda_1 \\ \lambda_2 \\ \lambda_3 \end{bmatrix} = \begin{bmatrix} 0 \\ 0 \\ 0 \end{bmatrix}$$

$R_2 \to R_2 - 2R_1, \qquad R_3 \to R_3 - 2R_1$

$$\begin{bmatrix} 1 & 1 & 2 \\ 0 & 1 & -2 \\ 0 & 0 & -2 \end{bmatrix} \begin{bmatrix} \lambda_1 \\ \lambda_2 \\ \lambda_3 \end{bmatrix} = \begin{bmatrix} 0 \\ 0 \\ 0 \end{bmatrix}$$

$\Rightarrow \qquad \lambda_3 = 0, \lambda_2 = 0, \lambda_1 = 0$

Thus, all scalars are zero. Therefore the given system of vectors is linearly independent (not dependent).

2.28 PARTITIONING OF MATRICES

Sub-Matrix : A matrix obtained by deleting some of the rows and columns of a matrix 'A' is said to be Sub-matrix.

e.g. $A = \begin{bmatrix} 4 & 1 & 0 \\ 5 & 2 & 1 \\ 6 & 3 & 4 \end{bmatrix}$, then $\begin{bmatrix} 4 & 1 \\ 5 & 2 \end{bmatrix}, \begin{bmatrix} 1 & 0 \\ 2 & 1 \end{bmatrix}, \begin{bmatrix} 5 & 2 \\ 0 & 3 \end{bmatrix}, \begin{bmatrix} 2 & 1 \\ 3 & 4 \end{bmatrix}$ are the sub-matrices.

Partitioning: A matrix may be subdivided into sub-matrices by drawing lines parallel to its rows and columns. These sub-matrices may be considered as the elements of the original matrix.

e.g. $$\left[\begin{array}{cc|ccc} 2 & 1 & 0 & 4 & 1 \\ 1 & 0 & 2 & 3 & 4 \\ \hline 4 & 5 & 1 & 6 & 5 \end{array}\right]$$

$$A_{11} = \begin{bmatrix} 2 & 1 \\ 1 & 0 \end{bmatrix}, \quad A_{12} = \begin{bmatrix} 0 & 4 & 1 \\ 2 & 3 & 4 \end{bmatrix}$$

$$A_{21} = [4 \ \ 5] \qquad A_{22} = [1 \ \ 6 \ \ 5]$$

Then, we may write $A = \begin{bmatrix} A_{11} & A_{12} \\ A_{21} & A_{22} \end{bmatrix}$

We then say the matrix is partitioned. The dotted lines divide the matrix into sub-matrices. A_{11}, A_{12}, A_{21}, A_{22} but behave like elements of the original matrix A. The matrix A can be partitioned in several ways.

Exercise 2.6

1. Prove that the set of vectors (0, 2, –4), (1, – 2, – 1), (1, –4, 3) is linearly dependent.
2. Determine whether the following sets of vectors are linearly dependent or linearly independent.
   (a) (1, –2, 1), (2, 1, –1), (7, –4, 1)
   (b) (1, 1, 1), (0, 4, 1), (3, 0, 1)
   (c) (2, 3, 1), (–1, 4, –2), (1, 18, –4)
3. Prove that the set of four vectors V_1 = (1, 0, –1), V_2 = (–1, 0, 0), V_3 = (1, 0, 1) and V_4 = (2, 1, 3) is linearly dependent.
4. Find 'a' if the vectors $\begin{bmatrix} 1 \\ -1 \\ 3 \end{bmatrix}, \begin{bmatrix} 1 \\ 2 \\ -3 \end{bmatrix}, \begin{bmatrix} a \\ 0 \\ 1 \end{bmatrix}$ are linearly dependent.
5. Find 'P' if the vectors $\begin{bmatrix} 1 \\ -1 \\ 3 \end{bmatrix}, \begin{bmatrix} 1 \\ P \\ -3 \end{bmatrix}$ *and* $\begin{bmatrix} 1 \\ 0 \\ 1 \end{bmatrix}$ are linearly dependent.

Answers

2. (a) Linearly dependent
   (b) Linearly independent
   (c) Linearly dependent
4. a = 1
5. P = 2

2.29 (A) LINEAR TRANSFORMATION

Let us consider the following relations

$$y_1 = a_{11} x_1 + a_{12} x_2 + a_{13} x_3 + \dots\dots\dots + a_{1n} x_n$$
$$y_2 = a_{21} x_1 + a_{22} x_2 + a_{23} x_3 + \dots\dots\dots + a_{2n} x_n$$
$$y_3 = a_{31} x_1 + a_{32} x_2 + a_{33} x_3 + \dots\dots\dots + a_{3n} x_n$$

....

....

$y_n = a_{n1} x_1 + a_{n2} x_2 + a_{n3} x_3 + ... + a_{nn} x_n$

These can be put in the matrix form as

$$\begin{bmatrix} y_1 \\ y_2 \\ y_3 \\ \vdots \\ \vdots \\ y_n \end{bmatrix} = \begin{bmatrix} a_{11} & a_{12} & a_{13} & \cdots & a_{1n} \\ a_{21} & a_{22} & a_{23} & \cdots & a_{2n} \\ a_{31} & a_{32} & a_{33} & \cdots & a_{3n} \\ \cdots & \cdots & \cdots & \cdots & \cdots \\ \cdots & \cdots & \cdots & \cdots & \cdots \\ a_{n1} & a_{n2} & a_{n3} & \cdots & a_{nn} \end{bmatrix} \begin{bmatrix} x_1 \\ x_2 \\ x_3 \\ \vdots \\ \vdots \\ x_n \end{bmatrix}$$

or Y = A X

which gives a linear transformation from X to Y by the matrix A

(b) Orthogonal Transformation

The linear transformation Y = AX is called orthogonal if it transforms $x_1^2 + x_2^2 + x_3^2 + ... + x_n^2$ and $y_1^2 + y_2^2 + y_3^2 + ... y_n^2$ and the matrix A as this transformation is called orthogonal matrix.

Now, $$X'X = [x_1\ x_2\ x_3\ ....\ x_n] \begin{bmatrix} x_1 \\ x_2 \\ x_3 \\ \vdots \\ x_n \end{bmatrix} = x_1^2 + x_2^2 + x_3^2 + ... + x_n^2$$

$$Y'Y = [y_1\ y_2\ y_3\ ....\ y_n] \begin{bmatrix} y_1 \\ y_2 \\ y_3 \\ \vdots \\ y_n \end{bmatrix} = y_1^2 + y_2^2 + y_3^2 + ... + y_n^2$$

If Y = A X is an orthogonal transformation

$$X'X = x_1^2 + x_2^2 + x_3^2 + ... + x_n^2 = y_1^2 + y_2^2 + y_3^2 + ... + y_n^2 = Y'Y$$

= (AX)' (AX)

= (X'A')(AX) = X'(A'A) X

= X'X if AA' = I

But $A^{-1}A = I$ ∴ $A^{-1} = A'$, I is unit matrix

Illustrative Examples

Example 1 If A is orthogonal then A', A^{-1} is also orthogonal

Proof: Since A is orthogonal $A' = A^{-1}$

A' is orthogonal if $y'\ (A')' = (A')^{-1}$

or $A = (A^{-1})'$

$= (A')'$

$= A$ Hence proved

A^{-1} is orthogonal if $(A^{-1})' = (A^{-1})^{-1}$

$(A')^{-1} = (A')^{-1}$

Hence the result

Note: A matrix 'A' is orthogonal if $AA' = I$ where I is a unit matrix.

Example 2 Show that the transformation

$$y_1 = x_1 + x_2 + 2x_3$$
$$y_2 = 2x_1 + x_3$$
$$y_3 = -x_1 + x_2 + 3x_2$$

is non-singular. Find the inverse transformation

Solution: The given system can be written as follows:

$$Y = AX \quad \text{where } y = \begin{bmatrix} y_1 \\ y_2 \\ y_3 \end{bmatrix}, \; X = \begin{bmatrix} x_1 \\ x_2 \\ x_3 \end{bmatrix} \text{ and } A = \begin{bmatrix} 1 & 1 & 2 \\ 2 & 0 & 1 \\ -1 & 1 & 3 \end{bmatrix}.$$

Now, $|A| = -4 \neq 0$, the matrix is non-singular and the given transformation is non-singular.

The inverse of A can be found by using any suitable method

$$A^{-1} = \begin{bmatrix} 1/4 & 1/4 & -1/4 \\ 7/4 & -5/4 & -3/4 \\ -1/2 & 1/2 & 1/2 \end{bmatrix}$$

Inverse transformation is $X = A^{-1}Y$

$$\begin{bmatrix} x_1 \\ x_2 \\ x_3 \end{bmatrix} = \begin{bmatrix} 1/4 & 1/4 & -1/4 \\ 7/4 & -5/4 & -3/4 \\ -1/2 & 1/2 & 1/2 \end{bmatrix} \begin{bmatrix} y_1 \\ y_2 \\ y_3 \end{bmatrix}$$

is
$$x_1 = \frac{1}{4}y_1 + \frac{1}{4}y_2 - \frac{1}{4}y_3$$
$$x_2 = \frac{1}{4}y_1 - \frac{5}{4}y_2 - \frac{3}{4}y_3$$
$$x_3 = -\frac{1}{2}y_1 + \frac{1}{2}y_2 + \frac{1}{2}y_3$$

Example 3 Consider the transformation

$$y_1 = -\frac{2}{3}x_1 + \frac{1}{3}x_2 + \frac{2}{3}x_3$$
$$y_2 = \frac{2}{3}x_1 + \frac{2}{3}x_2 + \frac{1}{3}x_3$$
$$y_3 = \frac{1}{3}x_1 - \frac{2}{3}x_2 + \frac{2}{3}x_3$$

Verify whether it can transform

$x_1^2 + x_2^2 + x_3^2$ to $y_1^2 + y_2^2 + y_3^2$, write down the inverse transformation.

Ans. Transformation is orthogonal

2.30 CHARACTERISTIC MATRIX

Let A be any nth order square matrix and λ be some scalar, then the matrix

$$A - \lambda I = \begin{bmatrix} a_{11}-\lambda & a_{12} & \cdots & a_{1n} \\ a_{21} & a_{22-\lambda} & \cdots & a_{2n} \\ \vdots & \vdots & \vdots & \vdots \\ a_{n1} & a_{n2} & \cdots & a_{nn-\lambda} \end{bmatrix}$$

is called the characteristic matrix of A [where I is an identity matrix of order n].

2.31 CHARACTERISTIC POLYNOMIAL

Let A be any nth order square matrix and λ be some scalar, then the determinant

$$|A - \lambda| = \begin{vmatrix} a_{11}-\lambda & a_{12} & \cdots & a_{1n} \\ a_{21} & a_{22}-\lambda & \cdots & a_{2n} \\ \vdots & \vdots & & \vdots \\ a_{n1} & a_{n2} & \cdots & a_{nn}-\lambda \end{vmatrix}$$

When expanded yield a polynomial $\phi(\lambda)$ of degree n in λ and is called the characteristic polynomial or function of the matrix A.

2.32 CHARACTERISTIC EQUATION

Let A be any nth order square matrix and λ be some scalar, then the equation $|A - \lambda I| = 0$ is called the characteristic equation of matrix A.

2.33 CHARACTERISTIC ROOTS

Let A be any nth order square matrix and λ be some scalar, then the roots of a charactristic equations $|A - \lambda I| = 0$ viz, $\lambda_1, \lambda_2, \ldots \lambda_n$ (say) are called characteristic roots of the matrix A.

Characteristic roots are also called latent roots, Eigen values or proper values.

2.34 SPECTRUM

The set of characteristic roots of a matrix A is called spectrum of a matrix A.

Illustrative Examples

Example 1 Find the Characteristic polynomial of matrix $A = \begin{bmatrix} 2 & -1 & 1 \\ 1 & 2 & 3 \\ -2 & 1 & 2 \end{bmatrix}$

Solution: Characteristic matrix of the matrix A is

$$A-\lambda I=\begin{bmatrix}2&-1&1\\1&2&3\\-2&1&2\end{bmatrix}-\lambda\begin{bmatrix}1&0&0\\0&1&0\\0&0&1\end{bmatrix}$$

$$=\begin{bmatrix}2-\lambda&-1&1\\1&2-\lambda&3\\-2&1&2-\lambda\end{bmatrix}$$

The characteristic polynomial of matrix A is

$$|A-\lambda I|=\begin{vmatrix}2-\lambda&-1&1\\1&2-\lambda&3\\-2&1&2-\lambda\end{vmatrix}=-\lambda^3+6\lambda^2-12\lambda+15 \quad \text{Ans.}$$

Example 2 Find the characteristic roots of matrix $A=\begin{bmatrix}8&-6&2\\-6&7&-4\\2&-4&3\end{bmatrix}$

Solution: Characteristic matrix is

$$[A-\lambda]=\begin{bmatrix}8&-6&2\\-6&7&-4\\2&-4&3\end{bmatrix}-\lambda\begin{bmatrix}1&0&0\\0&1&0\\0&0&1\end{bmatrix}$$

$$\begin{bmatrix}8-\lambda&-6&2\\-6&7-\lambda&-4\\2&-4&3-\lambda\end{bmatrix}$$

$\Rightarrow$ characteristic polynomial is $|A-\lambda I| = |\quad|$

$\Rightarrow$ $(8-\lambda)((7-\lambda)(3-\lambda)-16)+6(-18+6\lambda+8)+2(24-(14-2\lambda))=0$

or $(8-\lambda)(21-3\lambda-3\lambda+\lambda^2-16)+6(-10+6\lambda)+2(24-14+2\lambda)=0$

or $(8-\lambda)(05-6\lambda+\lambda^2)+6(6\lambda-10)+2(10+2\lambda)=0$

or $(8-\lambda)(\lambda^2-6\lambda+5)+36\lambda-60+20+4\lambda=0$

or $8\lambda^2-48\lambda+40-\lambda^3+6\lambda^2-5\lambda+36\lambda-60+20+4\lambda=0$

or $-\lambda^3+18\lambda^2-45\lambda=0$

or $\lambda(\lambda-18\lambda^2-45)=0$

Hence characteristic polynomial is $-\lambda^3+14\lambda^2-11\lambda=0$

for characteristic values:

$$\lambda(\lambda^2-18\lambda+45)=0$$

$\lambda=0,\quad \lambda^2-15\lambda-3\lambda+45=0$

$$\lambda(\lambda-15)-3(\lambda-15)=0$$

$$(\lambda-3)(\lambda-15)=0$$

$\lambda=0,\quad \lambda=3,\quad \lambda=15$ Ans.

2.35 PROPERTIES OF EIGEN VALUES

(a) The eigen value of a square matrix A and its transpose A' are same.

(b) The sum of the eigen values of a matrix is the sum of the elements on the principal diagonal.

(c) The product of the eigen values of a matrix A is equal to $|A|$.

(d) If λ is an eigen value of a non singular matrix A, then $\frac{1}{\lambda}$ is an eigen value of A^{-1}.

(e) If λ is an eigen value of an orthogonal matrix, then $\frac{1}{\lambda}$ is also its eigen value.

(f) If $\lambda_1, \lambda_2 \ldots\ldots \lambda_n$ are the eigen values of a matrix A, then A^m has the eigen values $\lambda_1^m, \lambda_2^m \ldots\ldots \lambda_n^m$

(m being a positive integer)

(g) The eigen values of an idempotent matrix are either zero or unity.

- Proof of (d) if x be the given eigen vector corresponding to given value λ of A then

$$Ax = \lambda x, \quad \text{premultipliying both sides by } A^{-1}$$

$$A^{-1}(Ax) = A^{-1}(\lambda x) \Rightarrow \quad (A^{-1}A)\,x = \lambda A^{-1}x \Rightarrow \frac{1}{\lambda}x = A^{-1}x$$

$\Rightarrow$ $\frac{1}{\lambda}$ is an eigen value of A^{-1}

- Proof of (f) If x be the eigen vector of the eigen value λ of matrix A, then

$$Ax = \lambda x \text{ premultiplying by A both sides}$$

$$A(Ax) = A(\lambda x) \Rightarrow A^2x = \lambda\,(Ax) \Rightarrow A^2x = \lambda.\lambda x$$

$\Rightarrow$ $A^2x = \lambda^2 x$ Again $A(A^2x) = A(\lambda^2 x) = \lambda^2\,(Ax)$

Similarly $A^m x = \lambda^m x$ $\Rightarrow$ $A^3x = \lambda^2\,(\lambda x) = \lambda^3 x$

which shows that λ^m is the eigen value of A^m $(m > 0)$

In general

If $\lambda_1, \lambda_2, \ldots \lambda_n$ be the eigen values of A, then $\lambda_1^m, \lambda_2^m, \lambda_3^m \ldots\ldots \lambda_n^m$ be the eigen values of A^m

2.36 CHARACTERISTIC VECTOR

If λ is a characteristic root of an n-square matrix A, then any solution of the equation $AX = \lambda X$ other than $X = 0$ (i.e. $X = x_1, x_2 \ldots\ldots\ldots\ldots x_n$) is called a characteristic vector of A. The characteristic vector of matrix A is also called eigen vector or latent vector of matrix A.

Illustrative Examples

Example 3. Find the eigen value and eigen vector of matrix A.

$$A = \begin{bmatrix} 8 & -6 & 2 \\ -6 & 7 & -4 \\ 2 & -4 & 3 \end{bmatrix}$$

Solution: For the characteristic equation

$$|A - \lambda I| = 0$$

$$\begin{vmatrix} 8-\lambda & -6 & 2 \\ -6 & 7-\lambda & -4 \\ 2 & -4 & 3-\lambda \end{vmatrix} = 0$$

$\Rightarrow$ $(8-\lambda)((7-\lambda)(3-\lambda)-16) + 6(-6(3-\lambda) + 8) + 2(24 - 2(7-\lambda)) = 0$

$\Rightarrow$ $(8-\lambda)(\lambda^2-10\lambda+5) + 6(-10 + 6\lambda) + 2(10 + 2\lambda) = 0$

$\Rightarrow$ $8\lambda^2 - 80\lambda + 40 - \lambda^3 + 10\lambda^2 - 5\lambda - 60 + 36\lambda + 20 + 4\lambda = 0$

$\Rightarrow$ $-\lambda^3 + 18\lambda^2 - 45\lambda = 0$

$\Rightarrow$ $\lambda^3 - 18\lambda^2 + 45\lambda = 0$ [Characteristic equation]

$\Rightarrow$ $\lambda[\lambda^2 - 18\lambda + 45] = 0$

$\Rightarrow \lambda = 0,\ \lambda^2 - 18\lambda + 45 = 0$

$\Rightarrow$ $\lambda^2 - 15\lambda - 3\lambda + 45 = 0$

$\Rightarrow$ $\lambda(\lambda - 15) - 3(\lambda - 15) = 0$

$\Rightarrow$ $\lambda - 3 = 0;\ \ \lambda - 15 = 0$

$\Rightarrow$ $\lambda = 3;\ \lambda = 15;\ \lambda = 0$

Eigen values are $\lambda = 0, 3, 15$

To find eigen vector : $(A - \lambda I)X = 0$

$$\begin{bmatrix} 8-\lambda & -6 & 2 \\ -6 & 7-\lambda & -4 \\ 2 & -4 & 3-\lambda \end{bmatrix}\begin{bmatrix} x_1 \\ x_2 \\ x_3 \end{bmatrix} = 0$$

for $\lambda = 0$

$$\begin{bmatrix} 8 & -6 & 2 \\ -6 & 7 & -4 \\ 2 & -4 & 3 \end{bmatrix}\begin{bmatrix} x_1 \\ x_2 \\ x_3 \end{bmatrix} = 0$$

$\Rightarrow$ $8x_1 - 6x_2 + 2x_3 = 0$ ...(1)

$\Rightarrow$ $-6x_1 + 7x_2 - 4x_3 = 0$ ...(2)

$\Rightarrow$ $2x_1 - 4x_2 + 3x_3 = 0$ ...(3)

Multiply equation (1) by 2 and adding to (2)

$$\begin{array}{r} 16x_1 - 12x_2 + 4x_3 = 0 \\ -6x_1 + 7x_2 - 4x_3 = 0 \\ \hline 10x_1 - 5x_2 = 0 \end{array}$$

$\Rightarrow$ $10x_1 = 5x_2$

$\Rightarrow$ $\boxed{x_2 = 2x_1}$ ...(4)

Multiply eqⁿ (3) by 3 and adding to (2)

$$\begin{array}{r} 6x_1 - 12x_2 + 9x_3 = 0 \\ -6x_1 + 7x_2 - 4x_3 = 0 \\ \hline -5x_2 + 5x_3 = 0 \end{array}$$

$\Rightarrow \quad -5x_2 = -5x_3$

$\Rightarrow \quad \boxed{x_2 = x_3} \quad ...(5)$

from (4) and (5) we get

$$2x_1 = x_2 = x_3 \Rightarrow \frac{x_1}{1} = \frac{x_2}{2} = \frac{x_3}{2}$$

Eigen vector corresponding to $\lambda = 0$ are

$$\begin{bmatrix} 1 \\ 2 \\ 2 \end{bmatrix}$$

for $\lambda = 3$
$$\begin{bmatrix} 8-3 & -6 & 2 \\ -6 & 7-3 & -4 \\ 2 & -4 & 0 \end{bmatrix} \begin{bmatrix} x_1 \\ x_2 \\ x_3 \end{bmatrix} = 0$$

$$\begin{bmatrix} 5 & -6 & 2 \\ -6 & 4 & -4 \\ 2 & -4 & 0 \end{bmatrix} \begin{bmatrix} x_1 \\ x_2 \\ x_3 \end{bmatrix} = 0$$

$\Rightarrow \quad 5x_1 - 6x_2 + 2x_3 = 0 \quad ...(6)$

$\Rightarrow \quad -6x_1 + 4x_2 - 4x_3 = 0 \quad ...(7)$

$\Rightarrow \quad 2x_1 - 4x_2 + 0.x_3 = 0 \quad ...(8)$

from (8) $\quad 2x_1 = 4x_2$

$\Rightarrow \quad x_1 = 2x_2$

Multiply (6) by 2 and adding in (7)

$$\begin{array}{r} 10x_1 - 12x_2 + 4x_3 = 0 \\ -6x_1 + 4x_2 - 4x_3 = 0 \\ \hline 4x_1 - 8x_2 = 0 \end{array}$$

$\Rightarrow \quad x_1 = 2x_2$

from (8), put $x_1 = 2x_2$ in (7), we get

$-6x_1 + 4x_2 - 4x_3 = 0$

or $\quad -12x_2 + 4x_2 - 4x_3 = 0$

$\Rightarrow \quad -8x_2 = 4x_3$

$\Rightarrow \quad x_3 = -2x_2$

Thus, $\quad x_1 = 2x_2 = -x_3$

$$\frac{x_1}{2} = \frac{x_2}{1} = \frac{x_3}{-2}$$

eigen vectors are (2, 1, −2)

for $\lambda = 15$

$$\begin{bmatrix} 8-15 & -6 & 2 \\ -6 & 7-15 & -4 \\ 2 & -4 & 3-15 \end{bmatrix}\begin{bmatrix} x_1 \\ x_2 \\ x_3 \end{bmatrix} = 0$$

$$\begin{bmatrix} -7 & -6 & 2 \\ -6 & -8 & -4 \\ 2 & -4 & -12 \end{bmatrix}\begin{bmatrix} x_1 \\ x_2 \\ x_3 \end{bmatrix} = 0$$

$\Rightarrow$
$$-7x_1 - 6x_2 + 2x_3 = 0 \quad ...(9)$$
$$-6x_1 - 8x_2 - 4x_3 = 0 \quad ...(10)$$
$$2x_1 - 4x_2 - 12x_3 = 0 \quad ...(11)$$

Multiply eq^n (11) by 3 and adding in (10)

$$6x_1 - 12x_2 + 36x_3 = 0$$
$$-6x_1 - 8x_2 - 4x_3 = 0$$
$$-20x_2 - 40x_3 = 0$$

$\Rightarrow \quad \boxed{x_2 = -2x_3}$

Multiply eq^n (9) by 2 and adding in (10)

$$-14x_1 - 12x_2 + 4x_3 = 0$$
$$-6x_1 - 8x_2 - 4x_3 = 0$$
$$-20x_1 - 20x_2 = 0$$

$\Rightarrow \quad x_1 = -x_2$

$\therefore \quad x_1 = -x_2 = 2x_3$

$\Rightarrow \quad \dfrac{x_1}{2} = \dfrac{x_2}{-2} = \dfrac{x_3}{1}$

The eigen vectors are (2, −2, 1)

Example 4 Find the eigen value and vector of $\begin{bmatrix} -2 & 2 & -3 \\ 2 & 1 & -6 \\ -1 & -2 & 0 \end{bmatrix}$

Solution: Let $A = \begin{bmatrix} -2 & 2 & -3 \\ 2 & 1 & -6 \\ -1 & -2 & 0 \end{bmatrix}$

$$|A - \lambda I| = 0$$

$$\begin{vmatrix} -2-\lambda & 2 & -3 \\ 2 & 1-\lambda & -6 \\ -1 & -2 & 0-\lambda \end{vmatrix} = 0 \Rightarrow \begin{vmatrix} -2-\lambda & 2 & -3 \\ 2 & 1-\lambda & -6 \\ -1 & -2 & -\lambda \end{vmatrix} = 0$$

$\Rightarrow \quad (-2-\lambda)(-(1-\lambda)\lambda - 12) - 2(-2\lambda - 6) - 3(-4 + 1 - \lambda) = 0$

$\Rightarrow \quad (-2-\lambda)(-\lambda + \lambda^2 - 12) - 2(-2\lambda - 6) - 3(-3 - \lambda) = 0$

$\Rightarrow \quad \lambda^3 + \lambda^2 - 21\lambda - 45 = 0$

$\Rightarrow \quad (\lambda + 3)(\lambda^2 - 2\lambda - 15) = 0$

$\Rightarrow \quad (\lambda + 3)((\lambda^2 - 5\lambda + 3\lambda - 15) = 0$

$\Rightarrow \quad (\lambda + 3)(\lambda(\lambda - 5) + 3(\lambda - 5)) = 0$

$\Rightarrow \quad (\lambda + 3)(\lambda + 3)(\lambda - 5) = 0$

$\Rightarrow \quad \lambda = -3, -3, 5$

eigen values are $-3, -3, 5$

To find eigen vector

for $\quad \lambda = -3 \qquad AX = \lambda X$

$$\begin{bmatrix} -2 & 2 & -3 \\ 2 & 1 & -6 \\ 1 & -2 & 0 \end{bmatrix} \begin{bmatrix} x \\ y \\ z \end{bmatrix} = -3 \begin{bmatrix} x \\ y \\ z \end{bmatrix}$$

$$\begin{bmatrix} -2x + 2y - 3z \\ 2x + y - 6z \\ -x - 2y + 0.z \end{bmatrix} = \begin{bmatrix} -3x \\ -3y \\ -3z \end{bmatrix}$$

$\Rightarrow \quad x + 2y - 3z = 0$

$2x + 4y - 6z = 0$

$-x - 2y + 3z = 0$

$x + 2y - 3z = 0$ is only one independent equation,

From the above set, it can be seen that

since rank of coefficient matrix is 1.

so, let $z = 0$, we get

$x + 2y = 0$

$\Rightarrow \quad x = -2y$

$$\frac{x}{2} = \frac{y}{-1} \Rightarrow z = 0$$

Hence eigen vector is (+2, −1, 0)

for $\lambda = -3$

$$x + 2y - 3z = 0$$

Let $y = 0$, we get

$$x - 3z = 0$$

$$\Rightarrow \quad x = 3z$$

$$\Rightarrow \quad \frac{x}{3} = \frac{z}{1}$$

$$\Rightarrow \quad y = 0$$

so (3, 0, 1) is eigen vector

for $\lambda = 5$, we have

$$\begin{bmatrix} -2 & 2 & -3 \\ 2 & 1 & -6 \\ -1 & -2 & 0 \end{bmatrix} \begin{bmatrix} x \\ y \\ z \end{bmatrix} = 5 \begin{bmatrix} x \\ y \\ z \end{bmatrix}$$

$$-7x + 2y - 3z = 0 \qquad ...(1)$$

$$\Rightarrow \quad 2x - 4y - 6z = 0 \qquad ...(2)$$

$$-x - 2y - 5z = 0 \qquad ...(3)$$

Multiply equation (1) by 2 and adding in (2)

$$-14x + 4y - 6z = 0$$

$$2x - 4y - 6z = 0$$

$$\overline{-12x - 12z = 0}$$

$$-12x = 12z$$

$$\Rightarrow \quad x = -z$$

Multiply equation (3) by 2 and adding in (2)

$$-2x - 4y - 10z = 0$$

$$2x - 4y - 6z = 0$$

$$\overline{y = -2z}$$

Hence $\frac{x}{1} = \frac{y}{2} = \frac{z}{-1}$

So, (1, 2, −1) is eigen vector.

Example 5 Find the eigen value and vector of $\begin{bmatrix} 6 & -2 & 2 \\ -2 & 3 & -1 \\ 2 & -1 & 3 \end{bmatrix}$

Solution: Let $A = \begin{bmatrix} 2 & -2 & 2 \\ -2 & 3 & -1 \\ 2 & -1 & 3 \end{bmatrix}$

and $|A - \lambda I| = 0$ (characteristic equation)

$$\Rightarrow \begin{vmatrix} 6-\lambda & -2 & 2 \\ -2 & 3-\lambda & -1 \\ 2 & -1 & 3-\lambda \end{vmatrix} = 0$$

$\Rightarrow$ $(6-\lambda)((3-\lambda)^2-1) + 2(-2(3-\lambda)+2) + 2(2-2(3-\lambda)) = 0$

$\Rightarrow$ $(6-\lambda)(\lambda^2 - 6\lambda + 8) + 2(2\lambda - 4) + 2(2\lambda - 4) = 0$

$\Rightarrow$ $-\lambda^3 + 12\lambda^2 - 36\lambda + 32 = 0$

$\Rightarrow$ $\lambda^3 - 12\lambda^2 + 36\lambda - 32 = 0$

$\Rightarrow$ $(\lambda - 2)(\lambda^2 - 10\lambda + 16) = 0$

$\Rightarrow$ $(\lambda - 2)(\lambda^2 - 8\lambda - 2\lambda + 16) = 0$

$\Rightarrow$ $(\lambda - 2)(\lambda(\lambda - 8) - 2(\lambda - 8)) = 0$

$\Rightarrow$ $(\lambda - 2)(\lambda - 2)(\lambda - 8) = 0$

$\Rightarrow$ $\lambda = 2, 2, 8$

eigen values are 2, 2, 8

for $\lambda = 2$, eigen vector is,

$$AX = \lambda X$$

$$\begin{bmatrix} -6 & -2 & 2 \\ -2 & 3 & -1 \\ 2 & -1 & 3 \end{bmatrix}\begin{bmatrix} x \\ y \\ z \end{bmatrix} = 2\begin{bmatrix} x \\ y \\ z \end{bmatrix}$$

$\Rightarrow$
$$6x - 2y + 2z = 2x$$
$$-2x + 3y - z = 2y$$
$$2x - y + 3z = 2z$$

or
$$4x - 2y + 2z = 0$$
$$-2x + y - z = 0$$
$$2x - y + z = 0$$

here $2x - y + z = 0$ is the only one independent equation

Let take $z = 0$

$$2x - y = 0$$
$$2x = y$$
$$\frac{x}{1} = \frac{y}{2}, z = 0$$

Hence (1, 2, 0) is eigen-vector

eigen vector for $\lambda = 2$ is (1, 2, 0)

Again for $\lambda = 2$

Let us take $y = 0$

here $2x - y + z = 0$ becomes $2x + z = 0$

$$z = -2x$$

$$\frac{x}{1} = \frac{z}{-2}, y = 0$$

hence

(1, 0, −2) is eigen vector.

for $\lambda = 8$
$$\begin{bmatrix} 6 & -2 & 2 \\ -2 & 3 & -1 \\ 2 & -1 & 3 \end{bmatrix}\begin{bmatrix} x \\ y \\ z \end{bmatrix} = 8\begin{bmatrix} x \\ y \\ z \end{bmatrix}$$

$\Rightarrow$
$$-2x - 2y + 2z = 0 \quad ...(1)$$
$$-2x - 5y - z = 0 \quad ...(2)$$
$$2x - y - 5z = 0 \quad ...(3)$$

Multiply eq^n (2) by 2 and adding in (1)

$$-4x - 10y - 2z = 0$$
$$-2x - 2y + 2z = 0$$
$$-6x - 12y = 0$$

$$\Rightarrow \quad x = -2y$$

Adding (2) and (3)

$$-2x - 5y - z = 0$$
$$2x - y - 5z = 0$$
$$y = -z$$

$$\Rightarrow \quad \frac{x}{2} = \frac{y}{-1} = \frac{z}{1}$$

(2, −1, 1) is eigen vector.

Example 6 Find eigen value and vector of $A = \begin{bmatrix} 1 & 0 & -1 \\ 1 & 2 & 1 \\ 2 & 2 & 3 \end{bmatrix}$

Solution: Characteristic equation is $|A - \lambda I| = 0$

$$\begin{vmatrix} 1-\lambda & 0 & -1 \\ 1 & 2-\lambda & 1 \\ 2 & 2 & 3-\lambda \end{vmatrix} = 0$$

$\Rightarrow \quad (1-\lambda)((2-\lambda)(3-\lambda)-2)-0-1(2-4+2\lambda)=0$

or $\quad (1-\lambda)(6-2\lambda-3\lambda+\lambda^2-2)-1(-2+2\lambda)=0$

or $\quad (1-\lambda)(\lambda^2-5\lambda+4)-1(2\lambda-2)=0$

or $\quad \lambda^2-5\lambda+4-\lambda^3+5\lambda^2-4\lambda-2\lambda+2=0$

or $\quad -\lambda^3+6\lambda^2-11\lambda+6=0$

or $\quad \lambda^3-6\lambda^2+11\lambda-6=0$

or $\quad (\lambda-1)(\lambda^2-5\lambda+6)=0$

or $\quad (\lambda-1)(\lambda-2)(\lambda-3)=0$

Eigen values are 1, 2, 3.

for $\lambda = 1$ $\quad [A-\lambda I]X=0$

$$\begin{bmatrix} 1-\lambda & 0 & -1 \\ 1 & 2-\lambda & 1 \\ 2 & 2 & 3-\lambda \end{bmatrix}\begin{bmatrix} x_1 \\ x_2 \\ x_3 \end{bmatrix}=1\begin{bmatrix} 0 \\ 0 \\ 0 \end{bmatrix}$$

$$\begin{bmatrix} 0 & 0 & -1 \\ 1 & 1 & 1 \\ 2 & 2 & 2 \end{bmatrix}\begin{bmatrix} x_1 \\ x_2 \\ x_3 \end{bmatrix}=\begin{bmatrix} 0 \\ 0 \\ 0 \end{bmatrix}$$

$$-x_3 = 0 \quad \ldots(1)$$

$$x_1 + x_2 + x_3 = 0 \quad \ldots(2)$$

$$2x_1 + 2x_2 + 2x_3 = 0 \quad \ldots(3)$$

putting equation (1) in (2) and (3)

$$x_1 + x_2 = 0 \quad \Rightarrow \quad x_1 = -x_2$$

$$2x_1 + 2x_2 = 0$$

$$\Rightarrow \quad \frac{x_1}{1}=\frac{x_2}{-1}=\frac{x_3}{0}$$

Thus eigen vector is $\begin{bmatrix} 1 \\ -1 \\ 0 \end{bmatrix}$

for $\lambda = 2$

$$\begin{bmatrix} 1-2 & 0 & -1 \\ 1 & 2-2 & 1 \\ 2 & 2 & 3-2 \end{bmatrix}\begin{bmatrix} x_1 \\ x_2 \\ x_3 \end{bmatrix}=\begin{bmatrix} 0 \\ 0 \\ 0 \end{bmatrix}$$

$$\begin{bmatrix} -1 & 0 & -1 \\ 1 & 0 & 1 \\ 2 & 2 & 1 \end{bmatrix}\begin{bmatrix} x_1 \\ x_2 \\ x_3 \end{bmatrix}=\begin{bmatrix} 0 \\ 0 \\ 0 \end{bmatrix}$$

$$-x_1 - x_3 = 0 \quad \Rightarrow \quad x_1 = -x_3$$

$$x_1 + x_3 = 0$$
$$2x_1 + 2x_2 + x_3 = 0 \quad \Rightarrow \quad x_1 + 2x_2 = 0$$

$$\Rightarrow \quad \frac{x_1}{2} = \frac{x_2}{1}; x_3 = -x_1$$

eigen vector is $\begin{bmatrix} 2 \\ 1 \\ -2 \end{bmatrix}$

for $\lambda = 3$

$$\begin{bmatrix} 1-3 & 0 & -1 \\ 1 & 2-3 & 1 \\ 2 & 2 & 3-3 \end{bmatrix}\begin{bmatrix} x_1 \\ x_2 \\ x_3 \end{bmatrix} = \begin{bmatrix} 0 \\ 0 \\ 0 \end{bmatrix}$$

$$\begin{bmatrix} -2 & 0 & -1 \\ 1 & -1 & 1 \\ 2 & 2 & 2 \end{bmatrix}\begin{bmatrix} x_1 \\ x_2 \\ x_3 \end{bmatrix} = \begin{bmatrix} 0 \\ 0 \\ 0 \end{bmatrix}$$

$$-2x_1 + 0.x_2 - x_3 = 0 \quad \Rightarrow \quad -2x_1 = x_3 \qquad ...(1)$$
$$x_1 - x_2 - x_3 = 0 \qquad ...(3)$$
$$2x_1 + 2x_2 + 0x_3 = 0 \quad \Rightarrow \quad x_1 = -x_2 \qquad ...(2)$$

put $\quad x_1 = 1$

then $\quad x_2 = -1$

$\quad x_3 = -2(1) = -2$ (from equation (1))

eigen vector is $\begin{bmatrix} 1 \\ -1 \\ -2 \end{bmatrix}$

Example 7 Find eigen value and eigen vector of $A = \begin{bmatrix} -5 & 2 \\ 1 & -2 \end{bmatrix}$

Solution: (Characteristic equation is) $|A - \lambda I| = 0$

$$\begin{vmatrix} -5-\lambda & 2 \\ 2 & -2-\lambda \end{vmatrix} = 0$$

$$(-5 - \lambda)(-2 - \lambda) - 4 = 0$$
$$\lambda^2 + 7\lambda + 10 - 4 = 0$$
$$\Rightarrow \quad \lambda^2 + 7\lambda + 6 = 0$$
$$\lambda^2 + \lambda + 6\lambda + 6 = 0$$
$$\lambda(\lambda + 1) + 6(\lambda + 1) = 0$$
$$(\lambda + 6)(\lambda + 1) = 0$$

$$\lambda = -6, -1$$

eigen values are –6 and –1

for eigen vectors

$$(A - \lambda I)\,X = 0$$

for $\lambda = -1$

$$\begin{bmatrix} -5+1 & 2 \\ 2 & -2+1 \end{bmatrix}\begin{bmatrix} x_1 \\ x_2 \end{bmatrix} = \begin{bmatrix} 0 \\ 0 \end{bmatrix}$$

$$\begin{bmatrix} -4 & 2 \\ 2 & -1 \end{bmatrix}\begin{bmatrix} x_1 \\ x_2 \end{bmatrix} = \begin{bmatrix} 0 \\ 0 \end{bmatrix}$$

$$-4x_1 + 2x_2 = 0$$

$$2x_1 - x_2 = 0$$

since $2x_1 - x_2$ is only 1 independent equation

$$2x_1 - x_2 = 0$$

$\Rightarrow$ $$2x_1 = x_2$$

or $$\frac{x_1}{1} = \frac{x_2}{2}$$

eigen vector is $\begin{bmatrix} 1 \\ 2 \end{bmatrix}$

for $\lambda = -6$

$$\begin{bmatrix} -5+6 & 2 \\ 2 & -2+6 \end{bmatrix}\begin{bmatrix} x_1 \\ x_2 \end{bmatrix} = \begin{bmatrix} 0 \\ 0 \end{bmatrix}$$

$$\begin{bmatrix} 1 & 2 \\ 2 & 4 \end{bmatrix}\begin{bmatrix} x_1 \\ x_2 \end{bmatrix} = \begin{bmatrix} 0 \\ 0 \end{bmatrix}$$

$$x_1 + 2x_2 = 0$$

$$2x_1 + 4x_2 = 0$$

since $x_1 + 2x_2$ is only 1 independent equation.

Now, $x_1 + 2x_2 = 0$

$$x_1 = -2x_2$$

$$\frac{x_1}{2} = \frac{x_2}{-1}$$

eigen vector is $\begin{bmatrix} -1 \\ 2 \end{bmatrix}$

Example 8 Find the eigen values and eigen vectors of matrix $A = \begin{bmatrix} 3 & 1 & 4 \\ 0 & 2 & 6 \\ 0 & 0 & 5 \end{bmatrix}$

Solution: characteristic equation is $|A - \lambda I| = 0$

$$\begin{vmatrix} 3-\lambda & 1 & 4 \\ 0 & 2-\lambda & 6 \\ 0 & 0 & 5-\lambda \end{vmatrix} = 0$$

$(3 - \lambda)((2 - \lambda)(5 - \lambda) - 0) - 1(0) + 4(0) = 0$

$(3 - \lambda)(10 - 2\lambda - 5\lambda + \lambda^2 - 0) = 0$

$(3 - \lambda)(10 - 7\lambda + \lambda^2 - 6) = 0 \Rightarrow (3 - \lambda)(\lambda^2 - 7\lambda + 10) = 0$

$(3 - \lambda)(\lambda^2 - 5\lambda - 2\lambda + 10) = 0$

$(3 - \lambda)(\lambda - 5)(\lambda - 2) = 0$

$\lambda = 2, 3, 5$

eigen values are 2, 3, 5

for eigen vector.

$$(A - \lambda I)X = 0$$

for $\lambda = 2$

$$\begin{bmatrix} 3-\lambda & 1 & 4 \\ 0 & 2-\lambda & 6 \\ 0 & 0 & 5-\lambda \end{bmatrix}\begin{bmatrix} x_1 \\ x_2 \\ x_3 \end{bmatrix} = 0$$

$$\begin{bmatrix} 1 & 1 & 4 \\ 0 & 0 & 6 \\ 0 & 0 & 3 \end{bmatrix}\begin{bmatrix} x_1 \\ x_2 \\ x_3 \end{bmatrix} = 0$$

$x_1 + x_2 + 4x_3 = 0$ ...(1)

$6x_3 = 0$

$3x_3 = 0$

$\Rightarrow x_3 = 0$ ...(2)

putting (2) in (1)

$x_1 + x_2 = 0$

or $\dfrac{x_1}{1} = \dfrac{-x_2}{1}$

$\dfrac{x_1}{1} = \dfrac{x_2}{-1}$

eigen vector is $\begin{bmatrix} 1 \\ -1 \\ 0 \end{bmatrix}$

for $\lambda = 3$ $\begin{bmatrix} 0 & 1 & 4 \\ 0 & -1 & 6 \\ 0 & 0 & 2 \end{bmatrix}\begin{bmatrix} x_1 \\ x_2 \\ x_3 \end{bmatrix} = \begin{bmatrix} 0 \\ 0 \\ 0 \end{bmatrix}$

$$0x_1 + x_2 + 4x_3 = 0$$
$$-x_2 + 6x_3 = 0$$
$$2x_3 = 0 \Rightarrow \boxed{x_3 = 0}$$
$$-x_2 + 6x_3 = 0$$
$$\boxed{x_2 = 0}$$

Let $\boxed{x_1 = 1}$

$$\frac{x_1}{1} = \frac{x_2}{0} = \frac{x_3}{0}$$

eigen vector is $\begin{bmatrix} 1 \\ 0 \\ 0 \end{bmatrix}$

for $\lambda = 5$, we have

$$\begin{bmatrix} 3-5 & 1 & 4 \\ 0 & 2-5 & 6 \\ 0 & 0 & 5-5 \end{bmatrix} \begin{bmatrix} x_1 \\ x_2 \\ x_3 \end{bmatrix} = 0$$

$$\begin{bmatrix} -2 & 1 & 4 \\ 0 & -3 & 6 \\ 0 & 0 & 0 \end{bmatrix} \begin{bmatrix} x_1 \\ x_2 \\ x_3 \end{bmatrix} = 0$$

$$2x_1 - x_2 - 4x_3 = 0$$
$$3x_2 - 6x_3 = 0 \Rightarrow x_2 = 2x_3$$
$$2x_1 - 2x_3 - 4x_3 = 0$$
$$2x_1 - 6x_3 = 0$$

$\Rightarrow$ $x_1 = 3x_3$

$\Rightarrow$ $\dfrac{x_1}{3} = \dfrac{x_3}{1}$ ...(1)

Also $x_2 = 2x_3$

$\dfrac{x_2}{2} = \dfrac{x_3}{1}$ ...(2)

from (1) and (2) $\dfrac{x_1}{3} = \dfrac{x_2}{2} = \dfrac{x_3}{1}$ eigen vector is $\begin{bmatrix} 3 \\ 2 \\ 1 \end{bmatrix}$

Example 9 Find eigen value and eigen vector of $A = \begin{bmatrix} 2 & 2 \\ 5 & -1 \end{bmatrix}$

Solution: characteristic equation is $|A - \lambda I| = 0$

$$\Rightarrow \qquad \begin{vmatrix} 2-\lambda & 2 \\ 5 & -1-\lambda \end{vmatrix} = 0$$

$$(2 - \lambda)(-1 - \lambda) - 10 = 0$$

$$-2 - 2\lambda + \lambda + \lambda^2 - 10 = 0$$

$$\lambda^2 - \lambda - 12 = 0$$

$$\lambda^2 - 4\lambda + 3\lambda - 12 = 0$$

$$\lambda(\lambda - 4) + 3(\lambda - 4) = 0$$

$$\lambda + 3 = 0, \qquad \lambda - 4 = 0$$

$$\lambda = -3, \qquad \lambda = 4$$

eigen values are –3 and 4.

eigen vector $\quad AX = \lambda x$

$$\begin{bmatrix} 2 & 2 \\ 5 & -1 \end{bmatrix}\begin{bmatrix} x \\ y \end{bmatrix} = -3\begin{bmatrix} x \\ y \end{bmatrix}$$

$\Rightarrow \quad 2x + 2y = -3x$

$\quad 5x - y = -3y$

or $\quad 5x + 2y = 0$

$\quad 5x + 2y = 0$

since there is only 1 independent equation

Thus, we have $\quad 5x + 2y = 0$

$\Rightarrow \quad 5x = -2y$

$$\Rightarrow \quad \frac{x}{-2} = \frac{y}{5}$$

so eigen vector is $\begin{bmatrix} -2 \\ 5 \end{bmatrix}$

for $\quad \lambda = 4$

$$\begin{bmatrix} 2-4 & 2 \\ 5 & -1-4 \end{bmatrix}\begin{bmatrix} x_1 \\ x_2 \end{bmatrix} = 0$$

$$\begin{bmatrix} -2 & 2 \\ 5 & -5 \end{bmatrix}\begin{bmatrix} x \\ y \end{bmatrix} = 0$$

$$-2x + 2y = 0$$

$$5x - 5y = 0$$

again both are linearly dependent

we have $\quad x - y = 0$

$\Rightarrow \quad x = y$

or $\quad \dfrac{x}{1} = \dfrac{y}{1}$

so eigen vector is $\begin{bmatrix} 1 \\ 1 \end{bmatrix}$

Example 10 Find the eigen value and eigen vector of $A = \begin{bmatrix} 7 & 0 & -3 \\ -9 & -2 & 3 \\ 18 & 0 & -8 \end{bmatrix}$

Solution: Characteristic equation is $|A - \lambda I| = 0$

$$\begin{vmatrix} 7-\lambda & 0 & -3 \\ -9 & -2-\lambda & 3 \\ -18 & 0 & -8-\lambda \end{vmatrix} = 0$$

$$= -(\lambda + 2)^2 (\lambda - 1)$$

$$\lambda = -2, \quad \lambda = 1, \quad \lambda = -2$$

for $\lambda = -2$

$$AX = \lambda X$$

$$\begin{bmatrix} 7 & 0 & -3 \\ -9 & -2 & 3 \\ 18 & 0 & -8 \end{bmatrix} \begin{bmatrix} x \\ y \\ z \end{bmatrix} = -2 \begin{bmatrix} x \\ y \\ z \end{bmatrix}$$

$$7x - 3z = -2x$$

$$-9x - 2y + 3z = -2y$$

$$18x - 8z = -2z$$

$\Rightarrow$ $$9x - 3z = 0$$

$$-9x + 3z = 0$$

$$18x - 6z = 0$$

$\Rightarrow$ so, there is only 1 independent equation

$$3x - z = 0$$

$$3x = z$$

$\Rightarrow$ $$\frac{x}{1} = \frac{z}{3}$$

so eigen vector is $\begin{bmatrix} 1 \\ 0 \\ 3 \end{bmatrix}$

for $\lambda = 1$

$$AX = \lambda X$$

$$\begin{bmatrix} 7 & 0 & -3 \\ -9 & -2 & 3 \\ 18 & 0 & -8 \end{bmatrix} \begin{bmatrix} x \\ y \\ z \end{bmatrix} = 1 \begin{bmatrix} x \\ y \\ z \end{bmatrix}$$

$\Rightarrow \quad -6x - 3z = 0$

$\Rightarrow \quad -9x - 3y + 3z = 0$

$\Rightarrow \quad 18x - 9z = 0$

$2x - z = 0 \quad ...(1)$

$3x + y - z = 0 \quad ...(2)$

from equation (1)

$2x = z$

$3x + y - 2z = 0$

$x + y = 0$

$x = -y$

$$\frac{x}{1} = \frac{y}{-1} \quad ...(3)$$

$2x = z$

$$\frac{x}{1} = \frac{z}{2} \quad ...(4)$$

from (3) and (4) $\frac{x}{1} = \frac{y}{-1} = \frac{z}{2}$

so eigen vector is $\begin{bmatrix} 1 \\ -1 \\ 2 \end{bmatrix}$

Example 11 Find the eigen value and eigen vector of $A = \begin{bmatrix} 2 & 2 & 1 \\ 1 & 3 & 1 \\ 1 & 2 & 2 \end{bmatrix}$

Solution: Characteristic equation is $|A - \lambda I| = 0$

$$\begin{vmatrix} 2-\lambda & 2 & 1 \\ 1 & 3-\lambda & 1 \\ 1 & 2 & 2-\lambda \end{vmatrix} = 0$$

$(2 - \lambda) [(3-\lambda)(2-\lambda) - 2] - 2(2 - \lambda - 1) + 1(2 - (3 - \lambda)) = 0$

$(2 - \lambda)(6 - 3\lambda - 2\lambda + \lambda^2 - 2) - 4 + 2\lambda + 2 + 2 - 3 + 3\lambda = 0$

$(2-\lambda)(\lambda^2 - 5\lambda + 4) + 5\lambda - 3 = 0$

$2\lambda^2 - 10\lambda + 8 - \lambda^3 + 5\lambda^2 - 4\lambda + 5\lambda - 3 = 0$

$-\lambda^3 + 7\lambda^2 - 9 + 5 = 0$

$\lambda^3 - 7\lambda^2 + 4 = 0$

$(\lambda - 1)^2 (\lambda - 5) = 0$

$\lambda = 1, 1, 5.$

eigen values are $\quad 1, 1, 5$

eigen vector $(A - \lambda I)\, X = 0$

for $\lambda = 1$

$$\begin{bmatrix} 1 & 2 & 1 \\ 1 & 2 & 1 \\ 1 & 2 & 1 \end{bmatrix}\begin{bmatrix} x \\ y \\ z \end{bmatrix} = \begin{bmatrix} 0 \\ 0 \\ 0 \end{bmatrix}$$

$$x + 2y + z = 0$$
$$x + 2y + z = 0$$
$$x + 2y + z = 0$$

There is only one independent equation

$$x + 2y + z = 0$$

put $z = 0$

$$x + 2y = 0$$
$$x = -2y$$
$$\frac{x}{-2} = \frac{y}{1}$$

so eigen vector is $\begin{bmatrix} -2 \\ 1 \\ 0 \end{bmatrix}$

Again for $\lambda = 1$

$$x + 2y + z = 0$$

put $y = 0$

$$x + z = 0$$
$$\frac{x}{1} = \frac{z}{-1}$$

so eigen vector is $\begin{bmatrix} 1 \\ 0 \\ -1 \end{bmatrix}$

for $\lambda = 5$ $|A - 5I|\, x = 0$

$$\begin{bmatrix} -3 & 2 & 1 \\ 1 & -2 & 1 \\ 1 & 2 & -3 \end{bmatrix}\begin{bmatrix} x \\ y \\ z \end{bmatrix} = \begin{bmatrix} 0 \\ 0 \\ 0 \end{bmatrix}$$

$$-3x + 2y + z = 0 \quad \ldots(1)$$
$$x - 2y + z = 0 \quad \ldots(2)$$
$$x + 2y - z = 0 \quad \ldots(3)$$

add (1) and (2)

$$\begin{array}{r} -3x + 2y + z = 0 \\ x - 2y + z = 0 \\ \hline -2x + 2z = 0 \end{array}$$

$$\boxed{x = z}$$

subtract (2) from (3)

$$\begin{array}{r} x - 2y + z = 0 \\ x + 2y - 3z = 0 \\ - \quad\; - \quad\; + \quad\;\; \\ \hline -4y + 4z = 0 \end{array}$$

$$y = z$$

$$\frac{x}{1} = \frac{y}{1} = \frac{z}{1}$$

so eigen vector is $\begin{bmatrix} 1 \\ 1 \\ 1 \end{bmatrix}$

Example 12 Find the eigen value and eigen vector of $A = \begin{bmatrix} -9 & 2 & 6 \\ 5 & 0 & -3 \\ 16 & 4 & 11 \end{bmatrix}$

Solution: Characteristic equation is $|A - \lambda I| = 0$

$$\begin{vmatrix} -9-\lambda & 2 & 6 \\ 5 & 0-\lambda & -3 \\ 16 & 4 & 11-\lambda \end{vmatrix} = 0$$

$$(-9-\lambda)\,[-\lambda\,(11-\lambda) + 12] - 2\,[5\,(11-\lambda) - 48] + 6\,[20 - 16\lambda] = 0$$

$$\Rightarrow \quad (-9-\lambda)\,(11\lambda + \lambda^2 + 12) - 2\,[55 - 5\lambda - 48] + 6\,[20 + 16\lambda] = 0$$

$$(\lambda + 1)\,(\lambda - 1)\,(\lambda - 2) = 0$$

$$\lambda = 1, -1, 2.$$

eigen values are 1, −1, 2

for eigen vector

for $\lambda = 1$ $\qquad AX = \lambda X$

$$\begin{bmatrix} -9 & 2 & 6 \\ 5 & 0 & -3 \\ -16 & 4 & 11 \end{bmatrix} \begin{bmatrix} x \\ y \\ z \end{bmatrix} = 1 \begin{bmatrix} x \\ y \\ z \end{bmatrix}$$

$$-9x + 2y + 6z = x$$

$$5x - 3z = y$$

$$-16x + 4y + 11z = z$$

$\Rightarrow$ $$-10x + 2y + 6z = 0$$

$$5x - y - 3z = 0$$

$$-16x + 4y + 10z = 0$$

so, we have 2 independent equation

$$5x - y - 3z = 0$$

$$8x - 2y - 5z = 0$$

After multiply by 2, we get $10x - 2y + 6z = 0$

$$8x - 2y - 5z = 0$$

$$\underline{\quad - \qquad + \qquad - \quad}$$

$$2x - z = 0$$

$$2x = z$$

$$\frac{x}{1} = \frac{z}{2} \qquad \text{...(1)}$$

Now

$$5x - y - 3z = 0$$

$$5x - y - 3(2x) = 0$$

$$5x - 6x - y = 0$$

$$-x - y = 0$$

$$-x = y$$

$$\frac{x}{1} = \frac{y}{-1} \qquad \text{...(2)}$$

from (1) and (2)

$$\frac{x}{1} = \frac{y}{-1} = \frac{z}{2}$$

so eigen vector is $\begin{bmatrix} 1 \\ -1 \\ 2 \end{bmatrix}$

for $\lambda = -1$

$AX = \lambda X$

$$\begin{bmatrix} -9 & 2 & 6 \\ 5 & 0 & -3 \\ -16 & 4 & 11 \end{bmatrix} \begin{bmatrix} x \\ y \\ z \end{bmatrix} = -1 \begin{bmatrix} x \\ y \\ z \end{bmatrix}$$

$$-9x + 2y + 6z = -x$$

$$5x + 0y - 3z = -y$$

$$-16x + 4y + 11z = -z$$

Now $$-8x + 2y + 6z = 0$$

$$5x + y - 3z = 0$$
$$-16x + 4y + 12z = 0$$

We have two linearly independent solution.

$$4x - y - 3z = 0$$
$$5x + y - 3z = 0 \qquad \ldots(A)$$

on adding we have

$$9x - 6z = 0$$
$$\Rightarrow \quad 3x - 2z = 0$$
$$\text{or} \quad 3x = 2z$$
$$\text{or} \quad \frac{x}{2} = \frac{z}{3} \qquad \ldots(1)$$
$$z = \frac{3}{2}x$$

putting in (A)

$$5x + y - 3 \times \frac{3}{2}x = 0$$
$$5x - \frac{9}{2}x + y = 0$$
$$\frac{x}{2} + y = 0$$
$$\frac{x}{2} = -y$$
$$\frac{x}{2} = \frac{y}{-1} \qquad \ldots(2)$$

using (1) and (2)

$$\frac{x}{2} = \frac{z}{3} = \frac{y}{-1}$$

so eigen vector is $\begin{bmatrix} 2 \\ -1 \\ 3 \end{bmatrix}$

for $\lambda = 2$ $\quad (A - \lambda I) X = 0$

$$\begin{bmatrix} -9-2 & 2 & 6 \\ 5 & 0-2 & -3 \\ -16 & 4 & 11-2 \end{bmatrix} \begin{bmatrix} x \\ y \\ z \end{bmatrix} = 0$$

$$-11x + 2y + 6z = 0 \qquad \ldots(1)$$
$$5x - 2y - 3z = 0 \qquad \ldots(2)$$
$$-16x + 4y + 9z = 0 \qquad \ldots(3)$$

on adding (1) and (2) we get

$$-6x + 3z = 0$$

$$2x = z$$

$$\frac{x}{1} = \frac{z}{2} \quad \text{...(4)}$$

$$z = 2x$$

putting z in eq^n (3) we get

$$-16x + 4y + 18x = 0$$

$$\Rightarrow \quad 2x + 4y = 0$$

$$\Rightarrow \quad 2x = -4y$$

$$x = -2y$$

$$\frac{x}{-2} = \frac{y}{1}$$

Multiply by –2

$$\frac{x}{1} = \frac{y}{-1/2} = \frac{z}{2}$$

or $$\frac{x}{2} = \frac{y}{-1} = \frac{z}{4}$$

so eigen vector is $\begin{bmatrix} 2 \\ -1 \\ 4 \end{bmatrix}$

Example 13 Find the eigen values of $\begin{bmatrix} 3 & 0 \\ 8 & -1 \end{bmatrix}$. Hence find the eigen values of A^{25}, and $A + 2I$.

Solution: Characteristic equation is $|A - \lambda I| = 0$

$$\Rightarrow \quad \begin{vmatrix} 3-\lambda & 0 \\ 8 & -1-\lambda \end{vmatrix} = 0$$

$$\Rightarrow \quad (3 - \lambda)(-1 - \lambda) = 0$$

$$-3 - 3\lambda + \lambda + \lambda^2 = 0$$

$$\Rightarrow \quad \lambda^2 - 2\lambda - 3 = 0$$

$$\lambda^2 - 3\lambda + \lambda - 3 = 0$$

$$\lambda(\lambda-3) + 1(\lambda-3) = 0$$

$$(\lambda + 1)(\lambda - 3) = 0$$

$$\lambda = -1, \ \lambda = 3.$$

eigen values of A^{25}

i.e., $(-1)^{25} = -1$

and $(3)^{25} = 3^{25}$

eigen value of $A + 2I$

for $A = 3$ $= 3 + 2I$ $= 3 + 2 = 5$

for $A = (-1)$ $= -1 + 2I$ $= -1 + 2 = 1$

Complex eigen values:

Example 14 Show that if $0 < \theta < \pi$, then $A = \begin{bmatrix} \cos\theta & -\sin\theta \\ \sin\theta & \cos\theta \end{bmatrix}$ has no real eigen values and consequently no eigen vector.

Solution: The characteristic equation is, $|A - \lambda I| = 0$

$$\Rightarrow \begin{vmatrix} \cos\theta - \lambda & -\sin\theta \\ \sin\theta & \cos\theta - \lambda \end{vmatrix} = 0$$

$$(\cos\theta - \lambda)^2 + \sin^2\theta = 0$$

$$\cos^2\theta - 2\lambda\cos\theta + \lambda^2 + \sin^2\theta = 0$$

$$\lambda^2 - 2\lambda\cos\theta + 1 = 0$$

$$\lambda = \frac{2\cos\theta \pm \sqrt{4\cos^2\theta - 4}}{2}$$

$$= \frac{2\cos\theta \pm 2i\sqrt{1 - \cos^2\theta}}{2}$$

$$= \cos\theta \pm i\sin\theta$$

Hence the matrix A has no real eigen values and consequently no eigen vector.

Exercise - 2.7

Q. Find the characteristic roots of the following matrices:

1. $\begin{bmatrix} 1 & 1 & -2 \\ -1 & 2 & 1 \\ 0 & 1 & -1 \end{bmatrix}$

2. $\begin{bmatrix} 2 & 3 & 11 \\ 0 & 3 & 17 \\ 0 & 0 & -2 \end{bmatrix}$

3. $\begin{bmatrix} a & h & g \\ 0 & b & f \\ 0 & 0 & c \end{bmatrix}$

Q. Find the eigen values and corresponding eigen vectors for the following matrices:

4. $\begin{bmatrix} -3 & -9 & -12 \\ 1 & 3 & 4 \\ 0 & 0 & 1 \end{bmatrix}$

5. $\begin{bmatrix} 1 & 2 & 2 \\ 0 & 2 & 1 \\ -1 & 2 & 2 \end{bmatrix}$

6. $\begin{bmatrix} 2 & 1 & 0 \\ 0 & 2 & 1 \\ 0 & 0 & 2 \end{bmatrix}$

7. $\begin{bmatrix} 1 & 0 & 0 \\ 3 & 1 & 0 \\ -7 & 5 & 1 \end{bmatrix}$

8. $\begin{bmatrix} 1 & 0 & 2 \\ 0 & 2 & 1 \\ 2 & 0 & 1 \end{bmatrix}$

9. $\begin{bmatrix} 2 & 0 & 0 \\ 0 & -3 & 1 \\ 0 & 1 & -3 \end{bmatrix}$

Answers

1. −1, 1, 2
2. 2, 3, −2
3. a, b, c
4. 0,0,1; 0, (−3, 1, 0); 1, (12, −4, −1)
5. 1, (1, 1, −1); 2, (2, 1, 0)
6. 2 (1, 0, 0)
7. 1, 1, 1 ; (0, 0, 1)
8. −1(−3, −1, 3); 2(0, 1, 0); 3(1, 1, 1)
9. (1, 0, 0), $\left(0, \frac{1}{\sqrt{2}}, \frac{1}{\sqrt{2}}\right)$; $\left(0, \frac{1}{\sqrt{2}}, -\frac{1}{\sqrt{2}}\right)$.

2.37 CAYLEY HAMILTON THEOREM

Definition

Every square matrix satisfies its characteristic equation i.e. if the characteristic equation of the nth order of square matrix A is $|A - \lambda I| = (-1)^n \lambda^n + k_1 \lambda^{n-1} + k_2\lambda^{n-2} + \ldots\ldots\ldots\ldots + k_n = 0$ ....(1)

then $(-1)^n A^n + k_1 A^{n-1} + k_2 A^{n-2} + \ldots\ldots + k_n I = 0$

2.38 GENERAL PROOF

Let $P = \text{adj } (A - \lambda I)$

Since, the elements of $(A-\lambda I)$ are at most of first degree in λ, the elements of $P = \text{adj } (A - \lambda I)$ are polynomials in λ of degree (n-1) or less.

We can therefore, split up P into a number of matrices each containing the same power of λ and write

$$P = P_1\lambda^{n-1} + P_2\lambda^{n-2} + P_{n-1}\lambda + P_n \quad \ldots(2)$$

Also, we know that if M is a square matrix, then $M (\text{adj } M) = |M| \times I$

$$\therefore \quad (A - \lambda I) P = |A-\lambda I| \times I$$

By (1) and (2), we have

$$(A - \lambda I) (P_1\lambda^{n-1} + P_2 \lambda^{n-2} + \ldots\ldots\ldots + P_{n-1} \lambda + P_n)$$
$$= [(-1)^n \lambda^n + k_1\lambda^{n-1} + k_2 \lambda^{n-2} + \ldots\ldots\ldots\ldots + k_{n-1}\lambda + k_n] I$$

Equating the co-efficient of like power of λ on both sides, we get

$$-P_1 = (-1)^n I \qquad (\because I P_1 = P_1)$$

$$AP_1 - P_2 = K_1 I$$
$$AP_2 - P_3 = k_2 I$$
$$\vdots \quad \vdots \quad \vdots$$
$$\vdots \quad \vdots \quad \vdots$$
$$AP_{n-1} - P_n = k_{n-1} I$$
$$AP_n = k_n I$$

Pre - Multiplying these equation by

$A^n, A^{n-1}, A^{n-2}, \ldots\ldots\ldots A, I$ respectively and adding

we get

$$0 = (-1)^n A^n + k_1 A^{n-1} + k_2 A^{n-2} + \ldots\ldots\ldots\ldots + k_{n-1} A + k_n I$$

or $\quad (-1)^n A^n + k_1 A^{n-1} + k_2 A^{n-2} + \ldots\ldots\ldots\ldots + k_{n-1} A + k_n I = 0 \quad \ldots(3)$

Hence proved

NOTE 1.

Multiplying (3) by A^{-1}, we have $(-1)^n A^{n-1} + k_1 A^{n-2} + \ldots\ldots +k_{n-1} I + k_n A^{-1} = 0$

$$\Rightarrow \quad A^{-1} = \frac{-1}{k_n}[(-1)^n A^{n-1} + k_1 A^{n-2} + \ldots\ldots\ldots + k_{n-1} I]$$

Thus, Cayley-Hamilton theorem gives another method for computing the inverse of a matrix. Since this method expresses the inverse of a matrix of order n in terms of (n–1) powers of A it is suitable for computing the inverse of large matrices.

NOTE 2.

If m be a positive integer such that $m > n$, then multiplying (3) by A^{m-n}, we get $(-1)^n A^m + k_1 A^{m-1} + k_2 A^{m-2} + \ldots\ldots\ldots + k_{n-1} A^{m-n+1} + k_n A^{m-n} = 0$. Showing that only positive integral power A^m $(m > n)$ of A is linearly expressible in terms of those of lower degree.

Illustrative Examples

Example 1. Using Cayley- Hamilton theorem, find the inverse of

$$\begin{bmatrix} 1 & 2 \\ 2 & -1 \end{bmatrix}$$

Solution: The characteristic equation of A is

$$|A - \lambda I| = 0$$

Computing $\quad |A - \lambda I| = \begin{vmatrix} 1-\lambda & 2 \\ 2 & -1-\lambda \end{vmatrix}$

$$= -(1-\lambda)(1+\lambda) - 4$$
$$= -1 + \lambda^2 - 4$$
$$\Rightarrow \quad \lambda^2 - 5 = 0$$

By Cayley-Hamilton theorem, put A in place of λ, gives the characteristic equation.

To show $A^2 - 5I = 0$

$$A^2 = \begin{bmatrix} 1 & 2 \\ 2 & -1 \end{bmatrix}\begin{bmatrix} 1 & 2 \\ 2 & -1 \end{bmatrix} = \begin{bmatrix} 5 & 0 \\ 0 & 5 \end{bmatrix}$$

Hence $A^2 - 5I = \begin{bmatrix} 5 & 0 \\ 0 & 5 \end{bmatrix} - \begin{bmatrix} 5 & 0 \\ 0 & 5 \end{bmatrix} = 0$

$$A^2 - 5I = 0$$

Multiplying both side by A^{-1}

$$A - 5A^{-1} = 0$$

$$\begin{bmatrix} 1 & 2 \\ 2 & -1 \end{bmatrix} - 5\left[A^{-1}\right] = 0$$

$$A^{-1} = \frac{1}{5}\begin{bmatrix} 1 & 2 \\ 2 & -1 \end{bmatrix} \text{ Answer}$$

Example 2 Using Cayley-Hamilton theorem, find the inverse of $\begin{bmatrix} 5 & 3 \\ 3 & 2 \end{bmatrix}$

Solution: The characteristic equation of A is $|A-\lambda I| = 0$

Computing $|A-\lambda I| = \begin{vmatrix} 5-\lambda & 3 \\ 3 & 2-\lambda \end{vmatrix}$

$$= (5-\lambda)(2-\lambda) - 9$$
$$= 10 - 5\lambda - 2\lambda + \lambda^2 - 9$$
$$= \lambda^2 - 7\lambda + 1 = 0$$

To verify Cayley-Hamilton theorem, put A in place of λ, and show $A^2 - 7A + I = 0$

$$A^2 = \begin{bmatrix} 5 & 3 \\ 3 & 2 \end{bmatrix}\begin{bmatrix} 5 & 3 \\ 3 & 2 \end{bmatrix} = \begin{bmatrix} 25+9 & 15+6 \\ 15+6 & 9+4 \end{bmatrix}$$

$$= \begin{bmatrix} 34 & 21 \\ 21 & 13 \end{bmatrix}$$

putting this in equation

$$A^2 - 7A + I$$

$$= \begin{bmatrix} 34 & 21 \\ 21 & 13 \end{bmatrix} - 7\begin{bmatrix} 5 & 3 \\ 3 & 2 \end{bmatrix} + \begin{bmatrix} 1 & 0 \\ 0 & 1 \end{bmatrix}$$

$$= \begin{bmatrix} 34 & 21 \\ 21 & 13 \end{bmatrix} - \begin{bmatrix} 35 & 21 \\ 21 & 14 \end{bmatrix} + \begin{bmatrix} 1 & 0 \\ 0 & 1 \end{bmatrix}$$

$= 0$

This verifies Cayley- Hamilton theorem.

Now multiplying both sides by A^{-1}

$$A^2 - 7A + I = 0$$

$$A - 7I + A^{-1} = 0$$

$$A^{-1} = 7I - A$$

$$= \begin{bmatrix} 7 & 0 \\ 0 & 7 \end{bmatrix} - \begin{bmatrix} 5 & 3 \\ 3 & 2 \end{bmatrix}$$

$$A^{-1} = \begin{bmatrix} 2 & -3 \\ -3 & 5 \end{bmatrix} \quad \text{Answer}$$

Example 3 If $A = \begin{bmatrix} 1 & 2 \\ -1 & 3 \end{bmatrix}$, express $A^6 - 4A^5 + 8A^4 - 12A^3 + 14A^2$ as a linear polynomial in A

Solution: Let $A = \begin{bmatrix} 1 & 2 \\ -1 & 3 \end{bmatrix}$

The characteristic equation of A is

$$|A - \lambda I| = 0$$

$$\begin{vmatrix} 1-\lambda & 2 \\ -1 & 3-\lambda \end{vmatrix} = 0$$

$$= (1-\lambda)(3-\lambda) + 2 = 0$$

$$= 3 - 3\lambda - \lambda + \lambda^2 + 2 = 0$$

$$= \lambda^2 - 4\lambda + 5 = 0$$

By Cayley-Hamilton theorem, we have

$$A^2 - 4A + 5I = 0 \quad ...(1)$$

Now $A^6 - 4A^5 + 8A^4 - 12A^3 + 14A^2$

$$= A^6 - 4A^5 + 5A^4 + 3A^4 - 12A^3 + 14A^2$$

$$= A^4(A^2 - 4A + 5I) + 3A^4 - 12A^3 + 14A^2$$

$$= 0 + 3A^4 - 12A^3 + 14A^2 \text{ (By (1))}$$

$$= 3A^4 - 12A^3 + 15A^2 - A^2$$

$$= 3A^2(A^2 - 4A + 5I) - A^2$$

$$= 0 - A^2$$

$$= + 4A - 5I \quad \text{By (I)}$$

Hence $A^6 - 4A^5 + 8A^4 - 12A^3 + 14A^2 = 4A - 5I$ Answer.

Example 4 Verify Cayley-Hamilton theorem for the matrix

$$A = \begin{bmatrix} 2 & -1 & 1 \\ -1 & 2 & -1 \\ 1 & -1 & 2 \end{bmatrix}$$ Hence, compute A^{-1}

Solution: The Characteristic equation of A is

$$|A-\lambda I| = 0$$

i.e
$$\begin{vmatrix} 2-\lambda & -1 & 1 \\ -1 & 2-\lambda & -1 \\ 1 & -1 & 2-\lambda \end{vmatrix} = 0$$

$$= (2-\lambda)\,[(2-\lambda)(2-\lambda) + 1] + 1\,[(-2 + \lambda) + 1] + 1\,[1 - 2+\lambda]$$

$$= \lambda^3 - 6\lambda^2 + 9\lambda - 4 = 0$$

To verify Cayley-Hamilton theorem, we have to show that $A^3 - 6A^2 + 9A - 4I = 0$

Now,
$$A^2 = \begin{bmatrix} 2 & -1 & 1 \\ -1 & 2 & -1 \\ 1 & -1 & 2 \end{bmatrix}\begin{bmatrix} 2 & -1 & 1 \\ -1 & 2 & -1 \\ 1 & -1 & 2 \end{bmatrix}$$

$$= \begin{bmatrix} 6 & -5 & 5 \\ -5 & 6 & -5 \\ 5 & -5 & 6 \end{bmatrix}$$

$$A^3 = \begin{bmatrix} 6 & -5 & 5 \\ -5 & 6 & -5 \\ 5 & -5 & 6 \end{bmatrix}\begin{bmatrix} 2 & -1 & 1 \\ -1 & 2 & -1 \\ 1 & -1 & 2 \end{bmatrix}$$

$$= \begin{bmatrix} 22 & -21 & 21 \\ -21 & 22 & -21 \\ 21 & -21 & 22 \end{bmatrix}$$

$\therefore$
$$A^3 - 6A^2 + 9A - 4I \qquad \text{...(1)}$$

$$= \begin{bmatrix} 22 & -21 & 21 \\ -21 & 22 & -21 \\ 21 & -21 & 22 \end{bmatrix} - 6\begin{bmatrix} 6 & -5 & 5 \\ -5 & 6 & -5 \\ 5 & -5 & 6 \end{bmatrix} + 9\begin{bmatrix} 2 & -1 & 1 \\ -1 & 2 & -1 \\ 1 & -1 & 1 \end{bmatrix} - 4\begin{bmatrix} 1 & 0 & 0 \\ 0 & 1 & 0 \\ 0 & 0 & 1 \end{bmatrix}$$

$$= \begin{bmatrix} 0 & 0 & 0 \\ 0 & 0 & 0 \\ 0 & 0 & 0 \end{bmatrix} = 0$$

This verifies Cayley-Hamilton theorem.

Now, multiplying both sides of (1) by A^{-1}

we have

$$A^3A^{-1} - 6A^2A^{-1} + 9AA^{-1} - 4I\,A^{-1} = 0$$

$$A^2 - 6A + 9I = 4A^{-1}$$

$$4A^{-1} = \begin{bmatrix} 6 & -5 & 5 \\ -5 & 6 & -5 \\ 5 & -5 & 6 \end{bmatrix} - 6\begin{bmatrix} 2 & -1 & 1 \\ -1 & 2 & -1 \\ 1 & -1 & 2 \end{bmatrix} + 9\begin{bmatrix} 1 & 0 & 0 \\ 0 & 1 & 0 \\ 0 & 0 & 1 \end{bmatrix}$$

$$4A^{-1} = \begin{bmatrix} 3 & 1 & -1 \\ 1 & 3 & 1 \\ -1 & 1 & 3 \end{bmatrix}$$

or $$A^{-1} = \frac{1}{4}\begin{bmatrix} 3 & 1 & -1 \\ 1 & 3 & 1 \\ -1 & 1 & 3 \end{bmatrix}$$ Answer.

Example 5 Using Cayley-Hamilton theorem. Find A^8 if $A = \begin{bmatrix} 1 & 2 \\ 2 & -1 \end{bmatrix}$

Solution: The characteristic equation of A is

$$|A-\lambda I| = 0$$

i.e.,
$$\begin{vmatrix} 1-\lambda & 2 \\ 2 & -1-\lambda \end{vmatrix} = 0$$

$$= -(1-\lambda^2) - 4 = 0$$

$$= \lambda^2 - 5 = 0$$

$$= \lambda^2 = 5 \qquad ...(1)$$

By Cayley-Hamilton theorem, A satisfies its characteristic equation ...(1)

$\therefore$
$$A^2 = 5I$$

$$= (A^2)^4 = (5I)^4$$

$\Rightarrow$
$$A^8 = 625\, I^4$$

But,
$$I^4 = I$$

So,
$$\boxed{A^8 = 625\, I}$$ Answer

Example 6 Using Cayley-Hamilton theorem for the matrix

$$A = \begin{bmatrix} 4 & 3 & 1 \\ 2 & 1 & -2 \\ 1 & 2 & 1 \end{bmatrix}, \text{ Find } A^{-1}$$

Solution: The characteristic equation of A is $|A-\lambda I| = 0$

i.e.
$$\begin{vmatrix} 4-\lambda & 3 & 1 \\ 2 & 1-\lambda & -2 \\ 1 & 2 & 1-\lambda \end{vmatrix} = 0$$

$$= (4-\lambda)[1 + \lambda^2 - 2\lambda + 4] - 3[2 - 2\lambda + 2] + 1[4 - 1 + \lambda] = 0$$

$$= (4-\lambda)(\lambda^2 - 2\lambda + 5) + 6\lambda - 12 + \lambda + 3 = 0$$

$$= 4\lambda^2 - 8\lambda + 20 - \lambda^3 + 2\lambda^2 - 5\lambda + 7\lambda - 9 = 0$$

$$= -\lambda^3 + 6\lambda^2 - 6\lambda + 11 = 0$$

$$= \lambda^3 - 6\lambda^2 + 6\lambda - 11 = 0$$

Hence By Cayley-Hamilton theorem, we get characteristic equation

$$A^3 - 6A^2 + 6A - 11 = 0$$

Multiplying the equation by A^{-1}

$$A^2 - 6A + 6I - 11A^{-1} = 0$$

or $$11A^{-1} = A^2 - 6A + 6I$$

$$A^2 = \begin{bmatrix} 4 & 3 & 1 \\ 2 & 1 & -2 \\ 1 & 2 & 1 \end{bmatrix}\begin{bmatrix} 4 & 3 & 1 \\ 2 & 1 & -2 \\ 1 & 2 & 1 \end{bmatrix}$$

$$= \begin{bmatrix} 16+6+1 & 12+3+2 & 4-6+1 \\ 8+2-2 & 6+1-4 & 2-2-2 \\ 4+4+1 & 3+2+2 & 1-4+1 \end{bmatrix}$$

$$= \begin{bmatrix} 23 & 17 & -1 \\ 8 & 3 & -2 \\ 9 & 7 & -2 \end{bmatrix}$$

$$11A^{-1} = \begin{bmatrix} 23 & 17 & -1 \\ 8 & 3 & -2 \\ 9 & 7 & -2 \end{bmatrix} - 6\begin{bmatrix} 4 & 3 & 1 \\ 2 & 1 & -2 \\ 1 & 2 & 1 \end{bmatrix} + \begin{bmatrix} 6 & 0 & 0 \\ 0 & 6 & 0 \\ 0 & 0 & 6 \end{bmatrix}$$

$$= \begin{bmatrix} 23-24+6 & 17-18+0 & -1-6+0 \\ 8-12+0 & 3-6+6 & -2+12+0 \\ 9-6+0 & 7-12+0 & -2-6+6 \end{bmatrix}$$

$$= \begin{bmatrix} 5 & -1 & -7 \\ -4 & 3 & 10 \\ 3 & -5 & -2 \end{bmatrix}$$

$$A^{-1} = \frac{1}{11}\begin{bmatrix} 5 & -1 & -7 \\ -4 & 3 & 10 \\ 3 & -5 & -2 \end{bmatrix} \quad \text{Answer.}$$

Example 7 Find the characteristic equation of the matrix $A = \begin{bmatrix} 2 & 1 & 1 \\ 0 & 1 & 0 \\ 1 & 1 & 2 \end{bmatrix}$ and, hence, find the matrix represented by $A^8 - 5A^7 + 7A^6 - 3A^5 - A^4 - 5A^3 + 8A^2 - 2A + I$

Solution: The characteristic equation of A is $|A - \lambda I| = 0$

$$\Rightarrow \quad \begin{vmatrix} 2-\lambda & 1 & 1 \\ 0 & 1-\lambda & 0 \\ 1 & 1 & 2-\lambda \end{vmatrix} = 0$$

or $$(2-\lambda)[(1-\lambda)(2-\lambda)] + 1[\lambda - 1] = 0$$

$\Rightarrow \quad (2-\lambda)^2(1-\lambda)+(\lambda-1)=0$

$\Rightarrow \quad \lambda^3-5\lambda^2+7\lambda-3=0$

By Cayley-Hamilton theorem,

$$A^3-5A^2+7A-3I=0 \quad \ldots(1)$$

Now $\quad A^8-5A^7+7A^6-3A^5+A^4-5A^3+8A^2-2A+I$

$= A^5(A^3-5A^2+7A-3I)+A(A^3-5A^2+7A-3I)+A^2-A+I$

$= 0+0+A^2+A+I \quad$ [By (1)]

$= A^2+A+I$

$$A^2=\begin{bmatrix}2&1&1\\0&1&0\\1&1&2\end{bmatrix}\begin{bmatrix}2&1&1\\0&1&0\\1&1&2\end{bmatrix}$$

$$=\begin{bmatrix}5&4&4\\0&1&0\\4&4&5\end{bmatrix}$$

$$\Rightarrow \quad A^2+A+I=\begin{bmatrix}5&4&4\\0&1&0\\4&4&5\end{bmatrix}+\begin{bmatrix}2&1&1\\0&1&0\\1&1&2\end{bmatrix}+\begin{bmatrix}1&0&0\\0&1&0\\0&0&1\end{bmatrix}$$

$$=\begin{bmatrix}8&5&5\\0&3&0\\5&5&3\end{bmatrix} \text{ Answer.}$$

Example 8 Verify Cayley-Hamilton theorem for the matrix $A=\begin{bmatrix}1&2&-2\\1&1&1\\1&3&-1\end{bmatrix}$, Hence compute A^{-1}

Solution: The characteristic equation of A is $|A-\lambda I|=0$

$$\Rightarrow \quad \begin{vmatrix}1-\lambda&2&-2\\1&1-\lambda&1\\1&3&-1-\lambda\end{vmatrix}=0$$

$\Rightarrow \quad (1-\lambda)[(1-\lambda)(-1-\lambda)-3]-2(-1-\lambda-1)-2[3-(1-\lambda)]=0$

$\Rightarrow \quad (1-\lambda)(-1-\lambda+\lambda+\lambda^2-3)-2(-\lambda-2)-2(3-1-\lambda)=0$

$\Rightarrow \quad (1-\lambda)(\lambda^2-4)+2\lambda+4-4+2\lambda=0$

$\Rightarrow \quad \lambda^2-4-\lambda^3+4\lambda+2\lambda+4-4+2\lambda=0$

$\Rightarrow \quad \lambda^3-\lambda^2-4\lambda+4=0$

By Cayley Hamilton theorem

$$A^3-A^2-4A+4I=0 \quad \ldots(1)$$

Now $$A^2 = \begin{bmatrix} 1 & 2 & -2 \\ 1 & 1 & 1 \\ 1 & 3 & -1 \end{bmatrix}\begin{bmatrix} 1 & 2 & -2 \\ 1 & 1 & 1 \\ 1 & 3 & -1 \end{bmatrix}$$

$$= \begin{bmatrix} 1+2-2 & 2+2-6 & -2+2+2 \\ 1+1+1 & 2+1+3 & -2+1-1 \\ 1+3-1 & 2+3-3 & -2+3+1 \end{bmatrix}$$

$$= \begin{bmatrix} 1 & -2 & 2 \\ 3 & 6 & -2 \\ 3 & 2 & 2 \end{bmatrix}$$

$$A^3 = \begin{bmatrix} 1 & -2 & 2 \\ 3 & 6 & -2 \\ 3 & 2 & 2 \end{bmatrix}\begin{bmatrix} 1 & 2 & -2 \\ 1 & 1 & 1 \\ 1 & 3 & -1 \end{bmatrix}$$

$$= \begin{bmatrix} 1-2+2 & 2-2+6 & -2-2-2 \\ 3+6-2 & 6+6-6 & -6+6+2 \\ 3+2+2 & 6+2+6 & -6+2-2 \end{bmatrix}$$

$$= \begin{bmatrix} 1 & 6 & -6 \\ 7 & 6 & 2 \\ 7 & 14 & 6 \end{bmatrix}$$

Putting these in eq ...(1)

$$A^3 - A^2 - 4A + 4I = \begin{bmatrix} 1 & 6 & -6 \\ 7 & 6 & 2 \\ 7 & 14 & 6 \end{bmatrix} - \begin{bmatrix} 1 & -2 & 2 \\ 3 & 6 & -2 \\ 3 & 2 & 2 \end{bmatrix} - \begin{bmatrix} 4 & 8 & -8 \\ 4 & 4 & 4 \\ 4 & 12 & -4 \end{bmatrix} + \begin{bmatrix} 4 & 0 & 0 \\ 0 & 4 & 0 \\ 0 & 0 & 4 \end{bmatrix}$$

$$= \begin{bmatrix} 0 & 0 & 0 \\ 0 & 0 & 0 \\ 0 & 0 & 0 \end{bmatrix} = 0$$

This verifies the Cayley Hamilton theorem Now, multiplying both side of eq (1) by A^{-1} we have

$$A^3 A^{-1} - A^2 A^{-1} - 4AA^{-1} + 4A^{-1} = 0$$

$$A^2 - A + 4A^{-1} - 4I = 0$$

$$A^2 - A - 4I = -4A^{-1}$$

$$4A^{-1} = -A^2 + A + 4I$$

$$4A^{-1} = \begin{bmatrix} -1 & 2 & -2 \\ -3 & -6 & 2 \\ -3 & -2 & -2 \end{bmatrix} + \begin{bmatrix} 1 & 2 & -2 \\ 1 & 1 & 1 \\ 1 & 3 & -1 \end{bmatrix} + \begin{bmatrix} 4 & 0 & 0 \\ 0 & 4 & 0 \\ 0 & 0 & 4 \end{bmatrix}$$

$$4A^{-1} = \begin{bmatrix} 4 & 4 & -4 \\ -2 & -1 & 3 \\ -2 & 1 & 1 \end{bmatrix}$$

$$A^{-1} = \frac{1}{4}\begin{bmatrix} 4 & 4 & -4 \\ -2 & -1 & 3 \\ -2 & 1 & 1 \end{bmatrix} \text{ Answer.}$$

Example 9 Using Cayley-Hamilton theorem, find A^6 if $A = \begin{bmatrix} 2 & 1 \\ 5 & -2 \end{bmatrix}$.

Solution: The characteristic equation of A is $|A-\lambda I| = 0$

$$\Rightarrow \begin{vmatrix} 2-\lambda & 1 \\ 5 & -2-\lambda \end{vmatrix} = 0$$

$$\Rightarrow \left[-(2-\lambda)(2+\lambda) - 5\right] = 0$$

$$\Rightarrow \left[-\left(2^2 - \lambda^2\right) - 5\right] = 0$$

$$\Rightarrow \left[-4 + \lambda^2 - 5\right] = 0$$

$$\Rightarrow \lambda^2 - 9 = 0$$

$$\Rightarrow \lambda^2 = 9 \qquad \ldots(1)$$

By Cayley Hamilton theorem A satisfies the eq - (1)

$$A^2 - 9I = 0$$

$$\Rightarrow A^2 = 9I$$

$$\Rightarrow (A^2)^3 = (9I)^3$$

$$\Rightarrow \boxed{A^6 = 729\,I} \text{ Answer.}$$

Example 10 Show that the matrix $A = \begin{bmatrix} 0 & r & -q \\ -r & 0 & p \\ q & -p & 0 \end{bmatrix}$ satisfies Cayley-Hamilton theorem.

Solution:

$$A = \begin{bmatrix} 0 & r & -q \\ -r & 0 & p \\ q & -p & 0 \end{bmatrix}$$

the characteristic equation of A is

$$|A - \lambda I| = \begin{vmatrix} 0-\lambda & r & -q \\ -r & 0-\lambda & p \\ q & -p & 0-\lambda \end{vmatrix} = 0$$

$$\Rightarrow \quad -\lambda(\lambda^2 + p^2) - r(r\lambda - pq) - q(rp + q\lambda) = 0$$

$\Rightarrow \quad -\lambda^3 - \lambda p^2 - r^2\lambda + rpq - rpq - q^2\lambda = 0$

$\Rightarrow \quad -\lambda^3 - \lambda p^2 - r^2\lambda - q^2\lambda = 0$

$\Rightarrow \quad \lambda^3 + \lambda p^2 + r^2\lambda + q^2\lambda = 0 \quad ...(1)$

By Cayley - Hamilton theorem.

$$A^3 + Ap^2 + r^2A + q^2A = 0 \quad ...(2)$$

$$A^2 = \begin{bmatrix} 0 & r & -q \\ -r & o & p \\ q & -p & o \end{bmatrix} \begin{bmatrix} o & r & -q \\ -r & o & p \\ q & -p & o \end{bmatrix}$$

$$= \begin{bmatrix} -r^2-q^2 & qp & rp \\ pq & -r^2-p^2 & rq \\ pr & qr & -q^2-p^2 \end{bmatrix}$$

$$A^3 = \begin{bmatrix} -r^2-q^2 & qp & rp \\ pq & -r^2-p^2 & rq \\ pr & qr & -q^2-p^2 \end{bmatrix} \begin{bmatrix} o & r & -q \\ -r & o & p \\ q & -p & o \end{bmatrix}$$

$$= \begin{bmatrix} -pqr+pqr & -r^3-rq^2-rp^2 & r^2q+q^3+qp^2 \\ r^3+rp^2+rq^2 & rpq-pqr & -pq^2-r^2p-p^3 \\ -qr^2-q^3-qp^2 & pr^2+pq^2+p^3 & -pqr+pqr \end{bmatrix}$$

$$= \begin{bmatrix} 0 & -r\left(r^2+q^2+p^2\right) & q\left(r^2+q^2+p^2\right) \\ r\left(r^2+p^2+q^2\right) & 0 & -p\left(r^2+q^2+p^2\right) \\ -q\left(r^2+p^2+q^2\right) & p\left(r^2+q^2+p^2\right) & 0 \end{bmatrix}$$

$$= \left(r^2+p^2+q^2\right) \begin{bmatrix} 0 & -r & q \\ r & 0 & -p \\ -q & p & 0 \end{bmatrix}$$

$$= -\left(r^2+p^2+q^2\right) \begin{bmatrix} 0 & r & -q \\ -r & 0 & p \\ q & -p & 0 \end{bmatrix}$$

$$= -\left(r^2+p^2+q^2\right) A$$

from the eq^n - (1)

we have $\quad A^3 + Ap^2 + r^2A + q^2A$

$$= -\left(r^2+p^2+q^2\right)A + \left(p^2+r^2+q^2\right)A$$

= 0 Hence it satisfies Cayley Hamilton theorem

Example 11 Find the characteristic equation of the matrix $A = \begin{bmatrix} 1 & 4 \\ 2 & 3 \end{bmatrix}$ and use it to find the matrix represented by $A^5 + 5A^4 - 6A^3 + 2A^2 - 4A + 7I$

Also express $A^5 - 4A^4 - 7A^3 + 11A^2 - A - 10I$ as a linear polynomial in A.

Solution: The Characteristic equation is

$$|A - \lambda I| = 0$$

$$\Rightarrow \quad \begin{vmatrix} 1-\lambda & 4 \\ 2 & 3-\lambda \end{vmatrix} = 0$$

$$\Rightarrow \quad (1-\lambda)(3-\lambda) - 8 = 0$$

$$\Rightarrow \quad 3 - \lambda - 3\lambda + \lambda^2 - 8 = 0$$

$$\text{or} \quad \lambda^2 - 4\lambda - 5 = 0$$

By Cayley-Hamilton theorem, we have

$$A^2 - 4A - 5I = 0 \quad \text{...(1)}$$

Now

$$A^5 + 5A^4 - 6A^3 + 2A^2 - 4A + 7I$$

$$= A^3 [A^2 + 5A - 6] + 2(A^2 - 2A) + 7I \quad \text{(By (1))}$$

$$= A^3 [4A + 5 + 5A - 6] + 2(4A + 5 - 2A) + 7I$$

$$= A^3 [9A - 1] + 2[2A + 5] + 7I$$

$$= A^2 (9A^2 - A) + 4A + 10I + 7I$$

$$= (4A + 5I)(36A + 45I - A) + 4A + 17I$$

$$= (4A + 5I)(35A + 45I) + 4A + 17I$$

$$= 5(4A + 5)(7A + 9) + 4A + 17I$$

$$= 5(28A^2 + 36A + 35A + 45) + 4A + 17I$$

$$= 5(28A^2 + 71A + 45I) + 4A + 17I)$$

$$= 5(112A + 140 + 71A + 45I) + 4A + 17I$$

$$= 5(183A + 185I) + 4A + 17$$

$$= 919A + 942I$$

$$= 919 \begin{bmatrix} 1 & 4 \\ 2 & 3 \end{bmatrix} + 942 \begin{bmatrix} 1 & 0 \\ 0 & 1 \end{bmatrix}$$

$$= \begin{bmatrix} 919 & 3676 \\ 1838 & 2757 \end{bmatrix} + \begin{bmatrix} 942 & 0 \\ 0 & 942 \end{bmatrix}$$

$$= \begin{bmatrix} 1861 & 3676 \\ 1838 & 3699 \end{bmatrix} \quad \text{Answer.}$$

Now

$$A^5 - 4A^4 - 7A^3 + 11A^2 - A - 10I$$

$$= A^5 - 4A^4 - 5A^3 - 2A^3 + 11A^2 - A - 10I$$

$$= A^3 [A^2 - 4A - 5I] - 8A^2 - 10A + 11A^2 - A - 10I$$

$$= 0 + 3A^2 - 11A - 10I$$

$$= 12A + 15I - 11A - 10I$$

$$= \boxed{A + 5I} \text{ Answer.}$$

Example 12 Find the characteristic equation of the matrix $A = \begin{bmatrix} 1 & 3 & 7 \\ 4 & 2 & 3 \\ 1 & 2 & 1 \end{bmatrix}$ and hence obtain the inverse of given matrix.

Solution: The characteristic equation is $|A - \lambda I| = 0$

$$\Rightarrow \begin{vmatrix} 1-\lambda & 3 & 7 \\ 4 & 2-\lambda & 3 \\ 1 & 2 & 1-\lambda \end{vmatrix} = 0$$

$$\Rightarrow (1 - \lambda)[(2 - \lambda)(1 - \lambda) - 6] - 3[4 - 4\lambda - 3] + 7[8 - 2 + \lambda] = 0$$

$$(1 - \lambda)[2 - 2\lambda - \lambda + \lambda^2 - 6] - 3[-4\lambda + 1] + 7[\lambda + 6] = 0$$

$$(1 - \lambda)[\lambda^2 - 3\lambda - 4] + 7\lambda - 3 + 7\lambda + 6 = 0$$

$$= \lambda^2 - 3\lambda - 4 - \lambda^3 + 3\lambda^2 + 4\lambda + 7\lambda - 3 + 7\lambda + 6 = 0$$

$$-\lambda^3 + 4\lambda^2 + 20\lambda + 35 = 0$$

$$\lambda^3 - 4\lambda^2 - 20\lambda - 35 = 0$$

By Cayley - Hamilton theorem, A satisfies the above characteristic equation.

$$\Rightarrow A^3 - 4A^2 - 20A - 35I = 0$$

Multiplying both side by A^{-1}

$$\Rightarrow A^2 - 4A - 20I - 35A^{-1} = 0$$

$$\Rightarrow 35A^{-1} = A^2 - 4A - 20I$$

$$\Rightarrow 35A^{-1} = \begin{bmatrix} 1 & 3 & 7 \\ 4 & 2 & 3 \\ 1 & 2 & 1 \end{bmatrix}\begin{bmatrix} 1 & 3 & 7 \\ 4 & 2 & 3 \\ 1 & 2 & 1 \end{bmatrix}$$

$$- 4\begin{bmatrix} 1 & 3 & 7 \\ 4 & 2 & 3 \\ 1 & 2 & 1 \end{bmatrix} - \begin{bmatrix} 20 & 0 & 0 \\ 0 & 20 & 0 \\ 0 & 0 & 20 \end{bmatrix}$$

$$\Rightarrow 35A^{-1} = \begin{bmatrix} -4 & +11 & -5 \\ -1 & -6 & 25 \\ 6 & 1 & -10 \end{bmatrix}$$

$$\Rightarrow A^{-1} = \frac{1}{35}\begin{bmatrix} -4 & 11 & -5 \\ -1 & -6 & 25 \\ 6 & 1 & -10 \end{bmatrix} \text{ Answer}$$

Exercise 2.8

Q. Verify Cayley Hamilton theorem for the following matrices:

1. $\begin{bmatrix} 1 & 2 & 3 \\ 2 & 3 & 1 \\ 3 & 1 & 2 \end{bmatrix}$ 2. $\begin{bmatrix} 1 & 0 & 2 \\ 0 & 2 & 4 \\ 2 & 0 & 3 \end{bmatrix}$

Q. Verify Cayley Hamilton theorem for the following matrices and compute A^{-1} also:

3. $\begin{bmatrix} 2 & 1 & 2 \\ 5 & 3 & 3 \\ -1 & 0 & -2 \end{bmatrix}$ 4. $\begin{bmatrix} 1 & 0 & 2 \\ 0 & 2 & 1 \\ 2 & 0 & 3 \end{bmatrix}$

Q.5. Find the characteristic equation of the matrix $A = \begin{bmatrix} 1 & 1 & 3 \\ 5 & 2 & 6 \\ -2 & -1 & -3 \end{bmatrix}$ and show that A satisfies this equation.

Q.6. If $A = \begin{bmatrix} 3 & 1 \\ -1 & 2 \end{bmatrix}$, use Cayley-Hamilton theorem to express $2A^5 - 3A^4 + A^2 - 4I$ as a linear polynomial in A.

Answers

3. $\begin{bmatrix} -6 & 2 & -3 \\ 7 & -2 & 4 \\ 3 & -1 & 1 \end{bmatrix}$ 4. $\begin{bmatrix} -3 & 0 & 2 \\ -1 & 1/2 & 1/2 \\ 2 & 0 & -1 \end{bmatrix}$ 5. $\lambda^3 = 0$ 6. $138A - 403I$

CHAPTER 3

Expansion of Functions

3.1 INTRODUCTION

This chapter deals with the expansion of various functions by using Taylor's and Maclaurin's series, Leibnitz's rule for successive differentiation.

Finite and Infinite series

A set of terms arranged by some definite law and connected by positive or negative sign is called a series. There are two kinds of series; finite and infinite. A series is said to be finite when the number of terms it contains is definite i.e. countable, for example, 1 + 2 + 3 + 4 + 5, is a finite series with 5 terms.

A series is said to be infinite when the number of terms it contains is infinite i.e., uncountable.

For example $1+\frac{1}{2}+\frac{1}{3}+\frac{1}{4}+\ldots\ldots\ldots\ldots$, is an infinite series with infinite terms.

3.2 TAYLOR'S INFINITE SERIES

If a function f(x) possesses derivatives of all orders in the interval [a, a + h], then for every integer h, however small, there corresponds a Taylor's expansion with Lagrange's form of remainder i.e.,

$$f(a+h) = f(a) + hf'(a) + \frac{h^2}{2!}f''(a) + \frac{h^3}{3!}f'''(a) + \ldots\ldots\ldots + \frac{h^{n-1}}{(n-1)!}f^{n-1}(a) + \frac{h^n}{n!}f^n(a+ph)$$

where $0 < p < 1$

or $f(a+h) = S_n + R_n$

where $S_n = f(a) + h\,f'(a) + \frac{h^2}{2!} f''(a) + \frac{h^3}{3!} f'''(a) + \dots\dots\dots + \frac{h^{n-1}}{(n-1)!} f^{n-1}(a)$

and $R_n = \frac{h^n}{n!} f^n(a + ph)$

considering $R_n \to 0$, as $n \to \infty$ then $S_n = f(a + h)$ as $n \to \infty$

Hence $f(a + h) = f(a) + h\,f'(a) + \frac{h^2}{2!} f''(a) + \dots\dots\dots + \frac{h^{n-1}}{(n-1)!} f^{n-1}(a) + \dots\dots\dots$

is convergent. Thus if

(i) f(x) possesses derivatives of all orders in [a, a + h]

(ii) $R_n = \frac{h^n}{n!} f^n(a + ph)$ where $0 < p < 1$ tends to zero as $n \to \infty$

Then $f(a + h) = f(a) + h\,f'(a) + \frac{h^2}{2!} f''(a) + \frac{h^3}{3!} f'''(a) + \dots\dots\dots + \frac{h^{n-1}}{(n-1)!} f^{n-1}(a) + \frac{h^n}{n!} f^n(a) + \dots\dots(i)$

Hence this is an expansion of f(a + h) as an infinite series in ascending integral powers of 'h' and is called Taylor's infinite series.

Note: To express f(x) in ascending integral powers of (x –a), change (a + h) to x i.e. h = x –a, in equation (i), then

$f(x) = f[a + (x - a)]$, we get

$$f(x) = f(a) + (x-a)\,f'(a) + \frac{(x-a)^2}{2!} f''(a) + \frac{(x-a)^3}{3!} f'''(a) + \dots\dots\dots + \frac{(x-a)^n}{n!} f^n(a), \dots\dots$$

3.3 Working Method for expansion as Taylor's infinite series

To expand f(x + h) as Taylor's series, we take following steps:-

Step I Let f(x+h) be a given function

Step II put h = 0 and find f(x)

Step III Differentiate f(x) a number of times and obtain f'(x), f''(x), f'''(x), ...

Step IV Substitute the values of f(x), f'(x), f''(x), f'''(x) in

$$f(x + h) = f(x) + h f'(x) + \frac{h^2}{2!} f''(x) + ... + \frac{h^n}{n!} f^n(x) + ...$$

3.4 Failure of Taylor's Series

Taylo-'s series fails to expand f(x + h) as an infinite series in the following situations:

(a) If any of functions f(x), f'(x), f''(x), ... becomes infinite or does not exist, i.e., indeterminate form for any value of 'x' in the interval under consideration.

(b) If R_n i.e. nth term does not tend to zero as $n \to \infty$.

Illustrative Examples

Example 1 Using Taylor's series, prove that

(i) $$e^{x+h} = e^x \left(1 + h + \frac{h^2}{2!} + \frac{h^3}{3!} + ...\right)$$

Solution: Let $f(x+h) = e^{x+h}$

putting $h = 0$, $f(x) = e^x$

$f'(x) = e^x$, $f''(x) = e^x$, $f'''(x) = e^x$, ...

According to Taylor's series,

$$f(x + h) = f(x) + hf'(x) + \frac{h^2}{2!} f''(x) + \frac{h^3}{3!} f'''(x) + ...$$

$$e^{x+h} = e^x + e^x . h + e^x . \frac{h^2}{2!} + \frac{h^3}{3!} e^x + ...$$

$$= e^x \left(1 + h + \frac{h^2}{2!} + \frac{h^3}{3!} + ...\right)$$

Proved.

Example 2 Prove that $a^{x+h} = a^x \left(1 + h \log a + \frac{h^2}{2!} (\log a)^2 + \frac{h^3}{3!} (\log a)^3 + ...\right.$

Solution: Here $f(x+h) = a^{x+h}$

putting $h = 0$

$f(x) = a^x$

$f'(x) = a^x \log a$

$f''(x) = a^x (\log a)^2$

$f'''(x) = a^x (\log a)^3$

According to Taylor' series $f(x+h) = f(x) + hf'(x) + \frac{h^2}{2!} f''(x) + ...$

$$f(x + h) = a^x + h\, a^x \log a + \frac{h^2}{2!} a^x (\log a)^2 + \frac{h^3}{3!} a^x (\log a)^3 \ ...$$

$$= a^x \{1 + h \log a + \frac{h^2}{2!} (\log a)^2 + \frac{h^3}{3!} (\log a)^3 \} \ ...... \text{ Proved.}$$

Example 3 Prove that

$$\log \cos(x+h) = \log \cos x - h \tan x - \frac{h^2}{2} \sec^2 x - \frac{h^3}{3} \sec^2 x. \tan x + ...$$

Solution:

Here $f(x+h) = \log \cos (x+h)$

putting $h = 0$

$f(x) = \log \cos x$

$$f'(x) = \frac{1}{\cos x}(-\sin x) = -\tan x$$

$f''(x) = -\sec^2 x$

$f'''(x) = -2 \sec x. \sec x \tan x$

$= -2 \sec^2 x. \tan x$

....

....

Acc. to Taylor's series,

$$f(x+h) = f(x) + h f'(x) + \frac{h^2}{2!} f''(x) + \frac{h^3}{3!} f'''(x) + ...$$

$$\log \cos (x+h) = \log \cos x + h. (-\tan x) + \frac{h^2}{2!}\left(-\sec^2 x\right) + \frac{h^3}{3!}(-2 \sec^2 x. \tan x) + ...$$

$$= \log \cos x - h \tan x - \frac{h^2}{2} \sec^2 x - \frac{h^3}{3} \sec^2 x. \tan x + ... \quad \text{Proved.}$$

Example 4 Prove that $\sin^{-1}(x+h) = \sin^{-1} x + \dfrac{h}{\sqrt{1-x^2}} + \dfrac{x}{\left(1-x^2\right)^{3/2}} \dfrac{h^2}{2!} + ...$

Solution: Here $f(x+h) = \sin^{-1}(x+h)$

putting $h = 0,$

$f(x) = \sin^{-1} x$

$$f'(x) = \frac{1}{\sqrt{1-x^2}} = \left(1-x^2\right)^{-1/2}$$

$$f''(x) = \frac{-1}{2}\left(1-x^2\right)^{-3/2}(-2x)$$

$$= \frac{x}{\left(1-x^2\right)^{3/2}}$$

Acc. to Taylor's series,

$$f(x+h) = f(x) + h f'(x) + \frac{h^2}{2!} f''(x) + \frac{h^3}{3!} f'''(x) + ...$$

$$\sin^{-1}(x+h) = \sin^{-1}x + \frac{h}{\sqrt{1-x^2}} + \frac{h^2}{2!}.\frac{x}{\left(1-x^2\right)^{3/2}} + ...$$ Proved.

Example 5 Prove that $\tan^{-1}(x+h) = \tan^{-1}x + \frac{h}{1+x^2} - \frac{xh^2}{\left(1+x^2\right)^2} + ...$

Solution: Here $f(x+h) = \tan^{-1}(x+h)$

putting $h = 0$

$f(x) = \tan^{-1}x$

$$f'(x) = \frac{1}{1+x^2} = \left(1+x^2\right)^{-1}$$

$$f''(x) = (-1)\left(1+x^2\right)^{-2}.(2x)$$

$$= \frac{-2x}{\left(1+x^2\right)^2}$$

Acc. to Taylor's series,

$$f(x+h) = f(x) + h f'(x) + \frac{h^2}{2!} f''(x) + \frac{h^3}{3!} f'''(x) + ...$$

$$\tan^{-1}(x+h) = \tan^{-1}x + \frac{h}{1+x^2} - \frac{h^2.x}{\left(1+x^2\right)^2} + ...$$ Proved.

Example 6 Prove that $\tan(x+h) = \tan x + h\sec^2 x + h^2\sec^2 x \tan x + \frac{h^3}{3}\left(1+3\tan^2 x\right).\sec^2 x + ...$

Solution: Here $f(x+h) = \tan(x+h)$

putting $h = 0$

$$f(x) = \tan x$$

$$f'(x) = \sec^2 x$$

$$f''(x) = 2 \sec x . \sec x \tan x$$

$$= 2 \sec^2 x \tan x$$

$$f'''(x) = 2(\sec^2 x . \sec^2 x + \tan x . 2 \sec x . \sec x \tan x)$$

$$= 2(\sec^2 x . \sec^2 x + 2 \sec^2 x . \tan^2 x)$$

$$= 2 \sec^2 x (\sec^2 x + 2 \tan^2 x)$$

$$= 2 \sec^2 x (1 + \tan^2 x + 2\tan^2 x)$$

$$= 2 \sec^2 x (1 + 3\tan^2 x).$$

....

....

Acc. to Taylor's series,

$$f(x+h) = f(x) + h f'(x) + \frac{h^2}{2!} f''(x) + \frac{h^3}{3!} f'''(x) + ...$$

$$\tan(x+h) = \tan x + h . \sec^2 x + \frac{h^2}{2!} . 2 \sec^2 x . \tan x + \frac{h^3}{3!} .2 \sec^2 x \left(1 + 3\tan^2 x\right) + ...$$

$$= \tan x + h \sec^2 x + h^2 \sec^2 x \tan x + \frac{h^3}{3} \sec^2 x (1 + 3 \tan^2 x) + ...$$ Proved

Example 7 Prove that $$\frac{f(x+h) + f(x-h)}{2} = f(x) + \frac{h^2}{2!} f'(x) + \frac{h^4}{4!} f^{iv}(x) + ...$$

Solution: Acc. to Taylor's series, we get $f(x+h) = f(x) + hf'(x) + \frac{h^2}{2!} f''(x) + \frac{h^3}{3!} f'''(x) + ...$

$$f(x-h) = f(x) - hf'(x) + \frac{h^2}{2!} f''(x) - \frac{h^3}{3!} f'''(x) + ...$$

$$\frac{f(x+h) + f(x-h)}{2} = \frac{[f(x) + h f'(x) + \frac{h^2}{2!} f''(x) + \frac{h^3}{3!} f'''(x) + ...] + [f(x) - h f'(x) + \frac{h^2}{2!} f''(x) - \frac{h^3}{3!} f'''(x) + ...]}{2}$$

$$= \frac{2f(x) + 2.\frac{h^2}{2!} f''(x) + 2.\frac{h^4}{4!} f^{iv}(x) + ...}{2}$$

$$= f(x) + \frac{h^2}{2!} f''(x) + \frac{h^4}{4!} f^{iv}(x) + \ldots\ldots \quad \text{Proved}$$

Example 8 Expand sin (x+h) in powers of h and deduce that sin (x+h) = sinx cos h + cos x sin h.

Solution: Here f(x+h) = sin (x+h)

putting h = 0

$f(x) = \sin x$

$f^{i}(x) = \cos x$

$f^{ii}(x) = -\sin x$

$f^{iii}(x) = -\cos x$

$f^{iv}(x) = \sin x$

$f^{v}(x) = \cos x$

⋮ ⋮

⋮ ⋮

Acc. to Taylor's series

$$f(x+h) = f(x) + hf'(x) + \frac{h^2}{2!} f''(x) + \frac{h^3}{3!} f'''(x) + \ldots\ldots\ldots$$

$$\sin(x+h) = \sin x + h\cos x - \frac{h^2}{2!}\sin x - \frac{h^3}{3!}\cos x + \frac{h^4}{4!}\sin x + \frac{h^5}{5!}\cos x + \ldots.$$

$$= \sin x\left(1 - \frac{h^2}{2!} + \frac{h^4}{4!} + \ldots..\right) + \cos x\left(h - \frac{h^3}{3!} + \frac{h^5}{5!} - \ldots\ldots\right)$$

$$= \sin x \cos h + \cos x. \sin h.$$

Example 9 Using Taylor's series, expand

(i) e^x in powers of (x – 2)

Solution: $f(x) = e^x \quad = e^{2+(x-2)} = e^{2+h}$, where h = x – 2

$f'(x) = e^x \quad , f(2) = e^2$

$f''(x) = e^x \quad , f'(2) = e^2$

$f'''(x) = e^x \quad , f''(2) = e^2$

$f'''(2) = e^2$

Acc. to Taylor's series,

$$f(x) = f(a) + (x-a)f'(a) + \frac{(x-a)^2}{2!} f''(a) + \ldots\ldots.$$

$$= e^2 + (x-2).e^2 + \frac{(x-2)^2}{2!}.e^2 + \frac{(x-2)^3}{3!}e^2 + \ldots$$

$$= e^2\left[1+(x-2)+\frac{(x-2)^2}{2!}+\frac{(x-2)^3}{3!}+\ldots\ldots\right] \quad \text{Answer}$$

Example 10 Expand sin x in powers of (x–π/2). Hence find the value of sin 91° correct to four decimal places.

Solution: $f(x) = \sin x$

$$= \sin\left[\pi/2 + (x-\pi/2)\right]$$

$$= \sin[\pi/2 + h] \quad , h = x - \pi/2$$

Acc. to Taylor's series,

$$= f(\pi/2) + h f'(\pi/2) + \frac{h^2}{2!} f''(\pi/2) + \frac{h^3}{3!} f'''(\pi/2) + \ldots\ldots$$

$f(x) = \sin x$, $f(\pi/2) = \sin \pi/2 = 1$

$f'(x) = \cos x$, $f'(\pi/2) = 0$

$f''(x) = -\sin x$, $f''(\pi/2) = -1$

$f'''(x) = -\cos x$, $f'''(\pi/2) = 0$

$f^{iv}(x) = \sin x$, $f^{iv}(\pi/2) = 1$

$$f(x) = 1 + h.0 + \frac{h^2}{2!}(-1) + \frac{h^3}{3!}(0) + \frac{h^4}{4!}(1) + \ldots.$$

$$\sin x = 1 - \frac{h^2}{2!} + \frac{h^4}{4!} \ldots\ldots.$$

$$= 1 - \frac{(x-\pi/2)^2}{2!} + \frac{(x-\pi/2)^4}{4!} + \ldots\ldots \quad \text{Proved Part I.}$$

Part II Find the value of sin 91°

Solution: let $f(x) = \sin x$

According to Taylor's series, $f(x+h) = f(x) + h\, f'(x) + \frac{h^2}{2!} f''(x) + \frac{h^3}{3!} f'''(x) + \ldots\ldots\ldots\ldots$

Here, let $x = 90°$, $h = 1°$.

$$\sin(x+h) = \sin x + 1°.\cos x + \frac{(1°)^2}{2!}(-\sin x) + \frac{(1°)^3}{3!}.(-\cos x) + \ldots\ldots.$$

$$\sin 91° = \sin 90° + \left(\frac{\pi}{180}\right).\cos 90° + \left(\frac{\pi}{180}\right)^2.\frac{1}{2}(-\sin 90°)$$

$$+\left(\frac{\pi}{180}\right)^3 \cdot \frac{1}{3!}(-\cos 90°)+............$$

$$=1-\frac{1}{2}\left(\frac{\pi}{180}\right)^2+\frac{1}{4!}\left(\frac{\pi}{180}\right)^4 ....$$

$$= 1 - 0.00015 + 0.0000000039$$

$$= 1 - 0.00015 = 0.9985$$ Answer.

Example 11 Find $\log_e x$ in powers of $(x-1)$. Hence evaluate $\log_e 1.1$ correct to 4 decimal places.

Solution: Here $f(x+h) = \log_e (x+h)$

$$= \log_e (1+(x-1))$$

$$= \log_e (x+h) \quad , \quad h = x-1$$

$f(x) = \log_e x$, $f(1) = \log_e 1 = 0$

$f'(x) = \frac{1}{x}$, $f'(1) = 1$

$f''(x) = \frac{-1}{x^2}$, $f''(1) = -1$

$f'''(x) = \frac{2}{x^3}$, $f'''(1) = 2$

$f^{iv}(x) = \frac{-6}{x^4}$, $f^{iv}(1) = -6$

Acc. to Taylor's series,

$$f(x+h) = f(x) + h f'(x) + \frac{h^2}{2!} f''(x) + \frac{h^3}{3!} f'''(x) + ...$$

$$f(1+(x-1)) = f(1) + (x-1) f'(1) + \frac{(x-1)^2}{2!} f''(1) + \frac{(x-1)^3}{3!} f'''(1) + ...$$

$$f(x) = \log_e x = 0 + (x-1).1 + \frac{(x-1)^2}{2!}(-1) + \frac{(x-1)^3}{3!}(2) + \frac{(x-1)^4}{4!}(-6) + ...$$

$$\log_e x = (x-1) - \frac{(x-1)^2}{2} + \frac{(x-1)^3}{3} - \frac{(x-1)^4}{4} + ...$$ Proved I Part

Now, $\log_e (1.1) = \log_e (1 + 0.1)$, $x = 1$, $h = 0.1$

$$= \log_e (x + h),$$

$\log_e (x + h) = f(x+h)$

$$= f(x) + hf'(x)\ \frac{h^2}{2!} + f''(x) + \ldots\ldots\ldots\ldots$$

$$= f(1) + 0.1\ f'(1) + \frac{(0.1)^2}{2!}\ f''(1) \ldots\ldots\ldots$$

$$= \log_e(1) + (0.1)\ .1 + \frac{(0.1)^2}{2!}\ (-1) + \frac{(0.1)^3}{3!}\ (2) + \ldots\ldots\ldots$$

$$= 0 + (0.1).1 + \frac{(0.1)^2}{2!}(-1) + \frac{(0.1)^3}{3!}(2) + \frac{(0.1)^4}{4!}(-6) + \ldots\ldots$$

$$= 0.1 - \frac{0.01}{2} + \frac{0.001}{3} - \frac{0.0001}{4} + \ldots\ldots\ldots$$

$$= 0.1 - 0.005 + 0.0003$$

$= 0.0953.$ Answer

Example 12 Expand tan x in powers of $(x-\pi/4)$.

Solution: Let $\tan x = \tan(\pi/4 + (x-\pi/4)) = \tan(x + h)$, $h = x - \pi/4$

$\tan(x+h) = f(x+h)$ $x = \pi/4$

putting $h = 0$

$$f(x) = \tan x$$

$$f'(x) = \sec^2 x$$

$$f''(x) = 2 \sec x.\ \sec x \tan x$$

$$= 2\sec^2 x\ .\ \tan x$$

$$f'''(x) = 2[\sec^2 x.\ \sec^2 x + \tan x.\ 2 \sec x.\ \sec x.\ \tan x]$$

$$= 2[\sec^2 x.\ \sec^2 x + 2\sec^2 x.\ \tan^2 x]$$

$$= 2 \sec^2 x\ [\sec^2 x + 2 \tan^2 x]$$

$$= 2 \sec^2 x\ [1 + \tan^2 x + 2 \tan^2 x]$$

$$= 2 \sec^2 x\ [1 + 3 \tan^2 x]$$

$$f(\pi/4) = 1$$

$$f(\pi/4) = \sec^2 \pi/4 = \tan^2 \pi/4 + 1 = 2$$

$$f'(\pi/4) = 2(1 + \tan^2 \pi/4).\ \tan \pi/4$$

$$= 2(1 + 1).\ 1 = 4$$

$$f'''(\pi/4) = 2(1 + \tan^2 \pi/4)(1 + 3 \tan^2 \pi/4)$$

$$= 2(1 + 1)(1 + 3) = 16$$

According to Taylor's series,

$$f(x) = f(\pi/4) + h f'(\pi/4) + \frac{h^2}{2!} f''(\pi/4) + \frac{h^3}{3!} f'''(\pi/4) + \ldots\ldots.$$

$$= 1 + \left(x - \frac{\pi}{4}\right).2 + \frac{(x-\pi/4)^2}{2!}(4) + \frac{\left(x - \frac{\pi}{4}\right)^3}{3!}(16) + ...$$

$$= 1 + 2\left(x - \frac{\pi}{4}\right) + 2\left(x - \frac{\pi}{4}\right)^2 + \frac{8}{3}\left(x - \frac{\pi}{4}\right)^3 + ... \quad \text{Answer}$$

Example 13 Expand $\tan^{-1}x$ in powers of $(x-1)$

Solution:

$f(x) = \tan^{-1}[1 + (x-1)]$

$= \tan^{-1}[x+h] \quad , \quad h = x-1$

$= f(x+h) \quad \quad x = 1$

$f(x) = \tan^{-1}x \quad , \quad \text{putting } h = 0$

$$f'(x) = \frac{1}{1+x^2} = \left(1+x^2\right)^{-1}$$

$$f''(x) = (-1)\left(1+x^2\right)^{-2}(2x)$$

$$= \frac{-2x}{\left(1+x^2\right)^2} = -2x\left(1+x^2\right)^{-2}$$

$$f'''(x) = -2\left[x.(-2)\left(1+x^2\right)^{-3}(2x) + \left(1+x^2\right)^{-2}(1)\right]$$

$$= 8x^2\left(1+x^2\right)^{-3} - 2\left(1+x^2\right)^{-2}$$

....

Now, we have

$f(1) = \pi/4 \quad , \quad f'(1) = \frac{1}{2}$

$f''(1) = \frac{-2}{4} = -\frac{1}{2}, \quad f'''(1) = \frac{8}{8} - \frac{2}{4} = 1 - \frac{1}{2} = \frac{1}{2}$

....

$$f(x) = f(1) + h\,f'(1) + \frac{h^2}{2!}f''(1) + ...$$

$$= \frac{\pi}{4} + (x-1).\frac{1}{2} + \frac{(x-1)^2}{2!}\left(-\frac{1}{2}\right) + \frac{(x-1)^3}{3!}\left(\frac{1}{2}\right) + ...$$

$$= \frac{\pi}{4} + \frac{(x-1)}{2} - \frac{(x-1)^2}{4} + \frac{(x-1)^3}{12} \ldots\ldots\ldots$$

$$= \frac{\pi}{4} + \frac{(x-1)}{2} - \frac{(x-1)^2}{4} + \frac{(x-1)^3}{12} \ldots\ldots\ldots \text{Answer}$$

Example 14 Expand $2+x^2-3x^5+7x^6$ in powers of $(x-1)$.

Solution: we have $f(x) = f(1+(x-1)) = f(x+h)$

$h = x-1$

$x = 1$

$f(x) = 2+x^2-3x^5+7x^6$, $f(1) = 2+1-3+7 = 7$

$f'(x) = 2x-15x^4+42x^5$, $f'(1) = 2-15+42 = 29$

$f''(x) = 2-60x^3+210x^4$, $f''(1) = 2-60+210 = 152$

$f'''(x) = -180x^2+840x^3$, $f'''(1) = -180+840 = 660$

$f^{iv}(x) = -360x+2520x^2$, $f^{iv}(1) = -360+2520 = 2160$

$f^{v}(x) = -360+5040x$, $f^{v}(1) = -360+5040 = 4680$

$f^{vi}(x) = 5040$

Acc. to Taylor's series,

$$f(x) = f(1) + h f'(1) + \frac{h^2}{2!} f''(1) + \frac{h^3}{3!} f'''(x) + \ldots$$

$$= 7 + (x-1).29 + \frac{(x-1)^2}{2!}.152 + \frac{(x-1)^3}{3!}.660 + \ldots\ldots \quad \text{Answer}$$

Example 15 Expand log cosx about the point $\pi/3$

Solution: We have $f(x) = \log\cos\left(\frac{\pi}{3} + \left(x - \frac{\pi}{3}\right)\right)$

$= \log\cos(x+h) = f(x+h)$ $\quad h = x - \pi/3$

$x = \pi/3$

$f(x) = \log\cos x$

$$f'(x) = \frac{1}{\cos x}(-\sin x) = -\tan x$$

$$f''(x) = -\sec^2 x$$

$$f'''(x) = -2\sec x.\sec x.\tan x$$

$$= -2\sec^2 x.\tan x$$

$\vdots \qquad \vdots$

$$f(\pi/3) = \log\cos\pi/3$$

$$f'(\pi/3) = -\tan\pi/3 = -\sqrt{3}$$

$$f''(\pi/3) = -\sec^2\pi/3 = -4$$

$$f'''(\pi/3) = -2\sec^2\pi/3.\tan\pi/3$$

$$= -2.(-4)\left(-\sqrt{3}\right) = -8\sqrt{3}$$

$$f(x) = f(\pi/3) + hf'(\pi/3) + \frac{h^2}{2!}f''(\pi/3) +$$

$$= \log\cos\frac{\pi}{3} + \left(x - \frac{\pi}{3}\right)\left(-\sqrt{3}\right) + \frac{\left(x - \frac{\pi}{3}\right)^2}{2!}(-4) + \frac{\left(x - \frac{\pi}{3}\right)^3}{3!}(-8\sqrt{3}) + \ldots.$$

$$= \log(0.5) - \sqrt{3}\left(x - \frac{\pi}{3}\right) - 2\left(x - \frac{\pi}{3}\right)^2 - \frac{4\sqrt{3}}{3}\left(x - \frac{\pi}{3}\right)^3 + \ldots \qquad \text{Answer}$$

Example 16 Using Taylor's series, compute the value of sin 31° to four decimal places.

Solution: We have $f(x+h) = f(x) + hf'(x) + \frac{h^2}{2!}f''(x) + \ldots\ldots\ldots$

$$\sin(30° + 1°) = \sin(x+h) = f(x+h)$$

$$f(x) = \sin x \quad , \qquad f(30°) = \sin 30° = \frac{1}{2}$$

$$f'(x) = \cos x \quad , \qquad f'(30°) = \frac{\sqrt{3}}{2}$$

$$f''(x) = -\sin x \quad , \qquad f''(30°) = -\frac{1}{2}$$

$$f'''(x) = -\cos x \quad , \qquad f'''(30°) = -\frac{\sqrt{3}}{2}$$

$f^{iv}(x) = \sin x$, $\quad f^{iv}(30°) = \frac{1}{2}$

$\vdots \qquad \vdots$

$$h = 1° = \frac{\pi}{180} = 0.175$$

$$\sin 31° = \sin(30°) + (0.175)\cos 30° + \left(\frac{0.175}{21}\right)^2 \sin 30°$$

$$+ \frac{(0.175)^3}{3!}\cos 30° + ...$$

$$= \frac{1}{2} + (0.175).\frac{\sqrt{3}}{2} - \frac{1}{2}\frac{(0.175)^2}{2!} - \frac{\sqrt{3}}{2}\frac{(0.175)^3}{3!} + ...$$

$= 0.5 + (0.175)(0.866) - (0.175)^2(0.25) -$ ....

$= 0.5 + 0.1515 - 0.0000765$

$= 0.5151$ Answer

Example 17 If $f(x) = x^3 + 8x^2 + 15x - 24$, calculate the value of $f\left(\frac{11}{10}\right)$ by using Taylor's series.

Solution: $f\left(\frac{11}{10}\right) = f\left(1 + \frac{1}{10}\right)$

$= f(x+h)$, $\quad h = \frac{1}{10} = 0.1$

$x = 1$

$f(x) = x^3 + 8x^2 + 15x - 24$, $\quad f(1) = 1 + 8 + 15 - 24 = 0$

$f'(x) = 3x^2 + 16x + 15$, $\quad f'(1) = 3 + 16 + 15 = 34$

$f''(x) = 6x + 16$, $\quad f''(1) = 6 + 16 = 22$

$f'''(x) = 6$, $\quad f'''(1) = 6$

$$f(x+h) = f(x) + hf'(x) + \frac{h^2}{2!}f''(x) + \frac{h^3}{3!}f'''(x) + ...$$

$$f\left(\frac{11}{10}\right) = f(1) + 0.1\ f'(1) + \frac{(0.1)^2}{2!}f''(1) + \frac{(0.1)^3}{3!}f'''(1) + ...$$

$$= 0 + (0.1)(34) + \frac{(0.1)^2}{2!}(22) + \frac{(0.1)^3}{3!}(6) + ...$$

$$f\left(\frac{11}{10}\right) = 3.4 + 0.011 + 0.001 + \ldots$$

$= 3.511.$ Answer

Example 18 Calculate the approximate value of $\sqrt{17}$ to four decimal places by taking the first four terms of an appropriate Taylor's series.

Solution: We have $\sqrt{17} = \sqrt{x+h} = \sqrt{16+1} = f(x+h)$

$$f(x+h) = \sqrt{x+h}, \qquad x = 16,\ h = 1$$

$$f(x) = \sqrt{x} \qquad f(16) = \sqrt{16} = 4$$

$$f'(x) = \frac{1}{2\sqrt{x}} \qquad f'(16) = \frac{1}{2\sqrt{16}} = \frac{1}{8}$$

$$f''(x) = -\frac{1}{4x^{3/2}} \qquad f''(16) = -\frac{1}{4.(16)^{3/2}} = -\frac{1}{4\times 64} = -\frac{1}{256}$$

$$f'''(x) = \frac{3}{8}.\frac{1}{x^{5/2}} \qquad f'''(16) = \frac{3}{8}.\frac{1}{(16)^{5/2}} = \frac{3}{8}.\frac{1}{1024} = \frac{3}{8192}$$

$$f(x+h) = f(x) + h f'(x) + \frac{h^2}{2!} f''(x) + \frac{h^3}{3!} f'''(x) + \ldots.$$

$$= 4 + \frac{1}{8} - \frac{1}{256} + \ldots..$$

$= 4.123.$ Answer

Example 19 Prove that (i)f(ax) = f(x) + (a–1)x f'(x) + $\frac{(a-1)^2 .x^2}{2!}$ f''(x) + ...

Solution: Acc. to Taylor's series,

$$f(x+h) = f(x) + hf'(x) + \frac{h^2}{2!} f''(x) + \frac{h^3}{3!} f'''(x) + \ldots...$$

$$f(ax) = f(x + (a-1)x)$$

$$= f(x+h) \quad , \quad h = (a-1)x$$

$$x = x$$

Thus,

$$f(ax) = f(x) + (a-1)x f'(x) + \frac{(a-1)^2 x^2}{2!} f''(x) + \frac{(a-1)^3 .x^3}{3!} f'''(x) + ... \quad \text{Proved}$$

(ii) Prove that

$$f\left(\frac{x^2}{1+x}\right) = f(x) - \frac{x}{1+x} f'(x) + \left(\frac{x}{1+x}\right)^2 \frac{f''(x)}{2!} - \left(\frac{x}{1+x}\right)^3 \frac{f'''(x)}{3!} + ...$$

$$f\left(\frac{x^2}{1+x}\right) = f\left(x - \frac{x}{1+x}\right),$$

$$= f(x+h)$$

$x = x$, $h = \frac{-x}{1+x}$, put values in Taylor's series,

$$f\left(\frac{x^2}{1+x}\right) = f(x) - \frac{x}{1+x} f'(x) + \left(\frac{x}{1+x}\right)^2 . \frac{f''(x)}{2!} - \left(\frac{x}{1+x}\right)^3 . \frac{f'''(x)}{3!} + ... \quad \text{Proved}$$

Exercise - 3.1

1. Using Taylor's series, prove that

(a) $a^{x+h} = a^x \left[1 + h \log a + \frac{h^2}{2!}(\log a)^2 + \frac{h^3}{3!}(\log a)^3 + ...\right]$

(b) $\frac{1}{x+h} = \frac{1}{x} - \frac{h}{x^2} + \frac{h^2}{x^3} - \frac{h^3}{x^4} + ...$

(c) $\sec^{-1}(x+h) = \sec^{-1} x + \frac{h}{x\sqrt{x^2-1}} - \frac{2x^2-1}{x^2(x^2-1)^{3/2}} \frac{h^2}{2!} + ...$

2. Using Taylor's series,

(a) Find the value of $f\left(\frac{21}{20}\right)$, if $f(x) = x^3 - 6x^2 + 7$

(b) Calculate the approximate value of $\sqrt{10}$ and $\sqrt{26}$ to four decimal places by taking first three terms of an appropriate Taylor's expansion.

(c) Calculate the value of $f\left(\frac{9}{10}\right)$ by the application of Taylor's series for $f(x+h)$, if

$f(x) = x^3 + 2x^2 - 5x + 11$

3. Expand log sin x in ascending power of (x – 3).
4. Calculate approximately $\log_{10} 404$, if $\log_{10} 4 = 0.6021$, using Taylor's series.
5. By use of Taylor's series, compute the value of cos 32°.
6. Use Taylor's Theorem to express the polynomial $2x^3 + 7x^2 + x - 6$ in power of (x–2).
7. Find the expansion of log sin (x+h) by applying Taylor's series.

Answers

2. (a) 1.5426 (b) 2.1623, 5.099 (c) 3.849
3. $\log \sin(3) + (x-3)\cot(3) - \dfrac{(x-3)^2}{2}\operatorname{cosec}^2(3) + \dfrac{(x-3)^3}{3}\operatorname{cosec}^3(3)\cot(3) + \ldots\ldots$
4. $\log_{10} 404 = 2.6064$
5. 0.8461
6. $40 + 53(x-2) + 19(x-2)^2 + 2(x-3)^3$
7. $\log \sin x + h \cot x - \dfrac{h^2}{2}\operatorname{cosec}^2 x + \dfrac{h^3}{3}\operatorname{cosec}^2 x \cot x + \ldots\ldots$

3.5 MACLAURIN'S INFINITE SERIES

If a = 0 and h = x in Taylor's infinite series such that:

(i) f(x) possesses derivatives of all order in the interval [0, x] and

(ii) Maclaurin's remainder $R_n = \dfrac{x^n}{n!} f^n(a + ph)$ tends to zero as $n \to \infty$, then

$$f(x) = f(0) + x f'(o) + \frac{x^2}{2!} f''(0) + \frac{x^3}{3!} f'''(0) + \ldots + \frac{x^n}{n!} f^n(0) + \ldots$$

which is known as Maclaurin's infinite series.

Note: If the function f(x) is denote by y, then Maclaurin's infinite series may be written as

$$y = y(0) + x y_1(0) + \frac{x^2}{2!} y_2(0) + \frac{x^3}{3!} y_3(0) + \ldots\ldots + \frac{x^n}{n!} y_n(0) + \ldots..$$

where $y(0)$, $y_1(0)$, $y_2(0)$, $y_3(0)$, ... $y_n(0)$ etc. denote the values of y, y_1, y_2, y_3, y_n respectively for x = 0.

3.6 WORKING METHOD OF EXPANSION AS MACLAURIN'S SERIES

Step-I put f(x) = given function

Step-II Differentiate f(x), a number of times, i.e., f(x), f'(x), f''(x), f'''(x) .... and so on.

step-III put x = 0 in the result of step-II and find f(0), f'(0), f''(0), f'''(o) .... and so on.

Step-IV On substituting the values of f(0), f'(0), f''(0), f'''(0), ... in Maclaurin's series

we get $f(x) = f(0) + x f'(0) + \frac{x^2}{2!} f''(0) + \frac{x^3}{3!} f'''(0) + \ldots\ldots + \frac{x^n}{n!} f^n(0) + \ldots\ldots$

Failure of Maclaurin's series

Maclaurin's series fails to expand f(x) in an infinite series in the following situation

(a) If any of f(x), f'(x), f''(x) .... becomes infinite or does not exist in the closed interval [0, x]

(b) If R_n does not tend to zero as $n \to \infty$.

3.7 SOME USEFUL EXPANSIONS

$$(1+x)^n = 1 + nx + \frac{n(n-1)}{2}x^2 + \frac{n(n-1)(n-2)}{3!}x^3 + \ldots\ldots \qquad (x < 1)$$

$$e^x = 1 + x + \frac{x^2}{2!} + \frac{x^3}{3!} + \frac{x^4}{4!} + \ldots\ldots$$

$$a^x = 1 + x(\log a) + \frac{x^2}{2!}(\log a)^2 + \frac{x^3}{3!}(\log a)^3 + \ldots\ldots \qquad (a > 0)$$

$$\log(1+x) = x - \frac{x^2}{2} + \frac{x^3}{3} - \frac{x^4}{4} + \ldots\ldots$$

$$\sin x = x - \frac{x^3}{3!} + \frac{x^5}{5!} - \frac{x^7}{7!} + \ldots\ldots$$

$$\cos x = 1 - \frac{x^2}{2!} + \frac{x^4}{4!} - \frac{x^6}{6!} + \ldots\ldots$$

$$\tan x = x + \frac{x^3}{3!} + \frac{2x^5}{15} + \ldots\ldots$$

$$\sin h\,x = x + \frac{x^3}{3!} + \frac{x^5}{5!} + \frac{x^7}{7!} + \ldots\ldots$$

$$\cos h\,x = 1 + \frac{x^2}{2!} + \frac{x^4}{4!} + \frac{x^6}{6!} + \ldots\ldots$$

$$\sin^{-1} x = x + \frac{x^3}{3} \times \frac{1}{2} + \frac{x^5}{5} \times \frac{3}{4} \times \frac{1}{2} + \frac{x^7}{7} \times \frac{5}{6} \times \frac{3}{4} \times \frac{1}{2} + ...$$

$$\cos^{-1} x = 1 + x + \frac{x^3}{6} + \frac{3x^5}{40} + ...$$

$$\tan^{-1} x = x - \frac{x^3}{3} + \frac{x^5}{5} - \frac{x^7}{7} + ...$$

Illustrative Examples

Expand the following functions, using Maclaurin's series:

(i) $f(x) = e^x$, $f(0) = e^0 = 1$

$f'(x) = e^x$, $f'(0) = e^0 = 1$

$f''(x) = e^x$, $f''(0) = 1$

$f'''(x) = e^x$, $f'''(0) = 1$

$f^{iv}(x) = e^x$, $f^{iv}(0) = 1$

$f^{v}(x) = e^x$, $f^{v}(0) = 1$

Using Maclaurin's series,

$$f(x) = f(0) + x f'(0) + \frac{x^2}{2!} f''(0) + \frac{x^3}{3!} f'''(0) + ...$$

$$= 1 + x + \frac{x^2}{2!} + \frac{x^3}{3!} + \frac{x^4}{4!} + ... \text{ Answer}$$

(ii) $f(x) = a^x \;\; a > 0$, $f(0) = a^0 = 1$

$f'(x) = a^x \log a$, $f'(0) = \log a$

$f''(x) = a^x (\log a)^2$, $f''(0) = (\log a)^2$

$f'''(x) = a^x.(\log a)^3$ $\quad f'''(0) = (\log a)^3$

$f^{v}(x) = a^x. (\log a)^4$ $\quad f^{v}(0) = (\log a)^4$

$\vdots \quad \vdots \quad \vdots \quad \vdots$

Using Maclaurin's series,

$$f(x) = f(0) + x f'(0) + \frac{x^2}{2!} f''(0) + \frac{x^3}{3!} f'''(0) + ...$$

$$= 1 + x \log a + \frac{x^2}{2!} (\log a)^2 + \frac{x^3}{3!} (\log a)^3 + ... \text{ Answer}$$

(iii) $f(x) = \sin x$, $f(0) = 0$

$f'(x) = \cos x$, $f'(0) = 1$

$f'(x) = -\sin x \quad , \quad f'(0) = 0$

$f''(x) = -\cos x \quad , \quad f''(0) = -1$

$f^{iv}(x) = \sin x \quad , \quad f^{iv}(0) = 0$

$f^{v}(x) = \cos x \quad , \quad f^{v}(0) = 1$

$f^{vi}(x) = -\sin x \quad , \quad f^{vi}(0) = 0$

Using Maclaurin's series,

$$f(x) = f(0) + x f'(0) + \frac{x^2}{2!} f''(0) + \ldots\ldots\ldots$$

$$= 0 + x + \frac{x^2}{2!}(0) + \frac{x^3}{3!}(-1) + \frac{x^4}{4!}(0) + \ldots\ldots$$

$$\sin x = x - \frac{x^3}{3!} + \frac{x^5}{5!} \ldots\ldots\ldots \quad \text{Answer}$$

(iv) $f(x) = \cos x \qquad f^{iv}(x) = \cos x$

$f'(x) = -\sin x \qquad f^{v}(x) = -\sin x$

$f''(x) = -\cos x \qquad f^{vi}(x) = -\cos x$

$f'''(x) = \sin x$

$f(0) = 1 \quad , \quad f''(0) = -1 \quad , \quad f^{iv}(0) = 1$

$f'(0) = 0 \qquad f'''(0) = 0 \quad , \quad f^{v}(0) = 0$

$f^{vi}(0) = -1$

Using Maclaurin's series,

$$f(x) = f(0) + x f'(0) + \frac{x^2}{2!} f''(0) + \frac{x^3}{3!} f'''(0) \ldots\ldots\ldots$$

$$= 1 + x.0 + \frac{x^2}{2!}(-1) + \frac{x^3}{3!}(0) + \frac{x^4}{4!}(1) + \ldots\ldots\ldots$$

$$\cos x = 1 - \frac{x^2}{2!} + \frac{x^4}{4!} - \frac{x^6}{6!} + \ldots\ldots \quad \text{Answer}$$

(v) $f(x) = \sec x \quad , \quad f(0) = 1$

$f'(x) = \sec x \tan x \quad , \quad f'(0) = 0$

$f''(x) = \sec x . \sec^2 x + \tan x . \sec x \tan x$

$= \sec^3 x + \sec x . \tan^2 x, \quad \boxed{\sec^2 x = 1 + \tan^2 x}$

$= \sec^3 x + \sec x(-1 + \sec^2 x)$

$= \sec^3 x - \sec x + \sec^3 x$

$= 2\sec^3 x - \sec x \qquad f''(0) = 2-1 = 1$

$f'''(x) = 6\sec^2 x.\ \sec x \tan x - \sec x.\ \tan x \qquad f'''(0) = 0$

Using Maclaurin's series,

$$f(x) = f(0) + x f'(0) + \frac{x^2}{2!} f''(0) + \frac{x^3}{3!} f'''(0) + \ldots\ldots$$

$$= 1 + x.(0) + \frac{x^2}{2!}(1) + \frac{x^3}{3!}(0) + \ldots\ldots\ldots$$

$$= 1 + \frac{x^2}{2!} + \frac{5x^4}{4!} + \ldots\ldots\ \text{Answer}$$

(vi) $f(x) = \sin h\, x \quad , \quad f(0) = 0$

$f'(x) = \cos h\, x \quad , \quad f'(0) = 1$

$f''(x) = \sin h\, x \quad , \quad f''(0) = 0$

$f'''(x) = \cos h\, x \quad , \quad f'''(0) = 1$

$f^{iv}(x) = \sin h\, x \quad , \quad f^{iv}(0) = 0$

$f^{v}(x) = \cos h\, x \quad , \quad f^{v}(0) = 1$

Acc. to Maclaurin's series,

$$f(x) = f(0) + x f'(0) + \frac{x^2}{2!} f''(0) + \frac{x^3}{3!} f'''(0) + \ldots.$$

$$= x + \frac{x^3}{3!} + \frac{x^5}{5!} + \ldots.\ \text{Answer}$$

(vii) $f(x) = \cos h\, x \quad , \quad f(0) = 1$

$f'(x) = \sin h\, x \quad , \quad f'(0) = 0$

$f''(x) = \cos h\, x \quad , \quad f''(0) = 1$

$f'''(x) = \sin h\, x \quad , \quad f'''(0) = 0$

$f^{iv}(x) = \cos h\, x \quad , \quad f^{iv}(0) = 1$

$f^{v}(x) = \sin h\, x \quad , \quad f^{v}(0) = 0$

....

....

Using Maclaurin's series,

$$f(x) = f(0) + x f'(0) + \frac{x^2}{2!} f''(0) + \frac{x^3}{3!} f'''(0) + \ldots\ldots$$

$$= 1 + \frac{x^2}{2!} + \frac{x^4}{4!} + \ldots\ldots\ \text{Answer}$$

(viii) $f(x) = \log(1-x)$, $f(0) = 0$

$$f'(x) = \frac{1}{1-x}(-1) = \frac{-1}{1-x} , \quad f'(0) = \frac{-1}{1} = -1$$

$$f''(x) = \frac{-1}{(1-x)^2} , \quad f''(0) = -1$$

$$f'''(x) = \frac{2}{(1-x)^3}(-1) = \frac{-2}{(1-x)^3} , \quad f'''(0) = -2$$

$$f^{iv}(x) = \frac{+6}{(1-x)^4}(-1) = \frac{-6}{(1-x)^4} , \quad f^{iv}(0) = -6$$

⋮ ⋮ ⋮ ⋮ ⋮

Acc. to Maclaurin's series

$$f(x) = f(0) + x f'(0) + \frac{x^2}{2!} f''(0) + \frac{x^3}{3!} f'''(0) + \ldots\ldots\ldots$$

$$= 0 + x(-1) + \frac{x^2}{2!}(-1) + \frac{x^3}{3!}(-2) + \frac{x^4}{4!}(-6) + \ldots\ldots\ldots$$

$$= -x - \frac{x^2}{2!} - \frac{x^3}{3} - \frac{x^4}{4} \ldots\ldots$$

Example 2 Prove that

(i) $$e^{\sin x} = 1 + x + \frac{x^2}{2} - \frac{x^4}{8} + \ldots..$$

Solution: $f(x) = e^{\sin x}$

$f'(x) = e^{\sin x}. \cos x$

$f''(x) = e^{\sin x}(-\sin x) + \cos x. e^{\sin x}. \cos x$

$= e^{\sin x}(\cos^2 x - \sin x)$

$f'''(x) = e^{\sin x}[2 \cos x(-\sin x) - \cos x] + (\cos^2 x - \sin x). e^{\sin x}. \cos x$

$= e^{\sin x}[-2 \sin x \cos x - \cos x + \cos^3 x - \sin x \cos x]$

....

$f(0) = e^0 = 1$, $f'(0) = 1$

$f''(0) = 1$, $f'''(0) = 0$, $f^{iv}(0) = -3$

Acc. to Maclaurin's series,

$$f(x) = f(0) + x f'(0) + \frac{x^2}{2!} f''(0) + \frac{x^3}{3!} f'''(0) + ...$$

$$= 1 + x + \frac{x^2}{2!} + \frac{x^4}{4!}(-3) + ...$$

$$= 1 + x + \frac{x^2}{2!} - \frac{x^4}{8} + ...\quad \text{Proved}$$

(ii) Prove that

$$e^{x \cos x} = 1 + x + \frac{x^2}{2} - \frac{x^3}{3} + ...$$

Solution:

$$f(x) = e^{x \cos x}$$

$$f'(x) = e^{x \cos x}(-x \sin x + \cos x)$$

$$f''(x) = e^{x \cos x}[-x \cos x - \sin x - \sin x] + [-x \sin x + \cos x].\, e^{x \cos x}[-x \sin x + \cos x]$$

$$= e^{x \cos x}[-2 \sin x - x \cos x] + (\cos x - x \sin x)^2.\, e^{x \cos x}$$

$$f'''(x) = e^{x \cos x}[-2 \cos x + x \sin x - \cos x] + 2[\cos x - x \sin x] (-\sin x - x \cos x - \sin x).\, e^{x \cos x}$$

$f(0) = 1$, $\quad f'(0) = 1$, $\quad f''(0) = 1$, $\quad f'''(0) = -2$,

Acc. to Maclaurin's series,

$$f(x) = 1 + x + \frac{x^2}{2!} - \frac{x^3}{3!}(-2) + ...$$

$$= 1 + x + \frac{x^2}{2} - \frac{x^3}{3} + ...\quad \text{Proved}$$

Another way

$$e^{x \cos x} = 1 + x + \frac{x^2}{2} - \frac{x^3}{3} + ...$$

using expansion of e^x,

$$e^x = 1 + x + \frac{x^2}{2!} + \frac{x^3}{3!} + ...$$

$$e^{x \cos x} = 1 + x \cos x + \frac{(x \cos x)^2}{2!} + \frac{(x \cos x)^3}{3!} + ...$$

$$= 1 + x\left(1 - \frac{x^2}{2!} + \frac{x^4}{4!} + \ldots\ldots\ldots\right) +$$

$$\left(\frac{x^2}{2!}\right)\left(1 - \frac{x^2}{2!} + \frac{x^4}{4!} \ldots\ldots\ldots\right)^2 +$$

$$\frac{x^3}{3!}\left(1 - \frac{x^2}{2!} + \frac{x^4}{4!} \ldots\ldots\right)^3 + \ldots\ldots\ldots\ldots\ldots$$

$$= 1 + \left(x - \frac{x^3}{2!} + \frac{x^5}{4!} \ldots\ldots\ldots\right) + \frac{x^2}{2}\left(1 + \frac{x^4}{4} \ldots\ldots\ldots\right) +$$

$$\frac{x^3}{3!}\left(1 + \frac{x^6}{8} \ldots\ldots\ldots\right)$$

$$= 1 + x - \frac{x^3}{2!} + \frac{x^5}{4!} \ldots\ldots\ldots + \frac{x^2}{2} + \frac{x^6}{8} - \frac{x^4}{2} \ldots\ldots\ldots + \frac{x^3}{6} \ldots\ldots\ldots$$

$$= 1 + x + \frac{x^2}{2} - \frac{x^3}{3} + \frac{11}{24}x^4 + \ldots\ldots\ldots \text{ Proved}$$

(iii) Prove that $\frac{x}{\sin x} = 1 + \frac{x^2}{6} + \frac{7}{360}x^4 + \ldots\ldots$

Solution: $f(x) = \frac{x}{\sin x}$

$$= \frac{x}{\left(x - \frac{x^3}{3!} + \frac{x^5}{5!} - \frac{x^7}{7!} \ldots\ldots\right)}$$

$$= \frac{x}{x\left[1 - \frac{x^2}{3!} + \frac{x^4}{5!} - \frac{x^6}{7!} \ldots\ldots\right]}$$

$$= \left[1 - \frac{x^2}{3!} + \frac{x^4}{5!} - \frac{x^6}{7!} \ldots\ldots\right]^{-1}$$

Using Binomial,

$$(1+x)^n = 1 + nx + \frac{n(n-1)}{2!}x^2 + \frac{n(n-1)(n-2)}{3!}x^3 + \ldots\ldots$$

$$= (1+z)^{-1} \quad , \quad z = \frac{-x^2}{3!} + \frac{x^4}{5!} - \frac{x^6}{7!} \ldots\ldots$$

$$= 1 + (-1)\,z + \frac{(-1)(-1-1)}{2!}z^2 + \ldots\ldots$$

$$= 1 + (-1)\left(\frac{-x^2}{3!} + \frac{x^4}{5!} - \frac{x^6}{7!} \ldots\ldots\right) + \left(\frac{-x^2}{3!} + \frac{x^4}{5!} - \frac{x^6}{7!} \ldots\ldots\right)^2 + \ldots\ldots$$

$$= 1 + \frac{x^2}{6} - \frac{x^4}{120} + \frac{x^6}{7!} + \frac{x^4}{36} + \ldots\ldots$$

$$= 1 + \frac{x^2}{6} + \frac{x^4}{36} - \frac{x^4}{120} + \ldots\ldots$$

$$= 1 + \frac{x^2}{6} + \frac{7x^4}{360} + \ldots\ldots \text{ Proved}$$

(iv) Prove that

$$\frac{e^x}{\cos x} = 1 + x + \frac{2x^2}{2!} + \frac{4x^3}{3!} + \ldots\ldots$$

Solution: ***I- Method***

$$e^x \sec x = \frac{e^x}{\cos x}$$

$f(x) = e^x \sec x$

$f'(x) = e^x . \sec x \tan x + \sec x . e^x$

$= e^x \sec x \;[1 + \tan x] = f(x) . (1 + \tan x)$

$f''(x) = f(x)\,(\sec^2 x) + (1 + \tan x) . f'(x)$

$= \sec^2 x . e^x . \sec x + (1 + \tan x) . e^x \sec x\,(1 + \tan x)$

$= e^x . \sec^3 x + e^x \sec x\,(1 + \tan x)^2$

$= e^x \sec x\,[\sec^2 x + (1 + \tan x)^2]$

$f'''(x) = e^x \sec x\,[2 \sec x . \sec x \tan x + 2(1 + \tan x) . \sec^2 x] +$

$[\sec^2 x + (1 + \tan x)^2] . e^x \sec x \tan x + \sec x. e^x$

....

$f(0) = 1$, $f'(0) = 2$

$f(0) = 1$, $f''(0) = 4$

Using Maclaurin's series

$$f(x) = 1 + x + \frac{x^2}{2!}.2 + \frac{x^3}{3!}4 + \ldots\ldots \text{Proved}$$

II-Method

$$\frac{e^x}{\cos x} = \left(1 + x + \frac{x^2}{2!} + \frac{x^3}{3!}\ldots\right).\left(1 - \frac{x^2}{2!} + \frac{x^4}{4!}\ldots\right)^{-1}$$

$$= \left(1 + x + \frac{x^2}{2!} + \frac{x^3}{3!}\ldots\ldots\right)(1+z)^{-1}$$

$$= \left(1 + x + \frac{x^2}{2!} + \frac{x^3}{3!}\ldots\right)(1 - z + z^2 \ldots)$$

$$= \left(1 + x + \frac{x^2}{2!} + \frac{x^3}{3!}\ldots\right)\left(1 + \frac{x^2}{2!} - \frac{x^4}{4!}\ldots + \frac{x^4}{4}\ldots\ldots\right)$$

$$= \left(1 + x + \frac{x^2}{2!} + \frac{x^3}{3!}\ldots\right)\left(1 + \frac{x^2}{2!}\ldots\right)$$

$$= 1 + x + \frac{x^2}{2!} + \frac{x^3}{3!} + \frac{x^2}{2!} + \frac{x^3}{2!} + \frac{x^4}{4}\ldots$$

$$= 1 + x + \frac{2x^2}{2!} + \frac{4x^3}{3!} + \ldots\ldots \text{Proved}$$

(v) Prove that

$$\log \sec x = \frac{x^2}{2} + \frac{x^4}{12} + \frac{x^6}{45} + \ldots$$

Solution: $f(x) = \log \sec x$

$$f'(x) = \frac{1}{\sec x} \sec x.\tan x = \tan x$$

$f'(x) = \sec^2 x$

$f'''(x) = 2\sec^2 x \,.\, \tan x$

$f^{iv}(x) = 4\sec^2 x \tan^2 x + 2\sec^4 x$

$= 2\sec^2 x + 6\tan^2 x \sec^2 x$

$f^{v}(x) = 4\sec^2 x \tan x + 12\tan x \sec^4 x + 6\tan^3 x\,(2\sec^2 x)$

$= 4\sec^2 x \tan x + 12\tan x \sec^4 x + 12\tan^3 x \sec^2 x$

$f^{vi}(x) = 8\sec^2 x \tan^2 x + 4\sec^4 x + 12\sec^6 x + 48\tan^2 x.\sec^4 x$

$+ 24\tan^2 x \sec^4 x + 24\tan^4 x \sec^2 x$

Now,

$f(0) = \log\sec 0 = 0$

$f'(0) = \tan 0 = 0$

$f''(0) = \sec^2 0 = 1$

$f'''(0) = 0$

$f^{iv}(0) = 2$

$f^{v}(0) = 16$

Using Maclaurin's series

$$f(x) = f(0) + x f'(0) + \frac{x^2}{2!} f''(0) + \frac{x^3}{3!} f'''(0) + ...$$

$$\log\sec x = \frac{x^2}{2!} + (2)\frac{x^4}{4!} + \frac{x^6}{6!}(16) + ...$$

$$= \frac{x^2}{2} + \frac{x^4}{12} + \frac{x^6}{45} + ... \text{ Proved}$$

(vi) Prove that

$$\cos^2 x = 1 - x^2 + \frac{x^4}{3} ...$$

Solution:

$f(x) = \cos^2 x$,	$f(0) = \cos^2 0 = 1$
$f'(x) = -2\cos x \sin x = -\sin 2x$	$f'(0) = 2\sin 0 = 0$
$f''(x) = -2\cos 2x$	$f''(0) = -2\cos 0 = -2$
$f'''(x) = 4\sin 2x$	$f'''(0) = 4\sin 0 = 0$
$f^{iv}(x) = 8\cos 2x$	$f^{iv}(0) = 8\cos 0 = 8$
$f^{v}(x) = -16\sin 2x$	$f^{v}(0) = -16\sin 0 = 0$

Using Maclaurin's series, we get

$$\cos^2 x = 1 - \frac{2x^2}{2!} + \frac{8x^4}{4!} + ...$$

$$= 1 - x^2 + \frac{x^4}{3} \text{.........} \text{ Proved}$$

(vii) Prove that

$$\tan^{-1} x = x - \frac{x^3}{3} + \frac{x^5}{5} - \frac{x^7}{7} + \ldots..$$

Solution: $f(x) = \tan^{-1}x$

$$f'(x) = \frac{1}{1+x^2}$$

$$f''(x) = \frac{-1(2x)}{\left(1+x^2\right)^2} = \frac{-2x}{\left(1+x^2\right)^2}$$

$$f'''(x) = \frac{\left(1+x^2\right)^2(-2) + 2x.\,2\left(1+x^2\right)(2x)}{\left(1+x^2\right)^4}$$

$$= \frac{(-2)\left(1+x^2\right)\left[\left(1+x^2\right) - 4x^2\right]}{\left(1+x^2\right)^4}$$

$$= \frac{(-2)\left(1-3x^2\right)}{\left(1+x^2\right)^3}$$

$$f^{iv}(x) = \frac{-2\left[(-6x)\left(1+x^2\right)^3 - \left(1-3x^2\right).\,3\left(1+x^2\right)^2.\,2x\right]}{\left(1+x^2\right)^6}$$

$$= \frac{-2\left(1+x^2\right)^2\left[-6x\left(1+x^2\right) - \left(1-3x^2\right)(6x)\right]}{\left(1+x^2\right)^6}$$

$$= \frac{-12x\left(1+x^2+1-3x^2\right)}{\left(1+x^2\right)^4}$$

$$= \frac{-12x\left(2-2x^2\right)}{\left(1+x^2\right)^4}$$

$$= \frac{24\left(x^3-x\right)}{\left(1+x^2\right)^4}$$

$$f^{v}(x) = \frac{24\left[\left(3x^2-1\right)\left(1+x^2\right)^4-\left(x^3-x\right).4\left(1+x^2\right)^3.2x\right]}{\left(1+x^2\right)^8}$$

$$= \frac{24\left(1+x^2\right)^3\left[\left(3x^2-1\right)\left(1+x^2\right)-8x\left(x^3-x\right)\right]}{\left(1+x^2\right)^8}$$

$$= \frac{24\left[3x^2+3x^4-1-x^2-8x^4+8x^2\right]}{\left(1+x^2\right)^5}$$

$$= \frac{24\left[10x^2-5x^4-1\right]}{\left(1+x^2\right)^5}$$

$f(x) = \tan^{-1}x$

$f(0) = \tan^{-1} 0 = 0$

$f'(0) = 1$

$f'(0) = 0$

$f''(0) = -2$

$f^{iv}(0) = 0$

$f^{iv}(0) = 24$

Using Maclaurin's series,

$$\tan^{-1}x = x - \frac{2}{3!}x^3 + \frac{24}{5!}x^5 \;....$$

$$= x - \frac{x^3}{3} + \frac{x^5}{5} - \frac{x^7}{7} \text{...... Proved}$$

(viii) Prove that $\tan^{-1}(1+x) = \frac{\pi}{4} + \frac{x}{2} - \frac{x^2}{4} + \frac{x^3}{12}$ .....

Solution: $f(x) = \tan^{-1}(1 + x)$

$$f'(x) = \frac{1}{1+(1+x)^2}$$

$$f''(x) = \frac{-1.2\,(1+x)}{\left[1+(1+x)^2\right]^2}$$

$$f'''(x) = \frac{-2\left[\left\{1+(1+x)^2\right\}^2(1) - (1+x).\,2\left(1+(1+x)^2\right).\,2(1+x)\right]}{\left[(1+x)^2+1\right]^4}$$

⋮

$f(0) = \tan^{-1} 1 = \pi/4$

$f(0) = 1/2$

$f'(0) = -2/4 = -1/2$

$f''(0) = 1/2$

⋮

Using Maclaurin's series,

$$\tan^{-1}(1+x) = \pi/4 + x\left(\frac{1}{2}\right) + \frac{x^2}{2!}\left(-\frac{1}{2}\right) + \frac{x^3}{3!}\left(\frac{1}{2}\right) + \ldots$$

$$= \frac{\pi}{4} + \frac{x}{2} - \frac{x^2}{4} + \frac{x^3}{12} \text{..... Proved}$$

Example 3 Prove that

(i) $$\log(1+\sin x) = x - \frac{x^2}{2} + \frac{x^3}{6} - \frac{x^4}{12} + \ldots..$$

Solution: Let $f(x) = \log(1 + \sin x)$

$$f'(x) = \frac{1}{1+\sin x}.\cos x = \frac{\cos x}{1+\sin x}$$

$$f''(x) = \frac{(1+\sin x)(-\sin x) - \cos x\,(\cos x)}{(1+\sin x)^2}$$

$$= \frac{-\sin x - \sin^2 x - \cos^2 x}{(1+\sin x)^2}$$

$$= \frac{-1}{1+\sin x}$$

$$f'''(x) = \frac{\cos x}{(1+\sin x)^2}$$

$$f^{iv}(x) = \frac{-\sin x\,(1+\sin x)^2 - \cos x.\,2(1+\sin x).\cos x}{(1+\sin x)^4}$$

$$= \frac{(1+\sin x)\left[-\sin x - \sin^2 x - 2\cos^2 x\right]}{(1+\sin x)^4}$$

$$= \frac{(1+\sin x)\left(-\sin x - 1 - \cos^2 x\right)}{(1+\sin x)^4}$$

$f(0) = 0$, $f'''(0) = 1$

$f'(0) = 1$ $f^{iv}(0) = -2$

$f''(0) = -1$

After putting all values in the Maclaurin's series, we get

$$\log(1 + \sin x) = x - \frac{x^2}{2} + \frac{x^3}{6} - \frac{x^4}{12} + \ldots\ldots \quad \text{Proved.}$$

(ii) Prove that $\log\left(1+e^x\right) = \log 2 + \frac{x}{2} + \frac{x^2}{8} - \frac{x^4}{192} + \ldots..$

Solution: $f(x) = \log(1 + e^x)$

$$f'(x) = \frac{1}{1+e^x}.\,e^x = \frac{e^x}{1+e^x}$$

$$f''(x) = \frac{e^x\left(1+e^x\right) - e^x\left(e^x\right)}{\left(1+e^x\right)^2}$$

$$= \frac{e^x}{\left(1+e^x\right)^2}$$

$$f'''(x) = \frac{\left(1+e^x\right)^2 e^x - 2e^x\left(1+e^x\right).e^x}{\left(1+e^x\right)^4}$$

$$= \frac{\left(1+e^x\right)\left[e^x\left(1+e^x\right) - 2e^{2x}\right]}{\left(1+e^x\right)^4}$$

$$= \frac{e^x + e^{2x} - 2e^{2x}}{\left(1+e^x\right)^3} = \frac{-e^{2x} + e^x}{\left(1+e^x\right)^3}$$

$$f^{iv}(x) = \frac{\left(1+e^x\right)^3\left(-2e^{2x} + e^x\right) - \left(-e^{2x} + e^x\right).3\left(1+e^x\right)^2\left(e^x\right)}{\left(1+e^x\right)^6}$$

$$= \frac{\left(1+e^x\right)^2\left[\left(1+e^x\right)\left(-2e^{2x} + e^x\right) - \left(-e^{2x} + e^x\right).3e^x\right]}{\left(1+e^x\right)^6}$$

$$= \frac{\left(1+e^x\right)\left(-2e^{2x} + e^x\right) - 3e^x\left(-e^{2x} + e^x\right)}{\left(1+e^x\right)^4}$$

Thus, $f(0) = \log 2$

$f(0) = 1/2$

$f'(0) = 1/4$

$f''(0) = 0$

$f^{iv}(0) = -1/8$

Using Maclaurin's series,

$$\log\left(1+e^x\right) = \log 2 + \frac{x}{2} + \frac{x^2}{8} - \frac{x^4}{192} + ... \quad \text{Proved}$$

(iii) Prove that

$$\frac{e^x}{1+e^x} = \frac{1}{2} + \frac{x}{4} - \frac{x^3}{48} + \text{.........}$$

Solution: $$f(x) = \frac{e^x}{1+e^x}$$

$$f'(x) = \frac{(1+e^x)(e^x) - e^x(e^x)}{(1+e^x)^2} = \frac{e^x}{(1+e^x)^2}$$

$$f''(x) = \frac{(1+e^x)^2(e^x) - e^x \cdot 2(1+e^x) \cdot e^x}{(1+e^x)^4}$$

$$= \frac{(1+e^x)\left[e^x(1+e^x) - 2e^{2x}\right]}{(1+e^x)^4}$$

$$= \frac{e^x + e^{2x} - 2e^{2x}}{(1+e^x)^3} = \frac{e^x - e^{2x}}{(1+e^x)^3}$$

$$f'''(x) = \frac{(1+e^x)(-2e^{2x} + e^x) - 3e^x(e^x - e^{2x})}{(1+e^x)^4}$$

$$f(0) = \frac{e^0}{1+e^0} = \frac{1}{2}$$, $f'(0) = 1/4$ $\quad f''(0) = 0$ $\quad f'''(0) = -1/8$

Using Maclaurin's series,

$$\frac{e^x}{1+e^x} = \frac{1}{2} + \frac{x}{4} - \frac{x^3}{48} + \ldots \quad \text{Proved.}$$

Example 4 Prove that

(i) $$e^{ax} \sin bx = bx + abx^2 + \frac{b(3a^2 - b^2)}{3!} x^3 + \ldots$$

Solution: $f(x) = e^{ax} \sin bx$

$f'(x) = ae^{ax} \sin bx + e^{ax} \,.\, b \cos bx$

$f''(x) = a^2e^{ax} \sin bx + a\, e^{ax} \,.\, b \cos bx + ab\, e^{ax} \cos bx - b^2 e^{ab} \sin bx$

$= a^2 e^{ax} \sin bx + 2ab\, e^{ax} \cos bx - b^2 e^{ax} \sin bx$

$= e^{ax} (a^2 \sin bx + 2ab \cos bx - b^2 \sin bx)$

$f'''(x) = e^{ax}(a^2b\cos bx - 2ab^2\sin bx - b^3\cos bx) +$

$(a^2\sin bx + 2ab\cos bx - b^2\sin bx)(a.e^{ax})$

$f(0) = 0, \quad f'(0) = b \quad , \quad f''(0) = 2ab$

$f'''(0) = 2a^2b + a^2b - b^3 = 3a^2b - b^3$

Using Maclaurin's series,

$$e^{ax}\sin bx = bx + abx^2 + \frac{b(3a^2 - b^2)}{3!}x^3 + \ldots\ldots \qquad \text{Proved}$$

Example 5 Prove that

$$e^x \cos x = 1 + x - \frac{2x^3}{3!} - \frac{2^2.x^4}{4!} \ldots\ldots\ldots$$

Solution:

$f(x) = e^x.\cos x$

$f'(x) = e^x(-\sin x) + \cos x.\,e^x$

$f''(x) = e^x(-\cos x) + (-\sin x).\,e^x + e^x(-\sin x) + \cos x.\,e^x$

$= -2e^x\sin x$

$f'''(x) = -2[e^x.\cos x + \sin x.e^x]$

$= -2e^x[\cos x + \sin x]$

$f^{v}(x) = -2e^x(-\sin x + \cos x) + (\sin x + \cos x)(-2e^x)$

$= -2e^x[-\sin x + \cos x + \sin x + \cos x]$

$= -2e^x.\,2\cos x = -4e^x\cos x$

$f(0) = e^0.\cos 0 = 1 \quad , \quad f'(0) = e^0\cos 0 - e^0\sin 0 = 1$

$f''(0) = 2e^0.\sin 0 = 0 \quad , \quad f'''(0) = -2, \quad f^{iv}(0) = -4$

Using Maclaurin's series

$$e^x.\cos x = 1 + x - \frac{2x^3}{3!} - \frac{2^2.x^4}{4!} + \ldots.. \quad \text{Proved.}$$

Example 6 Assuming the possibility of expansion, show that

(i) $$\sin(e^x - 1) = x + \frac{x^2}{2} - \frac{5x^4}{24} + \ldots$$

Solution:

$f(x) = \sin(e^x - 1)$

$f'(x) = \cos(e^x - 1).\,e^x$

$f''(x) = e^x.\cos(e^x - 1) - e^x.\sin(e^x - 1).\,e^x$

$= e^x\cos(e^x - 1) - e^{2x}\sin(e^x - 1)$

$f'''(x) = e^x\cos(e^x - 1) - e^x\sin(e^x - 1).\,e^x - e^{2x}\cos(e^x - 1).\,e^x - \sin(e^x - 1).\,2e^{2x}$

$= e^x\cos(e^x - 1) - e^{2x}\sin(e^x - 1) - e^{3x}\cos(e^x - 1) - 2e^{2x}\sin(e^x - 1)$

$f^{iv}(x) = e^x\cos(e^x - 1) - e^x\sin(e^x - 1).\,e^x - e^{2x}\cos(e^x - 1).\,e^x - \sin(e^x - 1).\,2e^{2x} +$

$e^{3x}\sin(e^x - 1).\,e^x - \cos(e^x - 1).\,3e^{3x} - 2e^{2x}\cos(e^x - 1).\,e^x - 4e^{2x}\sin(e^x - 1)$

$$= e^x \cos(e^x - 1) - e^{2x} \sin(e^x - 1) - 6e^{2x} \sin(e^x - 1) - 3e^{2x} \cos(e^x - 1)$$
$$- 3e^{3x} \cos(e^x - 1) + e^{3x} \sin(e^x - 1)$$

$f(0) = 0,$ $f'(0) = 1$, $f''(0) = 1$

$f'''(0) = 1 - 1 = 0$, $f^{iv}(0) = 1 - 3 - 3 = -5$

Using Maclaurin's series,

$$\sin\left(e^x - 1\right) = x + \frac{x^2}{2} - \frac{5}{24}x^4 + \ldots\ldots \quad \text{Proved.}$$

Example 7 Prove that

$$\log\left(1 - x + x^2\right) = -x + \frac{x^2}{2} + \frac{2}{3}x^3 + \frac{x^4}{4} \ldots\ldots$$

Solution: $f(x) = \log(1 - x + x^2)$

$$f'(x) = \frac{1}{1 - x + x^2}(-1 + 2x) = \frac{2x - 1}{1 - x + x^2}$$

$$f''(x) = \frac{\left(1 - x + x^2\right)(2) - (2x - 1)(-1 + 2x)}{\left(1 - x + x^2\right)^2}$$

$$= \frac{\left(2 - 2x + 2x^2\right) - \left(4x^2 + 1 - 4x\right)}{\left(1 - x + x^2\right)^2}$$

$$= \frac{2 - 2x + 2x^2 - 4x^2 - 1 + 4x}{\left(1 - x + x^2\right)^2}$$

$$= \frac{2x - 2x^2 + 1}{\left(1 - x + x^2\right)^2}$$

$$f'''(x) = \frac{\left(1 - x + x^2\right)^2 (2 - 4x) - \left(2x - 2x^2 + 1\right) 2(-1 + 2x)\left(1 - x + x^2\right)}{\left(1 - x + x^2\right)^4}$$

$$= \frac{\left(1 - x + x^2\right)\left[(2 - 4x)\left(1 - x + x^2\right) - 2(2x - 1)\left(1 + 2x - 2x^2\right)\right]}{\left(1 - x + x^2\right)^4}$$

$$=\frac{2-2x+2x^2-4x+4x^2-4x^3-(4x-2)(1+2x-2x^2)}{(1-x+x^2)^3}$$

$$f(0) = \log 1 = 0 \quad , \quad f(0) = \frac{-1}{1} = -1, \quad f'(0) = 1 \quad f''(0) = 4$$

Using Maclaurin's series,

$$\log(1-x+x^2) = -x + \frac{x^2}{2} + \frac{4}{3!}x^3 + \ldots.$$

$$= -x + \frac{x^2}{2} + \frac{2}{3}x^3 + \ldots. \quad \text{Proved.}$$

Example 8 Prove that

$$\log\frac{\tan x}{x} = \frac{x^2}{3} + \frac{7}{90}x^4 + \ldots\ldots$$

Solution: Let $f(x) = \log\frac{\tan x}{x}$

$$= \log\frac{\left[x + \frac{x^3}{3} + \frac{2}{15}x^5 + \ldots\ldots\right]}{x}$$

$$= \log\left[1 + \frac{x^2}{3} + \frac{2}{15}x^4 + \ldots\right]$$

$$= \left(\frac{x^2}{3} + \frac{2}{15}x^4 + \ldots.\right) - \frac{1}{2}\left(\frac{x^2}{3} + \frac{2}{15}x^4 + \ldots.\right)^2 + \ldots\ldots..$$

$$= \frac{x^2}{3} + \frac{2}{15}x^4 - \frac{x^4}{18} + \ldots$$

$$= \frac{x^2}{3} + \frac{7}{90}x^4 + \ldots \quad \text{Proved.}$$

Example 9 Prove that

$$\log\frac{\sin x}{x} = -\left(\frac{x^2}{6} + \frac{x^4}{180} + \frac{x^6}{2835} + \ldots\ldots..\right)$$

Solution: Let $$f(x) = \log \frac{\sin x}{x}$$

$$= \log \left[\frac{x - \frac{x^3}{3!} + \frac{x^5}{5!} - \frac{x^7}{7!} + \dots\dots}{x} \right]$$

$$= \log \left[1 - \frac{x^2}{3!} + \frac{x^4}{5!} - \frac{x^6}{7!} + \dots\dots \right]$$

$$= \left(\frac{-x^2}{3!} + \frac{x^4}{5!} - \frac{x^6}{7!} \dots \right) - \frac{1}{2}\left(\frac{-x^2}{3!} + \frac{x^4}{5!} - \frac{x^6}{7!} \dots \right)^2 + \frac{1}{3}\left(\frac{-x^2}{3!} + \frac{x^4}{5!} - \frac{x^6}{7!} \dots \right)^3 + \dots\dots$$

$$= \left(\frac{-x^2}{3!} + \frac{x^4}{5!} - \frac{x^6}{7!} \dots \right) - \frac{1}{2}\left(\frac{x^4}{36} - \frac{2x^6}{720} + \dots\dots \right) + \frac{1}{3}\left(\frac{-x^6}{216} \right) + \dots\dots$$

$$= -\left[\frac{x^2}{6} + \frac{x^4}{180} + \frac{x^6}{2835} + \dots\dots \right] \quad \text{Proved}$$

Example 10 Prove that

$$\sin x \cos hx = x + \frac{x^3}{3} - \frac{x^5}{30} + \dots\dots$$

Solution: Let $f(x) = \sin x \cos hx$

Using $\sin x = x - \frac{x^3}{3!} + \frac{x^5}{5!} - \frac{x^7}{7!} + \dots\dots$

$$\cos hx = 1 + \frac{x^2}{2!} + \frac{x^4}{4!} + \frac{x^6}{6!} + \dots\dots$$

$$f(x) = \left(x - \frac{x^3}{3!} + \frac{x^5}{5!} - \frac{x^7}{7!} \dots \right)\left(1 + \frac{x^2}{2!} + \frac{x^4}{4!} + \frac{x^6}{6!} \dots \right)$$

$$= x + \frac{x^3}{2!} + \frac{x^5}{4!} + \frac{x^7}{6!} \dots - \frac{x^3}{3!} - \frac{x^5}{3!2!} - \frac{x^7}{3!4!} \dots$$

$$- \frac{x^3 x^6}{3!6!} \dots\dots + \frac{x^5}{5!} + \frac{x^7}{2!5!} + \frac{x^9}{4!5!} +$$

$$\frac{x^{11}}{5!\,6!}\;.....\;\frac{x^7}{7!}-\frac{x^9}{2!\,7!}-\frac{x^{11}}{4!\,7!}-\frac{x^{13}}{6!\,7!}\;....$$

$$= x + x^3\left[\frac{1}{2}-\frac{1}{6}\right] + x^5\left[\frac{1}{24}-\frac{1}{12}+\frac{1}{120}\right] + ...$$

$$= x + \frac{x^3}{3} - \frac{x^5}{30} + ... \quad \text{Proved.}$$

Example 11 Prove that

(i) $\sin^{-1}\left(\frac{2x}{1+x^2}\right) = 2\left(x - \frac{x^3}{3} + \frac{x^5}{5} - \frac{x^7}{7} + ...\right)$

Solution: Let $f(x) = \sin^{-1}\left(\frac{2x}{1+x^2}\right)$ put $x = \tan\theta$ we have

$$= \sin^{-1}\left(\frac{2\tan\theta}{1+\tan^2\theta}\right)$$

$$= \sin^{-1}(\sin 2\theta) = 2\theta$$

$$f(x) = 2\tan^{-1}x$$

$$f'(x) = \frac{2}{1+x^2}$$

$$f''(x) = \frac{-2(2x)}{(1+x^2)^2} = \frac{-4x}{(1+x^2)^2}$$

$$f'''(x) = \frac{(1+x^2)^2(-4) + 4x.2(1+x^2)(2x)}{(1+x^2)^4}$$

$$= \frac{(1+x^2)\left[-4(1+x^2)+16x^2\right]}{(1+x^2)^4}$$

$$= \frac{16x^2 - 4 - 4x^2}{(1+x^2)^3} = \frac{12x^2 - 4}{(1+x^2)^3}$$

$$f^{iv}(x) = \frac{(1+x^2)^3 (24x) - (12x^2 - 4).3(1+x^2)^2 (2x)}{(1+x^2)^6}$$

$$= \frac{(1+x^2)^2 \left[(24x)(1+x^2) - 6x(12x^2 - 4)\right]}{(1+x^2)^6}$$

$$= \frac{24x + 24x^3 - 72x^3 + 24x}{(1+x^2)^4}$$

$$= \frac{48x - 48x^3}{(1+x^2)^6} = \frac{48x(1-x^2)}{(1+x^2)^6}$$

$f(0) = 2\tan^{-1} 0 = 0$

$f'(0) = 2$, $f''(0) = 0$

$f'''(0) = -4$, $f^{iv}(0) = 0$

Using Maclaurin's series,

$$\sin^{-1}\left(\frac{2x}{1+x^2}\right) = 2x - \frac{4x^3}{3!} + \ldots$$

$$= 2\left(x - \frac{x^3}{3} + \ldots\right)$$

(ii) Prove that

$$\tan^{-1}\frac{\sqrt{1+x^2}-1}{x} = \frac{1}{2}\left(x - \frac{x^3}{3} + \frac{x^5}{5} - \frac{x^7}{7} + \ldots\right)$$

Solution: Let $f(x) = \tan^{-1}\frac{\sqrt{1+x^2}-1}{x}$

put $x = \tan\theta$

$$= \tan^{-1}\frac{\sqrt{1+\tan^2\theta}-1}{\tan\theta}$$

$$= \frac{\theta}{2} = \frac{1}{2}\tan^{-1}x$$

using $\tan^{-1}x = x - \frac{x^3}{3} + \frac{x^5}{5} \ldots\ldots$

we get $f(x) = \frac{1}{2}\left[x - \frac{x^3}{3} + \frac{x^5}{5} \ldots\ldots\right]$ Proved.

(iii) Prove that

$$\tan^{-1}\frac{x}{\sqrt{1-x^2}} = x + \frac{1^2}{3!}x^3 + \frac{1^2\,3^2}{5!}x^5 + \ldots$$

Solution: Let $f(x) = \tan^{-1}\frac{x}{\sqrt{1-x^2}}$

put $x = \sin\theta$

$$= \tan^{-1}\left(\frac{\sin\theta}{\sqrt{1-\sin^2\theta}}\right) = \tan^{-1}\left(\frac{\sin\theta}{\cos\theta}\right) = \tan^{-1}\tan\theta$$

$$= \theta$$

$$= \sin^{-1}x$$

$$f'(x) = \frac{1}{\sqrt{1-x^2}} = \left(1-x^2\right)^{-1/2}$$

$$f''(x) = 1 + \frac{1}{2}x^2 + \frac{(1/2)(3/2)}{2!}x^4 + \frac{(1/2)(3/2)(5/2)}{3!}x^6 + \ldots.$$

$$f''(x) = x + \frac{3}{2}x^3 + \frac{15}{8}x^5 + \ldots$$

$$f'''(x) = 1 + \frac{9}{2}x^2 + \frac{75}{8}x^4 + \ldots$$

$$f^{iv}(x) = 9x + \frac{75}{2}x^3 + \ldots.$$

$$f^{v}(x) = 9 + \frac{75 \times 3}{2}x^2 + \ldots..$$

$$f^{vi}(x) = 225x + \ldots\ldots$$

$f^{vii}(x) = 225 + \ldots\ldots..$

$f(0) = 0,\quad f'(0) = 1,\quad f''(0) = 0,\quad f'''(0) = 1$

$f^{iv}(0) = 0,\quad f^{v}(0) = 9,\quad f^{vi}(0) = 0,\quad f^{vii}(0) = 225$

Using Maclaurin's series,

$$\tan^{-1}\frac{x}{\sqrt{1-x^2}} = x + \frac{1}{3!}x^3 + \frac{9}{5!}x^5 + \ldots.$$

$$= x + \frac{1^2}{3!}x^3 + \frac{1^2 \cdot 3^2}{5!}x^5 + \ldots \text{ Proved.}$$

(iv) Prove that $\sin^{-1}\left(3x - 4x^3\right) = 3\left(x + \frac{x^3}{2} + \frac{3x^5}{40} + \ldots.\right)$

Solution:

Let $f(x) = \sin^{-1}(3x - 4x^3)$

put $x = \sin\theta$

$= \sin^{-1}(3\sin\theta - 4\sin 3\theta)$

$= \sin^{-1}(\sin 3\theta) = 3\theta = 3\sin^{-1}x$

$$f'(x) = \frac{3}{\sqrt{1-x^2}} = 3\left(1-x^2\right)^{-1/2}$$

$$= 3\left[1 + \frac{1}{2}x^2 + \frac{(1/2)(3/2)}{2!}x^4 + \frac{1/2 \times 3/2 \times 5/2}{3!}x^6 + \ldots\right]$$

$$f'(x) = 3\left[1 + \frac{1}{2}x^2 + \frac{3}{8}x^4 + \frac{5}{16}x^6 + \ldots\right]$$

$$f''(x) = 3\left[x + \frac{3}{2}x^3 + \frac{15}{8}x^5 + \ldots.\right]$$

$$f'''(x) = 3\left[1 + \frac{9}{2}x^2 + \frac{75}{8}x^4 + \ldots.\right]$$

$$f^{iv}(x) = 3\left[9x + \frac{75}{2}x^3 + \ldots.\right]$$

$$f^{v}(x) = 3\left[9 + \frac{225}{2}x^2 + \ldots.\right]$$

$$f^{vi}(x) = 3[225\,x...]$$

$$f^{vii}(x) = 3 \times 225$$

$f(0) = 0,$ $f'(0) = 3,$ $f''(0) = 0$ $f'''(0) = 3,$

$f^{iv}(0) = 0,$ $f^{v}(0) = 3(9),$ $f^{vi}(0) = 0,$ $f^{vii}(0) = 3 \times 225$

Using Maclaurin's series, we get

$$\sin^{-1}\left(3x - 4x^3\right) = 3\left[x + \frac{x^3}{6} + \frac{3}{40}x^5 + \ldots\ldots\right].$$

3.8 SUCCESSIVE DIFFERENTIATION

Let y = f(x) be a function of x, which is differentiable. Then $\frac{dy}{dx}$ is known as first differential coefficient of y w.r.t. x which may be again differentiated. The term $\frac{d}{dx}\left(\frac{dy}{dx}\right)$ is called the second differential coefficient of y w.r.t. x and is denoted by $\frac{d^2y}{dx^2}$ and so on. In general, the nth differential coefficient of y w.r.t. x is denoted by $\frac{d^ny}{dx^n}$. Usually the differential operator $\frac{d}{dx}$ is represented by the symbol D such that $\frac{dy}{dx} = Dy$, similarly $\frac{d^ny}{dx^n} = D^ny$.

Note: If y = f(x), then y_r is the rth differentiation of y, the nth differential coefficient of y_r is (n+r)th differential coefficient of y i.e. $D^{n+r}y = y_{n+r}$ and $D^ny_2 = D^{n+2}y = y_{n+2}$

3.9 LEIBNITZ'S RULE

This rule help us to find nth differential coefficient of the product of two function i.e., if u and v are any two function of 'x' such that all their desired differential coefficient exist, then the nth differential coefficient of their product is given by

$$D^n(uv) = D^nu(v) + {}^nC_1\,D^{n-1}u\,(Dv) + {}^nC_2D^{n-2}u\left(D^2v\right) + \ldots\ldots + {}^nC_r\,D^{n-r}\,u\left(D^rv\right) + \ldots + u\left(D^nv\right)$$

Note: (i) Differentiation of $(1-x^2)\,y_2$ for n times by Leibnitz's rule is given by

$$D^n\,[y_2\,(1-x^2)] = D^ny_2\,(1-x^2) + {}^nC_1D^{n-1}\,y_2D(1-x^2) + {}^nC_2\,D^{n-2}\,y_2\,D^2\,(1-x^2)$$

$$= y_{n+2}\,(1-x^2) + {}^nC_1\,y_{n+1}\,(-2x) + {}^nC_2\,y_n(-2)$$

(ii) Differentiation of xy_1 for n times by Leibnitz's rule is given by

$$D^n[y_1 x] = D^n y_1(x) + {}^nC_1 Y_1(Dx)$$

$$= y_{n+1}(x) + {}^nC_1\, y_n(1)$$

Illustrative Examples

Example 1 By forming a differential eqn., prove that

(i) $\left(\sin^{-1}x\right)^2 = 2.\dfrac{x^2}{2!} + 2.2^2.\dfrac{x^4}{4!} + 2.2^2.4^2.\dfrac{x^6}{6!} + ...$

Solution: Let $f(x) = \left(\sin^{-1}x\right)^2$

$$y_1 = f'(x) = 2.\frac{\sin^{-1}x}{\sqrt{1-x^2}}$$

$$\sqrt{1-x^2}\, y_1 = 2\sin^{-1}x$$

$y(0) = 0$

$y_1(0) = 0$

$y_2(0) = 2$

squaring both sides

$(1-x^2)\, y_1^{\,2} = 4(\sin^{-1}x)^2$

$(1-x^2)\, y_1^{\,2} = 4y$

$y_1^{\,2}(-2x) + (1-x^2).\, 2y_1y_2 = 4y_1$

$-2xy_1 + 2y_2(1-x^2) = 4$

$(1-x^2)\, y_2 - xy_1 - 2 = 0$

differentiating 'n' times

$$y_{n+2}(1-x^2) + ny_{n+1}(-2x) + \frac{n(n-1)}{2} y_n(-2) - [y_{n+1}.\, x + ny_n] = 0$$

at $x = 0$

$y_{n+2}(0) - n(n-1)\, y_n(0) - n\, y_n(0) = 0$

$y_{n+2}(0) - n^2 y_n(0) = 0$

$y_{n+2}(0) = n^2 y_n(0)$

at $n = 1$

$y_3(0) = y_1(0) = 0$

at $n = 2$

$y_4(0) = 2^2.2$

at $n = 3$

$y_5(0) = 0$

at n = 4

$y_6(n) = 2.2^2.4^2$

Using Maclaurin's series,

$$\left(\sin^{-1} x\right)^2 = 2.\frac{x^2}{2!} + 2.2^2.\frac{x^4}{4!} + 2.2^2.4^2\,\frac{x^6}{6!} + \ldots \quad \text{Proved.}$$

Example 2 Prove that $\log\left(x + \sqrt{1+x^2}\right) = x - \frac{1}{2}.\frac{x^3}{3} + \frac{1.3}{2.4}.\frac{x^5}{5} + \ldots.$

Solution: Let $y = \log\left(x + \sqrt{1+x^2}\right)$

$$y_1 = \frac{1}{x + \sqrt{1+x^2}}\left(1 + \frac{x}{\sqrt{1+x^2}}\right)$$

$$y_1 = \frac{1}{\sqrt{x^2+1}}$$

squaring and cross multiplying both sides,

$$\left\{\begin{aligned} & y_1^2\left(x^2+1\right) = 1 \\ & 2y_1y_2\left(x^2+1\right) + y_1^2\, 2x = 0 \\ & y_2\left(x^2+1\right) + y_1\, x = 0 \end{aligned}\right. \quad \longleftrightarrow \quad \left.\begin{aligned} & y(0) = 0 \\ & y_1(0) = 1 \\ & y_2(0) = 0 \end{aligned}\right\} \text{ on differentiation}$$

Differentiating n times

$$y_{n+2}\left(x^2+1\right) + ny_{n+1}2x + \frac{n(n-1)}{2!}y_n 2 + y_{n+1}x + xy_n = 0$$

$$y_{n+2}\left(x^2+1\right) + (2n+1)\,xy_{n+1} + x^2y_n = 0$$

$$y_{n+2}(0) + x^2y_n(0) = 0$$

put n = 1

$$y_3(0) + y_1(0) = 0,\; y_3(0) = -1$$

put n = 2

$$y_4(0) + 4y_2(0) = 0,\; y_4(0) = 0$$

put n = 3

$$y_5(0) + 9y_3(0) = y_5(0) = -9$$

Apply Maclaurin's theorem, we get

$$y(x) = y(0) + xy'(0) + \frac{x^2}{2!}y''(0) + \frac{x^3}{3!}y'''(0) + ...$$

$$= 0 + x + \frac{x^2}{2!}(0) + \frac{x^3}{3!}(-1) + ...$$

$$= x - \frac{x^3}{2.3} + \frac{1.3}{2.4}\frac{x^5}{5}... \quad \text{Proved.}$$

Example 3 Prove that

$$\cos(m\sin^{-1}x) = 1 - \frac{m^2}{2!}x^2 - \frac{m^2(2^2 - m^2)}{4!}x^4 - \frac{m^2(2^2 - m^2)(4^2 - m^2)}{6!}x^6 + ...$$

Solution: Let $y = \cos(m \sin^{-1}x)$

$$y_1 = \frac{-\sin(m\sin^{-1}x).m}{\sqrt{1-x^2}} \qquad y_1(0) = 0$$

squaring and cross-multiplying

$$(1-x^2)\, y_1^2 = m^2 \sin^2(m \sin^{-1}x)$$

$$(1-x^2)\, y_1^2 = m^2(1 - y^2)$$

on differentiating

$$(1 - x^2).\, 2y_1y_2 + y_1^2(-2x) = m^2(-2y\, y_1)$$

$$(1 - x^2)y_2 - xy_1 + m^2y = 0,\ y_2(0) = -m^2$$

differentiating n times using Leibnitz's theorem

$$y_{n+2}(1 - x^2) + n\, y_{n+1}(-2x) + \frac{n(n-1)}{2}y_n(-2) -$$

$$[y_{n+1}.x + ny_n] + m^2y_n = 0$$

$$(1 - x^2)\, y_{n+2} - 2nxy_{n+1} - n(n-1)y_n - xy_{n+1} - xy_n + m^2y_n = 0$$

$$(1-x^2)y_{n+2} - (2nx + x)y_{n+1} - n^2y_n + ny_n - ny_n + m^2y_n = 0$$

$$(1-x^2)y_{n+2} - (2n+1)\, xy_{n+1} - (n^2 - m^2)\, y_{n-2} = 0$$

put $x = 0$

$$y_{n+2} = (n^2 - m^2)\, y_n$$

at $n = 1$, $y_3 = (1^2 - m^2)y_1$

at $n = 2$, $y_4 = (2^2 - m^2)y_2$

at $n = 3$, $y_5 = (3^2 - m^2)y_3$

After putting all these values, we will get the required result.

Example 4 Prove that $$e^{m\cos^{-1}x} = e^{\frac{m\pi}{2}}\left[1 - mx + \frac{m^2x^2}{2!} - \frac{m\left(1^2+m^2\right)}{3!}x^3 + \frac{m^2\left(2^2+m^2\right)x^4}{4!} - m^2\left(1^2+m^2\right)\left(3^2+m^2\right)\frac{x^5}{5!} + \ldots\right.$$

Solution:

Let $y = e^{m\cos^{-1}x}$

$$y_1 = \frac{-m}{\sqrt{1-x^2}}\,y$$

$$y_1\sqrt{1-x^2} = -my$$

Squaring both sides

$$y_1^2\left(1-x^2\right) = m^2y^2$$

Differentiating w.r.t. x

$$2y_1y_2\left(1-x^2\right) + y_1^2(-2x) = m^2.2y\,y_1$$

$$2\,y_2(1-x^2) - 2xy_1 = 2m^2y$$

$y_2\,(1-x^2) - xy_1 = m^2y$, Differentiating n times

$$y_{n+2}(1-x^2) + {}^nC_1\,y_{n+1}\,(-2x) + {}^nC_2\,y_n(-2) - [y_{n+1}\,x + {}^nC_1\,y_n] = m^2y_n$$

$$(1-x^2)\,y_{n+2} - 2nx\,y_{n+1} - n(n-1)\,y_n - ny_n - y_{n+1}x = m^2y_n$$

put $x = 0$

$$y_{n+2} - n^2y_n - m^2y_n = 0$$

$$y_{n+2} = (n^2 + m^2)y_n$$

put $n = 1$, $y_3(0) = (1^2 + m^2)\,(me^{m\pi/2})$

$n=2$, $y_4(0) = (2^2 + m^2)\,m^2\,e^{m\pi/2}$

And putting y_1, y_2, y_3 in the Maclaurin's series, we get the required result. Proved

Exercise 3.2

1. Expand log (1 + x) in power of x
2. Expand $\sin^{-1}x$ up to four term.
3. Using Maclaurin's series, show that

$$e^x \sec x = 1 + x + \frac{2x^2}{2!} + \frac{4x^3}{3!} + ...$$

4. Using Maclaurin's series, prove that

$$\log(\sec x + \tan x) = x + \frac{x^3}{6} + \frac{x^5}{24} + ...$$

5. Prove that

$$\tan^{-1}\frac{2x^2}{1-x^2} = 2\left(x - \frac{x^3}{3} + \frac{x^5}{5} .....\right)$$

6. Using Maclaurin's series, find the expansion of the following
   (a) e^x
   (b) $\tan x$
   (c) $\cos^{-1}x$
7. If $y = (\sin^{-1}x)^2$, prove that

(i) $\left(1-x^2\right)\frac{d^2y}{dx^2} - x\frac{dy}{dx} - 2 = 0$

(ii) $\left(1-x^2\right)y_{n+2} - (2n+1)xy_{n+1} - n^2y_n = 0$

8. By forming a differential equation, prove that

$$\sin\left(m \sin^{-1}x\right) = mx + \frac{m\left(1^2 - m^2\right)}{3!}x^3 + \frac{m\left(1^2 - m^2\right)\left(3^2 - m^2\right)}{5!}x^5 + ...$$

9. By forming a differential equation, expand $e^{m\cos^{-1}x}$ *upto* x^5 *terms.*
10. By forming a second order differential equation, prove that

$$e^{m \sin^{-1}x} = 1 + mx + \frac{m^2x^2}{2!} + \frac{m\left(1^2 + m^2\right)}{3!}x^3 + \frac{m^2\left(2^2 + m^2\right)}{4!}x^4 + ...$$

Answers

1. $\log(1+x) = x - \frac{x^2}{2} + \frac{x^3}{3} - \frac{x^4}{4} + ...$ 2. $\sin^{-1}x = x + \frac{x^3}{3!} + \frac{x^5}{5!} + \frac{x^7}{7!} + ...$

6. (a) $1 + x + \frac{x^2}{2!} + \frac{x^3}{3!} + \frac{x^4}{4!} + ...$ (b) $x + \frac{x^3}{6} + \frac{2}{15}x^5 + ...$

(c) $1 + x + \frac{x^3}{6} + \frac{3x^5}{40} + \dots$

9. $$e^{m\cos^{-1}x} = e^{m\pi/2}\left[1 - mx + \frac{m^2x^2}{2!} - m\left(1^2 + m^2\right)\frac{x^3}{3!} + m^2\left(2^2 + m^2\right)\frac{x^4}{4!}\right.$$

$$\left. - m^2\left(1^2 + m^2\right)\left(3^2 + m^2\right)\frac{x^5}{5!} + \dots\right]$$

CHAPTER 4

Asymptotes

4.1 INTRODUCTION

This chapter is devoted to the study of asymptotes and its types, in respect of both polar and cartesian co-ordinates.

Asymptote is mainly connected to the infinite branches of the curve.

Branch of a curve

Suppose that a curve is prescribed by the equation y = f(x) such that 'y' has two or more values for every value of x. Corresponding to these distinct values of y, we shall get different branches of the curve

Thus, if $x^2 + y^2 = a^2$, be a curve, then

$$y^2 = a^2 - x^2 \quad \text{or} \quad y = \pm\sqrt{a^2 - x^2}$$

Hence, corresponding to $y = \sqrt{a^2 - x^2}$ and $y = -\sqrt{a^2 - x^2}$, there are two branches of the curve i.e., circle $x^2 + y^2 = a^2$. There branches are referred to as the upper half and lower half of the circle.

Asymptote

An asymptote of a curve is a straight line at a finite distance from the origin, to which a tangent to the curve tends as the point of contact recedes to infinity. Hence when a point 'p' on the infinite branch of the curve tends to infinity, its perpendicular distance from the asymptotes tends to zero.

Note: Asymptotes may be parallel to either to x-axis or to y-axis and accordingly they are called horizontal and vertical asymptotes. If an asymptote is not parallel to either of the co-ordinate axes, then it is called an oblique asymptote.

Asymptote parallel to co-ordinate axis

(a) **Asymptotes parallel to y-axis:** Let $y = f(x)$ be any curve and let $x = c$ be an asymptote of the curve. For estimating the value of parameter 'c', consider a point c (x, y) on the curve AB and let 'd' be its distance from the line $x = c$ i.e.,

$$d = |x - c|$$

Then, by definition of an asymptote $d \to 0$ as $C(x, y) \to \infty$ along the curve AB.

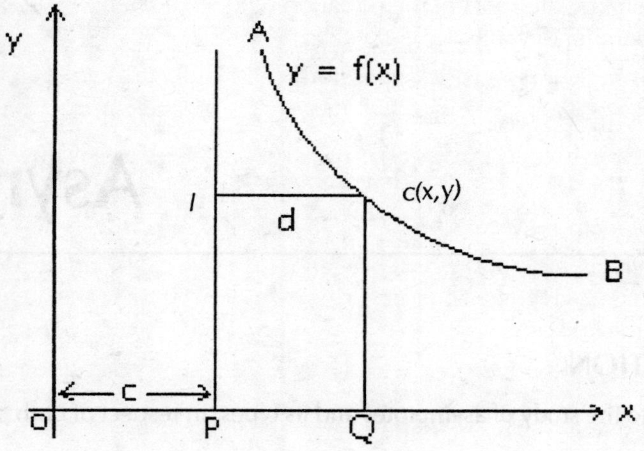

Therefore as $c(x, y) \to \infty$ or as $y \to \infty$ along an infinite branch of the curve $y = f(x)$, then $d \to 0$

$$\therefore \quad d = |x - c| \to 0 \text{ as } y \to \infty \text{ or } x \to c \text{ as } y \to \infty$$

$\therefore \quad \lim_{y \to k} x = c$, then $x = c$ is the asymptote parallel to y-axis.

(b) **Asymptote parallel to x-axis:** Similarly to obtain the asymptote parallel to x-axis, find from the given equation of the curve the value 'k' to which y tends as $x \to \infty$, then $y = k$ is an asymptote parallel to x-axis.

Asymptotes parallel to co-ordinate axes for f(x, y) = 0 i.e., Algebraic curve.

Case - I: Asymptote parallel to y-axis: Let an algebraic equation $f(x, y) = 0$ of nth order of the curve be given by

$$y^n \phi_0(x) + y^{n-1} \phi_1(x) + y^{n-2}\phi_2(x) + \ldots\ldots\ldots\ldots\ldots\ldots + \phi_n(x) = 0 \qquad \ldots(i)$$

where $\phi_0(x), \phi_1(x), \phi_2(x) \ldots\ldots \phi_n(x)$ are polynomials in x.

dividing (i) by y^n on both side, weget

$$\phi_0(x)+\frac{1}{y}\phi_1(x)+\frac{1}{y^2}\phi_2(x)+\dots\dots\dots+\frac{1}{y^n}\phi_n(x)=0$$

If $x = k$ is an asymptote of the curve parallel to y-axis of equation (i), then $k = \lim_{y\to\infty} x$. Therefore taking $y \to \infty$, and $x \to k$

$$\lim_{y\to\infty} \phi_0(x) \Rightarrow \phi_0(k) = 0$$

$$x \to k$$

$\therefore$ k is the root of the equation

$$\phi_0(x) = 0 \qquad \text{...(ii)}$$

If $k_1, k_2, \dots\dots\dots k_r$ etc. be the roots of the equation $\phi_0(x) = 0$. Then $x = k_1, x = k_2 \dots\dots\dots\dots x = k_r$ etc. are the asymptotes parallel to y-axis.

Since $\phi_0(x) = 0$ {from (ii)} has a root k, therefore x - k will be a factor of $\phi_0(x)$.

Also $\phi_0(x)$ is the coefficient of y^n.

Note: If the coefficient of the highest power of y is a constant, then there is no asymptote parallel to y-axis.

Case II: ***Asymptotes parallel to x-axis:***

In case (I) interchanging y and x and proceed as before, then a working method for finding asymptotes parallel to x-axis is given by equating to zero the coefficient of the highest power of x present in the given equation of the curve.

Note: If the coefficient of highest power of x is a constant, then there is no asymptotes parallel to x-axis.

Illustrative Examples

Example 1 Find the asymtotes parallel to axes:

$$\frac{a^2}{x^2}-\frac{b^2}{y^2}=1$$

Solution: $y^2a^2 - x^2b^2 = x^2y^2$

$$x^2y^2 - y^2a^2 + x^2b^2 = 0$$

Asymtotes parallel to x-axis

when highest degree coefficient of x is equated to zero, we get asymtotes parallel to x-axis.

$$\therefore \quad y^2 + b^2 = 0$$

$$y^2 = -b^2$$

As y is imaginary, we will not consider it as asymptote.

Asymtotes parallel to y-axis

when highest degree coefficient of y is equated to zero, we get asymtotes parallel to y-axis.

$$\therefore \quad x^2 - a^2 = 0$$

$\therefore \quad x^2 = a^2$

$\Rightarrow \quad x = \pm a$

Thus, it has two asymtotes parallel to y-axis

i.e., $x = \pm a$ Proved

Example 2 Find asymptotes parallel to axes

$$x^2y^2 - y^2 - 2 = 0$$

Solution: $x^2y^2 - y^2 - 2 = 0$

Asymtotes parallel to x-axis.

$$y^2 = 0 \Rightarrow y = 0$$

Asymtotes parallel to y-axis.

$$x^2 - 1 = 0$$

$$x^2 = 1 \Rightarrow x = \pm 1$$

$\therefore$ It has asymptote parallel to x-axis and also parallel to y-axis.

$$y = 0, x = \pm 1.$$

Example 3 Find asymptotes parallel to the axes.

$$x^3 - xy^2 = x^2 + 1$$

Solution: $x^3 - xy^2 = x^2 + 1$

Asymtotes parallel to x-axis

There is no asymptote parallel to x-axis as coefficient of highest degree term of x is a constant.

Asymptotes parallel to y-axis

$$-x = 0$$

$$\Rightarrow x = 0$$

$\therefore$ There is only one asymptote parallel to y-axis

i.e., $x = 0$, i.e., y - axis

Example 4 Find asymptotes parallel to the axes.

$$x^2y^2 - a^2\left(x^2 + y^2\right) - a^3(x + y) + a^4 = 0$$

Solution: $x^2y^2 - a^2\left(x^2 + y^2\right) - a^3(x + y) + a^4 + 0$

$$x^2y^2 - a^2x^2 - a^2y^2 - a^3x - a^3y + a^4 = 0$$

Asymptotes parallel to x-axis

$$y^2 - a^2 + 0$$

$\therefore \quad y^2 = a^2 \Rightarrow y = \pm a$

Asymptotes parallel to y-axis

$x^2 - a^2 = 0$

∴ $x^2 = a^2 \Rightarrow x = \pm a$

∴ It has 4 asymptotes. i.e., $y = \pm a$, $x = \pm a$

Example 5 Find asymptotes parallel to axes.

$$\left(x^2 - a^2\right)\left(y^2 - b^2\right) = a^2b^2$$

Solution: $x^2y^2 - x^2b^2 - a^2y^2 + a^2 b^2 = a^2 b^2$

∴ $x^2y^2 - x^2b^2 - a^2y^2 = 0$

Asymptotes parallel to x-axis.

$y^2 - b^2 = 0$

∴ $y^2 = b^2 \Rightarrow y = \pm b$

Asymptotes parallel to y-axis.

$x^2 - a^2 = 0$

∴ $x^2 = a^2 \Rightarrow x = \pm a$

∴ Asymptotes are

$y = \pm b$, $x = \pm a$.

Example 6 Find asymptotes parallel to axes.

$$x^2y^2 - x^2y - xy^2 + x + y + 1 = 0$$

Solution: $x^2y^2 - x^2y - xy^2 + x + y + 1 = 0$

Asymptotes parallel to x-axis.

$y^2 - y = 0$

$y(y - 1) = 0 \Rightarrow y = 0, y = 1$

Asymptotes parallel to y-axis.

$x^2 - x = 0$

$x(x-1) = 0 \Rightarrow x = 0, x = 1$

∴ Asymptotes are

$x = 0, x = 1, y = 0, y = 1$

Example 7 Show that asymptotes of $x^2y^2 = a^2 (x^2 + y^2)$ form a square of side 2a.

Solution: $x^2y^2 = a^2\left(x^2 + y^2\right)$

$x^2y^2 - a^2x^2 - a^2y^2 = 0$

Asymptotes parallel to x-axis.

$y^2 - a^2 = 0$

∴ $y^2 = a^2 \Rightarrow y = \pm a$

Asymptotes parallel to y-axis.

$$x^2 - a^2 = 0$$

$$\therefore \quad x^2 = a^2 \Rightarrow x = \pm a$$

As all the sides equals to a, hence the figure clearly shows that the asymtotes of the curve $x^2y^2 = a^2(x^2 + y^2)$ i.e., $x = \pm a$, $y = \pm a$ form a square of side

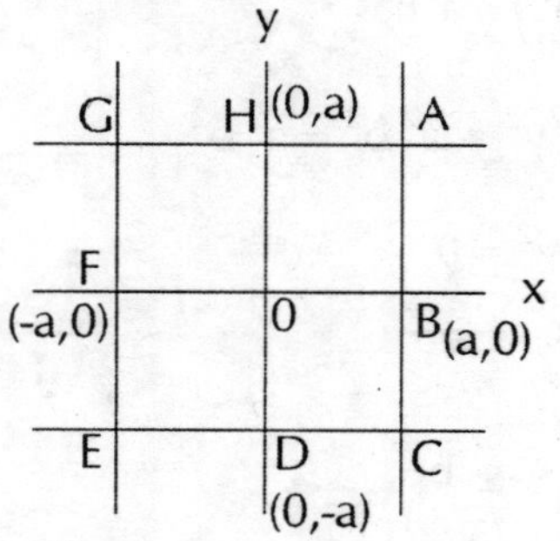

AB = a and BC = a $\therefore$ AC = 2a

$\therefore$ All sides equals = 2a and

Thus forms a square. Answer

Exercise 4.1

Find the asymptotes parallel to axes of the following curves

1. $xy^3 - x^3 - ax^2 - ay^2 = 0$
2. $y^3 - xy^2 = x^2 + 1$
3. $y^3 - xy^2 + x^2 + 1 = 0$
4. $x^2y^2 - a^2(x^2 + y^2) - a^3(x + y) + a^4 = 0$

Answers

1. x = 0 is the only asymptotes
2. No asymptotes
3. No asymptotes
4. $x = \pm a$, $y = \pm a$

4.2 OBLIQUE ASYMPTOTES

The asymptote which is not parallel to either of the co-ordinate axes is called an oblique asymptote, which is given by the equation y = mx + c, where 'm' and 'c' are the finite constants.

If y = mx + c is an asymptote to the curve f(x, y) = 0, prove that.

$$m = \lim_{x\to\infty}\left(\frac{y}{x}\right) \text{ and } c = \lim_{x\to\infty}(y - mx).$$

Suppose P(x, y) be any point on the infinite branch of the curve f(x, y) = 0 and p = PM be the perpendicular distance of the point P(x, y) from the line y = mx + c then $p = \dfrac{y - mx - c}{\sqrt{1+m^2}}$ ...(i)

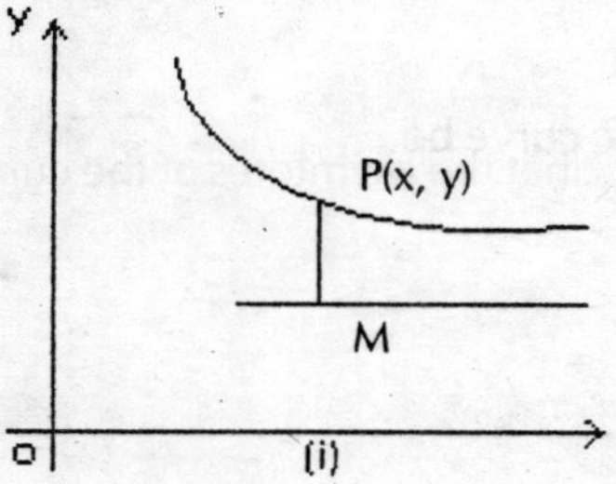

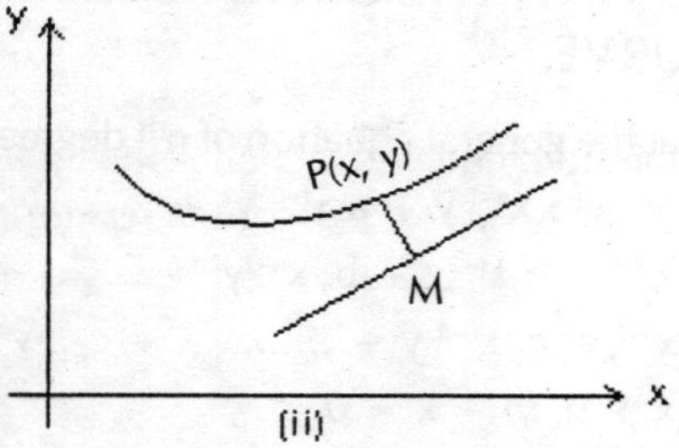

In the first figure the abscissa x tends to infinity as $p(x, y) \to \infty$ along the infinite branch of the curve.

Equation (i) can be written as

$$y - mx = c + p\sqrt{1+m^2}$$

dividing by x on both sides

$$\frac{y}{x} = m + \frac{c}{x} + \frac{p}{x}\sqrt{1+m^2} \qquad \ldots\text{(ii)}$$

Since $y = mx + c$ is an asymptote, so by definition, $p \to 0$ as $x \to \infty$, then equation (ii) becomes

$$\lim_{x\to\infty}\left(\frac{y}{x}\right) = \lim_{y\to\infty}\left(m + \frac{c}{x} + \frac{p}{x}\sqrt{1+m^2}\right) = m$$

Hence $\lim_{x\to\infty}\left(\frac{y}{x}\right) = m.$

Again from (i)

$$y - mx = c + p\sqrt{1+m^2}$$

$$\therefore \quad \lim_{x\to\infty}(y - mx) = \lim_{x\to\infty}\left(c + p\sqrt{1+m^2}\right) = \lim_{x\to\infty} c + \lim_{x\to\infty} p\sqrt{1+m^2}$$

$$\lim_{x\to\infty}(y - mx) = c + 0 \qquad \left(\because \lim_{x\to\infty} p = 0\right)$$

Hence $\lim_{x\to\infty}(y - mx) = c$

Therefore Asymptote is

$y = mx + c$

where $m = \lim_{x\to\infty}\left(\frac{y}{x}\right)$ and $c = \lim_{x\to\infty}(y - mx)$

4.3 TO FIND THE OBLIQUE ASYMPTOTES OF THE GENERAL ALGEBRAIC CURVE.

Suppose that the general equation of nth degree of an algebraic curve be

$$(a_0 x^n + a_1 x^{n-1} y + a_2 x^{n-2} y^2 + \ldots\ldots\ldots + a_n y^n) +$$
$$(b_0 x^{n-1} + b_1 x^{n-2} y + b_2 x^{n-3} y^2 + \ldots\ldots\ldots + b_{n-1} y^{n-1}) +$$
$$(c_0 x^{n-2} + c_1 x^{n-3} y + \ldots\ldots\ldots\ldots + c_{n-2} y^{n-2}) + \ldots\ldots\ldots\ldots +$$
$$(p_0 x + p_1 y) + k = 0$$

The above equation can be written as

$$x^n \phi_n\left(\frac{y}{x}\right) + x^{n-1} \phi_{n-1}\left(\frac{y}{x}\right) + x^{n-2} \phi_{n-2}\left(\frac{y}{x}\right) + \ldots\ldots + x\,\phi_1\left(\frac{y}{x}\right) + \phi_0\left(\frac{y}{x}\right) = 0 \qquad \ldots\text{(i)}$$

where $\phi_n\left(\frac{y}{x}\right), \phi_{n-1}\left(\frac{y}{x}\right), \phi_{n-2}\left(\frac{y}{x}\right) \ldots\ldots, \phi_0\left(\frac{y}{x}\right)$ are the polynomials in $\frac{y}{x}$ of degree n, n–1, n–2, ... 0, at the most.

Let y = mx + c be the asymptote of the curve, here m and c are finite and are to be found out.

To find m,

divide (i) by x^n on both sides, weget

$$\phi_n\left(\frac{y}{x}\right) + \frac{1}{x}\phi_{n-1}\left(\frac{y}{x}\right) + \frac{1}{x^2}\phi_{n-2}\left(\frac{y}{x}\right) + \ldots\ldots +$$

$$\frac{1}{x^{n-1}}\phi_1\left(\frac{y}{x}\right) + \frac{1}{x^n}\phi_0\left(\frac{y}{x}\right) = 0 \qquad \ldots\text{(ii)}$$

taking limit as $x \to \infty$ such that

$$\lim_{x\to\infty} \frac{y}{x} = m, \text{ we get}$$

$$\phi_n(m) + 0 + 0 + 0 \ldots\ldots + 0 + 0 = 0$$

$\therefore \quad \phi_n(m) = 0$ which gives 'n' values of m say m_1, m_2, m_3 etc.

To find 'c' corresponding to the values of 'm'

let us take y–mx = p, so that as $x \to \infty$, $p \to c$ because $\lim_{x\to\infty}(y - mx) = c$

Now, $\quad y = mx + p$

$$\therefore \quad \frac{y}{x} = m + \frac{p}{x}$$

On putting $\frac{y}{x} = m + \frac{p}{x}$ in (i), we get

$$x^n\phi_n\left(m+\frac{p}{x}\right)+x^{n-1}\,\phi_{n-1}\left(m+\frac{p}{x}\right)+x^{n-2}\phi_n\left(m+\frac{p}{x}\right)+\text{..........}\quad=0 \qquad \text{...(iii)}$$

On expanding each term by Taylor's series, we get

$$x^n\left[\phi_n(m)+\frac{p}{x}\phi_n{}'(m)+\frac{p^2}{2!x^2}\phi_n''(m)+\text{........}\right]+$$

$$x^{n-1}\left[\phi_{n-1}(m)+\frac{p}{x}\phi'_{n-1}(m)+\text{...........}\right]+x^{n-2}\left[\phi_{n-2}(m)+\frac{p}{x}\phi'_{n-2}(m)+\text{......}\right]+\text{.....}=0$$

on putting $\phi_n(m) = 0$ and arranging the rest of the terms in desending power of x i.e.,

$$x^{n-1}\left[p\,\phi_n{}'(m)+\phi_{n-1}(m)\right]+x^{n-2}\left[\frac{p^2}{2!}\phi_n''(m)+p\phi'_{n-1}(m)+\phi_{n-2}(m)\right]+\text{......}=0 \qquad \text{...... (iv)}$$

dividing by x^{n-1} on both sides, we get

$$\left[p\,\phi_n{}'(m)+\phi_{n-1}(m)\right]+\frac{1}{x}\left[\frac{p^2}{2!}\phi_n''(m)+p\phi_{n-1}{}'(m)+\text{.....}\right]+\text{......}=0 \qquad \text{...(v)}$$

Taking $x \to \infty$ such that $p \to c$, then

$$c\,\phi_n'(m)+\phi_{n-1}(m)=0 \qquad \text{...(vi)}$$

If $\phi_n'(m) \neq 0$ and $\phi_n(m) = 0$ has real and distinct values of m, then 'c' is given by

$$c=-\frac{\phi_{n-1}(m)}{\phi_n'(m)} \qquad \text{...(vii)}$$

put $\quad m = m_1, m_2, m_3,$ ..... then we get

$c = c_1, c_2, c_3$ .... and the asymptotes

are $\quad y = m_1x + c_1$, $y = m_2 x + c_2$, ...

Note:

(i) When $\phi_n'(m) = 0$ and $\phi_{n-1}(m) \neq 0$ then from equation (vii), 'c' is infinite, so there is no asymptote.

(ii) When $\phi_n(m) = 0$ and m has two real and equal values i.e., $m = m_1, m_1$ then $\phi_n'(m) = 0$ and (vii) becomes $c.0 + 0 = 0$ i.e., then it is not possible to find c.

Case of Parallel Asymptotes:

When $\phi_n'(m)$ and $\phi_{n-1}(m) = 0$, then equation (iv) reduces to

$$x^{n-2}\left[\frac{p^2}{2}\phi_n''(m)+p\,\phi'_{n-1}(m)+\phi_{n-2}(m)\right]+$$

$$x^{n-3}\left[\frac{p^3}{3!}\phi_n'''(m)+\frac{p^2}{2}\phi''_{n-1}(m)+p\,\phi'_{n-2}(m)+\phi_{n-3}(m)\right]+\ldots=0$$

on dividing by x^{n-2} on both sides, we get

$$\frac{p^2}{2!}\phi_n''(m)+p\,\phi'_{n-1}(m)+\phi_{n-2}(m)+$$

$$\frac{1}{x}\left[\frac{p^3}{3!}\phi_n'''(m)+\frac{p^2}{2}\phi''_{n-1}(m)+p\,\phi'_{n-2}(m)+\phi_{n-3}(m)\right]+\ldots=0$$

Taking limit as $x\to\infty$,

$p\to c$, we get

$$\frac{c^2}{2!}\phi_n''(m)+c\,\phi'_{n-1}(m)+\phi_{n-2}(m)=0$$

If $\phi_n''(m)\neq 0$, then we get two values of c i.e., c_1 and c_2 corresponding to the two equal values of m. i.e., m_1, m_1 and therefore asymptotes are given by

$$y = m_1 x + c_1 \text{ and } y = m_1 x + c_2$$

when $\phi_n''(m)=0$, $\phi'_{n-1}(m)=0$ and $\phi_{n-2}(m)=0$ i.e. when m has three real and equal values then equation (iv) reduce to

$$x^{n-3}\left[\frac{p^3}{3!}\phi_n'''(m)+\frac{p^2}{2!}\phi''_{n-1}(m)+p\,\phi'_{n-2}(m)+\phi_{n-3}(m)\right]+x^{n-4}\,[\ldots\ldots]+\ldots\ldots=0$$

on dividing by x^{n-3} on both sides, we get

$$\frac{p^3}{3!}\phi_n'''(m)+\frac{p^2}{2!}\phi''_{n-1}(m)+p\,\phi'_{n-2}(m)+\phi_{n-3}(m)+\frac{1}{x}[\ldots\ldots]+=0$$

Taking limit as $x\to 0$, $p\to c$, we get

or $$\frac{c^3}{3!}\phi_n'''(m)+\frac{c^2}{2!}\phi''_{n-1}(m)+c\phi'_{n-2}(m)+\phi_{n-3}(m)=0$$

If $\phi_n'''(m)\neq 0$ then we get three values of c i.e., c_1, c_2 and c_3 corresponding to the three equal values of m i.e., m_1, m_1, m_1 and therefore asymptotes are given by

$$y = m_1 x + c_1,\quad y = m_1 x + c_2,\quad y = m_1 x + c_3$$

Note: Total number of asymptotes of a curve cannot exceed the degree of a curve.

4.4 WORKING RULE FOR FINDING OBLIQUE ASYMPTOTES

Asymptotes of an algebraic curve of nth degree.

(1) **To find $\phi_n(m)$:** It can be obtained by putting x = 1 and y = m in the highest degree term of the given equation of the curve.

(2) **To find the value of m:** Putting $\phi_n(m) = 0$, solve for m, let $m = m_1, m_2, m_3$ .... be the roots.

(3) **To find $\phi_{n-1}(m)$:** It can be obtained by putting x = 1, y = m in the next lower degree terms of the given equation of the curve.

Case I: When m has real and distinct values i.e., $m = m_1, m_2, m_3, \ldots$ then the values of 'c' corresponding to the values of m's are given by $c = -\dfrac{\phi_{n-1}(m)}{\phi_n'(m)}$ provided $\phi_n'(m) \neq 0$

The required asymptote are given by $y = m_1x + c_1$; $y = m_2x + c_2, \ldots$

Case II: When m has two real and equal values then the value of 'c' corresponding to m's are given by

$$\frac{c^2}{2!}\phi_n''(m) + c\,\phi_{n-1}'(m) + \phi_{n-2}(m) = 0$$

provided $\phi_n''(m) \neq 0$

Now the term ϕ_{n-2} (m) can be found out by putting x = 1, y = m in the next lower degree term i.e., ϕ_{n-2} (m) then required asymptotes are given by

$$y = m_1x + c_1; \quad y = m_2x + c_2, \ldots$$

Case III: When m has three real and equal values then the value of 'c' corresponding to m's are given by

$$\frac{c^3}{3!}\phi_n'''(m) + \frac{c^2}{2!}\phi_{n-1}''(m) + c\,\phi_{n-2}'(m) + \phi_{n-3}(m) = 0,$$

provided $\phi_n'''(m) \neq 0$

Now the term $\phi_{n-3}(m)$ can be found out by putting x = 1, y = m in the next lower degree term i.e., ϕ_{n-3} (m), then the required asymptotes are given by

$$y = m_1x + c_1; \quad y = m_1x + c_2; \quad y = m_1x + c_3 \ldots$$

Illustrative Examples

Example 1. Find all the asymptotes of the following curves:

$$y^3 = x^3 + ax^2$$

Solution: we have $x^3 - y^3 + ax^2 = 0$

Since the coefficient of highest degree term i.e., x^3, y^3 are constants, so no asymptote llal to x and y-axis.

To find the oblique asymptote

put x = 1, and y = m in the highest degree term

$\phi_3(m) = 1 - m^3$, $\quad \phi_3'(m) = -3m^2$

$\phi_2(m) = a$

To find the value of m, equate

$\phi_3(m) = 0 \quad \Rightarrow \quad 1 - m^3 = 0$

$(1 - m)(m^2 - m + 1) = 0$

m = 1 is the only real root.

To find 'c', corresponding to m = 1

$$c = -\frac{\phi_{n-1}(m)}{\phi_n'(m)}$$

$$= \frac{\phi_2(m)}{\phi_3'(m)} = \frac{-a}{-3m^2} = \frac{a}{3}$$

Equation of asymptote is,

$$y = mx + c \quad \Rightarrow \quad y = -x + \frac{a}{3}$$

$$\Rightarrow \boxed{3x + 3y = a}$$

Example 2 Find all the asymptotes of the curve

$$x^2y + xy^2 + xy + y^2 + 3x = 0 \quad ...(1)$$

Solution: To find parallel asymptotes.

co-efficient of highest degree of x equated to zero i.e. co-eff of x^2,

$\boxed{y = 0}$ is an asymptote i.e., x –axis. Asymptote ||al to y-axis, co-eff. of y^2, i.e., x + 1 = 0

$\Rightarrow \boxed{x = -1}$

To find oblique asymptote

put x = 1 and y = m in the third and second degree terms in (1)

$\phi_3(m) = m + m^2 \quad , \quad \phi_3'(m) = 1 + 2m$

$\phi_2(m) = m + m^2$

$\phi_1(m) = 3$

$\phi_3(m) = 0 \Rightarrow \quad m + m^2 = 0 \Rightarrow m(m + 1) = 0$

m = 0, – 1 (all real and different)

∴ for m = 0 , c is given by

$$c = -\frac{\phi_2(m)}{\phi_3'(m)} = -\frac{m + m^2}{1 + 2m} = 0$$

for m = –1

$$c = -\frac{m + m^2}{1 + 2m} = -\frac{(-1) + 1}{1 - 2} = 0$$

Equation of asymptotes are

$y = mx + c$ for $m = 0,\ c = 0$

$\boxed{y = 0}$

for $m = -1,\ c = 0$

$\boxed{y = -x}$

so, the asymptotes are,

$\boxed{y = 0}$, $\boxed{x = -1}$ and $\boxed{y = -x}$ Answer

Example 3 Find all the asymptotes of the curve

(a) $y^2(x - 2a) = x^3 - a^3$

Solution: We have

$$x^3 - a^3 - xy^2 + 2ay^2 = 0$$

Asymptote parallel to x-axis

co-eff of highest power of x i.e., x^3 is constant. ∴ there is no asymptote llal to x – axis, Asymptote llal to y-axis, coefficient of highest power of y equated to zero,

We get $2a - x = 0 \Rightarrow \boxed{2a = x}$

To find oblique asymptote

put $x = 1,\ y = m$ is 3rd and 2nd degree terms,

$\phi_3(m) = 1 - m^2$, $\phi_3'(m) = -2m$

$\phi_2(m) = 2am^2$

$\phi_3(m) = 0 \Rightarrow 1 - m^2 = 0$

$m = \pm 1$ (real and different)

so, c is given by, $c = -\dfrac{\phi_{n-1}(m)}{\phi_n'(m)} = -\dfrac{\phi_2(m)}{\phi_3'(m)}$

$$= -\frac{2am^2}{-2m} = am$$

for $m = 1,\ c = a$

for $m = -1,\ c = -a$

so asymptotes are, $y = mx + c$,

$\boxed{y = x + a}$ $\boxed{y = -x - a}$ Answer.

Example 4 Find all the asymptotes of the curve $6x^2 + xy - 2y^2 + x + 2y + 1 + 0$.

Solution: Asymptotes parallel to the axes.

As the co-eff. of highest power of 'x' and 'y' are constant, there is no asymptote parallel to 'x' and 'y' axis.

For oblique asymptote put x = 1, y = m in 2nd and 1st degree term of the given curve.

$\phi_2(m) = 6 + m - 2m^2$

$\phi_1(m) = 1 + 2m$

$\phi_0(m) = 1$

$\phi_2'(m) = -4m + 1$

Put $\phi_2(m) = 0 \Rightarrow 2m^2 - m - 6 = 0$

$$\Rightarrow \quad m = \frac{1 \pm \sqrt{1+48}}{4}$$

$$= \frac{1 \pm 7}{4} = 2, -3/2$$

Let $m_1 = 2$, $m_2 = -3/2$

$$c = -\frac{\phi_1(m)}{\phi_2'(m)} = -\frac{1+2m}{1-4m}$$

for $m = 2$

$$c = -\frac{1+4}{1-8} = -\frac{5}{-7} = 5/7$$

for $m = -3/2$

$$c = -\frac{1 + 2 \times -3/2}{1 - 4^2 \times -3/2} = -\frac{1-3}{1+6} = +\frac{+2}{7} = 2/7$$

Asymptotes are given by, $y = mx + c$

$$\Rightarrow \quad y = 2x + 5/7 \Rightarrow \boxed{7y = 14x + 5}$$

and

$$y = \frac{-3}{2}x + \frac{2}{7} \Rightarrow 14y = -21x + 4$$

$$\Rightarrow \quad \boxed{21x + 14y - 4 = 0} \text{ Answer.}$$

Example 5 Find all the asymptotes of the curve

$$y^3 - 2xy^2 - x^2y + 2x^3 + 3y^2 - 7xy + 2x^2 + 2y + 2x + 1 = 0$$

Solution:

Since the co-eff. of highest degree term i.e., x^3 and y^3 are constants

$\therefore$ there is no asymptote parallel to x and y-axis.

To find oblique asymptote, put x = 1, y = m in 3rd and 2nd degree term.

$\phi_3(m) = m^3 - 2m^2 - m + 2$

$\phi_2(m) = 3m^2 - 7m + 2$

$\phi_1(m) = 2m + 2$

$\phi_0(m) = 1$

$\phi_3'(m) = 3m^2 - 4m - 1$

$$\phi_3(m) = 0 \Rightarrow \quad m^3 - 2m^2 - m + 2 = 0$$

$$m^2(m-2) - (m-2) = 0$$

$$(m-2)(m^2-1) = 0$$

$$m = 2, \pm 1 \qquad \text{(all real and distinct)}$$

$$c = -\frac{\phi_{n-1}(m)}{\phi_n'(m)} = -\frac{3m^2 - 7m + 2}{3m^2 - 4m - 1}$$

for m = 2,
$$c = -\left[\frac{3(2)^2 - 7(2) + 2}{3(2)^2 - 4(2) - 1}\right] = 0$$

for m = 1,
$$c = -\left[\frac{3(1)^2 - 7(1) + 2}{3(1)^2 - 4(1) - 1}\right] = -1$$

for m = −1,
$$c = \left[\frac{3(-1)^2 - 7(-1) + 2}{3(-1)^2 - 4(-1) - 1}\right] = -2$$

Equations of asymptotes are,

$\boxed{y = 2x}$, $\boxed{y = x - 1}$, $\boxed{y = -x - 2}$ Answer

Example 6 Find all the asymptotes of the curve

$$3x^3 + 2x^2y - 7xy^2 + 2y^3 - 14xy + 7y^2 + 4x + 5y = 0$$

Solution:

No asymptotes llal to x and y-axis, as the co-eff. of highest power of x i.e., x^3 and y i.e., y^3 is constant.

Oblique Asymptote.

put x = 1, y = m in 3rd and 2nd degree term,

$\phi_3(m) = 3 + 2m - 7m^2 + 2m^3$

$\phi_2(m) = -14m + 7m^2$

$\phi_1(m) = 4 + 5m$

$\phi_3'(m) = 2 - 14m + 6m^2$

$\phi_3(m) = 0 \Rightarrow 2m^3 - 7m^2 + 2m + 3 = 0$

or $2m^3 - 7m^2 + 2m + 3 = 0$

$\Rightarrow$ m = 1 is one root by inspection, we find other two roots

By synthetic division,

1	2	−7	2	3
		2	−5	−3
	2	−5	−3	0

$$\Rightarrow \quad 2m^2 - 5m - 3 = 0$$

$$m = \frac{5 \pm \sqrt{25 + 24}}{4}$$

$$= \frac{5 \pm 7}{4} = 3, -1/2$$

$$\Rightarrow \quad m = 1, -1/2, 3$$

$$c = -\frac{\phi_2(m)}{\phi_3'(m)} = -\frac{7m^2 - 14m}{6m^2 - 14m + 2}$$

for $m = 1$, $$c = -\frac{7 - 14}{6 - 14 + 2} = -\frac{-7}{-6} = -7/6$$

for $m = -1/2$, $$c = -\frac{7\left(-\frac{1}{2}\right)^2 - 14(-1/2)}{6\left(-\frac{1}{2}\right)^2 - 14\left(-\frac{1}{2}\right) + 2}$$

$$= -\frac{\frac{7}{4} + \frac{14}{2}}{\frac{+6}{4} + \frac{14}{2} + 2} = -\frac{28 + 7}{+6 + 28 + 8}$$

$$= -\frac{35}{42} = -\frac{5}{6}$$

for $m = 3$

$$c = -\frac{7(3)^2 - 14(3)}{6(3)^2 - 14(3) + 2}$$

$$= -\frac{7\times 9-42}{54-42+2} = -\frac{63-42}{56-42}$$

$$= -\frac{21}{14} = -\frac{3}{2}$$

asymptotes are

$\boxed{y = x - \frac{7}{6}}$ $\boxed{y = -\frac{1}{2}x - \frac{35}{42}}$ $\boxed{y = 3x - \frac{3}{2}}$ Answer

Example 7 Find the asymptotes of the curve

$$y^3 - 3x^2y + xy^2 - 3x^3 + 2y^2 + 2xy + 4x + 5y + 6 = 0$$

Solution:

Since the co-eff. of highest power of x and y i.e. x^3 and y^3 are constant

∴ there is no asymptote llal to x and y-axis.

For oblique asymptotes, put $x = 1$, $y = m$ in the 3rd and 2nd degree term,

$$\phi_3(m) = m^3 - 3m + m^2 - 3$$

$$\phi_2(m) = 2m^2 + 2m$$

$$\phi_1(m) = 4 + 5m$$

$$\phi_0(m) = 6$$

$$\phi_3'(m) = 3m^2 - 3 + 2m$$

$$\phi_3(m) = 0 \Rightarrow m^3 + m^2 - 3m - 3 = 0$$

$$m^2(m+1) - 3(m+1) = 0$$

$$(m^2-3)(m+1) = 0 \Rightarrow m = -1, \pm\sqrt{3}$$

$$c = -\frac{\phi_{n-1}(m)}{\phi_n'(m)} = -\frac{\phi_2(m)}{\phi_3'(m)}$$

$$= -\frac{2m^2+2m}{3m^2+2m-3}$$

for $m = -1$, $$c = -\frac{2(-1)^2+2(-1)}{3(-1)^2+2(-1)-3}$$

$$= -\frac{2-2}{3-2-3} = 0$$

for $m = +\sqrt{3}$, $c = -\dfrac{2\left(\sqrt{3}\right)^2 + 2\sqrt{3}}{3\left(\sqrt{3}\right)^2 + 2\sqrt{3} - 3}$

$$= -\frac{6 + 2\sqrt{3}}{9 + 2\sqrt{3} - 3} = -\frac{6 + 2\sqrt{3}}{6 + 2\sqrt{3}} = -1$$

$$= -1$$

for $m = -\sqrt{3}$, $c = -\dfrac{2\left(-\sqrt{3}\right)^2 + 2\left(-\sqrt{3}\right)}{3\left(-\sqrt{3}\right)^2 + 2\left(-\sqrt{3}\right) - 3}$

$$= -\frac{2 \times 3 - 2\sqrt{3}}{3 \times 3 - 2\sqrt{3} - 3} = -\frac{6 - 2\sqrt{3}}{6 - 2\sqrt{3}} = -1$$

Asymptotes are,

$\boxed{y = -x}$, $\boxed{y = \sqrt{3}x - 1}$, $\boxed{y = -\sqrt{3}x - 1}$

or

$\boxed{x + y = 0}$, $\boxed{\sqrt{3}x - y - 1 = 0}$, $\boxed{\sqrt{3}\,x + y + 1 = 0}$ Answer

Example 8 Find all the asymptotes of the curve:

$$x^3 + 3x^2y - xy^2 - 3y^3 + x^2 - 2xy + 3y^2 + 4x + 7 = 0$$

Solution: No asymptote parallel to x and y-axis as the co-eff. of highest power of x and y are constant.

For oblique Asymptote,

put $x = 1,\ y = m$

$\phi_3(m) = 1 + 3m - m^2 - 3m^3$

$\phi_2(m) = 1 - 2m + 3m^2$

$\phi_1(m) = 4$

$\phi_0(m) = 7$

$\phi_3'(m) = 3 - 2m - 9m^2$

$\phi_3(m) = 0 \Rightarrow 1 + 3m - m^2 - 3m^3 = 0$

$(1 + 3m) - m^2(1 + 3m) = 0$

$(1 + 3m)(1 - m^2) = 0$

$\Rightarrow m = -1/3,\ m = \pm 1$

$$c = -\frac{\phi_2(m)}{\phi_3'(m)} = -\frac{1 - 2m + 3m^2}{3 - 2m + 9m^2}$$

for m = 1, $c = -\frac{1-2+3}{3-2+9} = -\frac{2}{10} = -1/5$

for m = –1, $c = -\frac{1-2(-1)+3(-1)^2}{3-2(-1)+9(-1)^2} = -\frac{1+2+3}{3+2+9}$

$$= -\frac{6}{14} = -\frac{3}{7}$$

for m = –1/3, $c = -\frac{1-2\left(\frac{-1}{3}\right)+3\left(-\frac{1}{3}\right)^2}{3-2\left(-\frac{1}{3}\right)+9\left(-\frac{1}{3}\right)^2}$

$$= -\frac{1+\frac{2}{3}+\frac{3}{9}}{3+\frac{2}{3}+\frac{9}{9}} = -\frac{1+\frac{2}{3}+\frac{1}{3}}{3+\frac{2}{3}+1}$$

$$= -\frac{\frac{3+2+1}{3}}{\frac{9+2+3}{3}} = -\frac{6}{14} = -\frac{3}{7}$$

Asymptotes are,

$$y = x - \frac{1}{5} \Rightarrow \boxed{5y - 5x + 1 = 0}$$

$$y = -x - \frac{3}{7} \Rightarrow \boxed{7y + 7x + 3 = 0}$$

$$y = -\frac{1}{3}x - \frac{3}{7} \Rightarrow \boxed{21y + 7x + 9 = 0}$$ Answer

Example 9 Find all asymptotes of the curve:

$$y^3 - xy^2 - x^2y + x^3 + x^2 - y^2 = 0$$

Solution: No asymptote parallel to x and y axis, as co-efficient of highest power of x and y are constant in the given equation.

For oblique asymptote put x = 1, y = m, in the given equation

$\phi_3(m) = m^3 - m^2 - m + 1$

$\phi_2(m) = 1 - m^2$

$\phi_3'(m) = 3m^2 - 2m - 1$

$\phi_3(m) = 0 \Rightarrow m^3 - m^2 - m + 1 = 0$

$\Rightarrow m^2(m-1) - (m-1) = 0$

$(m-1)(m^2-1) = 0$

$m = 1, \quad m = \pm 1$

or $m = 1, 1, -1$

$$c = -\frac{1-m^2}{m^3 - m^2 - m + 1}$$

for $m = -1$,
$$c = -\frac{1-(-1)^2}{(-1)^3 - (-1)^2 - (-1) + 1} = 0$$

for m = 1, 1 (repeated roots), $\therefore$ c is given by

$$\frac{c^2}{2!}\phi_3''(m) + c\,\phi_2'(m) + \phi_1(m) = 0$$

$$\frac{c^2}{2!}(6m-2) + c(-2m) = 0$$

for $m = 1$,
$$\frac{c^2}{2!}(6-2) + c(-2) = 0$$

$$= \frac{4c^2}{2} - 2c = 0$$

$= 2c^2 - 2c = 0 \Rightarrow 2c(c-1) = 0$

$= c = 0, \; c = 1$

Asymptotes are,

$\boxed{y = -x}$, $\boxed{y = x}$, $\boxed{y = x + 1}$ Answer

Example 10 Find asymptotes of the curve

$4x^2(y-x) + y(y-2)(x-y) = 4x + 4y - 7$

Solution: we have $4x^2y - 4x^3 + (y^2 - 2y)(x - y) = 4x + 4y - 7$

$4x^2y - 4x^3 + xy^2 - 2xy - y^3 + 2y^2 - 4x - 4y + 7 = 0$

No asymptote parallel to x and y-axis, as the co-eff. of highest power of x and y are constant.

For oblique asymptote, put x = 1, y = m in the 3rd and 2nd degree term

$\phi_3(m) = 4m - 4 + m^2 - m^3$

$\phi_2(m) = -2m + 2m^2$

$\phi_1(m) = -4 - 4m$

$\phi_0(m) = 7$

$\phi_3'(m) = 4 + 2m - 3m^2$

$\phi_3(m) = 0 \Rightarrow 4m - 4 + m^2 - m^3 = 0$

or $m^3 - m^2 - 4m + 4 = 0$

$m^2(m-1) - 4(m-1) = 0$

$(m^2 - 4)\quad(m-1) = 0$

$\Rightarrow\quad m = 1, \quad m = \pm 2$ (real and distinct)

$$c = -\frac{\phi_2(m)}{\phi_3'(m)} = -\frac{-2m + 2m^2}{4 + 2m - 3m^2}$$

for $m = 1$,
$$c = -\frac{-2(1) + 2(1)^2}{4 + 2(1) - 3(1)^2} = -\frac{-2 + 2}{4 + 2 - 3} = 0$$

for $m = 2$,
$$c = -\frac{-2(2) + 2(2)^2}{4 + 2(2) - 3(2)^2}$$

$$= -\frac{-4 + 8}{4 + 4 - 12} = -\frac{4}{-4} = 1$$

for $m = -2$,
$$c = -\frac{-2(-2) + 2(-2)^2}{4 + 2(-2) - 3(-2)^2}$$

$$= -\frac{4 + 8}{4 - 4 - 12} = -\frac{12}{-12} = 1$$

Asymptotes are,

$\boxed{y = x}$, $\boxed{y = 2x + 1}$, $\boxed{y = -2x + 1}$ Answer

Example 11 Find asymptotes of the curve

$$x^3 + 3x^2y - 4y^3 - x + y + 3 = 0$$

Solution: No asymptote parallel to x and y-axis

For oblique asymptote, put $x = 1$, $y = m$, in 3rd and 2nd degree term.

$\phi_3(m) = 1 + 3m - 4m^3$

$\phi_2(m) = 0$

$\phi_1(m) = -1 + m$

$\phi_3'(m) = 3 - 12m^2$

$\phi_3(m) = 0 \Rightarrow 1 + 3m - 4m^3 = 0$

$4m^3 - 3m - 1 = 0$

$m = 1$ is one root by inspection

By synthetic division

$$\begin{array}{c|cccc} 1 & 4 & 0 & -3 & -1 \\ & & 4 & 4 & 1 \\ \hline & 4 & 4 & 1 & |0 \end{array}$$

$4m^2 + 4m + 1 = 0$

$$\Rightarrow \quad m = \frac{-4 \pm \sqrt{16-16}}{8} = \frac{-1}{2}, \frac{-1}{2}$$

for $m = 1$, $\quad c = -\dfrac{\phi_2(m)}{\phi_3'(m)} = 0$

for $m = -1/2, -1/2$

$$\frac{c^2}{2!}\phi_3''(m) + c\,\phi_2'(m) + \phi_1(m) = 0$$

$$\frac{c^2}{2!}(-24m) + (m-1) = 0$$

$$-12mc^2 + m - 1 = 0$$

$$-12c^2\left(-\frac{1}{2}\right) - \frac{1}{2} - 1 = 0 \quad \Rightarrow \quad (\text{for } m = -\frac{1}{2})$$

$$\Rightarrow \quad 6c^2 - \frac{3}{2} = 0$$

or $\quad 6c^2 = 3/2$

or $\quad c^2 = 1/4 \Rightarrow \quad c = \pm\dfrac{1}{2}$

Asymptotes are

$$\boxed{y = x}, \quad \boxed{y = \frac{-1}{2}x + \frac{1}{2}}, \quad \boxed{y = \frac{-1}{2}x - \frac{1}{2}} \quad \text{Answer.}$$

Example 12 Find asymptotes of the curve:

$$x^3 - x^2y - xy^2 + y^3 + 2x^2 - 4y^2 + 2xy + x + y + 1 = 0$$

Solution: No asymptote parallel to x and y-axis as the co-eff. of highest degree of x and y are constant.

For oblique asymptotes, put $x = 1$, $\quad y = m$ in 3rd and 2nd degree term of the given equation.

$\phi_3(m) = 1 - m - m^2 + m^3$

$\phi_2(m) = 2 - 4m^2 + 2m$

$\phi_1(m) = 1 + m$

$\phi_0(m) = 1$

$\phi_3'(m) = -1 - 2m + 3m^2$

$\phi_3(m) = 0 \Rightarrow m^3 - m^2 - m + 1 = 0$

$m^2(m-1) - 1(m-1) = 0$

$(m^2 - 1)\quad (m-1) = 0$

$\Rightarrow m = \pm 1, m = 1$

for $m = -1$, $\quad c = -\dfrac{\phi_2(m)}{\phi_3'(m)}$

$$= -\frac{2 - 4m^2 + 2m}{-1 - 2m + 3m^2}$$

$$= -\frac{2 - 4(-1)^2 + 2(-1)}{-1 - 2(-1) + 3(-1)^2}$$

$$= -\frac{2-4-2}{-1+2+3} = -\frac{-4}{4} = 1$$

for $m = 1, 1$

$$\frac{c^2}{2!}\phi_3''(m) + c\,\phi_2'(m) + \phi_1(m) = 0$$

$$\frac{c^2}{2!}(-2 + 6m) + c(-8m + 2) + (1 + m) = 0$$

$m = 1$, $\quad \dfrac{c^2}{2!}(-2+6) + c(-6) + 2 = 0$

$$\frac{c^2}{2!}(4) - 6c + 2 = 0$$

$2c^2 - 6c + 2 = 0$

$c^2 - 3c + 1 = 0$

$$c = \frac{+3 \pm \sqrt{9-4}}{2}$$

$$= \frac{3 \pm \sqrt{5}}{2}$$

Asymptotes are,

$\boxed{y = -x + 1}$, $\boxed{y = x + \frac{3+\sqrt{5}}{2}}$, $\boxed{y = x + \frac{3-\sqrt{5}}{2}}$

or $\boxed{y + x - 1 = 0}$, $\boxed{2y = 2x + 3 \pm \sqrt{5}}$ Answer

Example 13 Find asymptotes of the curve:

$$y^3 - 5xy^2 + 8x^2y - 4x^3 - 3y^2 + 9xy - 6x^2 + 2y - 2x + 1 = 0$$

Solution: No asymptote llal to x and y-axis

For oblique Asymptote, put $x = 1, \quad y = m$

$\phi_3(m) = m^3 - 5m^2 + 8m - 4$

$\phi_2(m) = -3m^2 + 9m - 6$

$\phi_1(m) = 2m - 2$

$\phi_0(m) = 1$

$\phi_3'(m) = 3m^2 - 10m + 8$

$\phi_3(m) = 0 \Rightarrow m^3 - 5m^2 + 8m - 4 = 0$

$m = 1$ is one root by inspection

By Synthetic division

1	1	−5	8	−4
		1	−4	4
	1	−4	4	0

$m^2 - 4m + 4 = 0$

$$m = \frac{4 \pm \sqrt{16-16}}{2}$$

$= 2, 2$

for $m = 1$,

$$c = -\frac{\phi_2(m)}{\phi_3'(m)}$$

$$= -\frac{-3m^2 + 9m - 6}{3m^2 - 10m + 8}$$

$$= -\frac{-3(1)^2 + 9(1) - 6}{3(1)^2 - 10(1) + 8}$$

$$= -\frac{-3 + 9 - 6}{3 - 10 + 8} = 0$$

for $m = 2, 2$ (repeated roots)

$$\frac{c^2}{2!}\phi_3''(m) + c\,\phi_2'(m) + \phi_1(m) = 0$$

$$\frac{c^2}{2!}(6m-10) + c\,(-6m+9) + (2m-2) = 0$$

for $m = 2, 2$

$$\frac{c^2}{2!}(2) + c\,(-3) + (2) = 0$$

$$c^2 - 3c + 2 = 0$$

$$c = \frac{3 \pm \sqrt{9-8}}{2} = \frac{3 \pm 1}{2} = 2, 1$$

Eqn. of Asymptotes are,

$\boxed{y = x}$, $\boxed{y = 2x + 2}$, $\boxed{y = 2x + 1}$ Answer

Problem 14 Find all asymptotes of the curve:

$$x^2 (x - y)^2 - a^2 (x^2 + y^2) = 0$$

Solution: we have $x^2 (x^2 + y^2 - 2xy) - a^2x^2 - a^2y^2 = 0$

or $x^4 + x^2y^2 - 2x^3y - a^2x^2 - a^2y^2 = 0$

No asymptote parallel to x-axis,

asymptote parallel to y-axis

$x^2 - a^2 = 0 \quad \Rightarrow \quad \boxed{x = \pm a}$

For oblique Asymptote, put $x = 1, \quad y = m$

$\phi_4(m) = 1 + m^2 - 2m$

$\phi_3(m) = 0$

$\phi_2(m) = -a^2 - a^2m^2$

$\phi_4'(m) = 2m - 2$

$\phi_4(m) = 0 \quad \Rightarrow \quad 1 + m^2 - 2m = 0$

$m^2 - 2m + 1 = 0$

$$m = \frac{2 \pm \sqrt{4-4}}{2}$$

$= 1, 1$

for $m = 1, 1$ (repeated roots)

$$\frac{c^2}{2!}\phi_4''(m) + c\,\phi_3'(m) + \phi_2(m) = 0$$

$$\frac{c^2}{2!}(2)+c(0)+\left(-a^2-a^2(1)^2\right)=0$$

$$c^2-2a^2=0 \Rightarrow c=\pm\sqrt{2}\,a$$

Asymptotes are,

$\boxed{y=x+\sqrt{2}\,a}$, $\boxed{y=x-\sqrt{2}\,a}$ Answer

Example 15 Find all asymptotes of the curve:

$$x^2y^2(x^2-y^2)^2-(x^2+y^2)^3=0$$

Solution: we have $x^2y^2(x^4+y^4-2x^2y^2)-(x^6+y^6+3x^2y^2(x^2+y^2))=0$

or $x^6y^2+x^2y^6-2x^4y^4-x^6-y^6-3x^4y^2-3x^2y^4=0$

Asymptote parallel to x-axis,

co-eff. of highest power of x equate to zero,

$$y^2-1=0 \quad \Rightarrow \quad \boxed{y=\pm 1}$$

Asymptote llal to y-axis,

co-eff. of highest power of y i.e., y^6 equate to zero.

$$x^2-1=0 \quad \Rightarrow \quad \boxed{x=\pm 1}$$

For oblique Asymptote, put $x=1,\ y=m$

$\phi_8(m)=m^2+m^6-2m^4$

$\phi_7(m)=0$

$\phi_6(m)=-1-m^6-3m^2-3m^4$

$\phi_8'(m)=6m^5-8m^3+2m$

$\phi_8(m)=0 \Rightarrow m^6-2m^4+m^2=0$

$m^2(m^4-2m^2+1)=0$

$m^2(m^2-1)^2=0$

$\Rightarrow \quad m=0,0, \quad m^2-1=0, \quad m^2-1=0$

$\boxed{m=\pm 1}$, $\boxed{m=\pm 1}$

$\Rightarrow \quad m=1,1,-1,-1$

for $m=1,1$

$$\frac{c^2}{2!}\phi_8''(m)+c\,\phi_7'(m)+\phi_6(m)=0$$

$$\frac{c^2}{2!}\left(30m^4-24m^2+2\right)+c(0)+\left(-1-m^6-3m^2-3m^4\right)=0$$

m = 1, 1

$$\frac{c^2}{2!}\left(30(+1)^4-24(+1)^2+2\right)+c(0)+\left(-1-(1)^6-3(1)^2-3(1)^4\right)=0$$

$$\frac{c^2}{2!}(30-24+2)+(-1-1-3-3)=0$$

$$\frac{c^2}{2!}(8)+(8)=0$$

$$4c^2-8=0 \quad \Rightarrow \quad 4(c^2-2)=0$$

$$c^2=2 \quad \Rightarrow \quad c=\pm\sqrt{2}$$

for m = –1, –1, again,

$$c=\pm\sqrt{2}$$

eqns. of asymptotes are,

$\boxed{y=x\pm\sqrt{2}}$, $\boxed{y=-x\pm\sqrt{2}}$ Answer

Example 16 Find all Asymptotes of the curve

$$(x+y+1)(x^2+y^2-xy)-3xy+x^2+y^2+2x-3y+5=0$$

Solution: we have $x^3+\cancel{xy^2}-\cancel{x^2y}+\cancel{x^2y}+y^3-\cancel{xy^2}-xy+x^2+y^2$

$-3xy+x^2+y^2+2x-3y+5=0$

or $x^3+2x^2+2y^2+y^3-4xy+2x-3y+5=0$

No asymptote parallel to x and y-axis.

For oblique Asymptote, put x = 1, y = m

$\phi_3(m) = 1+m^3$

$\phi_2(m) = 2+2m^2-4m$

$\phi_1(m) = 2-3m$

$\phi_3'(m) = 3m^2$

$\phi_3(m) = 0 \Rightarrow 1+m^3=0$

$(m+1)\quad(m^2-m+1)=0$

m = –1 is the only real root,

$$c=-\frac{\phi_2(m)}{\phi_3'(m)}=-\frac{2+2m^2-4m}{3m^2}$$

$$m=-1, \qquad c=-\frac{2+2(-1)^2-4(-1)}{3(-1)^2}$$

$$= -\frac{2+2+4}{3} = \frac{-8}{3}$$

Eqn. of asymptote is,

$$y = -x - \frac{8}{3} \quad or \quad \boxed{3y + 3x + 8 = 0} \quad \text{Answer}$$

Example 17 Show that the parabola $y^2 = 4ax$ has no asymptote

Solution: We have $y^2 = 4ax$ or $y^2 - 4ax = 0$

or put $x = 1$, $y = m$ in the given equation

$\phi_2(m) = m^2$

$\phi_1(m) = -4a$

$\phi_2(m) = 0 \Rightarrow m^2 = 0, \quad m = 0, 0$

$\phi_2'(m) = 2m$

since co-eff. of highest power of 'y' and 'x' are constants, there is no asymptote llal to 'x' and 'y' axes

$$c = -\frac{\phi_1(m)}{\phi_2'(m)} = -\frac{-4a}{2m}$$

for $m = 0 \Rightarrow c = \frac{4a}{0} = \infty$

it means no asymptote for $m = 0$

$m = 0$ corresponds asymptote llal to x-axis, since we have shown above that we have no asymptote llal to 'x' and 'y' axis, thats why it contradicts our observation.

$\therefore$ There is no asymptote for parabola $\boxed{y^2 = 4ax}$ Proved.

Exercise 4.2

1. Find all the asymptotes of the following curves
   (a) $xy^2 + 4a^2(2a - x)$
   (b) $y^2(x - 2a) = x^3 - a^2$
   (c) $x^3 - 6x^2y + 11xy^2 - 6y^3 + x + y + 1 = 0$
   (d) $x(y - x)^2 = x(y - x) + 2$
2. Find all the asymptotes of the following curves
   (a) $(x - y)^2(x + 2y - 1) = 3x + y - 7$
   (b) $x^3 + 3x^2y - 4y^3 - x + y + 3 = 0$
   (c) $(x + y)^2(x + y + 2) = x + 9y - 2$
   (d) $y^3 - 2xy^2 - x^2y + 2x^3 + 3y^2 - 7xy + 2x^2 + 2y + 2x + 1 = 0$

3. Find all the asymptotes of the following curve

(a) $(x - y)^2 (x^2 + y^2) - 10(x - y) x^2 + 12y^2 + 2x + y = 0$

(b) $(y - x)(y - 2x)^2 + (y + 3x)(y - 2x) + 2x + 2y - 1 = 0$

(c) $x^3 + y^3 - 3axy = 0$

(d) $y^3 - 3x^2y + xy^2 - 3x^3 + 2y^2 + 2xy + 4x + 5y + 6 = 0$

4.5 INTERSECTION OF A CURVE AND ITS ASYMPTOTES

Let $y = mx + c$ be an asymptote of the curve ...(i)

$$x^n\phi_n\left(\frac{y}{x}\right)+x^{n-1}\,\phi_{n-1}\left(\frac{y}{x}\right)+x^{n-1}\,\phi_{n-2}\left(\frac{y}{x}\right)+\ldots\ldots=0 \text{ of nth degree} \quad \ldots\text{(ii)}$$

The points of intersection of the curve and the asymptotes are given by putting the values of y from (i) in (ii)

$$x^n\,\phi_n\left(m+\frac{c}{x}\right)+x^{n-1}\,\phi_{n-1}\left(m+\frac{c}{x}\right)+x^{n-2}\,\phi_{n-2}\left(m+\frac{c}{x}\right)+\ldots\ldots=0$$

or

$$x^n\phi_n(m)+(c\,\phi_n'(m)+\phi_{n-1}(m))\,x^{n-1}+$$

$$\left[\frac{1}{2!}c^2\,\phi_n''(m)+c\,\phi'_{n-1}(m)+\phi_{n-2}(m)\right]x^{n-2}+\ldots\ldots=0 \quad \ldots\text{(iii)}$$

since $y = mx + c$ is the asymptote of the curve (ii), then

$$\phi_n(m) = 0, \quad c\,\phi_n'(m) + \phi_{n-1}(m) = 0$$

The equation (iii) reduces to

$$\left[\frac{1}{2!}c^2\phi_n''(m)+c\,\phi'_{n-1}(m)=\phi_{n-2}(m)\right]x^{n-2}+\ldots\ldots=0 \quad \ldots\text{(iv)}$$

the coefficients of x^{n-2} will be non-zero provided there are no parallel asymptotes, since the equation (iv) is of degree (n–2) in x, there will be (n –2) points of intersection

Now for one asymptote we have (n–2) points of intersection. But there are at most 'n' asymptotes for the equation of n degree. Hence there will be at the most n(n–2) points of intersection.

4.6 ASYMPTOTES BY INSPECTION

If the equation of the curve can be put in the form $F_n + F_{n-2} = 0$ where F_n is of degree n (i.e., contains terms of degree n and may also contain lower degree terms) and F_{n-2} is of degree (n–2) at the most, then $F_n = 0$ will give the asymptotes and $F_{n-2} = 0$ will give the curve on which the points of intersection of the curve and asymptote lie.

Illustrative Examples

Example 1 Find asymptotes of the curve

$$\frac{x^2}{a^2}-\frac{y^2}{b^2}+1=0$$

Solution: It is of the form $F_n + F_{n-2} = 0$

asymptotes are $\frac{x^2}{a^2}-\frac{y^2}{b^2}=0$

$$y^2=\frac{b^2x^2}{a^2}$$

$$y=\pm\frac{b}{a}x$$

Example 2 Find asymptotes of the curve

$x^2y - xy^2 + xy - y^2 + x + y = 0$

Solution: It is of the form

$F_n + F_{n-2} = 0$

$F_n = x^2y - xy^2 + xy - y^2$

$F_{n-2} = x + y$

$F_n = 0 \Rightarrow x^2y - xy^2 + xy - y^2 = 0$

$xy(x - y) + y(x-y) = 0$

$(x-y)(xy + y) = 0$

$(x-y)(x + 1)y = 0$

$y = 0, x-y = 0, x + 1 = 0$ are the asymptotes. Answer

Example 3 Show that the points of intersection of the curve $xy(x^2 - y^2) + a^2y^2 + b^2x^2 - a^2b^2 = 0$ with its asymptotes lie on an ellipse,

Solution: The eqn. of curve is

$xy(x^2 - y^2) + a^2y^2 + b^2x^2 - a^2b^2 = 0$

equating to zero the co-eff. of highest power of x and y, we get

asymptote $y = 0$ parallel to x-axis

and $x = 0$ parallel to y-axis

Oblique asymptote

$\phi_4(m) = m(1- m^2)$, $\phi_3(m) = 0$,

$\phi_2(m) = a^2m^2 + b^2$

$\phi_4(m) = 0 \Rightarrow m(1 - m^2) = 0$

$m = 0, \quad m = \pm 1$

$\phi_4'(m) = 1 - 3m^2$

$c = 0$ since $\phi_3(m) = 0$ $\phi_4'(m) \neq 0$

so the asymptotes are

$$x = 0, \quad y = 0, \quad y = x, \quad y = -x$$

These asymptote cut the curve at 4(4 – 2) = 8 pts.

To find the equation of curve on which these 8 points. will lie, we find the joint eqn. of asymptote which is

$$xy(x + y)(x - y) = 0$$

$$xy(x^2 - y^2) = 0$$

$$F_n = 0$$

$F_n + F_{n-2} = 0$, the curve in which 8 points lie

or $\quad a^2y^2 + b^2x^2 - a^2b^2 = 0$

or $\quad \dfrac{x^2}{a^2} + \dfrac{y^2}{b^2} = 1$

which is an ellipse

alternatively

$$F_n + F_{n-2} = 0$$

$$xy(x^2 - y^2) + a^2y^2 + b^2x^2 - a^2b^2 = 0$$

$$F_n = 0$$

asymptotes are $\quad x = 0, \; y = 0, \; (x - y) = 0, \quad (x + y) = 0$

These four asymptotes lie on the conic

$F_{n-2} = 0$ i.e.,

$$a^2y^2 + b^2x^2 - a^2b^2 = 0$$

or $\quad \dfrac{x^2}{a^2} + \dfrac{y^2}{b^2} = 1 \quad$ which is an ellipse. Answer

Example 4 Find the equation of cubic which has the same asymptotes as the curve

$$x^3 - 6x^2y + 11xy^2 - 6y^3 + 4x + 5y + 7 = 0$$

and which passes through the points (0, 0), (0, 2) and (2, 0)

Solution: The equation of given curve

$$x^3 - 6x^2y + 11xy^2 - 6y^3 + 4x + 5y + 7 = 0$$

$$(x - y)(x - 2y)(x - 3y) + 4x + 5y + 7 = 0$$

which is of the form

$$F_3 + F_1 = 0$$

asymptotes are given by

$$F_3 = 0$$

$$\Rightarrow \quad x - y = 0, \qquad x - 2y = 0, \qquad x - 3y = 0$$

eqn. of cubic for which asymptotes are given by $F_3 = 0$ is of the form $F_3 + F_1 = 0$, where F_1 is of degree one at most. It will be of the form

$$(x - y)(x - 2y)(x - 3y) + ax + by + c = 0$$

a, b, c are constants. The cubic passes through (0,0) (0, 2), (2, 0), these points must satisfy the above eqn. with $a = -4$, $b = 24$, $c = 0$

substitute a, b, c, we get

$$(x - y)(x - 2y)(x - 3y) - 4x + 24y = 0$$

or $x^3 - 6x^2y + 11xy^2 - 6y^3 - 4x + 24y = 0$ is the required cubic. Answer

Example 5 Find the equation of hyperbola having $x + y - 1 = 0$ and $x - y + 2 = 0$ as its asymptotes and passing through the origin.

Solution: Joint eqn. of asymptote is $F_2 = (x + y - 1)(x - y + 2) = 0$

The eqn. of curve for which asymptotes are given by $F_n = 0$

is $F_n + F_{n-2} = 0$ If $n = 2$ then $F_{n-2} = F_0 =$ constant

$$\Rightarrow \quad (x + y - 1)\ (x - y + 2) + k = 0$$

it passes through the origin

(0, 0), if $(-1)(2) + K = 0$

$$\Rightarrow \boxed{k = 2}$$

Thus, $(x + y - 1)\ (x - y + 2) + 2 = 0$

or $x^2 - y^2 + x + 3y = 0$ Answer

Example 6 Find the asymptote of the curve $x^2y - xy^2 + xy + y^2 + x - y = 0$ and show that they cut the curve in three points which lie on the straight line $x + y = 0$

Solution: Eqn. of given curve is,

$$x^2y - xy^2 + xy + y^2 + x - y = 0 \qquad \text{(i)}$$

Asymptote ||al to x-axis

co-eff. of x^2, $\boxed{y = 0}$

Asymptote ||al to y-axis,

$$\boxed{1 - x = 0}$$

For oblique asymptote, put $x = 1$, $y = m$ in equation (i)

$$\phi_3(m) = m - m^2$$

$$\phi_2(m) = m + m^2$$

$$\phi_1(m) = 1 - m$$

$$\phi_3'(m) = 1 - 2m$$

$$\phi_3(m) = 0 \quad \Rightarrow \quad m - m^2 = 0$$

$$m(1 - m) = 0$$

$$m = 0, 1$$

for m = 0, 1 , 'c' is given by

$$c = -\frac{\phi_2(m)}{\phi_3'(m)} = -\frac{m+m^2}{1-2m}$$

for m = 0, c = 0

for m = 1, $c = -\frac{1+1}{1-2} = 2$

putting the values of 'c' and 'm' in

$y = mx + c$, the asymptotes are,

$\boxed{y=0}$, $\boxed{y=x+2}$

total asymptotes are,

$x - 1 = 0, \quad y = 0, \quad x - y + 2 = 0$

Thus, three asymptotes will cut the curve again in n(n–2) i.e. 3(3–2) = 3 points.

joint eqn. of asymptotes are,

$y(x - 1)(x - y + 2) = 0$

$(xy - y)(x - y + 2) = 0$

$x^2y - xy^2 + 2xy - xy + y^2 - 2y = 0$

Eqn. of given curve is,

$x^2y - xy^2 + xy + y^2 + x - y = 0$

which can be re-written as,

$$\left[x^2y - xy^2 + xy + y^2 - 2y\right] + [x + y] = 0$$

which is of the form,

$F_n + F_{n-2} = 0$

Hence, the points of intersection lie on the curve $F_{n-2} = 0$

i.e., $x + y = 0$, which is the required straight line Answer

Example 7 Find the equation of the straight line on which lie the three points of intersection of the curve $x^3 + 2x^2y - xy^2 - 2y^3 + 4y^2 + 2xy + y - 1 = 0$ and its asymptotes.

Solution: Eqn. of given curve is,

$x^3 + 2x^2y - xy^2 - 2y^3 + 4y^2 + 2xy + y - 1 = 0$

No asymptote parallel to x and y-axis, as the co-eff of highest power of x and y are constant.

For oblique asymptotes put x = 1, y = m in the 3rd and 2nd degree term

$\phi_3(m) = 1 + 2m - m^2 - 2m^3$

$\phi_2(m) = 4m^2 + 2m$

$\phi_1(m) = m$

$\phi_0(m) = -1$

$\phi_3'(m) = 2 - 2m - 6m^2$

$\phi_3(m) = 0 \quad \Rightarrow \quad 1 + 2m - m^2 - 2m^3 = 0$

or $\quad 2m^3 + m^2 - 2m - 1 = 0$

$m^2(2m + 1) - 1(2m + 1) = 0$

$(2m + 1)(m^2 - 1) = 0$

$m = -1/2, \quad m = \pm 1$

$$c = -\frac{\phi_2(m)}{\phi_3'(m)} = -\frac{4m^2 + 2m}{2 - 2m - 6m^2}$$

for $m = 1$

$$c = -\frac{4(1)^2 + 2(1)}{2 - 2(1) - 6(1)^2}$$

$$= -\frac{4+2}{2-2-6} = -\frac{6}{-6} = 1$$

for $m = -1$,

$$c = -\frac{4(-1)^2 + 2(-1)}{2 - 2(-1) - 6(-1)^2}$$

$$= -\frac{4-2}{2+2-6} = -\frac{2}{-2} = 1$$

for $m = -1/2$,

$$c = -\frac{4\left(-\frac{1}{2}\right)^2 + 2\left(-\frac{1}{2}\right)}{2 - 2\left(-\frac{1}{2}\right) - 6\left(-\frac{1}{2}\right)^2}$$

$$= \frac{\frac{4}{4} - \frac{2}{2}}{2 + \frac{2}{2} - \frac{6}{4}}$$

$$= \frac{1-1}{2 + \frac{3}{2}} = 0$$

putting the values of m and c in the eqn. $y = mx + c$

$\boxed{y = x + 1}$, $\quad \boxed{y = -x + 1}$, $\quad \boxed{y = -\frac{1}{2}x}$

These asymptotes will cut the curve in 3(3-2) = 3 points

Joint eqn. of asymptotes is

$$(x - y + 1)(x + y - 1)(x + 2y) = 0$$

or $$x^3 + 2x^2y - xy^2 - 2y^3 + 4y^2 + 2xy - 2y - x = 0$$

Eqn. of given curve can be written as,

$$(x^3 + 2x^2y - xy^2 - 2y^3 + 4y^2 + 2xy - 2y - x) + (x + 3y - 1) = 0$$

which is of the form

$$F_n + F_{n-2} = 0$$

points of intersection lie on the curve $F_{n-2} = 0$

i.e. $x + 3y - 1 = 0$ Answer

Example 8 Show that the asymptotes of the curve $x^3 - 2y^3 + xy(2x - y) + y(x - y) + 1 = 0$ cut the curve in three points which lie on the straight line $x - y + 1 = 0$

Solution: Eqn. of the given curve is,

$$x^3 - 2y^3 + 2x^2y - xy^2 + xy - y^2 + 1 = 0$$

This eqn. has no asymptote llal to x and y-axis, as the co-eff. of highest degree of x and y are constant.

For oblique asymptote, put $x = 1$, $y = m$ in the third and second degree term of the given curve.

$$\phi_3(m) = 1 - 2m^3 + 2m - m^2$$

$$\phi_2(m) = m - m^2$$

$$\phi_0(m) = 1$$

$$\phi_3'(m) = -6m^2 + 2 - 2m$$

$$\phi_3(m) = 0 \Rightarrow 2m^3 + m^2 - 2m - 1 = 0$$

$$m^2(2m + 1) - 1(2m + 1) = 0$$

$$(2m + 1)(m^2 - 1) = 0$$

$$\boxed{m = -\frac{1}{2}, \quad m = \pm 1}$$

$$c = -\frac{\phi_2(m)}{\phi_3'(m)} = \frac{m - m^2}{-6m^2 + 2 - 2m}$$

for $m = 1$, $$c = -\frac{(1) - (1)^2}{-6(1)^2 + 2 - 2(1)}$$

$$= \frac{1 - 1}{-6 + 2 - 2} = 0$$

for $m = -1$, $$c = -\frac{(-1) - (-1)^2}{-6(-1)^2 + 2 - 2(-1)}$$

$$= -\frac{-1-1}{-6+2+2} = -\frac{-2}{-2} = -1$$

for $m = -1/2$, $$c = -\frac{-\frac{1}{2}-\left(-\frac{1}{2}\right)^2}{-6\left(-\frac{1}{2}\right)^2+2-2\left(-\frac{1}{2}\right)}$$

$$= -\frac{-\frac{1}{2}-\frac{1}{4}}{-\frac{6}{4}+2+1}$$

$$= \frac{1}{2}$$

Asymptotes are,

$\boxed{y = x}$, $\boxed{y = -x-1}$, $\boxed{y = -\frac{1}{2}x + \frac{1}{2}}$

These three asymptote will cut the curve at $3(3-2) = 3$ points

Joint eqn. of the asymptote is,

$$(x-y)(x+y+1)(2y+x-1) = 0$$

$$x^3 - 2y^3 + 2x^2y - xy^2 + xy - y^2 - x + y = 0$$

eqn. of given curve can be rewritten as,

$$[x^3 - 2y^3 + 2x^2y - xy^2 + xy - y^2 - x + y] + [x - y + 1] = 0$$

which is of the form

$$F_n + F_{n-2} = 0$$

Hence, the pts. of intersection lie on the curve $F_{n-2} = 0$

i.e., $x - y + 1 = 0$ Answer

Example 9 Find the asymptote of the curve $xy(x^2 - y^2) + x^2 + y^2 = a^2$ and show that the eight points of intersection of the curve and its asymptotes lie on a circle whose center is at the origin.

Solution: Equation of given curve is

$$x^3y - xy^3 + x^2 + y^2 - a^2 = 0$$

Asymptote parallel to x-axis,

$\boxed{y = 0}$

Asymptote parallel to y-axis

$\boxed{x = 0}$

For oblique asymptote, put $x = 1$, $y = m$

$$\phi_4(m) = m - m^3$$
$$\phi_3(m) = 0$$
$$\phi_2(m) = 1 + m^2$$
$$\phi_0(m) = -a^2$$
$$\phi_4'(m) = 1 - 3m^2$$
$$\phi_4(m) = 0 \quad \Rightarrow \quad m - m^3 = 0$$
$$\Rightarrow \quad m(1 - m^2) = 0$$
$$\Rightarrow \quad m = 0, \qquad m = \pm 1$$

$$c = -\frac{\phi_3(m)}{\phi_4'(m)} = 0$$

for $m = 0, 1, -1,$ $\quad c = 0$

Eqn. of asymptotes are,

$\boxed{y = 0}$, $\boxed{y = x.}$, $\boxed{y = -x}$

Total asymptotes are

$\boxed{x = 0}$ $\boxed{y = 0}$, $\boxed{x - y = 0}$ $\boxed{x + y = 0}$

Joint eqn. of the asymptote is,

$$xy(x^2 - y^2) = 0$$
$$x^3y - xy^3 = 0$$

Eqn. of given curve can be re-written as,

$$(x^3y - xy^3) + (x^2 + y^2 - a^2) = 0$$

which is of the form

$$F_n + F_{n-2} = 0$$

Hence, the pts. of intersection lie on the curve $F_{n-2} = 0$

i.e., $x^2 + y^2 = a^2$, which is a circle of radius 'a' and center at origin.

The four asymptotes will cut the curve at $4(4 - 2) = 8$ points. Answer

Example 10 Show that the eight points of intersection of the curve $x^4 - 5x^2y^2 + 4y^4 + x^2 - y^2 + x + y + 1 = 0$ and its asymptotes lie on a rectangular hyperbola.

Solution: Eqn. of the given curve is,

$$x^4 - 5x^2y^2 + 4y^4 + x^2 - y^2 + x + y + 1 = 0$$

No asymptote parallel to x and y-axis

(co-effs. of highest power of x and y are constant)

put $x = 1$, $y = m$ in the given curve to find oblique asymptote

$$\phi_4(m) = 1 - 5m^2 + 4m^4$$
$$\phi_3(m) = 0$$
$$\phi_2(m) = 1 - m^2$$

$\phi_1(m) = 1 + m$

$\phi_4'(m) = -10m + 16m^3$

$\phi_4(m) = 0 \quad \Rightarrow \quad 4m^4 - 5m^2 + 1 = 0$

$$4m^4 - 4m^2 - m^2 + 1 = 0$$

$$4m^2(m^2 - 1) - 1(m^2 - 1) = 0$$

$$(m^2 - 1)(4m^2 - 1)$$

$$m = \pm 1, \quad m = \pm\frac{1}{2}$$

$$c = -\frac{\phi_3(m)}{\phi_4'(m)} = 0$$

Eqn. of asymptotes are,

$\boxed{y = x}$ $\boxed{y = -x}$, $\boxed{y = -\frac{1}{2}x}$, $\boxed{y = \frac{1}{2}x}$

The four asymptote will cut the curve in 4(4 – 2) = 8 points.

Joint eqn. of asymptotes is,

$$(x - y)(x + y)(2y + x)(2y - x) = 0$$

$$x^4 - 5x^2y^2 + 4y^4 = 0$$

Eqn. of given curve can be re-written as,

$$(x^4 - 5x^2y^2 + 4y^4) + (x^2 - y^2 + x + y + 1) = 0$$

which is of the form,

$$F_n + F_{n-2} = 0$$

Hence, the pts. of intersection lie on the curve $F_{n-2} = 0$

i.e., $x^2 - y^2 + x + y + 1 = 0$, which is a rectangular hyperbola. Answer

Example 11 Show that the four asymptotes of the curve $xy(x^2 - y^2) + 9x^2 + 25y^2 = 144$ cut the curve in eight points which lie on an ellipse of eccentricity 4/5.

Solution: Eqn. of given curve is,

$$x^3y - xy^3 + 9x^2 + 25y^2 - 144 = 0$$

Asymptote parallel to x-axis,

$\boxed{y = 0}$

Asymptote llal to y-axis

$\boxed{x = o}$

For oblique asymptote, put $x = 1, \quad y = m$

$$\phi_4(m) = m - m^3$$

$\phi_3(m) = 0$

$\phi_2(m) = 9 + 25m^2$

$\phi_4'(m) = 1 - 3m^2$

$\phi_4(m) = 0 \quad \Rightarrow \quad m - m^3 = 0$

$\Rightarrow \quad m(1 - m^2) = 0$

$\Rightarrow \quad m = 0, \quad m = \pm 1$

$$c = -\frac{\phi_3(m)}{\phi_4'(m)} = 0$$

$c = 0$ for all $m = 0, 1, -1$

Eqn. of asymptotes are,

$\boxed{y = 0}$, $\boxed{y = x}$, $\boxed{y = -x}$

Joint eqn. of asymptotes is

$xy(x - y)(x + y) = 0$

$x^3y - xy^3 = 0$

Eqn. of given curve can be re-written as,

$[x^3y - xy^3] + [9x^2 + 25y^2 - 144] = 0$

which is of the form

$F_n + F_{n-2} = 0$

Hence the pts. of intersection lie on the curve $F_{n-2} = 0$

i.e., $9x^2 + 25y^2 - 144 = 0$

or $$\frac{x^2}{25} + \frac{y^2}{9} = 1$$

$a = 5, \quad b = 3$

$$e = \sqrt{1 - \frac{b^2}{a^2}}$$

$$= \sqrt{1 - \frac{9}{25}} = \sqrt{\frac{16}{25}} = \frac{4}{5}$$

$e = 4/5$ Answer

Example 12 Find the equation of the cubic which has the same asymptotes as the curve $x^3 - 6x^2y + 11xy^2 - 6y^3 + 4x + 5y + 7 = 0$ and which passes through the points (0, 0), (0, 2) and (2,0)

Solution: No asymptote llal to x and y-axis, as the co-eff. of highest degree term in x and y-axis is constant.

For oblique asymptote, put $x = 1, \quad y = m,$

$$\phi_3(m) = 1 - 6m + 11m^2 - 6m^3$$
$$\phi_2(m) = 0$$
$$\phi_1(m) = 4 + 5m$$
$$\phi_0(m) = 7$$
$$\phi_3'(m) = -6 + 22m - 18m^2$$
$$\phi_3(m) = 0 \quad \Rightarrow \quad 6m^3 - 11m^2 + 6m - 1 = 0$$

m = 1 is one root by inspection

By synthetic division

1	6	−11	6	−1
		6	−5	1
	6	−5	1	0

$$6m^2 - 5m + 1 = 0$$

$$m = \frac{5 \pm \sqrt{25 - 24}}{12}$$

$$= \frac{5 \pm 1}{12} = \frac{1}{2}, \frac{1}{3}$$

for $m = 1, \frac{1}{2}, \frac{1}{3},$

$$c = -\frac{\phi_2(m)}{\phi_3'(m)} = 0$$

for every m, $c = 0$

Eqns. of asymptotes are

$\boxed{y = x}$, $\boxed{y = \frac{1}{2}x}$, $\boxed{y = \frac{1}{3}x}$

Joint eqn. of asymptotes is,

$$(x - y)(x - 2y)(x - 3y) = 0$$
$$F_n = 0$$
$$x^3 - 6x^2y + 11xy^2 - 6y^3 = 0$$

Eqn. of any cubic having asymptotes given by $F_n = 0$ is of the form

$$F_n + F_{n-2} = 0$$
$$F_3 + F_1 = 0$$

i.e., $(x - y)(x - 2y)(x - 3y) + ax + by + c = 0$

$ax + by + c$ is a general linear eqn. Since curve passes through

(0, 0), (0, 2) and (2, 0)

for (0, 0), $\Rightarrow$ $c = 0$

for (0, 2) $\Rightarrow$ $(-2)(-4)(-6) + 2b + c = 0$

$-48 + 2b = 0$

$2b = 48$

$\boxed{b = 24}$

for (2, 0) $\Rightarrow$ $(2)(2)(2) + 2a + c = 0$

$2a = -8$

$\boxed{a = -4}$

Eqn. of the curve is,

$(x-y)(x-2y)(x-3y) + (-4x + 24y + 0) = 0$

$x^3 - 6x^2y + 11xy^2 - 6y^3 - 4x + 24y = 0$ Answer

Example 13 Find the eqn. of the curve which has the same asymptotes as the curve $x^3 - 6x^2y + 11xy^2 - 6y^3 + x + y + 1 = 0$ and which touches the axis of y at the origin and goes through the pt. (3, 2).

Solution: Eqn. of the curve is,

$x^3 - 6x^2y + 11xy^2 - 6y^3 + x + y + 1 = 0$

Curve has no asymptote llal to x and y-axis, co-eff. of x^3 and y^3 are constant.

For oblique asymptotes, put $x = 1$, $y = m$,

$\phi_3(m) = 1 - 6m + 11m^2 - 6m^3$

$\phi_2(m) = 0$

$\phi_1(m) = 1 + m$

$\phi_3'(m) = -6 + 22m - 18m^2$

$\phi_3(m) = 0 \Rightarrow 6m^3 - 11m^2 + 6m - 1 = 0$

$m = 1, 1/2, 1/3$

$c = 0$ for all m.

The asymptotes are $x - y = 0$, $x - 2y = 0$ $x - 3y = 0$

Joint eqn. of asymptotes is

$(x-y)(x-2y)(x-3y) = 0$

$F_n = 0$

Eqn. of cubic having asymptotes of the form $F_n = 0$ is

$F_n + F_{n-2} = 0$

i.e., $F_3 + F_1 = 0$

$(x-y)(x-2y)(x-3y) + ax + by + c = 0$ ...(2)

$ax + by + c = 0$ is general linear eqn.

passes through the pt. (0, 0) and (3, 2)

for (0, 0) $\Rightarrow$ $c = 0$

for (3, 2), we get $(1)(-1)(-3) + 3a + 2b + 0 = 0$ {from (2)}

$$3a + 2b = -3 \qquad \ldots(3)$$

x = 0 is a tangent to the curve at (0, 0)

$$\Rightarrow \frac{dx}{dy} = 0$$

differentiating (2) w.r.t. y

$$\left(\frac{dx}{dy} - 1\right)(x - 2y)(x - 3y) + \left(\frac{dx}{dy} - 2\right)(x - y)(x - 3y) + \left(\frac{dx}{dy} - 3\right)(x - y)(x - 2y) +$$

$$a.\frac{dx}{dy} + b = 0$$

at (0, 0)

$$\frac{dx}{dy} = -b/a$$

given $\frac{dx}{dy} = 0 \quad \Rightarrow \quad -b/a = 0 \quad \Rightarrow \quad \boxed{b = 0}$

from (3)

$$3a + 2(0) = -3$$

$$\boxed{a = -1}$$

Eqn. of the curve is,

$$x^3 - 6x^2y + 11xy^2 - 6y^3 - x = 0$$

Exercise 4.3

1. Find the equation of the hyperbola with $x + y - 1 = 0, x + y - 2 = 0$ as its asymptotes pass through the origin.
2. Find the equation of the cubic which has the same asymptotes as the curve $x^3 - 6x^2y + 11xy^2 - 6y^3 + x + y + 1 = 0$ and which passes through the points (0, 0), (1, 0) and (0, 1).
3. Show that the four asymptotes of the curve $xy(x^2 - y^2) + 25y^2 + 9x^2 - 144 = 0$ cut it again in 8 points on an ellipse whose eccentricity is 4/5.
4. Find the equation of the bi-quadratic curve which has $x = 0, y = x, y = -x$ as asymptotes and which passes through (a, b) and which cuts the asymptotes in 8 points that lie on the circle $x^2 + y^2 = a^2$.
5. Find the asymptotes to the curve $x^3 + 3x^2y - 4y^3 - x + y + 3 = 0$ and show that they cut the curve in 8 points which lie on the line $x - y - 3 = 0$.

Answers

1. $(x + y - 1)(x + y - 2) - 2 = 0$
2. $x^3 - 6x^2y + 11xy^2 - 6y^3 - x + 6y = 0$
4. $bxy(y^2 - x^2) + a(a^2 - b^2)(x^2 + y^2 - a^2) = 0$
5. $y - x = 0, \quad 2y + x - 1 = 0, \quad 2y - x + 1 = 0$

4.7 ASYMPTOTES IN POLAR CO-ORDINATES

When a curve is given in the form of 'r' and 'θ' i.e., $r = f(\theta)$, it is called a polar curve.

Theorem - I The polar equation of any line is $p = r\cos(\theta - \alpha)$, where p is the length of perpendicular from the pole to the line and α is the angle which the perpendicular makes with the initial line (x –axis.)

Proof: Let O be the pole and OX be the initial line.

Let $p(r,\theta)$ be any point on the given line such that

$$OP = r \text{ and } \angle XOP = \theta$$

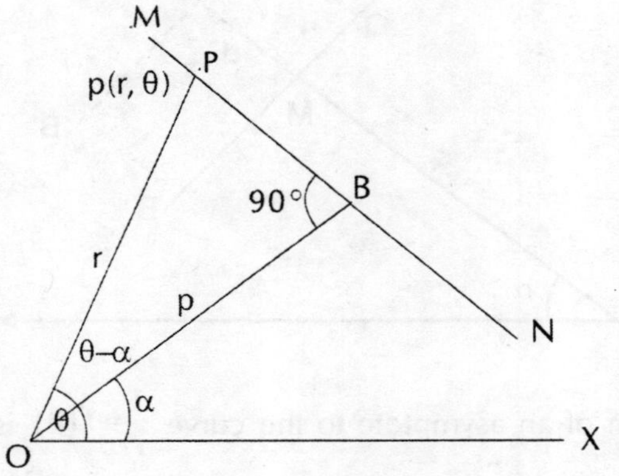

Let $OB = p$, be the length of perpendicular from pole to the given line MN and $\angle XOB = \alpha$

$\therefore$ $\angle BOP = \angle XOP - \angle XOB = \theta - \alpha$

Now in ΔOPB

$$\cos(\theta - \alpha) = \frac{OB}{OP} = \frac{p}{r}$$

$\therefore$ $p = r\cos(\theta - \alpha)$ which is the required equation of the line.

Theorem - II The length of the perpendicular from a point $p(r_1, \theta_1)$ to the line $p = r\cos(\theta - \alpha)$ is

$$\pm\{p - r_1\cos(\theta_1 - \alpha)\}$$

Proof: Let $p = r\cos(\theta - \alpha)$ be the equation of the line AB.

$\therefore$ $OM = p$ and $\angle XOM = \alpha$

Now through $p(r_1, \theta_1)$ draw a line A'B' parallel to AB and let $PQ = d$ be the length of perpendicular from A'B' to the line AB.

$\therefore$ $ON = OM + MN = OM + QP = p + d$ and $\angle XON = \alpha$

$\therefore$ equation of line A'B' is $p + d = r\cos(\theta - \alpha)$

It passes through $p(r_1, \theta_1)$

$\therefore$ $p + d = r_1\cos(\theta_1 - \alpha)$

or $d = -p + r_1\cos(\theta_1 - \alpha)$

$= -[p - r_1 \cos(\theta_1 - \alpha)$

Similarly by taking p below, we can see that $d = \{p - r_1 \cos(\theta_1 - \alpha)\}$

on combining the two; d, the perpendicular distance of $p(r_1, \theta_1)$ from the line $p = r\cos(\theta_1 - \alpha)$ is given by

$$d = \pm\{p - r_1 \cos(\theta_1 - \alpha)\}$$

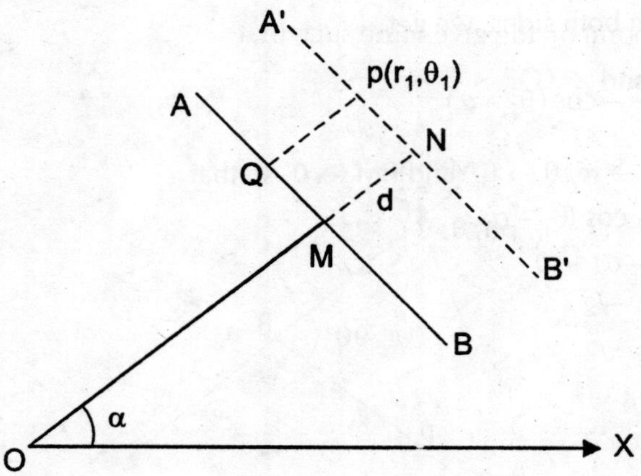

Theorem - III polar equation of an asymptote to the curve $\frac{1}{r} = f(\theta)$ is $p = r\sin(\theta_1 - \theta)$, where $p = \underset{\theta \to \theta_1}{lt.}\left(\frac{-d\theta}{du}\right)$, $u = \frac{1}{r}$ and θ_1 is a root of the equation obtained by putting $u = \frac{1}{r} = 0$, i.e., $f(\theta) = 0$.

Proof: Let $p = r\cos(\theta - \alpha)$ ...(i)

be an asymptote to the curve $r = f(\theta)$ ...(ii)

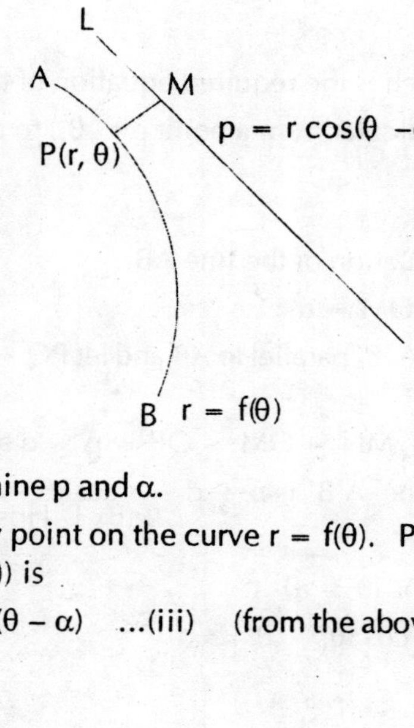

we are to determine p and α.

Let $P(r, \theta)$ be any point on the curve $r = f(\theta)$. PM, the perpendicular distance of $P(r, \theta)$ from the line $p = r(\cos(\theta - \alpha))$ is

$$PM = p - r\cos(\theta - \alpha) \quad ...(iii) \quad \text{(from the above theorem)}$$

$\because$ line $p = r\cos(\theta - \alpha)$ is an asymptote to the curve $r = f(\theta)$, therefore $PM \to 0$ as $P(r, \theta) \to \infty$ along the curve

As $P(r, \theta) \to \infty$ along the curve $r \to \infty$ and let θ tends to a finite value (say θ_1) also $r \to \infty$ implies $u \to 0$ $\quad \because r = \dfrac{1}{u}$

dividing (iii) by r on both sides, we get

$$\frac{PM}{r} = \frac{p}{r} - \cos(\theta_1 - \alpha)$$

taking limit as $r \to \infty$, $\theta \to \theta_1$, and $PM \to 0$, so that

$$0 = 0 - \cos(\theta_1 - \alpha)$$

or $\quad \cos(\theta_1 - \alpha) = 0$

$$\theta_1 - \alpha = \pi/2$$

$$\alpha = \theta_1 - \pi/2 \qquad \text{...(iv)}$$

Again taking limit in (iii) as

$PM \to 0$, $r \to \infty$ and $\theta \to \theta_1$

$$O = p - \underset{\substack{r\to\infty \\ \theta - \theta_1}}{\text{lt.}}\; r\cos(\theta - \alpha)$$

or $\quad p = \underset{\substack{r\to\infty \\ \theta\to\theta_1}}{\text{lt.}}\; r\cos(\theta - \alpha)$

from (iv) $p = \underset{\substack{r\to\infty \\ \theta\to\theta_1}}{\text{lt.}}\; r\cos(\theta - \theta_1 + \pi/2)$

or $\quad p = \underset{\substack{r\to\infty \\ \theta\to\theta_1}}{\text{lt.}}\; -r\sin(\theta - \theta_1)$ (it is of the form $\infty \times 0$)

or $\quad p = \underset{\substack{r\to\infty \\ \theta\to\theta_1}}{\text{lt.}}\; \dfrac{-\sin(\theta - \theta_1)}{(1/r)}$ $\qquad$ (form $\dfrac{0}{0}$)

or $\quad p = \underset{\substack{u\to 0 \\ \theta\to\theta_1}}{\text{lt.}}\; \dfrac{-\sin(\theta - \theta_1)}{u}$ $\qquad \left(\dfrac{0}{0}\right)$

or $\quad p = \underset{\substack{u\to 0 \\ \theta\to\theta_1}}{\text{lt.}}\; \dfrac{-\cos(\theta - \theta_1)}{\dfrac{du}{d\theta}}$ $\qquad$ (By L' Hospital's Rule)

$$= \underset{u\to 0}{\text{lt}} \frac{1}{\frac{du}{d\theta}} \lim_{\theta\to\theta_1} [-\cos(\theta-\theta_1)]$$

$$p = \lim_{u\to 0}\left(-\frac{d\theta}{du}\right) \qquad \left[\because \lim_{\theta\to\theta_1} \cos(\theta-\theta_1) = 1\right]$$

putting the value of α in (i) then the equation of asymptote is

$$p = r\cos\left(\theta-\theta_1+\frac{\pi}{2}\right) = r\cos\left[\frac{\pi}{2}-(\theta_1-\theta)\right]$$

$$p = r\sin(\theta_1-\theta)$$

$$p = \underset{\theta\to\theta_1}{\text{lt}}\left(-\frac{d\theta}{du}\right) = -\frac{1}{f'(\theta)}$$

$$p = -\frac{1}{f'(\theta)}$$

$\therefore$ asymptote is $r\sin(\theta_1-\theta) = \underset{u\to 0}{\text{lt}}\left(-\frac{d\theta}{du}\right) = -\frac{1}{f'(\theta)}$

$$\therefore \left(\begin{aligned} &\frac{1}{r} = f(\theta) \Rightarrow u = f(\theta) \\ &\frac{du}{d\theta} = f'(\theta) \Rightarrow \frac{d\theta}{du} = \frac{1}{f'(\theta)} \end{aligned}\right.$$

Working method for finding polar asymptotes

If α be the root of equation $f(\theta) = 0$ then $r\sin(\theta-\alpha) = \frac{1}{f'(\alpha)}$ is called the asymptotes of a polar curve $\frac{1}{r} = f(\theta)$. We adopt the following steps to find asymptotes of the above equation.

Step- I: Find the value of $\frac{1}{r}$ from the equation of curve and let it be $f(\theta)$. Also change all t-ratios if any into $\sin\theta$ and $\cos\theta$.

Step- II: Put $f(\theta) = 0$ and solve for θ and let $\theta = \alpha, \beta, \gamma$ ...... be the roots.

Step- III: Find $f'(\theta)$ i.e. differentiate $f(\theta)$ with respect to θ and find $f'(\theta)$ at $\theta = \alpha, \beta, \gamma$ ......

Step- IV: The asymptotes corresponding to $\theta = \alpha$ is

$$r \sin(\theta - \alpha) = \frac{1}{f'(\alpha)}, \text{ where } f'(\alpha) \neq 0$$

Note: If $\frac{1}{f'(\alpha)}$ at $\theta = \alpha$ is infinite, then the curve has no asymptotes corresponding to $\theta = \alpha$.

Remarks: Some helpful results of trigonometry are as follows:

1. f $\sin\theta = 0$ then $\theta = n\pi$, where n is any integer i.e $n \in I$
2. If $\cos\theta = 0$ then $\theta = (2n+1)\frac{\pi}{2}$, $n \in I$
3. If $\sin\theta = \sin\alpha$ then $\theta = n\pi + (-1)^n \alpha$, $n \in I$
4. If $\cos\theta = \cos\alpha$ then $\theta = 2n\pi \pm \alpha$, $n \in I$
5. If $\tan\theta = \tan\alpha$ then $\theta = n\pi + \alpha$, $n \in I$
6. $\frac{1}{(-1)^n} = (-1)^n$, $n \in I$
7. $\sin(n\pi + \theta) = (-1)^n \sin\theta$
8. $\cos(n\pi + \theta) = (-1)^n \cos\theta$
9. $\sin(n\pi - \theta) = -(-1)^n \sin\theta$

Illustrative Examples

Example 1 Prove that the curve $r = \frac{a}{1-\cos\theta}$ has no asymptotes.

Solution: Eqn. of curve is $r = \frac{a}{1-\cos\theta}$

$$\frac{1-\cos\theta}{a} = \frac{1}{r} = f(\theta)$$

Here, $f(\theta) = 0 \Rightarrow \frac{1-\cos\theta}{a} = 0$

$\Rightarrow$ $\cos\theta = 1$ $\Rightarrow$ $\theta = 2n\pi$, $n \in I$

$= \theta_1 =$ (say)

$$f'(\theta) = \frac{\sin\theta}{a} \text{ gives}$$

$$f'(2n\pi) = \frac{\sin(2n\pi)}{a} = 0$$

Eqn. of asymptote corresponding $\theta = \theta_1$ is

$$r \sin(\theta - 2n\pi) = \frac{1}{f'(\theta_1)} = \frac{1}{f'(2n\pi)} = \frac{1}{0}$$

$\Rightarrow$ no asymptote as $\frac{1}{f'(\theta_1)} = \infty$

Example 2 Find asymptotes of the curve

$$r \sin\theta = 2\cos 2\theta$$

Solution:

$$\frac{1}{r} = \frac{\sin\theta}{2\cos 2\theta} = f(\theta)$$

$$f(\theta) = 0 \Rightarrow \frac{\sin\theta}{2\cos 2\theta} = 0$$

$$\Rightarrow \quad \theta = n\pi = \theta_1 \text{ (say)}$$

$$f'(\theta) = \frac{1}{2}\left[\frac{\cos 2\theta(\cos\theta) + \sin\theta(2\sin 2\theta)}{\cos^2 2\theta}\right]$$

$$f'(\theta) = \left[\frac{\cos\theta.\cos 2\theta + 2\sin\theta \ \sin 2\theta}{2\cos^2 2\theta}\right]$$

$$f'(\theta_1) = f'(n\pi) = \frac{\cos n\pi.\cos 2n\pi}{2\cos^2 2n\pi} + \frac{2\sin n\pi.\sin 2n\pi}{2\cos^2 2n\pi}$$

$$= \frac{(-1)^n}{2}$$

$$r\sin(\theta - n\pi) = \frac{2}{(-1)^n} = 2(-1)^n$$

$$-r\sin(n\pi - \theta) = 2(-1)^n$$

$$-r\,\cancel{(-1)^n}\sin(-\theta) = 2\cancel{(-1)^n}$$

$$\cancel{-}(r)(\cancel{-}\sin\theta) = 2$$

$\boxed{r\sin\theta = 2}$ required asymptote **Answer**

Example 3 Find asymptote of the curve $r = a\tan\theta$

Solution:

$$\frac{1}{r} = \frac{1}{a\tan\theta} = f(\theta)$$

$$f(\theta) = 0 \Rightarrow \frac{1}{a\tan\theta} = 0 \Rightarrow \frac{\cot\theta}{a} = 0$$

$$\Rightarrow \cot\theta = 0 \Rightarrow \theta = (2n+1)\pi/2 = \theta_1$$

$$f'(\theta) = -\frac{\operatorname{cosec}^2\theta}{a}$$

$$f'\,(2n+1)\,\pi/2 = -\frac{\operatorname{cosec}^2\,(2n+1)\,\pi/2}{a}$$

$$= -\frac{1}{a}$$

asymptote is,

$$r\sin\left(\theta - (2n+1)\pi/2\right) = \frac{1}{-1/a} = -a$$

$$-r\sin\left(n\pi + \left(\frac{\theta-\pi}{2}\right)\right) = -a$$

$$(-1)^n\, r\sin(\theta - \pi/2) = -a$$

$$(-1)^n\, r\sin(\pi/2 - \theta) = a$$

$$(-1)^n\, r\cos\theta = a$$

n is even, $r\cos\theta = a$

n is odd $r\cos\theta = -a$

or $\boxed{r\cos\theta = \pm a}$ Answer

Example 4 Find the asymptote of the curve $r = \dfrac{a\,\theta}{\theta - 1}$

Solution: Eqn. can be written as,

$$\frac{1}{r} = \frac{\theta - 1}{a\theta} = f(\theta)\ say$$

put $f(\theta) = 0 \quad \Rightarrow \quad \dfrac{\theta - 1}{a\theta} = 0$

$$\Rightarrow \quad \theta = 1 = \theta_1$$

$$f'(\theta) = \frac{(a\theta).1 - (\theta - 1).a}{(a\theta)^2}$$

$$= \frac{a\theta - a\theta + a}{(a\theta)^2} = \frac{a}{a^2\,\theta^2} = \frac{1}{a\,\theta^2}$$

$$f'(\theta_1) \Rightarrow f'(1) = \frac{1}{a}$$

put in formula,

$$r\sin(\theta - \theta_1) = \frac{1}{f'(\theta_1)}$$

$$r\sin(\theta - 1) = \frac{1}{1/a}$$

$\Rightarrow$ $r\sin(\theta - 1) = a$

Example 5 Find the asymptotes of the curve $r(1 - e^\theta) = a$

Solution: Given eqn. can be written as,

$$\frac{1}{r} = \frac{1 - e^\theta}{a} = f(\theta) = (say)$$

putting $f(\theta) = 0 \Rightarrow \frac{1 - e^\theta}{a} = 0 \Rightarrow e^\theta = 1$

$$\boxed{\theta = 0} \Rightarrow \boxed{\theta_1 = 0}$$

$$f'(\theta) = \frac{-e^\theta}{a} = -\frac{1}{a} = f'(\theta_1) = f'(0)$$

Put the value in the formula,

$$r\sin(\theta - 0) = \frac{1}{-\frac{1}{a}}$$

$\Rightarrow$ $\boxed{r\sin\theta + a = 0}$ Answer

Example 6 Find the asymptotes of the curve $r\log\theta = a$

Solution:

$$\frac{1}{r} = \frac{\log\theta}{a} = f(\theta)\,(say)$$

put $f(\theta) = 0 \Rightarrow \frac{\log\theta}{a} = 0$

$\log\theta = 0$

$\Rightarrow \quad \boxed{\theta = 1}$

$\Rightarrow \quad \boxed{\theta_1 = 1}$

$$f'(\theta) = \frac{1}{a\theta}$$

$$f'(\theta_1) = f'(1) = \frac{1}{a}$$

put the values in the formula,

$$r \sin(\theta - \theta_1) = \frac{1}{f'(\theta_1)}$$

$$\Rightarrow \quad r \sin(\theta - 1) = \frac{1}{1/a}$$

$$\Rightarrow \quad \boxed{r \sin(\theta - 1) = a}$$

Example 7. Find the asymptote of the curve $r \sin n\theta = a$

Solution: Given eqn. can be written as

$$\frac{1}{r} = \frac{\sin n\theta}{a} = f(\theta)\ (say)$$

$$f(\theta) = 0 \quad \Rightarrow \quad \frac{\sin n\theta}{a} = 0$$

$\sin n\theta = 0$

$\Rightarrow \quad n\theta = m\pi$

$$\theta = \frac{m\pi}{n} = \theta_1 \text{ (say)}$$

$$f'(\theta) = \frac{n \cos n\theta}{a}$$

$$f'(\theta_1) = f'\left(\frac{m\pi}{n}\right) = \frac{\not{n} \cos \not{n}.\frac{m\pi}{\not{n}}}{a}$$

$$= \frac{n \cos m\pi}{a}$$

$$= \frac{n}{a}(-1)^m$$

put the values in the formula,

$$r\sin(\theta-\theta_1)=\frac{1}{f'(\theta_1)}$$

$$r\sin\left(\theta-\frac{m\pi}{n}\right)=\frac{1}{\frac{n}{a}(-1)^m}=\frac{a(-1)^m}{n}$$

$$r\sin\left(\theta-\frac{m\pi}{n}\right)=\frac{a(-1)^m}{n}$$ Answer

Example 8 Find the asymptote of the curve $r\sin\theta = ae^{\theta}$

Solution: Given eqn. can be written as

$$\frac{1}{r}=\frac{\sin\theta}{a e^{\theta}}=f(\theta)\ (say)$$

put $$f(\theta)=0 \Rightarrow \frac{\sin\theta}{a e^{\theta}}=0$$

$$\sin\theta=0 \Rightarrow \boxed{\theta=n\pi}$$

$$\theta_1=n\pi$$

$$f'(\theta)=\frac{1}{a}\left[\frac{e^{\theta}.\cos\theta-\sin\theta\, e^{\theta}}{e^{2\theta}}\right]$$

$$=\frac{e^{\theta}.(\cos\theta-\sin\theta)}{a\, e^{2\theta}}=\frac{\cos\theta-\sin\theta}{a\, e^{\theta}}$$

$$f'(\theta_1)=f'(n\pi)=\frac{\cos n\pi-\sin n\pi}{a\, e^{n\pi}}$$

$$=\frac{(-1)^n}{a e^{n\pi}}$$

put the values in the formula,

$$r\sin(\theta-n\pi)=\frac{1}{(-1)^n}\, a e^{n\pi}$$

$$-r\sin(n\pi-\theta)=\frac{a e^{n\pi}}{(-1)^n}$$

$$\Rightarrow \quad -(-1)^n r\sin(-\theta)=\frac{a.\, e^{n\pi}}{(-1)^n}$$

or $$(-1)^n r\sin\theta=\frac{a\, e^{n\pi}}{(-1)^n}$$

or $$(-1)^{2n} r\sin\theta=a\, e^{n\pi}$$

or $r\sin\theta = a\, e^{n\pi}$ Answer

Example 9 Find the asymptote of the curve $r = a(\sec\theta + \tan\theta)$

Solution: Given eqn. can be written as

$$r=\frac{a}{\cos\theta}+\frac{a\sin\theta}{\cos\theta}$$

$$=\frac{a+a\sin\theta}{\cos\theta}$$

$$\frac{1}{r}=\frac{\cos\theta}{a+a\sin\theta}=f(\theta)$$

put $f(\theta) = 0 \Rightarrow \cos\theta = 0$

$\theta = (2n+1)\pi/2 = \theta_1$

$$f'(\theta)=\frac{(a+a\sin\theta)(-\sin\theta)-\cos\theta\,(a\cos\theta)}{(a+a\sin\theta)^2}$$

$$=\frac{-a\sin\theta-a\sin^2\theta-a\cos^2\theta}{(a+a\sin\theta)^2}$$

$$=\frac{-a\sin\theta-a}{(a+a\sin\theta)^2}$$

$$=\frac{(-1)(a+a\sin\theta)}{(a+a\sin\theta)^2}=\frac{(-1)}{a+a\sin\theta}$$

$f'(\theta_1) = f'(2n+1)\pi/2$

$$=\frac{(-1)}{a+\sin(2n+1)\pi/2}=\frac{(-1)}{a+a(-1)^n}$$

put the values in the formula,

$$r \sin\left[\theta - (2n+1)\frac{\pi}{2}\right] = -a - a(-1)^n$$

$$-r \sin\left(n\pi + \frac{\pi}{2} - \theta\right) = -[a + a(-1)^n]$$

$$(-1)^n r \sin\left(\frac{\pi}{2} - \theta\right) = [a + a(-1)^n]$$

$$r \cos\theta = 0 \quad (\text{n is odd})$$

$$= 2a \quad (\text{n is even}) \quad \text{Answer}$$

Exercise 4.4

Find the asymptotes of the following polar curves:-

1. $r\theta = a$
2. $r\left(1 - e^{\theta}\right) = a$
3. $r\left(e^{\theta} - 1\right) = a\left(e^{\theta} + 1\right)$
4. $r = \dfrac{2a}{1 - 2\cos\theta}$
5. $r\cos 2\theta = a \sin 3\theta$
6. $r(2\theta^2 - 3\theta + 7) = 6\theta^2 + 5\theta - 1$
7. $r = a\sec\theta + b\tan\theta$
8. $r = \dfrac{3a\sin\theta\cos\theta}{\sin^3\theta + \cos^3\theta}$

Answers

1. $r\sin\theta = a$
2. $r\sin\theta + a = 0$
3. $r = a$
4. $r\sin\left(\dfrac{\pi}{3} \pm \theta\right) + \dfrac{2a}{\sqrt{3}} = 0$
5. $\dfrac{a}{2}\sin\left(\dfrac{n\pi}{2} + \dfrac{3\pi}{4}\right) = r\sin\left[(2n+1)\dfrac{\pi}{4} - \theta\right]$
6. $r = 3$
7. $r\cos\theta = a \pm b$
8. $r(\sin\theta + \cos\theta) + a = 0$

CHAPTER 5

Curvature

5.1 Introduction

In the figure a it can be seen that the curve PQ is more curved (i.e., bends more sharply) than the curve RS. In this section we shall find a formula to measure this property of being curved. The numerical measure of the sharpness of the bending of a curve at a point is called the curvature of the curve at that point.

Fig. a

Now from the fig .b, let the arc PQ = δs, where the arc AP = S.A, being the fixed point on the curve from where the length of the arc is measured. Let the tangent at *P* and Q make angles ψ and $\psi + \delta\psi$ with a fixed line say *OX*. Then the angle between the tangents at *P* and Q is $\delta\psi$.

The amount by which the curve bends & P can be described in terms of the rate at which the tangent line is turning as you travel along the curve at P, and this, in turn, can be measured by the rate at which the angle ψ is changing at P. We study the rate of change of the angle ψ with respect to the rate of change of arc length.

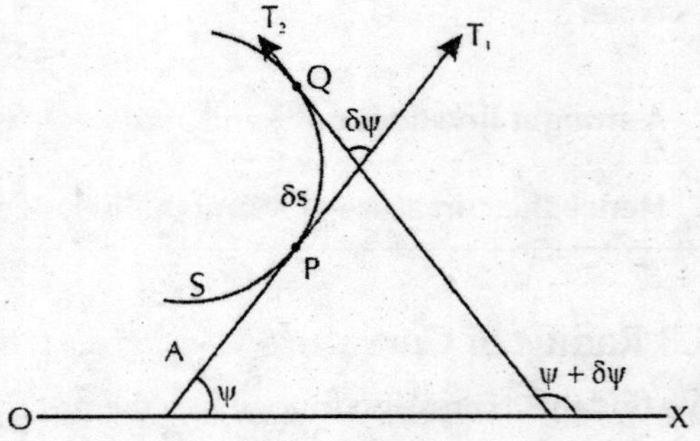

Fig. b

The ratio $\frac{\delta\psi}{\delta s}$ represents the average rate of changing in the angle ψ per unit of arc length along the curve and is called average curvature of arc PQ.

In general this ratio $\frac{\delta\psi}{\delta s}$ will approach a limit as $\delta s \to 0$ namely the derivative $\frac{d\psi}{ds}$.

Thus the curvature at point $$P = \lim_{Q \to P} \frac{\delta\psi}{\delta s} = \lim_{\delta s \to 0} \frac{\delta\psi}{\delta s} = \frac{d\psi}{ds}.$$

Definition:

The limiting value of average curvature $\frac{\delta\psi}{\delta s}$ as $\delta s \to 0$ is called the curvature at point P. Notation for curvature is κ (kappa).

and is given by $$\kappa = \left|\frac{d\psi}{ds}\right|$$

5.2 Radius of Curvature

The reciprocal of the curvature at any point is called the radius of curvature at that point. Obviously the curvature at any point should not be zero for defining radius of curvature at that point. It is usually denoted by ρ. Hence

$$\rho = \frac{1}{\kappa} = \frac{ds}{d\psi} \text{ at } P.$$

Graphically

For the curve CD, if ρ is the radius of curvature at point P, we draw normal at the point P and then O′ P is the distance equal to radius of curvature at P with O'.

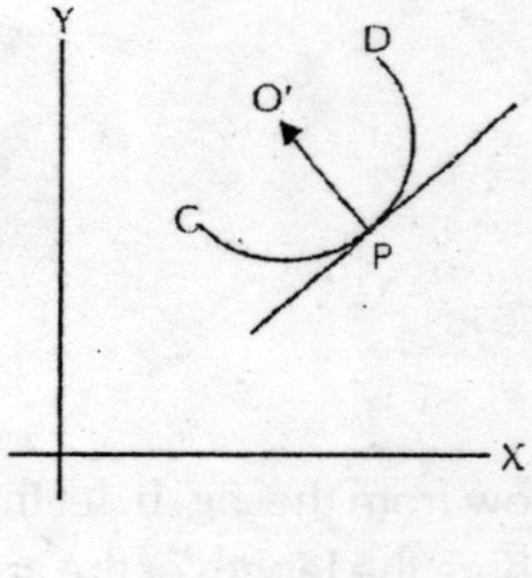

Note:

A straight line does not bend at all (as ψ is constant so $\frac{d\psi}{ds}$ is zero)

Hence the curvature of a straight line is zero.

5.3 Radius of Curvature for the Cartesian Curve

We find the expression for ρ, when the equation of curve is given in cartesian co-ordinates in different forms.

(A) To find ρ for the curve $y = f(x)$

If ψ is the angle which the tangent at P (x, y) on the curve makes with x- axis, then we have

$$\sin\psi = \frac{dy}{ds}, \cos\psi = \frac{dx}{ds} \text{ and } \tan\psi = \frac{dy}{dx}$$

So that from the last relation we have

$$\psi = \tan^{-1}(y_1) \text{ where } y_1 = \frac{dy}{dx}$$

Differentiating w.r.t x, we have

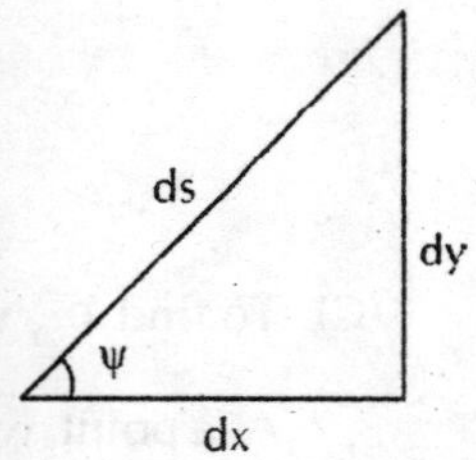

$$\frac{d\psi}{dx} = \frac{1}{1+y_1^2}\cdot y_2$$

$$\frac{d\psi}{dx} = \frac{d\psi}{dx}\cdot\frac{dx}{ds} = \frac{y_2}{1+y_1^2}\cdot\cos\psi$$

$$= \frac{y_2}{1+y_1^2}\cdot\frac{1}{\sqrt{1+y_1^2}}$$

$$\left[\because \cos\psi = \frac{1}{\sec\psi} = \frac{1}{\sqrt{1+\tan^2\psi}} = \frac{1}{\sqrt{1+(dy/dx)^2}} = \frac{1}{\sqrt{1+y_1^2}}\right]$$

$$\frac{1}{\rho} = \frac{y_2}{(1+y_1^2)^{\frac{3}{2}}}$$

$$\rho = \frac{[1+y_1^2]^{\frac{3}{2}}}{y_2} \quad \text{where } y_2 \neq 0$$

(B) To find ρ for the curve $x = f(t)$, $y = \phi(t)$ i.e., when the parametric equation of the curve is given.

Here $x = f(t)$ and $y = \phi(t)$, t being parameter

$$x' = \frac{dx}{dt}, \quad y' = \frac{dy}{dt}$$

$$x'' = \frac{d^2x}{dt^2}, \quad y'' = \frac{d^2y}{dt^2}$$

$$y_1 = \frac{dy}{dx} = \frac{\frac{dy}{dt}}{\frac{dx}{dt}} = \frac{y'}{x'} \qquad \text{...(1)}$$

Differentiating (1) w.r.t x we have

$$y_2 = \frac{x'y'' - y'x''}{x'^2} \cdot \frac{dt}{dx} = \frac{x'y'' - y'x''}{x'^3} \qquad \left[\because \frac{dt}{dx} = \frac{1}{x}\right]$$

$$\rho = \frac{\left[1 + y_1^2\right]^{\frac{3}{2}}}{y_2} = \frac{\left(x'^2 + y'^2\right)^{\frac{3}{2}}}{x'y'' - y'x''}$$

(C) To find ρ, when the equation of the curve is in implicit form i.e., $f(x,y)$.

At a point where $fy \neq 0$ we have

$$y_1 = \frac{dy}{dx} = \frac{-fx}{fy}$$

$$y_2 = \frac{d^2y}{dx^2} = -\frac{fxxfy^2 - 2fxyfxfy + fyyfx^2}{fy^3}$$

$$\rho = \frac{\left(1 + y_1^2\right)^{\frac{3}{2}}}{y_2}$$

$$= \frac{\left(fx^2 + fy^2\right)^{\frac{3}{2}}}{fxxfy^2 - 2fxyfxfy + fyyfx^2}.$$

(D) To find ρ when x and y are given as functions of s, the length of an arc, measured from a fixed point on the curve.

We have

$$\cos\psi = \frac{dx}{ds} \quad \text{...(1)}$$

$$\sin\psi = \frac{dy}{ds} \quad \text{...(2)}$$

Differentiating (1) and (2) w.r.t s, we get

$$-\sin\psi\frac{d\psi}{ds} = \frac{d^2x}{ds^2} \quad \text{...(3)}$$

$$\frac{-dy}{ds}\cdot\frac{1}{\rho} = \frac{d^2x}{ds^2}$$

$$\rho = \frac{-\frac{dy}{ds}}{\frac{d^2x}{ds^2}}$$

Again differentiating (2) w.r.t s, we get

$$\cos\psi\frac{d\psi}{ds} = \frac{d^2y}{ds^2} \quad \text{...(4)}$$

$$\rho = \frac{\frac{dx}{ds}}{\frac{d^2y}{ds^2}}$$

Squaring and adding (3) and (4) and putting $\frac{1}{\rho}$ for $\frac{d\psi}{ds}$, we get

$$\frac{1}{\rho^2} = \left(\frac{d^2x}{ds^2}\right)^2 + \left(\frac{d^2y}{ds^2}\right)^2$$

Again from (1)

$$\cos\psi = \frac{dx}{ds} = \frac{dx}{d\psi}\cdot\frac{d\psi}{ds} = \frac{1}{\rho}\cdot\frac{dx}{d\psi}$$

$$\rho\cos\psi = \frac{dx}{d\psi} \quad \text{...(5)}$$

Similarly from (2) we have

$$\rho \sin\psi = \frac{dy}{d\psi} \qquad \text{...(6)}$$

Squaring and adding (5) and (6) we get

$$\rho^2 = \left(\frac{dx}{d\psi}\right)^2 + \left(\frac{dy}{d\psi}\right)^2$$

Illustrative Examples

Example 1:

Find the radius of curvature at the given point of following curves :

(a) $y = 4\sin x - \sin 2x$ at $x = \frac{\pi}{2}$ (b) $\sqrt{x} + \sqrt{y} = 1$ at the point $\left(\frac{1}{4}, \frac{1}{4}\right)$

Solution :

(a)

$$y = 4\sin x - \sin 2x \qquad \text{...(1)}$$

Differentiate (1) w.r.t x

$$\frac{dy}{dx} = 4\cos x - 2\cos 2x$$

$$\frac{d^2y}{dx^2} = -\ 4\sin x + 4\sin 2x$$

$$\rho = \frac{\left[1+\left(\frac{dy}{dx}\right)^2\right]^{\frac{3}{2}}}{\frac{d^2y}{dx^2}} = \frac{\left[1+(4\cos x - 2\cos 2x)^2\right]^{\frac{3}{2}}}{-4\sin x + 4\sin 2x}$$

$$\rho \text{ at } \frac{\pi}{2} = \frac{\left[1+(0-2\cos\pi)^2\right]^{\frac{3}{2}}}{-4\sin\frac{\pi}{2}} = \frac{(1+4)^{\frac{3}{2}}}{-4} \qquad \text{(ignoring the negative sign)}$$

$$= \frac{5^{\frac{3}{2}}}{4} \quad \textbf{(Answer)}$$

(b) $$\sqrt{x}+\sqrt{y} = 1 \qquad ...(1)$$

Differentiating the equation (1) w.r.t x

$$\frac{1}{2\sqrt{x}}+\frac{1}{2\sqrt{y}}\frac{dy}{dx} = 0$$

$$\frac{dy}{dx} = -\frac{\frac{1}{\sqrt{x}}}{\frac{1}{\sqrt{y}}} = -\frac{\sqrt{y}}{\sqrt{x}}$$

$$= -1 \text{ at } \left(\frac{1}{4},\frac{1}{4}\right)$$

$$\frac{d^2y}{dx^2} = \frac{\left(\sqrt{x}\cdot\frac{1}{2\sqrt{y}}\frac{dy}{dx}-\sqrt{y}\cdot\frac{1}{2\sqrt{x}}\right)}{x}$$

$$= \frac{\left[\frac{\sqrt{x}}{2\sqrt{y}}\left(-\frac{\sqrt{y}}{\sqrt{x}}\right)-\frac{\sqrt{y}}{2\sqrt{x}}\right]}{x} = \frac{\left[-\frac{1}{2}-\frac{\sqrt{y}}{2\sqrt{x}}\right]}{x}$$

$$= \frac{\sqrt{x}+\sqrt{y}}{2x\sqrt{x}} = \frac{\frac{1}{2}+\frac{1}{2}}{2\cdot\frac{1}{4}\cdot\frac{1}{2}} \qquad \left(at\frac{1}{4},\frac{1}{4}\right)$$

$$= 4$$

$$p = \frac{\left[1+\left(\frac{dy}{dx}\right)^2\right]^{\frac{3}{2}}}{\frac{d^2y}{dx^2}} = \frac{(1+1)^{\frac{3}{2}}}{4} = \frac{2^{\frac{3}{2}}}{4}$$

$$= \frac{1}{\sqrt{2}} \quad \textbf{(Answer)}$$

Example 2: Find the least value of $|\rho|$ for $y = \log x$, $x > 0$.

Solution:

$$\text{Let } y = \log x \qquad .....(1)$$

Differentiating (1) w.r.t x, we have

$$\frac{dy}{dx} = \frac{1}{x} \text{ and } \frac{d^2y}{dx^2} = -\frac{1}{x^2}$$

$$\rho = \frac{\left[1+y_1^{\,2}\right]^{\frac{3}{2}}}{y_2} = \frac{\left(1+\frac{1}{x^2}\right)^{\frac{3}{2}}}{-\frac{1}{x^2}}$$

$$= \frac{-(x^2+1)^{\frac{3}{2}}}{x}$$

Let $$f(x) = \frac{(x^2+1)^{\frac{3}{2}}}{x}$$

To find the least value of $f(x)$, we find $f'(x)$

$$f'(x) = \frac{x\cdot\frac{3}{2}(x^2+1)^{\frac{1}{2}}\,2x-(x^2+1)^{\frac{3}{2}}}{x^2}$$

$$= \frac{3x^2\sqrt{x^2+1}-(x^2+1)^{\frac{3}{2}}}{x^2}$$

$$= \frac{\sqrt{x^2+1}\left(3x^2-x^2-1\right)}{x^2} = \frac{\sqrt{x^2+1}\left(2x^2-1\right)}{x^2}$$

Equate $f'(x) = 0$ to get $x = \pm\frac{1}{\sqrt{2}}$

and also $f''(x)$ is positive for $x = \frac{1}{\sqrt{2}}$

Hence $|\rho|$ is minimum for $x = \frac{1}{\sqrt{2}}$

$$\therefore \quad |\rho|_{min} = \left[\frac{(x^2+1)^{\frac{3}{2}}}{x}\right]_{x=\frac{1}{\sqrt{2}}} = \frac{\left(\frac{1}{2}+1\right)^{\frac{3}{2}}}{\frac{1}{\sqrt{2}}}$$

$$= \left(\frac{3}{2}\right)^{\frac{3}{2}} \sqrt{2} = \frac{3\sqrt{3}}{2}$$

$\therefore$ Minimum value of $|\rho|$ is $\frac{3\sqrt{3}}{2}$. Answer

Example 3:

Find the radius of curvature for Rectangular hyperbola $xy = c^2$ at the point (x, y)

Solution :

$$xy = c^2$$

$$y = \frac{c^2}{x} \qquad ...(1)$$

Differentiating (1) w.r.t. x, we get

$$\frac{dy}{dx} = \frac{-c^2}{x^2}$$

$$\frac{d^2y}{dx^2} = \frac{2c^2}{x^3}$$

$\therefore$ Radius of curvature at point (x, y) is

$$\rho = \frac{\left[1+\left(\frac{dy}{dx}\right)^2\right]^{\frac{3}{2}}}{\frac{d^2y}{dx^2}} = \frac{\left(1+\frac{c^4}{x^4}\right)^{\frac{3}{2}}}{\frac{2c^2}{x^3}}$$

$$= \frac{\left(x^4+c^4\right)^{\frac{3}{2}}}{x^6} \cdot \frac{x^3}{2c^2} = \frac{\left(x^4+c^4\right)^{\frac{3}{2}}}{2c^2x^3}$$

$$= \frac{\left(x^2+y^2\right)^{\frac{3}{2}}}{2c^2} \qquad \left(\because \text{ by using } y = \frac{c^2}{x}\right)$$

Example 4:

Find the radius of curvature at the origin of the two branches of the curve given by

$x = 1 - t^2,\ y = t - t^3$

Solution :

At origin (0, 0) two common value for t are 1 and -1.

Hence for two branches of the curves, value of t are 1 and –1

Differentiating w.r.t. 't', we get

$$\frac{dy}{dt} = 1 - 3t^2, \quad \frac{dx}{dt} = -2t$$

$$\frac{dy}{dx} = \frac{1-3t^2}{-2t} = -\frac{1}{2t} + \frac{3}{2}t \qquad \text{...(1)}$$

Again differentiating (1) w.r.t. 'x', we get

$$\frac{d^2y}{dx^2} = \left(\frac{1}{2t^2} + \frac{3}{2}\right) \cdot \frac{dt}{dx} = \frac{\left(\frac{1}{2t^2} + \frac{3}{2}\right)}{-2t}$$

$$= -\frac{1}{4t^3} - \frac{3}{4t}$$

$$\left(\frac{dy}{dx}\right)_{t=1} = -\frac{1}{2} + \frac{3}{2} = 1$$

$$\left(\frac{d^2y}{dx^2}\right)_{t=1} = -\frac{1}{4} - \frac{3}{4} = -1$$

$$\left(\frac{dy}{dx}\right)_{t=-1} = \frac{1}{2} - \frac{3}{2} = -1$$

$$\left(\frac{d^2y}{dx^2}\right)_{t=-1} = \frac{1}{4} + \frac{3}{4} = 1$$

$$(\rho)_{t=1} = \frac{\left[1 + \left(\frac{dy}{dx}\right)^2\right]^{3/2}}{\frac{d^2y}{dx^2}} = \frac{(1+1)^{3/2}}{-1} = -2\sqrt{2}$$

$$(\rho)_{t=-1} = \frac{\left[1 + \left(\frac{dy}{dx}\right)^2\right]^{3/2}}{\frac{d^2y}{dx^2}} = \frac{(1+1)^{3/2}}{1} = 2\sqrt{2} \qquad \text{Answer}$$

Example 5:

Prove that the radius of curvature for the catenary $y = c\cosh\frac{x}{c}$ is equal to the portion of the normal intercepted between the curve and the x-axis and that it varies as the square of the ordinate.

Solution :

$$y = c\cosh\frac{x}{c} \qquad \text{...(1)}$$

Differentiating (1) w.r.t. x, *we get*

$$\frac{dy}{dx} = c\sinh\left(\frac{x}{c}\right)\cdot\frac{1}{c}$$

$$= \sinh\left(\frac{x}{c}\right)$$

Again differentiating

$$\frac{d^2y}{dx^2} = \cosh\left(\frac{x}{c}\right)\cdot\frac{1}{c}$$

$$\rho = \frac{\left[1+\left(\frac{dy}{dx}\right)^2\right]^{3/2}}{\frac{d^2y}{dx^2}} = \frac{\left[1+\sinh^2\left(\frac{x}{c}\right)\right]^{3/2}}{\frac{1}{c}\cdot\cosh\left(\frac{x}{c}\right)}$$

$$\Rightarrow \quad = \frac{c\left(\cosh^2\frac{x}{c}\right)^{3/2}}{\cosh\left(\frac{x}{c}\right)}$$

$$= c\cosh^2\frac{x}{c}$$

Now for the length of the normal between the curve and the x-axis, we find the equation of the normal at the point (a, b) as

$$y-b = -\frac{1}{y_1}(x-a) \quad \text{where } y_1 = \frac{dy}{dx}$$

$\therefore$ x co-ordinate of the point where normal meets x-axis is obtained by substituting $y=0$

$\therefore \quad x = by_1 + a$

Hence the length of intercept between the point (a, b) and (x, 0) is

$$\sqrt{(by_1 + a - a)^2 + b^2} = \sqrt{b^2y_1^2 + b^2} = b\sqrt{1+y_1^2}$$

For curve $\quad y = c\cos h\frac{x}{c}$, the point (a, b) has

$$b = c\cos h\left(\frac{x}{c}\right)$$

$\therefore$ length of normal $= c\cos h\left(\frac{a}{c}\right)\sqrt{1+y_1^2}$

$$= c\cos h\left(\frac{a}{c}\right)\sqrt{1+\sin h^2\left(\frac{a}{c}\right)}$$

$$= c\cos h^2\left(\frac{a}{c}\right)$$

$\Rightarrow$ it varies as the square of ordinate. Proved

Example 6:

If ρ_1 and ρ_2 be the radii of curvature at the ends of focal chord of the parabola $y^2 = 4ax$, then prove that $\rho_1^{\frac{-2}{3}} + \rho_2^{\frac{-2}{3}} = (2a)^{\frac{-2}{3}}$

Solution :

The equation of parabola is $y^2 = 4ax$ which is in parametric form is $x = at^2, y = 2at$

If $(at_1^2, 2at_1)$ and $(at_2^2, 2at_2)$ are the two extremities of a focal chord of the parabola,

then $\quad t_1t_2 = -1$

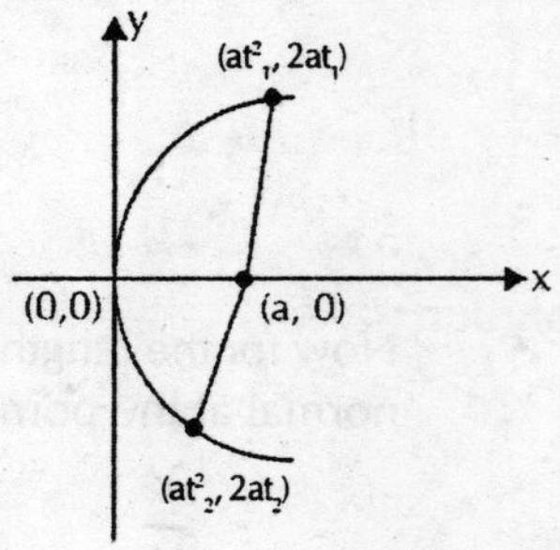

For $\quad x = at^2, \quad y = 2at$

We have, $\quad \frac{dx}{dt} = 2at \quad \frac{dy}{dt} = 2a$

$$\frac{dy}{dx} = \frac{1}{t}$$

$$\frac{d^2y}{dx^2} = -\frac{1}{t^2}\cdot\frac{dt}{dx} = -\frac{1}{2at^3}$$

$$\rho = \frac{\left[1+\left(\frac{dy}{dx}\right)^2\right]^{\frac{3}{2}}}{\frac{d^2y}{dx^2}} = \frac{\left[1+\frac{1}{t^2}\right]^{\frac{3}{2}}}{-\frac{1}{2at^3}}$$

$$= -2a\left(t^2+1\right)^{\frac{3}{2}}$$

$$\therefore \quad \rho \text{ at } \left(at_1^2, 2at_1\right) = -2a\left(t_1^2+1\right)^{\frac{3}{2}}$$

$$\text{and} \quad \rho \text{ at } \left(at_2^2, 2at_2\right) = -2a\left(t_2^2+1\right)^{\frac{3}{2}}$$

$$\therefore \quad (\rho_1)^{\frac{-2}{3}} + (\rho_2)^{\frac{-2}{3}} = \left[2a\left(t_1^2+1\right)^{\frac{3}{2}}\right]^{\frac{-2}{3}} + \left[2a\left(t_2^2+1\right)^{\frac{3}{2}}\right]^{\frac{-2}{3}}$$

$$= (2a)^{\frac{-2}{3}}\left[\frac{1}{t_1^2+1}+\frac{1}{t_2^2+1}\right]$$

$$= (2a)^{\frac{-2}{3}}\left[\frac{t_2^2+1+t_1^2+1}{\left(t_1^2+1\right)\left(t_2^2+1\right)}\right]$$

$$= (2a)^{\frac{-2}{3}}\left[\frac{t_1^2+t_2^2+2}{t_1^2+t_2^2+(-1)^2+1}\right] = (2a)^{\frac{-2}{3}} \qquad \textbf{Proved}$$

Example 7:

Show that the radius of curvature at the end of the major axis of an ellipse $\frac{x^2}{a^2}+\frac{y^2}{b^2} = 1$ is equal to the semi-latus rectum of the ellipse.

Solution :

Equation of an ellipse in parametric form is $x = a\cos t,\ y = b\sin t$

and the ends of major axis are $(\pm a, 0)$

Differentiating w.r.t. t, we get

$$\frac{dx}{dt} = -a\sin t\ , \quad \frac{dy}{dt} = b\cos t$$

$$\frac{dy}{dx} = -\frac{b}{a}\cot t$$

$$\frac{d^2y}{dx^2} = \frac{b}{a}\operatorname{cosec}^2 t\frac{dt}{dx} = \frac{b}{a}\operatorname{cosec}^2 t \times \frac{1}{-a\sin t}$$

$$= \frac{-b}{a^2}\operatorname{cosec}^3 t$$

Radius of curvature is

$$\rho = \frac{\left[1+\left(\frac{dy}{dx}\right)^2\right]^{\frac{3}{2}}}{\frac{d^2y}{dx^2}} = \frac{\left[1+\frac{b^2\cos^2 t}{a^2\sin^2 t}\right]^{\frac{3}{2}}}{\frac{-b}{a^2}\operatorname{cosec}^3 t}$$

$$= \frac{1}{ab}\left(a^2\sin^2 t + b^2\cos^2 t\right)^{\frac{3}{2}} \text{ (ignoring the negative sign)}$$

ρ at (a, 0) is $\frac{1}{ab}\left(a^2 \sin^2 0 + b^2 \cos^2 0\right)^{3/2}$

$$= \frac{b^2}{a} \text{ (Semi latus ractum of the ellipse)} \qquad \text{Proved}$$

Example 8:

Find radius of curvature for the curve $x = c\log\left(s+\sqrt{s^2+c^2}\right)$, $y = \sqrt{s^2+c^2}$

Solution :

$$x = c\log\left(s+\sqrt{s^2+c^2}\right)$$

Differentiating w.r.t. s, we get

$$\frac{dx}{ds} = \frac{c}{s+\sqrt{s^2+c^2}}\left(1+\frac{2s}{2\sqrt{s^2+c^2}}\right) = \frac{c}{\sqrt{s^2+c^2}}$$

$$\frac{dy}{ds} = \frac{2s}{2\sqrt{s^2+c^2}} = \frac{s}{\sqrt{s^2+c^2}}$$

$$\frac{dy}{dx} = \frac{\frac{dy}{ds}}{\frac{dx}{ds}} = \frac{s}{c} \quad \text{...(1)}$$

But $\quad \tan\psi = \frac{dy}{dx}$

$$s = c\tan\psi \qquad \text{[from (1)]}$$

$\therefore$ Radius of curvature $\frac{ds}{d\psi} = c\sec^2\psi$

$$= c\left(1+\tan^2\psi\right)$$

$$= c\left(1+\frac{s^2}{c^2}\right)$$

$$= \frac{c^2+s^2}{c} \qquad \text{Proved}$$

Example 9:

If ρ_1 and ρ_2 be the radii of curvature at the extremities of two conjugate semi – diameters of the ellipse $\frac{x^2}{a^2}+\frac{y^2}{b^2} = 1$, then prove that $\left(\rho_1^{\frac{2}{3}}+\rho_2^{\frac{2}{3}}\right)(ab)^{\frac{2}{3}} = a^2+b^2$

Solution :

Co-ordinates at the extremities of two conjugate semi-diameter are

at P (a cost, b sint) and

at D (–a sint , b cost)

Where CP and CD are conjugate semi – diameters with C as (0, 0)

Also ρ at $\quad P = \frac{1}{ab}\left(a^2\sin^2 t+b^2\cos^2 t\right)^{\frac{3}{2}}$

At D $\quad x = -a\,\text{sint} \qquad y = b\,\text{cost}$

$$\frac{dx}{dt} = -a\,\text{cost} \qquad \frac{dy}{dt} = -b\,\text{sint}$$

$$\frac{dy}{dx} = \frac{b}{a}\tan t$$

$$\frac{d^2y}{dx^2} = \frac{b}{a}\sec^2 t\frac{dt}{dx} = \frac{b}{a}\sec^2 t\left(\frac{1}{-a\cos t}\right)$$

$$= -\frac{b}{a^2}\sec^3 t$$

$$\rho \text{ at D} = \frac{\left[1+\left(\frac{dy}{dx}\right)^2\right]^{\frac{3}{2}}}{\frac{d^2y}{dx^2}} = \frac{\left[1+\frac{b^2}{a^2}\tan^2 t\right]^{\frac{3}{2}}}{\frac{-b}{a^2}\sec^3 t}$$

$$= \frac{1}{ab}(a^2\cos^2 t + b^2\sin^2 t)^{\frac{3}{2}} \text{ (ignoring negative sign)}$$

$$\left(\rho_1^{\frac{2}{3}} + \rho_2^{\frac{2}{3}}\right)(ab)^{\frac{2}{3}} = a^2\sin^2 t + b^2\cos^2 t + a^2\cos^2 t + b^2\sin^2 t$$

$$= a^2 + b^2$$

$$\Rightarrow \quad \left(\rho_1^{\frac{2}{3}} + \rho_2^{\frac{2}{3}}\right)(ab)^{\frac{2}{3}} = a^2 + b^2 \qquad \text{Proved}$$

Exercise 5.1

Find the radius of curvature at the given point of the following curves:

1. $\sqrt{x}+\sqrt{y} = \sqrt{a}$ at $\left(\frac{a}{4},\frac{a}{4}\right)$

2. $x^3+y^3 = 3axy$ at $\left(\frac{3a}{2},\frac{3a}{2}\right)$

3. $y^2 = \frac{a^2(a-x)}{x}$ at $(a, 0)$ (Equation of curve is $x = \frac{a^3}{y^2+a^2}$)

4. $x^2y = a(x^2+y^2)$ at $(-2a, 2a)$

5. Find the radius of curvature of the curve $y = e^x$ at the point where it crosses the y-axis.

6. Find the radius of curvature at any point $(0, c)$ of the catenary $y = c\cosh\frac{x}{c}$.

7. Show that for the parabola $y^2 = 4ax$, ρ^2 *varies as* $(SP)^3$, where ρ is the radius of curvature at any point P of the parabola and S is the focus of the parabola.

8. The tangent at two points P,Q on the cycloid $x = a(\theta-\sin\theta)$, $y = a(1-\cos\theta)$ are at right angles: show that if ρ_1, ρ_2 be the radii of curvature at these points, then $\rho_1^2+\rho_2^2 = 16a^2$.

9. Prove that the radius of curvature at any point of the astroid $x^{\frac{2}{3}}+y^{\frac{2}{3}} = a^{\frac{2}{3}}$ is three times the length of the perpendicular from the origin to the tangent at that point.

10. Prove that for the ellipse $\frac{x^2}{a^2}+\frac{y^2}{b^2} = 1$, $\rho = \frac{a^2b^2}{P^3}$ where P is the perpendicular from the center upon the tangent at (x, y).

11. Find the points on the parabola $y^2 = 8x$ at which the radius of curvature is $7\frac{13}{16}$.

12. Show that for the curve;

$x = a\cos\theta(1+\sin\theta)$, $y = a\sin\theta(1+\cos\theta)$, the radius of curvature at $\theta = \frac{-\pi}{4}$ is a.

Answers

1. $\frac{a}{\sqrt{2}}$ 2. $\frac{3a}{8\sqrt{2}}$ 3. $\frac{a}{2}$

4. $2a$ 5. $2\sqrt{2}$ 6. c

11. $\left(\frac{9}{8}, 3\right)$ and $\left(\frac{9}{8}, -3\right)$

5.4 Radius of Curvature for Polar Equations

(A) To find ρ for the curve $r = f(\theta)$ or $f(r, \theta) = 0$

Let ϕ be the angle which the tangent at $P(r, \theta)$ makes with OP, then we have

$$\left.\begin{aligned} \tan\phi &= \frac{rd\theta}{dr}, \quad \sin\phi = \frac{rd\theta}{ds} \\ \text{and} \quad \cos\phi &= \frac{dr}{ds} \end{aligned}\right\} \quad \ldots\ldots 1$$

Again if ψ be the angle which the tangent at $P(r, \theta)$ makes with OX,

then $\psi = \theta + \phi$

$$\therefore \quad \frac{d\psi}{ds} = \frac{d\theta}{ds} + \frac{d\phi}{ds}$$

$$= \frac{d\theta}{ds} + \frac{d\phi}{d\theta} \cdot \frac{d\theta}{ds}$$

$$= \frac{d\theta}{ds}\left[1 + \frac{d\phi}{d\theta}\right] \quad \ldots(2)$$

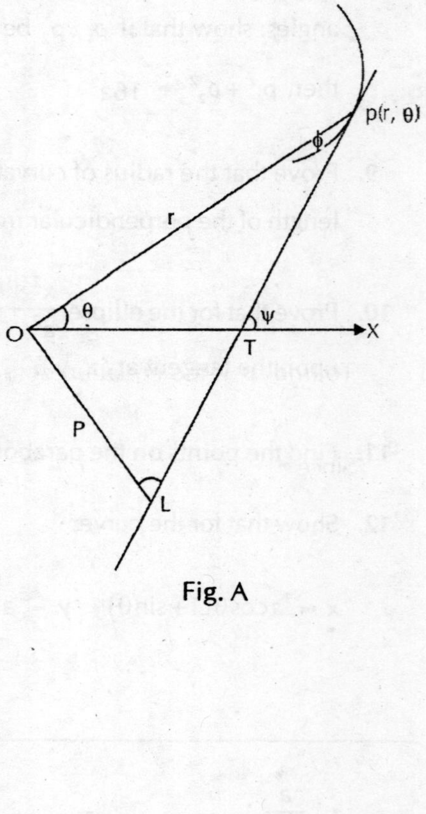

Fig. A

from 1, $\tan\phi = r \cdot \dfrac{d\theta}{dr} = \dfrac{r}{\dfrac{dr}{d\theta}}$

or $\tan\phi = \dfrac{r}{r_1}$ where $r_1 = \dfrac{dr}{d\theta}$

differentiating both side w.r.t θ, we have

$$\sec^2\phi \cdot \frac{d\phi}{d\theta} = \frac{r_1 \cdot r_1 - rr_2}{r_1^2}$$

$$\frac{d\phi}{d\theta} = \frac{r_1^2 - rr_2}{r_1^2} \cdot \frac{1}{\sec^2\phi}$$

$$= \frac{r_1^2 - rr_2}{r_1^2} \cdot \frac{1}{1 + \tan^2\phi}$$

$$\Rightarrow \quad \frac{r_1^2 - rr_2}{r_1^2} \cdot \frac{r_1^2}{r_1^2 + r^2} \qquad \left[\because \tan\phi = \frac{r}{r_1}\right]$$

$$= \frac{r_1^2 - rr_2}{r_1^2 + r^2} \qquad ...(3)$$

Again from 1, $r\frac{d\theta}{ds} = \sin\phi = \frac{1}{\operatorname{cosec}\phi}$

$$\frac{d\theta}{ds} = \frac{1}{r}\cdot\frac{1}{\sqrt{1+\cot^2\phi}} = \frac{1}{\sqrt{r^2 + r_1^2}} \qquad ...(4)$$

From (2), (3) and (4) we have

$$\frac{d\psi}{ds} = \frac{1}{\sqrt{r^2 + r_1^2}}\cdot\left[1 + \frac{r_1^2 - rr_2}{r_1^2 + r^2}\right]$$

$$= \frac{2r_1^2 + r^2 - rr_2}{\left(r^2 + r_1^2\right)^{\frac{3}{2}}}$$

$$\rho = \frac{ds}{d\psi} = \frac{\left(r^2 + r_1^2\right)^{\frac{3}{2}}}{2r_1^2 + r^2 - rr_2}$$

(B) To find ρ if the equation of the curve is in the form $u = f(\theta)$ *where* $u = \frac{1}{r}$.

Since $r = \frac{1}{u}$

$$r_1 = -\frac{1}{u^2}\cdot u_1 \text{ where } u_1 = \frac{du}{d\theta}$$

$$r_2 = -\frac{u^2 u_2 - u_1 2uu_1}{u^4} = \frac{-2u_1^2 - uu_2}{u^3}$$

$$\rho = \frac{\left(\frac{1}{u^2} + \frac{u_1^2}{u^4}\right)^{\frac{3}{2}}}{\frac{2u_1^2}{u^4} + \frac{1}{u^2} - \frac{2u_1^2 - uu_2}{u^4}}$$

$$\rho = \frac{\left(u^2 + u_1^2\right)^{\frac{3}{2}}}{u^3\left(u + u_2\right)}$$

(C) Radius of curvature for Pedal Equations :

To prove that for the curve $P = f(r)$, $\rho = r \cdot \dfrac{dr}{dP}$

Solution:

From the fig A we have

$$\psi = \theta + \phi$$

$$\frac{d\psi}{ds} = \frac{d\theta}{ds} + \frac{d\phi}{ds} \qquad \ldots\ldots 1$$

Also from the same fig. $P = r\sin\phi$

Differentiating both side of (2) w.r.t r, we have

$$\frac{dP}{dr} = r\cos\phi \cdot \frac{d\phi}{dr} + \sin\phi$$

$$= r \cdot \frac{dr}{ds} \cdot \frac{d\phi}{dr} + \frac{rd\theta}{ds}$$

$$= r\left(\frac{d\phi}{ds} + \frac{d\theta}{ds}\right) = r \cdot \frac{d\psi}{ds}$$

$$= r \cdot \frac{1}{\rho}$$

$$\rho = r\frac{dr}{dp} \qquad \text{Proved}$$

Illustrative Examples

Example 1:

Find the radius of curvature for the parabola $\dfrac{2a}{r} = 1 + \cos\theta$. Hence show that the square of the radius of curvature varies as the cube of the focal distance.

Solution:

Given equation of parabola is

$$\frac{2a}{r} = 1 + \cos\theta \qquad \ldots(1)$$

Putting $\qquad u = \frac{1}{r}$ in (1)

$$2au = 1+\cos\theta \qquad \text{...(2)}$$

differentiating (2) w.r.t θ, we have

$$2au_1 = -\sin\theta \qquad \text{...(3)}$$

differentiating (3) w.r.t. θ, we have

$$2au_2 = -\cos\theta$$

then, $$\rho = \frac{\left(u^2+u_1^2\right)^{\frac{3}{2}}}{u^3\left(u+u_2\right)}$$

$$= \frac{\left[\left(\frac{1+\cos\theta}{2a}\right)^2 + \frac{\sin^2\theta}{4a^2}\right]^{\frac{3}{2}}}{\left(\frac{1+\cos\theta}{2a}\right)^3\left[\frac{1+\cos\theta}{2}-\frac{\cos\theta}{2a}\right]}$$

$$= \frac{\frac{1}{8a^3}\left[2(1+\cos\theta)\right]^{\frac{3}{2}}}{\frac{1}{16a^4}(1+\cos\theta)^3} = \frac{4\sqrt{2a}}{(1+\cos\theta)^{\frac{3}{2}}}$$

$$= \frac{2}{\sqrt{a}}r^{\frac{3}{2}}$$

$$\therefore \qquad \rho^2 = \frac{4}{a}r^3$$

In the given equation of parabola $\frac{2a}{r} = 1+\cos\theta$, the pole is at the focus. Therefore the focal distance of the point on the parabola is *r*. Hence the square of the radius of curvature varies as the cube of focal distance.

Remark:

The radius of curvature for polar curves can also be found by first changing the polar equation of the curve to pedal form.

5.5 Rule to Convert Polar Equation to Pedal Equation

1. Use the relation $\tan\phi = r\frac{d\theta}{dr}$ and find ϕ.
2. Put this value of ϕ in $p = r\sin\phi$
3. Eliminate θ to get the equation in pedal form.

Example 2:

For any curve $r = f(\theta)$ prove that

$$\frac{r}{\rho} = \sin\phi\left(1+\frac{d\phi}{d\theta}\right)$$

Solution:

From the fig (B)

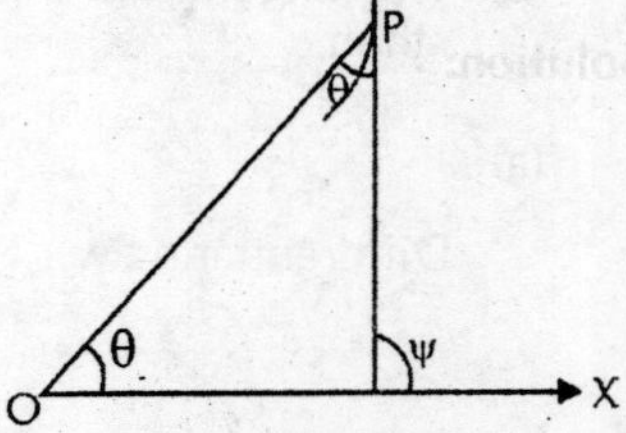

Fig. B

$$\psi = \theta+\phi$$

$$\frac{d\psi}{ds} = \frac{d\theta}{ds}+\frac{d\phi}{ds} \quad (s \text{ is the arc length })$$

$$= \frac{d\theta}{ds}+\frac{d\phi}{d\theta}\cdot\frac{d\theta}{ds}$$

$$= \left(1+\frac{d\phi}{d\theta}\right)\frac{d\theta}{ds} \qquad \text{...(i)}$$

But $\tan\phi = r\frac{d\theta}{dr}$ and $(ds)^2 = (rd\theta)^2+(dr)^2$ ...(ii)

Hence $\frac{dr}{d\theta} = r\cot\phi$ and $\left(\frac{ds}{d\theta}\right)^2 = r^2+\left(\frac{dr}{d\theta}\right)^2$

$$\therefore \quad \left(\frac{ds}{d\theta}\right)^2 = r^2+r^2\cot^2\phi = r^2\,\text{cosec}^2\phi$$

$$\frac{ds}{d\theta} = r\,\text{cosec}\,\phi$$

Substituting in (i) we get

$$\frac{d\psi}{ds} = \left(1+\frac{d\phi}{d\theta}\right)\cdot\frac{1}{r\,\text{cosec}\,\phi}$$

$$\rho = \frac{ds}{d\psi} = \frac{r\operatorname{cosec}\phi}{1+\dfrac{d\phi}{d\theta}}$$

$$\frac{r}{\rho} = \sin\phi\left(1+\frac{d\phi}{d\theta}\right) \quad \text{Proved}$$

Example 3:

Prove that in the curve $r^2 = a^2 \sin 2\theta$

(a) The curvature varies as the radius vector.

(b) The tangent runs three times as fast as the radius vector.

Solution:

(a) $$r^2 = a^2 \sin 2\theta$$

Differentiating w.r.t θ we get

$$2rr_1 = 2a^2 \cos 2\theta$$

$$r_1 = \frac{a^2 \cos 2\theta}{r}$$

$$r_2 = \frac{a^2\left[r(-2\sin 2\theta)-(\cos 2\theta)r_1\right]}{r^2}$$

$$= \frac{a^2\left[-2r\sin 2\theta - \cos 2\theta\left(\dfrac{a^2\cos 2\theta}{r}\right)\right]}{r^2}$$

$$= \frac{a^2}{r^2}\left[\frac{-2r^2 \sin 2\theta - a^2\cos^2 2\theta}{r}\right]$$

$$= \frac{a^2}{r^3}\left[-2a^2\sin^2 2\theta - a^2\cos^2 2\theta\right] = \frac{-a^4}{r^3}(1+\sin^2 2\theta)$$

Hence $$\rho = \frac{(r^2+r_1^2)^{\frac{3}{2}}}{r^2+2r_1^2-rr_2}$$

$$= \frac{\left[r^2+\dfrac{a^4\cos^2 2\theta}{r^2}\right]^{\frac{3}{2}}}{r^2+2\dfrac{a^4\cos^2 2\theta}{r^2}-r\left[\dfrac{-a^4}{r^3}(1+\sin^2 2\theta)\right]}$$

$$= \frac{\left(r^4 + a^4 \cos^2 2\theta\right)^{\frac{3}{2}}}{r^3} \cdot \frac{r^2}{r^4 + 2a^4 \cos^2 2\theta + a^4\left(1 + \sin^2 2\theta\right)}$$

$$= \frac{\left(r^4 + a^4 \cos^2 2\theta\right)^{\frac{3}{2}}}{r\left(r^4 + a^4 \cos^2 2\theta + 2a^4\right)} = \frac{\left(a^4 \sin^2 2\theta + a^4 \cos^2 2\theta\right)^{\frac{3}{2}}}{r\left(a^4 \sin^2 2\theta + a^4 \cos^2 2\theta + 2a^4\right)}$$

$$= \frac{a^6}{r\left(a^4 + 2a^4\right)} = \frac{a^2}{3r}$$

Hence curvature $k = \frac{1}{\rho} = \frac{3r}{a^2}$. Proved

(b) Since $\tan\phi = r\frac{d\theta}{dr}$ where ϕ is the angle.

$$\tan\phi = \frac{r}{r_1} = \frac{r}{\frac{a^2 \cos 2\theta}{r}}$$

$$= \frac{r^2}{a^2 \cos 2\theta} = \frac{a^2 \sin 2\theta}{a^2 \cos 2\theta} = \tan 2\theta$$

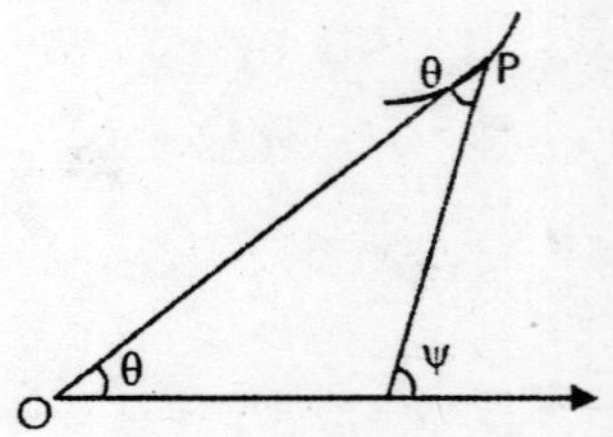

$$\phi = 2\theta$$

$$\psi = \theta + \phi$$

$$= \theta + 2\theta = 3\theta$$

$\therefore$ Tangent runs 3 times as fast as radius vector. Proved

Example 4:

If ρ_1 and ρ_2 are the radii of curvature at the extremities of any chord through the pole of the cardioid $r = a(1-\cos\theta)$, show that $\rho_1^2 + \rho_2^2 = \frac{16a^2}{9}$

Solution :

The cardioid $r = a(1-\cos\theta)$ is as shown in fig.

If the point P_1 is (r_1, θ) P_2 would be $(r_2, \theta+\pi)$ as P_1 and P_2 are the extremities of the chord through pole.

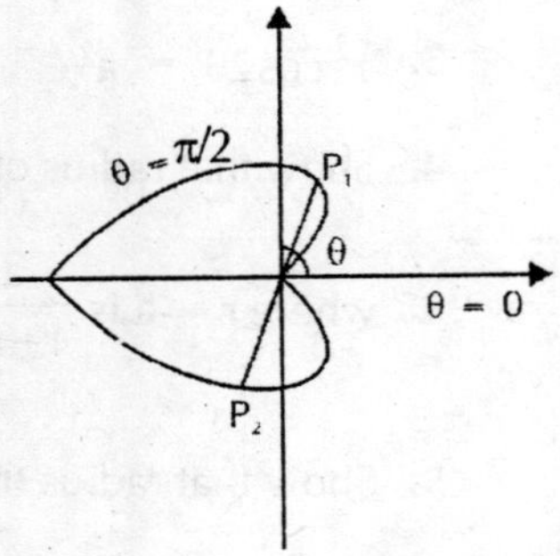

$$\frac{dr}{d\theta} = r_1 = a\sin\theta$$

$$\frac{d^2r}{d\theta^2} = r_2 = a\cos\theta$$

$$\therefore \quad \rho = \frac{\left(r^2 + r_1^2\right)^{\frac{3}{2}}}{r^2 + 2r_1^2 - rr_2}$$

$$= \frac{\left[a^2(1-\cos\theta)^2 + a^2\sin^2\theta\right]^{\frac{3}{2}}}{a^2(1-\cos\theta)^2 + 2a^2\sin^2\theta - a^2(1-\cos\theta)\cos\theta}$$

$$= \frac{a(2-2\cos\theta)^{\frac{3}{2}}}{3-3\cos\theta} = a\cdot\frac{2\sqrt{2}}{3}(1-\cos\theta)^{\frac{1}{2}}$$

$$\rho_1 = \frac{2\sqrt{2}}{3}a(1-\cos\theta)^{\frac{1}{2}}$$

$$\rho_2 = \frac{2\sqrt{2}}{3}a(1-\cos(\theta+\pi))^{\frac{1}{2}}$$

$$= \frac{2\sqrt{2}}{3}a(1+\cos\theta)^{\frac{1}{2}}$$

$$\therefore \quad \rho_1^2 + \rho_2^2 = \frac{8a^2}{9}(1-\cos\theta + 1 + \cos\theta)$$

$$= \frac{16a^2}{9}$$

$$\therefore \quad \rho_1^2 + \rho_2^2 = \frac{16a^2}{9}. \quad \text{Proved}$$

Exercise 5.2

Find the radius of curvature for the following curves :

1. $r = a\cos n\theta$.

2. $r^m = a^m \sin m\theta$.

3. $r^2 \cos 2\theta = a^2$.

4. Show that radius of curvature at any point to the curve $r = a\cos n\theta$ where $r = a$ is $\frac{a}{1+n^2}$.

5. Show that radius of curvature to $r = a\sin n\theta$ at the pole is $\frac{na}{2}$.

6. Show that radius of curvature for the curve $\theta = \frac{\sqrt{r^2-a^2}}{a} - \cos^{-1}\frac{a}{r}$ is $\sqrt{r^2-a^2}$.

7. Find the radius of curvature for the ellipse

$$\frac{1}{p^2} = \frac{1}{a^2}+\frac{1}{b^2}-\frac{r^2}{a^2b^2}.$$

8. Find the radius of curvature for

$$P^2 = \frac{r^4}{r^2+a^2}.$$

9. Show that the radius of curvature of the lemniscate $r^2 = a^2\cos 2\theta$ at the point where the tangent is parallel to x – axis is $\frac{\sqrt{2}}{3}a$.

Answers

1. $\frac{\left(r^2+a^2n^2-r^2n^2\right)^{\frac{3}{2}}}{r^2-r^2n^2+2a^2n^2}$. **2.** $\frac{a^m}{(m+1)r^{m-1}}$. **3.** $\frac{r^3}{a^2}$

7. $\frac{a^2b^2}{p^3}$ **8.** $\frac{\left(r^2+a^2\right)^{\frac{3}{2}}}{r^2+2a^2}$.

5.6 Radius of Curvature at the Origin

When the curve passes through the origin, then to find the radius of curvature at the origin, the following methods may be employed :

(a) **Method of Direct Substitution :**

The value of y_1 and y_2 at the origin is calculated and then these values substituted directly in the formula

$$\rho = \frac{\left(1+y_1^2\right)^{\frac{3}{2}}}{y_2}.$$

(b) **Method of Expansion :**

Let

$$y = f(x) = f(0) + xf'(0) + \frac{x^2}{2!}f''(0) + \frac{x^3}{3!}f'''(0) + \ldots$$

be Maclaurin's expansion of the equation of the curve. Since the curve passes through the origin(0, 0) therefore $f(0) = 0$,

So that the equation of the curve becomes

$$y = px + q\frac{x^2}{2!} + r\cdot\frac{x^3}{3!} + \ldots \qquad \ldots(1)$$

$$p = f'(0) = y_1(0)$$

$$q = f''(0) = y_2(0)$$

Differentiating (1) w.r.t x we have

$$y_1 = p + \frac{2qx}{2!} + \frac{3rx^2}{3!} + \ldots \qquad \ldots(2)$$

Differentiating (2) w.r.t x, we get

$$y_2 = \frac{2q}{2!} + \frac{6rx}{3!} + \ldots$$

$\therefore$ At the origin (0, 0)

$$y_1 = p \text{ and } y_2 = \frac{2q}{2!} = q$$

$$\therefore \quad \rho = \frac{\left(1+y_1^2\right)^{\frac{3}{2}}}{y_2} = \frac{\left(1+p^2\right)^{\frac{3}{2}}}{q}.$$

(c) **Newton's Method :**

(a) If x – axis is the tangent to the curve at the origin, then

$$\rho \text{ (at the origin)} = \lim_{\substack{x\to 0\\ y\to 0}} \frac{x^2}{2y}.$$

(b) If y – axis is a tangent to the curve at the origin, then

$$\rho \text{ (at the origin)} = \lim_{\substack{x\to 0\\ y\to 0}} \frac{y^2}{2x}.$$

Proof :

(a) Since x – axis is a tangent to the curve at (0, 0).

$$f'(0) = 0 \text{ i.e } y_1 = 0 \text{ or } p = 0$$

$$\rho \text{ (at the origin)} = \frac{(1+0^2)^{\frac{3}{2}}}{q} = \frac{1}{q}$$

Again because $p = 0$ the equation of the curve becomes

$$y = q\frac{x^2}{2!} + r\cdot\frac{x^3}{3!} + \ldots$$

$$\frac{2y}{x^2} = q + r\cdot\frac{x}{3} + \ldots$$

At the origin $x \to 0$, $y \to 0$ so that

$$q = \lim_{\substack{x\to 0\\ y\to 0}} \frac{2y}{x^2}$$

or

$$\rho = \frac{1}{q} = \lim_{\substack{x\to 0\\ y\to 0}} \frac{x^2}{2y}$$

(b) Similarly if y – axis is a tangent to the curve at the origin, then

$$\rho = \lim_{\substack{x\to 0\\ y\to 0}} \frac{y^2}{2x}$$

Curvature at the origin when polar equation of the curve is given :

When the initial line i.e $\theta = 0$ is a tangent to the curve $r = f(\theta)$ then

$$\rho \text{ (at the pole)} = \lim_{\substack{x\to 0\\ y\to 0}} \frac{x^2}{2y} = \lim_{\substack{\theta\to 0\\ r\to 0}} \frac{r^2\cos^2\theta}{2r\sin\theta} \quad \text{[changing to polar co-ordinates]}$$

$$= \lim_{\substack{\theta\to 0\\ r\to 0}} \left(\frac{r}{2\theta}\cdot\frac{\theta}{\sin\theta}\cdot\cos^2\theta\right)$$

$$= \lim_{\substack{\theta\to 0\\ r\to 0}} \frac{r}{2\theta} \quad \left[\because \lim_{\theta\to 0}\frac{\theta}{\sin\theta} = 1\right]$$

$$= \lim_{\theta\to\infty} \left(\frac{1}{2}\frac{dr}{d\theta}\right) \quad \text{(By L'Hospital Rule)}$$

> **Note :**
>
> **1. Newton's method for finding the radius of curvature at the origin should be used when *x* – axis or *y* – axis is the tangent at the origin.**
>
> **2. Tangents at the origin of an algebraic curve are obtained by equating to zero the lowest degree term.**

Illustrative Examples

Example 1:

Find the radius of curvature at the origin for the curve

$$2x^4 + 4x^3y + xy^2 + 6y^3 - 3x^2 - 2xy + y^2 - 4x = 0$$

Solution:

Since the constant term in the equation is absent, therefore the curve passes through origin. and the tangent at the origin is $x = 0$

i.e $y-$ axis is a tangent at the origin.

Thus by Newton's method

$$\rho \text{ (at the origin)} = \lim_{\substack{x \to 0 \\ y \to 0}} \frac{y^2}{2x}$$

Dividing both sides of the given equation by $2x$, we have

$$x^3 + 2x^2y + \frac{y^2}{2} + 6\cdot\left(\frac{y^2}{2x}\right)\cdot y - \frac{3}{2}x - y + \frac{y^2}{2x} - 2 = 0$$

Taking limit when $x \to 0,\ y \to 0$ and

$$\lim_{\substack{x \to 0 \\ y \to 0}} \frac{y^2}{2x} = \rho, \text{ we get}$$

$$\rho - 2 = 0$$

$$\rho = 2 \qquad \text{Proved}$$

Example 2:

Find the radius of curvature at the origin of the curve

$$x^2 + 6y^2 + 2x - y = 0$$

Solution:

Since the constant term in the equation is absent, therefore the curve passes through the origin and the tangent at the origin is

$$2x - y = 0$$

Let

$$y = px + q\frac{x^2}{2!} + r\frac{x^3}{3!} + ...$$

by the Maclaurin's expansion of the equation of the curve. Then

$$x^2 + 6\left(px + q \cdot \frac{x^2}{2!} + ...\right)^2 + 2x - \left(px + q\frac{x^2}{2!} + ...\right) = 0$$

is an identity. Equating the coefficients of x and x^2, we have

$$2 - p = 0 \quad ...(1)$$

$$p = 2$$

$$2 + 6p^2 + \frac{q}{2!} = 0 \quad ...(2)$$

Putting $p = 2$ in (2) we have $q = -50$

and $$\rho = \frac{(1+p^2)^{\frac{3}{2}}}{q} = \frac{(1+4)^{\frac{3}{2}}}{50}$$

$$= \frac{1}{2\sqrt{5}}$$ Proved

Example 3:

Find the radius of curvature at the origin of the two branches of the curve given by $x = 1 - t^2, y = t - t^3$.

Solution : We have

$$y = t - t^3 = t(1 - t^2) = tx$$

and $$t = \frac{y}{x}$$

$$x = 1 - \frac{y^2}{x^2} \Rightarrow x^3 = x^2 - y^2$$

which is cartesian equation of the curve . It clearly passes through the origin and the tangents at the origin are given by

$$x^2 - y^2 = 0$$

$$y = \pm x$$

Let

$$y = px + q\frac{x^2}{2!} + r\frac{x^3}{3!} + \ldots$$

be the Maclaurin's expansion.

$$x^3 = x^2 - \left(px + q\frac{x^2}{2!} + \ldots\right)^2$$

Equating the coefficient of x^2 and x^3 on both sides, we have

$$0 = 1 - p^2$$

$$p = \pm 1$$

$$1 = pq$$

So that when $p = 1, q = 1$ and when $p = -1, q = -1$

$$p = -1, q = -1$$

$$\rho = \frac{(1+p^2)^{\frac{3}{2}}}{q} = \frac{(1+1)^{\frac{3}{2}}}{\pm 1} = \pm 2\sqrt{2}$$

which are the radii of curvature at the origin to the two branches of the curve. Answer

Example 4:

Find the radius of curvature at the pole of the curve $r = a(\theta + \sin\theta)$.

Solution :

When $\theta = 0$ we get $r = 0$

$\therefore$ the curve passes through the pole and $\theta = 0$

i.e the initial line is the tangent at the pole.

$$\rho = \lim_{\theta \to 0} \frac{1}{2}\frac{dr}{d\theta}$$

$$= \lim_{\theta \to 0} a(1 + \cos\theta)$$

$$= 2a \quad \text{Answer}$$

Exercise 5.3

Find the radius of curvature at the origin for the following curves:

1. $x^3 - 2x^2y - 3xy^2 - 4y^3 + 5x^2 - 6xy + 7y^2 - 8y = 0$

2. $2x^4 + 3y^4 + 4x^2y + xy - y^2 + 2x = 0$

3. $x^3 + y^2 - 2x^2 + 6y = 0$

4. $2x^3 + 4x^2y + xy^2 + 5x^3 - x^2 - 2xy + y^2 + 4x = 0$

Find the radius of curvature at the origin of the following curves:

5. $y - x = x^2 + 2xy + y^2$

6. $y = 6x + 5x^2 + x^3$

7. $a_1x + a_2y + b_1x^2 + b_2xy + b_3y^2 + c_1x^3 + ... = 0$

Find the radius of curvature at the origin of the two branches of following curves:

8. $y^2 - 3xy + 2x^2 - x^3 + y^4 = 0$

9. $x^3 + y^3 - 3axy = 0$

10. $a(y^2 - x^2) = x^3$

11. $y^2 = x^2 \dfrac{a+x}{a-x}$

12. Apply Newton's formula to find the radius of curvature at the origin for the cycloid

$x = a(\theta + \sin\theta),\ y = a(1 - \cos\theta)$.

Answers

1. $\dfrac{4}{5}$	**2.** 1	**3.** $\dfrac{3}{2}$
4. 2	**5.** $\dfrac{1}{2\sqrt{2}}$	**6.** $\dfrac{37\sqrt{37}}{10}$
7. $\dfrac{1}{2}\dfrac{\left(a_1^2 + a_2^2\right)^{\frac{3}{2}}}{b_1a_2^2 - b_2a_1a_2 + b_3a_1^2}$	**8.** $\sqrt{2}, \dfrac{5\sqrt{2}}{2}$	**9.** $\dfrac{3a}{2}, \dfrac{3a}{2}$.
10. $\pm 2\sqrt{2}a$	**11.** $\pm\sqrt{2}a$	**12.** 4a

CHAPTER

6

Partial Differentiation

6.1 FUNCTIONS HAVING MORE THAN ONE VARIABLE

So far we have been concerned with functions of a single independent variable. We can however, think of any number of examples or quantities that depend upon two or more than two independent variables. For instance, the area of a triangle depends upon its base and altitude, the volume of a rectangular parallelopiped depends upon its length, breadth and height. In this chapter, we shall explain the limit, continuity and differentiation in relation to function of more than one variable.

If to each pair of values of x and y, there corresponds definite value of z, then z is said to be a function of two variables x and y and it is written as $z = f(x,y)$. Also, x and y are called independent variables and z is called the dependent variable. A similar definition can be given for function of more than two variables.

6.2 LIMIT AND CONTINUITY

The idea of the limit of a function of single variable can be extended to the case of the limit of a function of two variables.

Limit: A function f(x,y) is said to have the limit l as $x \to a$, $y \to b$, if for any arbitrarily choosen positive number 'ε' however small, a positive number 'δ' depending upon 'ε', can be determined such that $|f(x,y) - l| < \epsilon \ \forall \ x,y$ for $|x - a| < \delta$ and $|y - b| < \delta$. Note that we are in no way concerned with what happens when $x = a$, $y = b$. Further, the function f(x,y) must tend to same limit in whatever way, the point (x, y) tends to the point (a, b).

Remark: It should be presumed that

$$\lim_{x\to a}\left[\lim_{y\to b} f(x,y)\right] = \lim_{y\to b}\left[\lim_{x\to a} f(x,y)\right]$$

what ever be the function f(x,y). In general, it may or may not possible.

For example, if $f(x,y) = \dfrac{x^2-y^2}{x^2+y^2}$, then

$$\lim_{x\to 0}\left[\lim_{y\to 0}\frac{(x^2-y^2)}{(x^2+y^2)}\right] = \lim_{x\to 0}\left[\frac{x^2}{x^2}\right] = 1$$

and $$\lim_{y\to 0}\left[\lim_{x\to 0}\frac{(x^2-y^2)}{(x^2+y^2)}\right] = \lim_{y\to 0}\left[\frac{-y^2}{y^2}\right] = -1$$

Thus, the two repeated limits are not equal.

6.2.1 Continuity

A function f(x,y) is said to be continuous at a point (a, b) if for any arbitrarily chosen $\varepsilon > 0$, however small, there exists a positive number δ, depending upon $\in$ can be determined such that,

$$|f(x,y) - f(a,b)| < \in$$

for all values of x, y such that

$$|x-a| < \delta \text{ and } |y-b| < \delta.$$

The above definition of continuity implies that $f(x,y) \to f(a,b)$ when $x \to a$ and $y \to b$ in any manner. If f(x,y) does not tend to a unique limit, then it is said to be discontinuous at (a, b).

Note: In the following sections, the functions are assumed to be continuous at all points unless otherwise expressly stated.

6.3 PARTIAL DERIVATIVES

If z be a function of two independent variables x, y i.e. if z = f(x, y), the derivative of z w.r.t. x, when y is regarded as a constant, is called the partial derivative of z w.r.t. x and is denoted by the symbol $\dfrac{\partial z}{\partial x}$.

Similarly when x is regarded as constant, we may find the partial derivative of z with respect to y, which we shall denoted by the symbol $\dfrac{\partial z}{\partial y}$. Thus

$$\frac{\partial z}{\partial x} = \frac{\partial}{\partial x}[f(x,y)] = \lim_{\delta x\to 0}\frac{f(x+\delta x, y) - f(x,y)}{\delta x}$$

and $$\frac{\partial z}{\partial y} = \frac{\partial}{\partial y}[f(x,y)] = \lim_{\delta y \to 0} \frac{f(x, y+\delta y) - f(x,y)}{\delta y}$$

provided these limit exists.

The following symbols are commonly used to denote the partial derivatives of a function of two variables.

$$\frac{\partial z}{\partial x}, \frac{\partial f}{\partial x}, \frac{\partial}{\partial x}[f(x,y)], f_x(x,y),$$

$$\frac{\partial z}{\partial y}, \frac{\partial f}{\partial y}, \frac{\partial}{\partial y}[f(x,y)], f_y(x,y),$$

In general, if 'u' is a function of n independent variables $x_1, x_2, x_3, \ldots\ldots\ldots\ldots x_n$, then the partial derivatives of u with respect to any variable x_r, r = 1, 2, ... n, is obtained by differentiating u, w.r.t. x_r, treating all other variables independent of x_r, as constant.

6.3.1 Partial derivatives of higher Orders

If z = f(x,y), then the partial derivatives $\frac{\partial z}{\partial x}$ and $\frac{\partial z}{\partial y}$ are usually functions of both x and y (though one or both of them may be either constant or dependent on only one of the variables) they may posses partial derivatives with respect to x or y. These partial derivatives are called the second order partial derivatives of z.

These derivatives are commonly denoted by one of the following symbols:

$$\frac{\partial}{\partial x}\left(\frac{\partial z}{\partial x}\right) = \frac{\partial^2 z}{\partial x^2} = z_{xx} = \frac{\partial}{\partial x}\left(\frac{\partial f}{\partial x}\right) = \frac{\partial^2 f}{\partial x^2} = f_{xx};$$

$$\frac{\partial}{\partial y}\left(\frac{\partial z}{\partial x}\right) = \frac{\partial^2 z}{\partial y \partial x} = z_{xy} = \frac{\partial}{\partial y}\left(\frac{\partial f}{\partial x}\right) = \frac{\partial^2 f}{\partial y \partial x} = f_{xy};$$

$$\frac{\partial}{\partial y}\left(\frac{\partial z}{\partial y}\right) = \frac{\partial^2 z}{\partial y^2} = z_{yy} = \frac{\partial}{\partial y}\left(\frac{\partial f}{\partial y}\right) = \frac{\partial^2 f}{\partial y^2} = f_{yy};$$

$$\frac{\partial}{\partial x}\left(\frac{\partial z}{\partial y}\right) = \frac{\partial^2 z}{\partial x \partial y} = z_{yx} = \frac{\partial}{\partial x}\left(\frac{\partial f}{\partial y}\right) = \frac{\partial^2 f}{\partial x \partial y} = f_{yx};$$

Partial derivatives of higher order are defined in the similar manner.

Remark: It should be noted that $\frac{\partial^2 z}{\partial x \partial y}$ and $\frac{\partial^2 z}{\partial y \partial x}$ are, ingeneral, not equal. If however, the function possesses continuous partial derivatives of second order, then, $\frac{\partial^2 z}{\partial x \partial y} = \frac{\partial^2 z}{\partial y \partial x}$ and this property is known as the commutative property of the partial derivatives.

Illustrative Examples

Example 1 Find $\frac{\partial u}{\partial x}$ and $\frac{\partial u}{\partial y}$ when $u = \log_e\left(x^2 + y^2\right)$

Solution: Here $u = \log_e\left(x^2 + y^2\right)$ .....(i)

differentiating (i) partially w.r.t. x (treating y as constant),

we get $$\frac{\partial u}{\partial x} = \frac{1}{x^2 + y^2}\,2x = \frac{2x}{x^2 + y^2}$$

differentiating (i) partially w.r.t. y (treating x as constant),

we get $$\frac{\partial u}{\partial y} = \frac{1}{x^2 + y^2}.2y = \frac{2y}{x^2 + y^2}$$

Example 2 Verify that $\frac{\partial^2 u}{\partial x \partial y} = \frac{\partial^2 u}{\partial y \partial x}$, where $u = \log_e\left(\frac{x^2 + y^2}{xy}\right)$

Solution: Here $u = \log_e\left(\frac{x^2 + y^2}{xy}\right)$

or $u = \log_e(x^2 + y^2) - \log_e x - \log_e y$ ... (1)

differentiating (1) partially w.r.t. x, weget

$$\frac{\partial u}{\partial x} = \frac{1}{x^2 + y^2}.2x - \frac{1}{x} = \frac{2x}{x^2 + y^2} - \frac{1}{x} \quad ...(2)$$

differentiating (1) partially w.r.t. y, we get

$$\frac{\partial u}{\partial y} = \frac{1}{x^2 + y^2}.2y - \frac{1}{y} = \frac{2y}{x^2 + y^2} - \frac{1}{y} \quad ...(3)$$

differentiating (2) partially w.r.t. y, we get

$$\frac{\partial}{\partial y}\left(\frac{\partial u}{\partial x}\right) = \frac{\partial}{\partial y}\left(\frac{2x}{x^2 + y^2} - \frac{1}{x}\right) = 2x\frac{\partial}{\partial y}\left(\frac{1}{x^2 + y^2}\right)$$

$$= \frac{-2x}{\left(x^2 + y^2\right)^2}.2y$$

i.e., $$\frac{\partial^2 u}{\partial y \partial x} = \frac{-4xy}{\left(x^2 + y^2\right)^2} \quad ...(4)$$

differentiating (3) partially w.r.t. x, we get

$$\frac{\partial}{\partial x}\left(\frac{\partial u}{\partial y}\right)=\frac{\partial}{\partial x}\left(\frac{2y}{x^2+y^2}-\frac{1}{y}\right)=2y\frac{\partial}{\partial x}\left(\frac{1}{x^2+y^2}\right)$$

$$=\frac{-2y}{\left(x^2+y^2\right)^2}.2x$$

i.e.,
$$\frac{\partial^2 u}{\partial x\partial y}=\frac{-4xy}{\left(x^2+y^2\right)^2} \qquad ...(5)$$

∴ from (4) and (5), we get

$$\frac{\partial^2 u}{\partial y\partial x}=\frac{\partial^2 u}{\partial x\partial y} \qquad \text{Proved}$$

Example 3 If u (x+y) = (x² + y²), then show that

$$\left(\frac{\partial u}{\partial x}-\frac{\partial u}{\partial y}\right)^2=4\left(1-\frac{\partial u}{\partial x}-\frac{\partial u}{\partial y}\right)$$

Solution: Given that
$$u=\frac{x^2+y^2}{x+y} \qquad ...(1)$$

differentiating (1) partially w.r.t. x, we get

$$\frac{\partial u}{\partial x}=\frac{(x+y).\,2x-\left(x^2+y^2\right).1}{(x+y)^2}=\frac{x^2-y^2+2xy}{(x+y)^2} \qquad ...(2)$$

differentiating (1) partially w.r.t. y, we get

$$\frac{\partial u}{\partial y}=\frac{(x+y).\,2y-\left(x^2+y^2\right).1}{(x+y)^2}=\frac{y^2-x^2+2xy}{(x+y)^2} \qquad ...(3)$$

adding (2) and (3), we get

$$\frac{\partial u}{\partial x}+\frac{\partial u}{\partial y}=\frac{4xy}{(x+y)^2}$$

$$\therefore \quad \left(1-\frac{\partial u}{\partial x}-\frac{\partial u}{\partial y}\right)=1-\frac{4xy}{(x+y)^2}=\frac{(x-y)^2}{(x+y)^2} \qquad ...(4)$$

subtracting (3) from (2), we get

$$\frac{\partial u}{\partial x}-\frac{\partial u}{\partial y}=\frac{2(x^2-y^2)}{(x+y)^2}=\frac{2(x-y)}{(x+y)} \quad ...(5)$$

from (4) and (5), we get

$$\left[\left(\frac{\partial u}{\partial x}\right)-\left(\frac{\partial u}{\partial y}\right)\right]^2=\frac{4(x-y)^2}{(x+y)^2}=4\left[1-\frac{\partial u}{\partial x}-\frac{\partial u}{\partial y}\right] \quad \text{Proved}$$

Example 4 If $u = f(y + ax) + \phi(y - ax)$, prove that

$$\frac{\partial^2 u}{\partial x^2}-a^2\frac{\partial^2 u}{\partial y^2}=0$$

Solution: Given that

$$u = f(y + ax) + \phi(y - ax) \quad ...(1)$$

differentiating (1) partially w.r.t. x, we get

$$\frac{\partial u}{\partial x}=f'(y+ax).a+\phi'(y-ax)(-a)$$

$$=a f'(y+ax)-a\phi'(y-ax) \quad ...(2)$$

differentiating (2) partially w.r.t x, we get

$$\frac{\partial^2 u}{\partial x^2}=a^2 f''(y+ax)+a^2\phi''(y-ax) \quad ...(3)$$

differentiating (1) partially w.r.t. y, we get

$$\frac{\partial u}{\partial y}=f'(y+ax).1+\phi'(y-ax).1 \quad ...(4)$$

differentiating (4) partially w.r.t. y, we get

$$\frac{\partial^2 u}{\partial y^2}=f''(y+ax)+\phi''(y-ax) \quad ...(5)$$

from (3) and (5), we get

$$\frac{\partial^2 u}{\partial x^2}=a^2\frac{\partial^2 u}{\partial y^2} \Rightarrow \frac{\partial^2 u}{\partial x^2}-a^2\frac{\partial^2 u}{\partial y^2}=0 \quad \text{Proved.}$$

Example 5 If $u = \log_e(x^3 + y^3 + z^3 - 3xyz)$, then show that

$$\left(\frac{\partial}{\partial x}+\frac{\partial}{\partial y}+\frac{\partial}{\partial z}\right)^2 u=\frac{-9}{(x+y+z)^2}$$

Solution: Since $u = \log_e(x^3 + y^3 + z^3 - 3xyz)$

$$\therefore \quad \frac{\partial u}{\partial x} = \frac{1}{\left(x^3 + y^3 + z^3 - 3xyz\right)}.\left(3x^2 - 3yz\right) \qquad ...(1)$$

$$\frac{\partial u}{\partial y} = \frac{3y^2 - 3xz}{\left(x^3 + y^3 + z^3 - 3xyz\right)} \qquad ...(2)$$

and $$\frac{\partial u}{\partial z} = \frac{3z^2 - 3xy}{\left(x^3 + y^3 + z^3 - 3xyz\right)} \qquad ...(3)$$

adding (1), (2) and (3), we get

$$\left(\frac{\partial u}{\partial x} + \frac{\partial u}{\partial y} + \frac{\partial u}{\partial z}\right) = \frac{3\left(x^2 + y^2 + z^2 - yz - zx - xy\right)}{(x + y + z)\left(x^2 + y^2 + z^2 - yz - zx - xy\right)}$$

$$= \frac{3}{(x + y + z)} \qquad ...(4)$$

Now $$\left(\frac{\partial}{\partial x} + \frac{\partial}{\partial y} + \frac{\partial}{\partial z}\right)^2 u = \left(\frac{\partial}{\partial x} + \frac{\partial}{\partial y} + \frac{\partial}{\partial z}\right)\left(\frac{\partial u}{\partial x} + \frac{\partial u}{\partial y} + \frac{\partial u}{\partial z}\right)$$

$$= \left(\frac{\partial}{\partial x} + \frac{\partial}{\partial y} + \frac{\partial}{\partial z}\right)\left(\frac{3}{x + y + z}\right) \qquad \text{[by (4)]}$$

$$= \frac{-3}{(x + y + z)^2} - \frac{3}{(x + y + z)^2} - \frac{3}{(x + y + z)^2}$$

$$= \frac{-9}{(x + y + z)^2} \quad \text{Proved.}$$

Example 6 If $u = \sin^{-1}\left(\frac{x}{y}\right) + \tan^{-1}\left(\frac{y}{x}\right)$,

show that $$x\frac{\partial u}{\partial x} + y\frac{\partial u}{\partial y} = 0$$

Solution: Given $u = \sin^{-1}\left(\frac{x}{y}\right) + \tan^{-1}\left(\frac{y}{x}\right)$,

$$\therefore \quad \frac{\partial u}{\partial x} = \frac{1}{\sqrt{1 - \frac{x^2}{y^2}}}.\left(\frac{1}{y}\right) + \frac{1}{\left(1 + \frac{y^2}{x^2}\right)}\left(-\frac{y}{x^2}\right)$$

or $$x\frac{\partial u}{\partial x}=\frac{x}{\sqrt{y^2-x^2}}-\frac{xy}{x^2+y^2} \qquad ...(1)$$

and $$\frac{\partial u}{\partial y}=\frac{1}{\sqrt{1-\frac{x^2}{y^2}}}\left(-\frac{x}{y^2}\right)+\frac{1}{\left(1+\frac{y^2}{x^2}\right)}\cdot\frac{1}{x}$$

or $$y\frac{\partial u}{\partial y}=\frac{-x}{\sqrt{y^2-x^2}}+\frac{xy}{x^2+y^2} \qquad ...(2)$$

adding (1) and (2), we get

$$x\frac{\partial u}{\partial x}+y\frac{\partial u}{\partial y}=0$$ Proved.

Example 7 If $x^x\, y^y\, z^z=\lambda$, show that at x = y = z

$$\frac{\partial^2 z}{\partial x\partial y}=-\left(x\log_e xe\right)^{-1}$$

Solution: Given that

$x^x\, y^y\, z^z=\lambda$, where z is a function of x and y.

Taking logrithms on both sides, we get

$$x\log x+y\log y+z\log z=\log\lambda \qquad ...(1)$$

differentiating (1) partially w.r.t. x, we get

$$\left[\log x+x.\frac{1}{x}\right]+\left[\log z+z.\frac{1}{z}\right].\frac{\partial z}{\partial x}=0$$

or $$\frac{\partial z}{\partial x}=-\frac{(1+\log x)}{(1+\log z)} \qquad ...(2)$$

differentiating (1) partially w.r.t. y, we get

$$\left[\log y+y.\frac{1}{y}\right]+\left[\log z+z.\frac{1}{z}\right].\frac{\partial z}{\partial y}=0$$

or $$\frac{\partial z}{\partial y}=-\frac{(1+\log y)}{(1+\log z)} \qquad ...(3)$$

Now, $$\frac{\partial^2 z}{\partial x\partial y}=\frac{\partial}{\partial y}\left(\frac{\partial z}{\partial x}\right)=\frac{\partial}{\partial y}\left[-\frac{(1+\log x)}{(1+\log z)}\right]$$

$$= -(1+\log x)\frac{\partial}{\partial y}\left(\frac{1}{(1+\log z)}\right)$$

$$= -(1+\log x)\left[-\frac{1}{(1+\log z)^2}\frac{1}{z}\frac{\partial z}{\partial y}\right]$$

$$= \frac{1}{z}\frac{(1+\log x)}{(1+\log z)^2}\left[-\frac{(1+\log y)}{(1+\log z)}\right] \quad \text{[from 3]}$$

$$= -\frac{(1+\log x)(1+\log y)}{z.(1+\log z)^3}$$

$\therefore$ At $x = y = z$,

$$\frac{\partial^2 z}{\partial x \partial y} = -\frac{(1+\log x)^2}{x(1+\log x)^3} = -\frac{1}{x(1+\log x)}$$

$$= -\frac{1}{x[\log_e e + \log_e x]} = -\frac{1}{x \log_e x e}$$

i.e., $$\frac{\partial^2 z}{\partial x \partial y}(\text{at } x = y = z) = -[x \log_e x e]^{-1}$$ **Proved.**

Example 8 If $u = x^4 y^2 \sin^{-1}\left(\frac{x}{y}\right) + \log x - \log y$, then show that

$$x\frac{\partial z}{\partial x} + y\frac{\partial z}{\partial y} = 6x^4 z^2 \sin^{-1}\left(\frac{x}{y}\right).$$

Solution: Since $z = x^4 y^2 \sin^{-1}\left(\frac{x}{y}\right) + \log x - \log y$ ...(1)

differentiating (1) partially w.r.t. x, we get

$$\frac{\partial z}{\partial x} = x^4.y^2.\frac{1}{\sqrt{1-\frac{x^2}{y^2}}}.\frac{1}{y} + 4x^3y^2 \sin^{-1}\left(\frac{x}{y}\right) + \frac{1}{x}$$

$$x\frac{\partial z}{\partial x} = x^5y^2.\frac{1}{\sqrt{y^2 - x^2}} + 4x^4y^2 \sin^{-1}\left(\frac{x}{y}\right) + 1 \quad ...(2)$$

differentiating (1) partially w.r.t. y, we get

$$\frac{\partial z}{\partial y} = x^4y^2 \cdot \frac{1}{\sqrt{1-\frac{x^2}{y^2}}}\left(-\frac{x}{y^2}\right) + 2x^4y \sin^{-1}\left(\frac{x}{y}\right) - \frac{1}{y}$$

or
$$y\frac{\partial z}{\partial y} = -x^5y^2 \cdot \frac{1}{\sqrt{y^2-x^2}} + 2x^4y^2 \sin^{-1}\left(\frac{x}{y}\right) - 1 \qquad \text{...(3)}$$

Adding (2) and (3), we get

$$x\frac{\partial z}{\partial x} + y\frac{\partial z}{\partial y} = 6x^4y^2 \sin^{-1}\left(\frac{x}{y}\right) \qquad \text{Proved.}$$

Example 9 Find $\frac{\partial z}{\partial r}$ and $\frac{\partial z}{\partial \theta}$, if $z = e^{r\cos\theta}.\cos(r\sin\theta)$

Solution: Given that

$$z = e^{r\cos\theta}.\cos(r\sin\theta) \qquad \text{...(1)}$$

differentiating (1) partially w.r.t. r, we get

$$\frac{\partial z}{\partial r} = e^{r\cos\theta}\left[\{-\sin(r\sin\theta)\sin\theta\} + \{\cos(r\sin\theta).\cos\theta\}\right]$$

$$= e^{r\cos\theta}.\cos(r\sin\theta + \theta) \qquad \text{...(2)}$$

differentiating (1) partially w.r.t. θ, we get

$$\frac{\partial z}{\partial \theta} = e^{r\cos\theta}\left[-\sin(r\sin\theta)\,r\cos\theta\right] + \left[-r\sin\theta\, e^{r\cos\theta}\right].\cos(r\sin\theta)$$

$$= -r\,e^{r\cos\theta}\left[\sin(r\sin\theta).\cos\theta + \sin\theta\cos(r\sin\theta)\right]$$

$$= -r\,e^{r\cos\theta}\left[\sin(r\sin\theta + \theta)\right] \qquad \text{Answer}$$

Example 10 If $u = f(r)$, where $r^2 = x^2 + y^2$, show that

$$\frac{\partial^2 u}{\partial x^2} + \frac{\partial^2 u}{\partial y^2} = f''(r) + \frac{1}{r}f'(r)$$

Solution: Given that, $r^2 = x^2 + y^2$ ...(1)

differentiating (1) partially w.r.t. x, we get

$$2r\frac{\partial r}{\partial x} = 2x \Rightarrow \frac{\partial r}{\partial x} = \frac{x}{r} \qquad \text{...(2)}$$

Similarly
$$\frac{\partial r}{\partial y} = \frac{y}{r} \qquad \text{...(3)}$$

Now $u = f(r)$

$$\therefore \quad \frac{\partial u}{\partial x} = \frac{\partial u}{\partial r}.\frac{\partial r}{\partial x} = f'(r).\frac{x}{r} \qquad \text{(from (2)}$$

differentiating both sides partially w.r.t. x, we get

$$\frac{\partial^2 u}{\partial x^2} = \frac{\partial}{\partial x}\left(\frac{\partial u}{\partial x}\right) = \frac{\partial}{\partial x}\left(f'(r).\frac{x}{r}\right)$$

$$\frac{\partial^2 u}{\partial x^2} = \frac{r\left[f'(r).1 + x.f''(r).\frac{\partial r}{\partial x}\right] - xf'(r).\frac{\partial r}{\partial x}}{r^2}$$

$$\Rightarrow \quad \frac{\partial^2 u}{\partial x^2} = \frac{1}{r^2}\left[r\,f'(r) + x^2 f''(r) - \frac{x^2}{r} f'(r)\right] \qquad \text{[from (2)]}$$

Similarly,

$$\frac{\partial^2 u}{\partial y^2} = \frac{1}{r^2}\left[r\,f'(r) + y^2 f''(r) - \frac{y^2}{r} f'(r)\right]$$

$$\therefore \quad \frac{\partial^2 u}{\partial x^2} + \frac{\partial^2 u}{\partial y^2} = \frac{1}{r^2}\left[2r\,f'(r) + \left(x^2 + y^2\right) f''(r) - \frac{\left(x^2 + y^2\right)}{r} f'(r)\right]$$

$$= \frac{1}{r^2}\left[2r\,f'(r) + r^2 f''(r) - \frac{r^2}{r} f'(r)\right], \quad \text{(from (1))}$$

or

$$\frac{\partial^2 u}{\partial x^2} + \frac{\partial^2 u}{\partial y^2} = f''(r) + \frac{1}{r} f'(r) \qquad \text{Proved.}$$

Example 11 If $\dfrac{x^2}{a^2+u} + \dfrac{y^2}{b^2+u} + \dfrac{z^2}{c^2+u} = 1$ Prove that

$$\left(\frac{\partial u}{\partial x}\right)^2 + \left(\frac{\partial u}{\partial y}\right)^2 + \left(\frac{\partial u}{\partial z}\right)^2 = 2\left[x\frac{\partial u}{\partial x} + y\frac{\partial u}{\partial y} + z\frac{\partial u}{\partial z}\right]$$

Solution: Given that

$$\frac{x^2}{a^2+u} + \frac{y^2}{b^2+u} + \frac{z^2}{c^2+u} = 1 \qquad \text{...(1)}$$

where u is a function of x,y and z. Differentiating (1) both sides partially w.r.t. x, we get

$$\frac{2x}{a^2+u} - \left[\frac{x^2}{\left(a^2+u\right)^2} + \frac{y^2}{\left(b^2+u\right)^2} + \frac{z^2}{\left(c^2+u\right)^2}\right]\frac{\partial u}{\partial x} = 0$$

$$\therefore \qquad \frac{\partial u}{\partial x} = \frac{\dfrac{2x}{a^2+u}}{\Sigma\left(\dfrac{x^2}{(a^2+u)^2}\right)}$$

similarly,
$$\frac{\partial u}{\partial y} = \frac{\dfrac{2y}{b^2+u}}{\Sigma\left(\dfrac{x^2}{(a^2+u)^2}\right)}$$

and
$$\frac{\partial u}{\partial z} = \frac{\dfrac{2z}{c^2+u}}{\Sigma\left(\dfrac{x^2}{(a^2+u)^2}\right)}$$

$$\therefore \qquad \left(\frac{\partial u}{\partial x}\right)^2 + \left(\frac{\partial u}{\partial y}\right)^2 + \left(\frac{\partial u}{\partial z}\right)^2 = \frac{u\left(\Sigma\dfrac{x^2}{(a^2+u)^2}\right)}{\left[\Sigma\dfrac{x^2}{(a^2+u)^2}\right]^2}$$

$$= \frac{u}{\Sigma\left(\dfrac{x^2}{(a^2+u)^2}\right)} \qquad \text{...(2)}$$

Further,
$$x\frac{\partial u}{\partial x} + y\frac{\partial u}{\partial y} + z\frac{\partial u}{\partial z} =$$

$$\frac{2\,\Sigma\left(\dfrac{x^2}{(a^2+u)}\right)}{\Sigma\left(\dfrac{x^2}{(a^2+u)^2}\right)} = \frac{2}{\Sigma\left(\dfrac{x^2}{(a^2+u)^2}\right)} \qquad \text{(from (1))}$$

$$\therefore \quad 2\left[x\frac{\partial u}{\partial x}+y\frac{\partial u}{\partial y}+z\frac{\partial u}{\partial z}\right]=\frac{4}{\Sigma\left[\frac{x^2}{\left(a^2+u\right)^2}\right]} \quad \text{...(3)}$$

from (2) and (3), we get

$$\left(\frac{\partial u}{\partial x}\right)^2+\left(\frac{\partial u}{\partial y}\right)^2+\left(\frac{\partial u}{\partial z}\right)^2=2\left[x\frac{\partial u}{\partial x}+y\frac{\partial u}{\partial y}+z\frac{\partial u}{\partial z}\right] \text{ Proved.}$$

Example 12 If $\theta = t^n e^{-r^2/4t}$ for what value of 'n', we have $\frac{1}{r^2}\frac{\partial}{\partial r}\left(r^2\frac{\partial\theta}{\partial r}\right)=\frac{\partial\theta}{\partial t}$.

Solution: Let $\theta = t^n e^{-r^2/4t}$, then

$$\frac{\partial\theta}{\partial r}=t^n e^{-r^2/4t}\left(\frac{-2r}{4t}\right)=\frac{-r}{2}t^{n-1}e^{-r^2/4t}$$

$$\therefore \quad r^2\frac{\partial\theta}{\partial r}=\frac{-1}{2}r^3 t^{n-1}e^{-r^2/4t}$$

$$\therefore \quad \frac{\partial}{\partial r}\left(r^2\frac{\partial\theta}{\partial r}\right)=\frac{-3r^2}{2}t^{n-1}e^{-r^2/4t}-\frac{1}{2}r^3t^{n-1}e^{-r^2/4t}\left(\frac{-2r}{4t}\right)$$

$$=\frac{-3}{2}r^2t^{n-1}e^{-r^2/4t}+\frac{1}{4}r^4t^{n-2}e^{-r^2/4t}$$

$$\therefore \quad \frac{1}{r^2}\frac{\partial}{\partial r}\left(r^2\frac{\partial\theta}{\partial r}\right)=-\frac{3}{2}t^{n-1}e^{-r^2/4t}+\frac{1}{4}r^2t^{n-2}e^{-r^2/4t}$$

Also,
$$\frac{\partial\theta}{\partial t}=n\,t^{n-1}e^{-r^2/4t}+t^n e^{-r^2/4t}\left(\frac{r^2}{4t^2}\right)$$

$$=n\,t^{n-1}e^{-r^2/4t}+\frac{1}{4}r^2t^{n-2}e^{-r^2/4t}$$

Now,
$$\frac{1}{r^2}\frac{\partial u}{\partial r}\left(r^2\frac{\partial\theta}{\partial r}\right)=\frac{\partial\theta}{\partial t}$$

$$\Rightarrow \quad \frac{-3}{2}t^{n-1}e^{-r^2/4t}+\frac{1}{4}r^2t^{n-2}e^{-r^2/4t}=nt^{n-1}e^{-r^2/4t}+\frac{1}{4}r^2t^{n-2}e^{-r^2/4t}$$

$$\Rightarrow \quad \frac{-3}{2}t^{n-1}e^{-r^2/4t}=n\,t^{n-1}e^{-r^2/4t} \quad \text{(for all possible values of r and t)}$$

$\Rightarrow \qquad n = \dfrac{-3}{2}$

Example 13 Show that the function u = arc tan $\left(\dfrac{y}{x}\right)$ satisfies the Laplace equation i.e.,

$$\frac{\partial^2 u}{\partial x^2} + \frac{\partial^2 u}{\partial y^2} = 0$$

Solution: Here, u = arc tan $\left(\dfrac{y}{x}\right) = \tan^{-1}\left(\dfrac{y}{x}\right)$ ...(1)

differentiating (1) partially w.r.t. x, we get

$$\frac{\partial u}{\partial x} = \frac{1}{1+\dfrac{y^2}{x^2}}\left(\frac{-y}{x^2}\right)$$

or $$\frac{\partial u}{\partial x} = \frac{-y}{x^2+y^2} \qquad ...(2)$$

differentiating (1) partially w.r.t. y, we get

$$\frac{\partial u}{\partial y} = \frac{1}{1+\dfrac{y^2}{x^2}}\left(\frac{1}{x}\right)$$

$$\frac{\partial u}{\partial y} = \frac{x}{x^2+y^2} \qquad ...(3)$$

Again $$\frac{\partial^2 u}{\partial x^2} = \frac{\partial}{\partial x}\left(\frac{\partial u}{\partial x}\right)$$

$$\Rightarrow \qquad \frac{\partial}{\partial x}\left(\frac{-y}{x^2+y^2}\right) = -\left[\frac{(x^2+y^2).0 - y.(2x)}{(x^2+y^2)^2}\right]$$

$$\Rightarrow \qquad \frac{\partial^2 u}{\partial x^2} = \frac{2xy}{(x^2+y^2)^2} \qquad ...(4)$$

and $$\frac{\partial^2 u}{\partial y^2} = \frac{\partial}{\partial y}\left(\frac{x}{x^2+y^2}\right) = \left[\frac{(x^2+y^2).0 - x.\,2y}{(x^2+y^2)^2}\right]$$

$$\frac{\partial^2 u}{\partial y^2} = \frac{-2xy}{\left(x^2 + y^2\right)^2} \qquad ...(5)$$

adding (4) and (5), we get

$$\frac{\partial^2 u}{\partial x^2} + \frac{\partial^2 u}{\partial y^2} = 0 \qquad \text{Proved.}$$

Example 14 If $u = x^y$, then show that

$$\frac{\partial^3 u}{\partial x^2 \partial y} = \frac{\partial^3 u}{\partial x\, \partial y\, \partial x}$$

Solution: we have

$$u = x^y \qquad ...(1)$$

differentiating (1) partially w.r.t., 'y' , we get

$$\frac{\partial u}{\partial y} = x^y \log_e x \qquad ...(2)$$

differentiating (2) partially w.r.t. x, we get

$$\frac{\partial^2 u}{\partial x \partial y} = \frac{\partial}{\partial x}\left(\frac{\partial u}{\partial y}\right) = \frac{\partial}{\partial x}\left(x^y \log_e x\right)$$

$$= yx^{y-1} \log x + x^y . \frac{1}{x}$$

$$= x^{y-1}\left[y \log x + 1\right] \qquad ...(3)$$

Again differentiating (3) partially w.r.t. x, we get

$$\frac{\partial^3 u}{\partial x^2 \partial y} = \frac{\partial}{\partial x}\left(\frac{\partial^2 u}{\partial x \partial y}\right) = \frac{\partial}{\partial x}\left[x^{y-1}\left(y \log x + 1\right)\right] \qquad ...(4)$$

Also, $\dfrac{\partial u}{\partial x} = yx^{y-1}$

$$\therefore \quad \frac{\partial^2 u}{\partial y \partial x} = x^{y-1} . 1 + y . \frac{1}{x} x^y \log x$$

$$= x^{y-1}\left[1 + y \log x\right]$$

$$\frac{\partial^3 u}{\partial x \partial y \partial x} = \frac{\partial}{\partial x}\left(\frac{\partial^2 u}{\partial y \partial x}\right)$$

$$= \frac{\partial}{\partial x}\left[x^{y-1}(1+y\log x)\right] \qquad \ldots(5)$$

From (4) and (5)

$$\frac{\partial^3 u}{\partial x^2 \partial y} = \frac{\partial^3 u}{\partial x \partial y \partial x} \qquad \text{Proved.}$$

Example 15 If $u = e^{xyz}$, then show that

$$\frac{\partial^3 u}{\partial x \partial y \partial z} = \left(1 + 3xyz + x^2y^2z^2\right) e^{xyz}$$

Solution: Let $u = e^{xyz}$

$$\therefore \quad \frac{\partial u}{\partial z} = xye^{xyz}$$

Again differentiating partially w.r.t. y, we get

$$\frac{\partial^2 u}{\partial y \partial z} = \frac{\partial}{\partial y}\left(\frac{\partial u}{\partial z}\right) = x\frac{\partial}{\partial y}\left(y\, e^{xyz}\right)$$

$$= x\left[y.z.x.e^{xyz} + e^{xyz}\right]$$

$$= e^{xyz}\left[x^2yz + x\right]$$

Again differentiating partially w.r.t. x, we get

$$\frac{\partial^3 u}{\partial x \partial y \partial z} = \frac{\partial}{\partial x}\left[e^{xyz}\left(x^2yz + x\right)\right]$$

$$= e^{xyz}(2xyz + 1) + yz\, e^{xyz}\left(x^2yz + x\right)$$

$$= e^{xyz}\left(1 + 2xyz + x^2y^2z^2 + xyz\right)$$

$$= e^{xyz}\left(1 + 3xyz + x^2y^2z^2\right) \qquad \text{Proved.}$$

Exercise 6.1

1. Find $\frac{\partial u}{\partial x}$ and $\frac{\partial u}{\partial y}$, when

(i) $u = x \sin y + y \sin x$ (ii) $u = \frac{1}{\sqrt{x^2 + y^2}}$

(iii) $u = x^y$ (iv) $u = \sin^{-1} x/y$

2. If $u = x^2 \tan^{-1}\left(\frac{y}{x}\right) - y^2 \tan^{-1}\left(\frac{x}{y}\right)$, prove that

$$\frac{\partial^2 u}{\partial x \partial y} = \frac{\partial^2 u}{\partial y . \partial x}$$

3. If $u = f\left(\frac{y}{x}\right)$, then prove that

$$x\frac{\partial u}{\partial x} + y\frac{\partial u}{\partial y} = 0$$

4. If $u = \sin^{-1}\frac{\sqrt{x}-\sqrt{y}}{\sqrt{x}+\sqrt{y}}$, then show that $\frac{\partial u}{\partial x} = -\frac{y}{x}\frac{\partial u}{\partial y}$.

5. If $u = e^x (x \cos y - y \sin y)$, then show that $\frac{\partial^2 u}{\partial x^2} + \frac{\partial^2 u}{\partial y^2} = 0$

6. If $u = \tan^{-1}\left(\frac{xy}{\sqrt{1+x^2+y^2}}\right)$, then prove that

$$\frac{\partial^2 u}{\partial x \partial y} = \frac{1}{\left(1+x^2+y^2\right)^{3/2}}.$$

7. If $u = (1 - 2xy + y^2)^{-1/2}$, then prove that

$$\frac{\partial}{\partial x}\left\{\left(1-x^2\right)\frac{\partial u}{\partial x}\right\} + \frac{\partial}{\partial y}\left\{y^2\frac{\partial u}{\partial y}\right\} = 0$$

8. Find p and q, if $\begin{cases} x = \sqrt{a}\,(\sin u + \cos v) \\ y = \sqrt{a}\,(\cos u - \sin v) \\ z = 1 + \sin(u - v) \end{cases}$

where p and q are $\frac{\partial z}{\partial x}$ and $\frac{\partial z}{\partial y}$ respectively **Hint:** $z = \frac{1}{2a}\left(x^2+y^2\right)$

9. If $f(u) = \sin u$ and $u = \sqrt{x^2+y^2}$, then show that $\left(\frac{\partial f}{\partial x}\right)^2 + \left(\frac{\partial f}{\partial y}\right)^2 = \cos^2 u$.

10. Prove that, if $f(x,y)=\frac{1}{\sqrt{y}}e^{\frac{(x-a)^2}{4y}}$, then $f_{xy}(x, y) = f_{yx}(x, y)$

11. If $v = \log(x^2 + y^2) + \tan^{-1}\left(\frac{y}{x}\right)$, then showthat $\frac{\partial^2 v}{\partial x^2}+\frac{\partial^2 v}{\partial y^2}=0$

12. If $v=e^{xyz}+\left(\frac{xz}{y}\right)$, then prove that

$$x\frac{\partial v}{\partial x}+y\frac{\partial v}{\partial y}=2xyz, \quad y\frac{\partial v}{\partial y}+z\frac{\partial v}{\partial z}=2xyzv$$

Also deduce that $x\frac{\partial^2 v}{\partial z\partial x}\neq y\frac{\partial^2 v}{\partial z\partial y}$

13. If $u=x\,\phi\left(\frac{y}{x}\right)+\phi\left(\frac{y}{x}\right)$, then show that

(i) $x\frac{\partial u}{\partial x}+y\frac{\partial u}{\partial y}=x\,\phi\left(\frac{y}{x}\right)$

(ii) $x^2\frac{\partial^2 u}{\partial x^2}+2xy\frac{\partial^2 u}{\partial x\partial y}+y^2\frac{\partial^2 u}{\partial y^2}=0$

14. If $v = x^2\tan^{-1}\left(\frac{y}{x}\right)-y^2\tan^{-1}\left(\frac{x}{y}\right)$, then prove that

$\frac{\partial^2 v}{\partial x\partial y}=\frac{\partial^2 u}{\partial y\partial x}$, where v is a function of x and y.

Answers

1. (i) $\sin y + y\cos x;\ x\cos y + \sin x$ (ii) $\frac{-x}{(x^2+y^2)^{3/2}};\ \frac{-y}{(x^2+y^2)^{3/2}}$

(iii) $x^y.\frac{y}{x};\ x^y\log x$ (iv) $\frac{-1}{\sqrt{y^2-x^2}};\ -\frac{x}{\sqrt{y^2-x^2}}$

8. $p = x/a$ and $q = y/a$

6.4 HOMOGENEOUS FUNCTIONS

A function f(x,y) is said to be a homogeneous function of degree (or order) n, if each term in x and y is of same degree n.

Thus the expression

$$a_0x^n + a_1x^{n-1}y + a_2x^{n-2}y^2 + \dots\dots\dots + a_ny^n,$$

is a homogeneous function of degree n. This expression can also be written as

$x^n\left[a_0 + a_1\left(\frac{y}{x}\right) + a_2\left(\frac{y}{x}\right)^2 + \dots\dots\dots + a_n\left(\frac{y}{x}\right)^n\right]$, which can be denoted by the symbol $x^n f\left(\frac{y}{x}\right)$. Hence to enlarge the concept of homogeneity so as to bring even transcendental functions within its scope, we define a homogeneous function as follows.

A function of two variables x, y is said to be a homogeneous function of degree n in x and y if it can be expressible in the form $x^n f(y/x)$.

Thus, for example, the expression, $ax^2 + by^2 + 2hxy$, i.e. $x^2\left[a + 2h\left(\frac{y}{x}\right) + b\left(\frac{y}{x}\right)^2\right]$ is the homogeneous function of degree 2.

$$\frac{\sqrt{x}+\sqrt{y}}{x+y} = \frac{\sqrt{x}\left(1+\frac{\sqrt{y}}{\sqrt{x}}\right)}{x\,(1+y/x)} = x^{-1/2}\frac{\left(1+\frac{\sqrt{y}}{\sqrt{x}}\right)}{(1+y/x)}$$ is a homogeneous function of degree −1/2.

Similarly $\sin^{-1}\left(\frac{x^{1/2} - y^{1/2}}{x^{1/2} + y^{1/2}}\right)$ is a homogeneous function of degree zero.

6.5 EULER'S THEOREM

Statement: If u be a homogeneous function of degree n in x and y, then $x\frac{\partial u}{\partial x} + y\frac{\partial u}{\partial y} = nu$

Proof: Since u is a homogeneous function of degree n in x and y, it can be written as, $u = x^n f(y/x)$

$$\therefore \quad \frac{\partial u}{\partial x} = nx^{n-1} f(y/x) + x^n f'(y/x)\left(\frac{-y}{x^2}\right)$$

$$= nx^{n-1} f(y/x) - x^{n-2}y\, f'(y/x) \qquad \dots(1)$$

and $$\frac{\partial u}{\partial y} = x^n f'(y/x)\left(\frac{1}{x}\right)$$

$$= x^{n-1} f'(y/x) \qquad \dots(2)$$

Multiplying (1) and (2) by x and y respectively and adding, we obtain

$$x\frac{\partial u}{\partial x}+y\frac{\partial u}{\partial y}=nx^n f(y/x)=nu \qquad ...(3)$$

Hence the theorem.

This theorem can be generalized to homogeneous functions of any number of variables. Thus, if $u = f(x_1, x_2 .... x_m)$ be a function of degree n of variables $x_1, x_2, ... x_m$, then

$$x_1\frac{\partial u}{\partial x}+x_2\frac{\partial u}{\partial x_2}+..........+x_m\frac{\partial u}{\partial x_m}=nu$$

The proof is similar to that for two variables.

Deduction - I: If u is a homogeneous function of x and y of degree n and u = f (v), then

$$x\frac{\partial v}{\partial x}+y\frac{\partial v}{\partial y}=n\frac{f(v)}{f'(v)}$$

Proof: Let u be a homogeneous function of x and y of degree n, we have, on applying Euler's theorem,

$$x\frac{\partial u}{\partial x}+y\frac{\partial u}{\partial y}=nu \qquad ...(1)$$

Now given $u = f(v)$

$$\therefore \quad \frac{\partial u}{\partial x}=f'(v)\frac{\partial v}{\partial x} \text{ and } \frac{\partial u}{\partial y}=f'(v),\frac{\partial v}{\partial y}$$

Putting these values in equation (1), we get

$$x f'(v)\frac{\partial v}{\partial x}+y f'(v)\frac{\partial v}{\partial y}=n f(v)$$

or

$$x\frac{\partial v}{\partial x}+y\frac{\partial v}{\partial y}=n\frac{f(v)}{f'(v)}$$

Deduction - II: Prove that

$$x^2\frac{\partial^2 v}{\partial x^2}+2xy\frac{\partial^2 v}{\partial x\partial y}+y^2\frac{\partial^2 v}{\partial y^2}=g(v)\left[g'(v)-1\right]$$

where $g(v) = n\frac{f(v)}{f'(v)}$

Proof: Since v is a homogeneous function of x and y of degree n, by Euler's deduction I, we have

$$x\frac{\partial v}{\partial x}+y\frac{\partial v}{\partial y}=n\frac{f(v)}{f'(v)}$$

given that $g(v) = n\dfrac{f(v)}{f'(v)}$

Then $$x\frac{\partial v}{\partial x} + y\frac{\partial v}{\partial y} = g(v) \qquad ...(1)$$

differentiating (1) partially w.r.t. x, we get

$$\left[x\frac{\partial^2 v}{\partial x^2} + \frac{\partial v}{\partial x}.1\right] + y\frac{\partial^2 v}{\partial y \partial x} = g'(v).\frac{\partial v}{\partial x}$$

or $$x\frac{\partial^2 v}{\partial x^2} + y\frac{\partial^2 v}{\partial x \partial y} = [g'(v) - 1]\frac{\partial v}{\partial x} \qquad ...(2)$$

similarly differentiating (1) partially w.r.t. y, we get

$$y.\frac{\partial^2 v}{\partial y^2} + x\frac{\partial^2 v}{\partial y \partial x} = [g'(v) - 1]\frac{\partial v}{\partial y} \qquad ...(3)$$

Multiplying (2) by 'x' and (3) by 'y' and adding, we get

$$x^2\frac{\partial^2 v}{\partial x^2} + 2xy\frac{\partial^2 v}{\partial x \partial y} + y^2\frac{\partial^2 v}{\partial y^2} = [g'(v) - 1]\left[x\frac{\partial v}{\partial x} + y\frac{\partial v}{\partial y}\right]$$

$$= [g'(v) - 1]\, g(v) \qquad \text{[from (1)]}$$

or $$x^2\frac{\partial^2 v}{\partial x^2} + 2xy\frac{\partial^2 v}{\partial x \partial y} + y^2\frac{\partial^2 v}{\partial y^2} = g(v)[g'(v) - 1].$$ Proved.

Illustrative Examples

Example 1 Verify, Euler's theorem for the function $f(x,y) = ax^2 + 2hxy + by^2$

Solution: Here $f(x,y) = ax^2 + 2hxy + by^2$ ...(1)

clearly, f(x,y) is a homogeneous function of degree '2'.

∴ we have to prove that

$$x\frac{\partial f}{\partial x} + y\frac{\partial f}{\partial y} = 2f$$

differentiating (1) partially w.r.t. x, we get

$$\frac{\partial f}{\partial x} = 2ax + 2hy \Rightarrow x\frac{\partial f}{\partial x} = 2ax^2 + 2hxy \qquad ...(2)$$

differentiating (1) partially w.r.t y, we get

$$\frac{\partial f}{\partial y} = 2by + 2hx \Rightarrow y\frac{\partial f}{\partial y} = 2by^2 + 2hxy \qquad ...(3)$$

adding (2) and (3), we get

$$x\frac{\partial f}{\partial x}+y\frac{\partial f}{\partial y}=2ax^2+4h\,xy+2by^2=2f$$

or $$x\frac{\partial f}{\partial x}+y\frac{\partial f}{\partial y}=2f$$ which verifies Euler's theorem.

Example 2 If $u=\sin^{-1}\left(\frac{x+y}{\sqrt{x}+\sqrt{y}}\right)$, then show that

$$x\frac{\partial u}{\partial x}+y\frac{\partial u}{\partial y}=\frac{1}{2}\tan u$$

Solution: Here $u=\sin^{-1}\left(\frac{x+y}{\sqrt{x}+\sqrt{y}}\right)$

$$\therefore \quad \sin u=\frac{x+y}{\sqrt{x}+\sqrt{y}}=f(x,y) \text{ (say)}$$

Here sin u is a homogeneous function of degree $\left(1-\frac{1}{2}\right)=\left(\frac{1}{2}\right)$

Hence by Euler's theorem

$$x\frac{\partial}{\partial x}(\sin u)+y\frac{\partial}{\partial y}(\sin u)=\frac{1}{2}\sin u$$

or $$x\cos u\frac{\partial u}{\partial x}+y\cos u\frac{\partial u}{\partial y}=\frac{1}{2}\sin u$$

or $$x\frac{\partial u}{\partial x}+y\frac{\partial u}{\partial y}=\frac{1}{2}\frac{\sin u}{\cos u}=\frac{1}{2}\tan u$$ Proved.

Example 3 If $u=\sin^{-1}\left(\frac{x}{y}\right)+\tan^{-1}\left(\frac{y}{x}\right)$, then prove that

$$x\frac{\partial u}{\partial x}+y\frac{\partial u}{\partial y}=0$$

Solution: Let $u_1=\sin^{-1}\left(\frac{x}{y}\right)$ and $u_2=\tan^{-1}\left(\frac{y}{x}\right)$

$$\therefore \quad u=u_1+u_2 \qquad \text{...(1)}$$

Here, sin $u_1 = \frac{x}{y}$ is a homogeneous function of degree 0, therefore by Euler's theorem

$$x\frac{\partial}{\partial u}(\sin u_1) + y\frac{\partial}{\partial y}(\sin u_1) = 0.\ \sin u_1 = 0$$

or $$\left(x\frac{\partial u_1}{\partial x} + y\frac{\partial u_1}{\partial y}\right)\cos u_1 = 0$$

i.e., $$x\frac{\partial u_1}{\partial x} + y\frac{\partial u_1}{\partial y} = 0 \qquad ...(2)$$

Again tan $u_2 = \frac{y}{x}$ is a homogeneous function of degree 0, therefore by Euler's theorem, we get

$$x\frac{\partial}{\partial x}(\tan u_2) + y\frac{\partial}{\partial y}(\tan u_2) = 0.\ \tan u_2 = 0$$

or $$\left(x\frac{\partial u_2}{\partial x} + y\frac{\partial u_2}{\partial y}\right)\sec^2 u_2 = 0$$

i.e., $$x\frac{\partial u_2}{\partial x} + y\frac{\partial u_2}{\partial y} = 0 \qquad ...(3)$$

Adding (2) and (3), we get

$$x\frac{\partial}{\partial x}(u_1 + u_2) + y\frac{\partial}{\partial y}(u_1 + u_2) = 0$$

or $$x\frac{\partial u}{\partial x} + y\frac{\partial u}{\partial y} = 0 \qquad \text{[from (1)]}$$

Hence Proved.

Example 4 If $v = \log_e\left(\frac{x^4 + y^4}{x + y}\right)$, then show that

$$x\frac{\partial v}{\partial x} + y\frac{\partial v}{\partial y} = 3.$$

Solution: Let $v = \log_e\left(\frac{x^4 + y^4}{x + y}\right)$

Here v is not a homogeneous function, but

but if $$z = e^v = \frac{x^4 + y^4}{x + y} = \frac{x^4\left(1 + y^4/x^4\right)}{x(1 + y/x)}$$

$$= x^3 \phi\,(y/x)$$

Here z is a homogeneous function of degree 3. Hence by Euler's theorem,

$$x\frac{\partial z}{\partial x} + y\frac{\partial z}{\partial y} = nz$$

i.e., $$x\frac{\partial}{\partial x}\left(e^v\right) + y\frac{\partial}{\partial y}\left(e^v\right) = 3e^v$$

or $$x\,e^v\,\frac{\partial v}{\partial x} + y\,e^v\,\frac{\partial v}{\partial y} = 3e^v$$

or $$x\frac{\partial v}{\partial x} + y\frac{\partial v}{\partial y} = 3.$$ Proved

Example 5 If $u = \sin^{-1}\left(\dfrac{x + 2y + 3z}{\sqrt{x^8 + y^8 + z^8}}\right)$, show that

$$x\frac{\partial u}{\partial x} + y\frac{\partial u}{\partial y} + z\frac{\partial u}{\partial z} = -\,3\tan u$$

Solution: we have, $$u = \sin^{-1}\left(\frac{x + 2y + 3z}{\sqrt{x^8 + y^8 + z^8}}\right),$$

$$\therefore \quad \sin u = \frac{x + 2y + 3z}{\sqrt{x^8 + y^8 + z^8}}$$

Here sin u is a homogeneous function of degree 1 – 4 = –3. Hence by Euler's theorem

$$x\frac{\partial}{\partial x}(\sin u) + y\frac{\partial}{\partial y}(\sin u) + z\frac{\partial}{\partial z}(\sin u) = -\,3\sin u$$

$$\Rightarrow \quad x\cos u\frac{\partial u}{\partial x} + y\cos u\frac{\partial u}{\partial y} + z\cos u\frac{\partial u}{\partial z} = -\,3\sin u$$

or $$x\frac{\partial u}{\partial x} + y\frac{\partial u}{\partial y} + z\frac{\partial u}{\partial z} = -\,3\frac{\sin u}{\cos u} = -\,3\tan u$$

or $$x\frac{\partial u}{\partial x} + y\frac{\partial u}{\partial y} + z\frac{\partial u}{\partial z} + 3\tan u = 0$$ Proved.

Example 6 If $v = \tan^{-1}\left(\dfrac{x^3 + y^3}{x - y}\right)$, then show that

(i) $x\dfrac{\partial v}{\partial x} + y\dfrac{\partial v}{\partial y} = \sin 2v$

(ii) $x^2\dfrac{\partial^2 v}{\partial x^2} + 2xy\dfrac{\partial^2 v}{\partial x \partial y} + y^2\dfrac{\partial^2 v}{\partial y^2} = 2\cos 3v \sin v.$

Solution: Here v is not a homogeneous function

But $z = \tan v = \dfrac{x^3 + y^3}{x - y} = \dfrac{x^3\left(1 + y^3/x^3\right)}{x\left(1 - y/x\right)}$

$= f(x)$ (*say*)

$= x^2\,\phi\,(y/x)$

Therefore z is a homogeneous function of x, y of degree 2. Hence by Euler's deduction- I formula,

$$x\frac{\partial v}{\partial x} + y\frac{\partial v}{\partial y} = n\frac{f(v)}{f'(v)} = 2\frac{\tan v}{\sec^2 v} = \frac{2\sin v\cos^2 v}{\cos v}$$

$= 2\sin v\cos v = \sin 2v$

Also by Euler's deduction II formula

$$x^2\frac{\partial^2 v}{\partial x^2} + 2xy\frac{\partial^2 v}{\partial x \partial y} + y^2\frac{\partial^2 v}{\partial y^2} = g(v)\left[g'(v) - 1\right]$$

Here $g(v) = \sin 2v$

$$x^2\frac{\partial^2 v}{\partial x^2} + 2xy\frac{\partial^2 v}{\partial x \partial y} + y^2\frac{\partial^2 v}{\partial y^2} = \sin 2v\,(2\cos 2v - 1)$$

$= 2\sin 2v\cos 2v - \sin 2v$

$= \sin 4v - \sin 2v$

$= 2\sin v\cos 3v$ Proved

Example 7 If $u = \sin^{-1}\left(\dfrac{x^3 + y^3 + z^3}{ax + by + cz}\right)$, then prove that

$$x\frac{\partial u}{\partial x} + y\frac{\partial u}{\partial y} + z\frac{\partial u}{\partial z} = 2\tan u$$

Solution: we have

$$u = \sin^{-1}\left(\frac{x^3 + y^3 + z^3}{ax + by + cz}\right)$$

$$\therefore \quad \sin u = \frac{x^3 + y^3 + z^3}{ax + by + cz} = f(x, y)\,(say)$$

here sin u is a homogeneous function of degree 3–1 i.e. 2. Hence by Euler's theorem

$$x\frac{\partial f}{\partial x} + y\frac{\partial f}{\partial y} + z\frac{\partial f}{\partial z} = nf$$

$$\Rightarrow \quad x\frac{\partial}{\partial x}(\sin u) + y\frac{\partial}{\partial y}(\sin u) + z\frac{\partial}{\partial z}(\sin u) = 2\sin u$$

$$\Rightarrow \quad x\cos u\frac{\partial u}{\partial x} + y\cos u\frac{\partial u}{\partial y} + z\cos u\frac{\partial u}{\partial z} = 2\sin u$$

or

$$x\frac{\partial u}{\partial x} + y\frac{\partial u}{\partial y} + z\frac{\partial u}{\partial z} = 2\frac{\sin u}{\cos u} = 2\tan u$$

$$\Rightarrow \quad x\frac{\partial u}{\partial x} + y\frac{\partial u}{\partial y} + z\frac{\partial u}{\partial z} = 2\tan u \quad \text{Proved.}$$

Example 8 If $u = x\sin^{-1}\left(\frac{y}{x}\right)$, Prove that

$$x^2\frac{\partial^2 u}{\partial x^2} + 2xy\frac{\partial^2 u}{\partial x \partial y} + y^2\frac{\partial^2 u}{\partial y^2} = 0$$

Solution: Here u is a homogeneous function of x and y of degree 1 i.e. n = 1

Now, by using Euler's Formula,

$$x^2\frac{\partial^2 u}{\partial x^2} + 2xy\frac{\partial^2 u}{\partial x \partial y} + y^2\frac{\partial^2 u}{\partial y^2} = n(n-1)u$$

by putting n = 1, we get the required result.

Example 9 If $u = \frac{x^2 + y^2}{\sqrt{x + y}}$, then verify that Euler's theorem holds good.

Solution: Here $u = \frac{x^2 + y^2}{\sqrt{x + y}} = \frac{x^{3/2}\left(1 + y^2/x^2\right)}{(1 + y/x)^{1/2}}$

Here u is a homogeneous function of x and y of degree 3/2 i.e. n = 3/2
Then using Euler's formula,

$$x\frac{\partial u}{\partial x}+y\frac{\partial u}{\partial y}=nu$$

Then put n = 3/2, we get the required result.

Example 10 If $v=\sec^{-1}\left(\dfrac{x^3-y^3}{x+y}\right)$, then prove that

$$x\frac{\partial v}{\partial x}+y\frac{\partial v}{\partial y}=2\cot v,$$ then evaluate

$$x^2\frac{\partial^2 v}{\partial x^2}+2xy\frac{\partial^2 v}{\partial x\partial y}+y^2\frac{\partial^2 v}{\partial y^2}$$

Solution: Here $\sec v=\left(\dfrac{x^3-y^3}{x+y}\right)=f(x,y)$

Since f is a homogeneous function of x and y of degree 2 i.e. n = 2, Hence applying Euler's formula, we have

$$x\frac{\partial f}{\partial x}+y\frac{\partial f}{\partial y}=nf$$

$$x\sec v\tan v\frac{\partial v}{\partial x}+y\sec v.\tan v\frac{\partial v}{\partial y}=2\sec v$$

or $$x\frac{\partial v}{\partial x}+y\frac{\partial u}{\partial y}=2\cot v$$ I- part Proved.

Now, for IInd part, find the value of

$$x^2\frac{\partial^2 v}{\partial x^2}+2xy\frac{\partial^2 v}{\partial x\partial y}+y^2\frac{\partial^2 v}{\partial y^2}=n(n-1)v$$

and also $$x^2\frac{\partial^2 v}{\partial x^2}+2xy\frac{\partial^2 v}{\partial x\partial y}+y^2\frac{\partial^2 v}{\partial y^2}=[g'(v)-1]g(v)$$ we get the required result.

Example 11 If $u=\sin^{-1}\left[\dfrac{x^{1/3}+y^{1/3}}{x^{1/2}+y^{1/2}}\right]^{1/2}$, Prove that

$$x^2\frac{\partial^2 u}{\partial x^2}+2xy\frac{\partial^2 u}{\partial x\partial y}+y^2\frac{\partial^2 u}{\partial y^2}=\frac{\tan u}{144}\left(13+\tan^2 u\right)$$

Hint: Let $z = \sin u = \left[\dfrac{x^{1/3}+y^{1/3}}{x^{1/2}+y^{1/2}}\right]^{1/2}$, since z is a homogeneous function of x and y of degree −1/12 i.e., n = −1/12

Now applying Euler's formula,

$$x\frac{\partial u}{\partial x} + y\frac{\partial u}{\partial y} = \frac{-1}{12}\tan u$$

Again let $g(u) = \dfrac{-1}{12}\tan u$, then

Applying Euler's IInd deduction formula, we get

$$x^2\frac{\partial^2 u}{\partial x^2} + 2xy\frac{\partial^2 u}{\partial x \partial y} + y^2\frac{\partial^2 u}{\partial y^2} = g(u)\left[g'(u)-1\right]$$

$$= \frac{-1}{12}\tan u\left[\frac{-1}{12}\sec^2 u - 1\right]$$

$$= \frac{\tan u}{144}\left(\tan^2 u + 13\right).$$

Exercise 6.2

1. Verify Euler's theorem for the following functions:

   (i) $u = 3x^2yz + 4xy^2z + 5y^4$

   (ii) $u = \dfrac{x(x^3 - y^3)}{x^3 + y^3}$

   (iii) $u = x^2 \log\left(\dfrac{y}{x}\right)$

   (iv) $u = \dfrac{x^{1/4} + y^{1/4}}{x^{1/5} + y^{1/5}}$

   (v) $u = axy + byz + czx$

2. If $u = \dfrac{x^3 y^3}{x^3 + y^3}$, show that $x\dfrac{\partial u}{\partial x} + y\dfrac{\partial u}{\partial y} = 3u$

3. If $v = \log\left(\dfrac{x^3 + y^3}{x^2 + y^2}\right)$, prove that $x\dfrac{\partial u}{\partial x} + y\dfrac{\partial u}{\partial y} = 1$

4. If $u = x^4 y^2 \sin^{-1}\left(\frac{y}{x}\right)$, then find the value of $x\frac{\partial u}{\partial x} + y\frac{\partial u}{\partial y}$.

5. If $u = \sin^{-1}\left(\frac{\sqrt{x} - \sqrt{y}}{\sqrt{x} + \sqrt{y}}\right)$, show that

$$x\frac{\partial u}{\partial x} + y\frac{\partial u}{\partial y} = 0$$

6. If $u = \tan^{-1}\left(\frac{x^3 + y^3}{x - y}\right)$, Prove that

$$x\frac{\partial u}{\partial x} + y\frac{\partial u}{\partial y} = \sin 2u.$$

7. If f(x,y) is a homogeneous function of degree n, show that

$$x^2 f_{11} + xy\, f_{12} + xy\, f_{21} + y^2 f_{22} = n(n-1)f$$

State the continuity assumption for f(x, y), where the subscription of 'f' stands for the partial derivatives.

$$f_1 = \frac{\partial f}{\partial x}, f_{12} = \frac{\partial^2 f}{\partial x \partial y}, f_{21} = \frac{\partial^2 f}{\partial y \partial x}, f_{11} = \frac{\partial^2 f}{\partial x^2}, f_2 = \frac{\partial f}{\partial y}, \text{and } f_{22} = \frac{\partial^2 f}{\partial y^2}$$

8. If $u = \frac{1}{x^3} + \frac{1}{x^2 y} + \frac{1}{x^3 + 5y^3}$, Prove that

$$x\frac{\partial u}{\partial x} + y\frac{\partial u}{\partial y} + 3u = 0$$

9. State and Prove Euler's theorem and verify for $u = \frac{x}{y} + \frac{y}{z} + \frac{z}{x}$

Answers

4. 6u

6.6 TOTAL DIFFERENTIAL COFFICIENT

If u = f(x,y), where x = φ(t) and y = ψ(t), then u itself is a function of a single variable 't', and hence we can find the ordinary derivative $\frac{du}{dt}$. This , in reference to partial differentiation is called the total differentiation of u w.r.t t.

The problem now is to find $\frac{du}{dt}$ without actually substituting the values of x and y in terms of t in f (x, y). We can proceed as follows:

Let δx, δy and δu denote the small increments in x, y and u respectively corresponding to a small increment δt in t.

we have $x + \delta x = \phi(t + \delta t)$ and

$y + \delta y = \psi(t + \delta t)$

∴ $u + \delta u = f(x + \delta x, y + \delta y)$

Then by definition

$$\frac{du}{dt} = \lim_{\delta t \to 0} \frac{\delta u}{\delta t} = \lim_{\delta t \to 0}\left[\frac{f(x+\delta x, y+\delta y) - f(x,y)}{\delta t}\right]$$

$$= \lim_{\delta t \to 0}\left[\frac{f(x+\delta x, y+\delta y) - f(x,y+\delta y) + f(x, y+\delta y) - f(x,y)}{\delta t}\right]$$

$$= \lim_{\delta t \to 0}\left[\frac{f(x+\delta x, y+\delta y) - f(x,y+\delta y)}{\delta x}\cdot\frac{\delta x}{\delta t}\right]$$

As $\delta t \to 0$, $\delta x \to 0$ and $\delta y \to 0$

∴ $$= \lim_{\delta t \to 0} \frac{\delta x}{\delta t} = \frac{dx}{dt}; \quad \lim_{\delta t \to 0} \frac{\delta y}{\delta t} = \frac{dy}{dt};$$

∴ $$\frac{du}{dt} = \lim_{\delta x \to 0}\left[\frac{f(x+\delta x, y+\delta y) - f(x, y+\delta y)}{\delta x}\right]\cdot\left[\lim_{\delta t \to 0}\frac{\delta x}{\delta t}\right]$$

$$+\left[\lim_{\delta t \to 0}\frac{f(x, y+\delta y) - f(x,y)}{\delta y}\right]\cdot\left[\lim_{\delta t \to 0}\frac{\delta y}{\delta t}\right]$$

But by definition of partial derivatives

$$\lim_{\delta y \to 0}\frac{f(x, y+\delta y) - f(x,y)}{\delta y} = \frac{\partial f(x,y)}{\partial y} = \frac{\partial f}{\partial y}$$

and $$\lim_{\delta x \to 0}\frac{f(x+\delta x, y+\delta y) - f(x, y+\delta y)}{\delta x} = \frac{\partial f(x, y+\delta y)}{\partial x}$$

Now assuming that $\frac{\partial f(x,y)}{\partial x}$ is a continuous function of y,

$$\lim_{\delta y \to 0}\frac{\partial f(x, y+\delta y)}{\delta x} = \frac{\partial f(x,y)}{\partial x} = \frac{\partial f}{\partial x}$$

Hence $$\frac{du}{dt} = \frac{\partial f}{\partial x}.\frac{dx}{dt} + \frac{\partial f}{\partial y}.\frac{dy}{dt}$$

i.e. $$\frac{du}{dt} = \frac{\partial u}{\partial x}.\frac{dx}{dt} + \frac{\partial u}{\partial y}.\frac{dy}{dt} \quad ...(A)$$

The above result can be extended to a function of more than two variables, i.e.,

If $u = f(x_1, x_2, \ldots\ldots x_n)$ and $x_1\ x_2 \ldots\ldots\ldots\ x_n$ are functions of 't' only, then we can prove that

$$\frac{du}{dt} = \frac{\partial u}{\partial x_1}.\frac{dx_1}{dt} + \frac{\partial u}{\partial x_2}.\frac{dx_2}{dt} + \ldots\ldots\ldots + \frac{\partial u}{\partial x_n}.\frac{dx_n}{dt} \quad ...(B)$$

Deduction If $u = f(x, y)$, where $x = \phi(t_1, t_2)$ and $y = \psi(t_1, t_2)$, then from the above result,

we have $$\frac{\partial u}{\partial t_1} = \frac{\partial u}{\partial x}.\frac{\partial x}{\partial t_1} + \frac{\partial u}{\partial y}.\frac{\partial y}{\partial t_1}$$

and $$\frac{\partial u}{\partial t_2} = \frac{\partial u}{\partial x}.\frac{\partial x}{\partial t_2} + \frac{\partial u}{\partial y}.\frac{\partial y}{\partial t_2}$$

Also, from the given equations, we can obtain t_1, t_2 in terms of x, y i.e.,

$$t_1 = \phi_1(x, y) \text{ and } t_2 = \phi(x, y)$$

then $$\frac{\partial u}{\partial x} = \frac{\partial u}{\partial t_1}.\frac{\partial t_1}{\partial x} + \frac{\partial u}{\partial t_2}.\frac{\partial t_2}{\partial x},$$

and $$\frac{\partial u}{\partial y} = \frac{\partial u}{\partial t_1}.\frac{\partial t_1}{\partial y} + \frac{\partial u}{\partial t_2}.\frac{\partial t_2}{\partial y}.$$

By using these equations, we can change one system of independent variables to another system of independent variables.

The higher order partial derivatives of u can be obtained by a repeated application of the above results.

6.7 DIFFERENTIATION OF IMPLICIT FUNCTIONS

The results (A) of the preceding section can be used to find the differential coefficient of implicit functions.

If $u = f(x,y)$ and y is a function of x, then the total differential coefficient of u w.r.t. x is given by

$$\frac{du}{dx} = \frac{\partial u}{\partial x} + \frac{\partial u}{\partial y}.\frac{dy}{dx}$$

In particular, if we are given an implicit relation between x and y of the form, $u = f(x, y) = c$,

where 'c' is a constant, then $$0 = \frac{\partial u}{\partial x} + \frac{\partial u}{\partial y}.\frac{dy}{dx}$$

i.e.,
$$0 = \frac{\partial f}{\partial x} + \frac{\partial f}{\partial y} \cdot \frac{dy}{dx}$$

or,
$$\frac{dy}{dx} = -\frac{\partial f / \partial x}{\partial f / \partial y}$$

In general, if $u = f(x_1, x_2, x_3 \, ... \, x_n)$ and the variables $x_2, x_3 \, ... \, x_n$ are all functions of x, then from the result (B) of the preceding section, we obtain

$$\frac{du}{dx_1} = \frac{\partial u}{\partial x_1} + \frac{\partial u}{\partial x_2} \cdot \frac{dx_2}{dx_1} + \frac{\partial u}{\partial x_3} \cdot \frac{dx_3}{dx_1} + ... + \frac{\partial u}{\partial x_n} \cdot \frac{dxn}{dx_1}$$

6.8 SECOND ORDER DERIVATIVE OF AN IMPLICIT FUNCTION

If f(x,y) = c, we have $\dfrac{dy}{dx} = \dfrac{-\partial f / \partial x}{\partial f / \partial y}$

Now our aim is to find $\dfrac{d^2y}{dx^2}$ in terms of the partial derivatives of (x, y). For now, we shall denote the partial derivatives $\dfrac{\partial f}{\partial x}, \dfrac{\partial f}{\partial y}, \dfrac{\partial^2 f}{\partial x^2}, \dfrac{\partial^2 f}{\partial y \partial x}\left(= \dfrac{\partial^2 f}{\partial x \partial y}\right)$ and $\dfrac{\partial^2 f}{\partial y^2}$ by p, q, r, s and t respectively.

using these notations, we have

$$\frac{dy}{dx} = -p/q$$

differentiating this w.r.t. x, we get

$$\frac{d^2y}{dx^2} = \frac{-\left(q\dfrac{dp}{dx} - p\dfrac{dq}{dx}\right)}{q^2} \qquad ...(1)$$

Since p and q are functions of x and y,

$$\therefore \quad \frac{dp}{dx} = \frac{\partial p}{\partial x} + \frac{\partial p}{\partial y} \cdot \frac{dy}{dx} = \frac{\partial^2 f}{\partial x^2} + \frac{\partial^2 f}{\partial x \partial y} \cdot \frac{dy}{dx}$$

$$= r + s\left(\frac{-p}{q}\right) = \frac{qr - ps}{q};$$

substituting these values in (1), we get

$$\frac{d^2y}{dx^2} = -\left[\frac{q^2r - 2pqs + p^2t}{q^3}\right]$$

Illustrative Examples

Example 1 If $u = x^5y^4$, where $x = t^2$ and $y = t^3$, find $\frac{du}{dt}$,

Solution: Here u is a function of x and y, where x and y are the functions of t.

$\therefore$ u is also a function of single variable t.

$$\therefore \quad \frac{du}{dt} = \frac{\partial u}{\partial x}.\frac{dx}{dt} + \frac{\partial u}{\partial y}.\frac{dy}{dt}$$

$$= \left(5x^4y^4\right).\, 2t + \left(4x^5y^3\right).\, 3t^2$$

$$= 10x^4y^4t + 12x^5y^3t^2$$

$$= 10\, t^8.t^{12}.t + 12.t^{10}.t^9.t^2$$

$$= 22\, t^{21}$$

Example 2 If $u = f(x-y, y-z, z-x)$, prove that

$$\frac{\partial u}{\partial x} + \frac{\partial u}{\partial y} + \frac{\partial u}{\partial z} = 0$$

Solution: Let

$$x - y = X \qquad \text{...(1)}$$

$$y - z = Y \qquad \text{...(2)}$$

$$z - x = Z \qquad \text{...(3)}$$

$\therefore$ $u = f(X, Y, Z)$, where X, Y, Z, are the functions of x, y, z, then

$$\frac{\partial u}{\partial x} = \frac{\partial u}{\partial X}.\frac{\partial X}{\partial x} + \frac{\partial u}{\partial Y}.\frac{\partial Y}{\partial x} + \frac{\partial u}{\partial Z}.\frac{\partial Z}{\partial x}$$

$$= \frac{\partial u}{\partial X}(1) + \frac{\partial u}{\partial Y}.(0) + \frac{\partial u}{\partial Z}.(-1)$$

$$\therefore \quad \frac{\partial u}{\partial x} = \frac{\partial u}{\partial X} - \frac{\partial u}{\partial Z} \qquad \text{...(4)}$$

similarly,

$$\frac{\partial u}{\partial y} = \frac{\partial u}{\partial X}.\frac{\partial X}{\partial y} + \frac{\partial u}{\partial Y}.\frac{\partial Y}{\partial y} + \frac{\partial u}{\partial Z}.\frac{\partial Z}{\partial y}$$

$$= \frac{-\partial u}{\partial X} + \frac{\partial u}{\partial Y} \qquad \text{...(5)}$$

and

$$\frac{\partial u}{\partial z} = \frac{\partial u}{\partial X}.\frac{\partial X}{\partial z} + \frac{\partial u}{\partial Y}.\frac{\partial Y}{\partial z} + \frac{\partial u}{\partial Z}.\frac{\partial Z}{\partial z}$$

$$= \frac{-\partial u}{\partial Y} + \frac{\partial u}{\partial Z} \quad ...(6)$$

Adding (4), (5) and (6), we get

$$\frac{\partial u}{\partial x} + \frac{\partial u}{\partial y} + \frac{\partial u}{\partial z} + 0 \quad \text{Proved.}$$

Example 3 If $x = r\cos\theta$, $y = r\sin\theta$, then prove that

$$\frac{\partial^2 r}{\partial x^2} + \frac{\partial^2 r}{\partial y^2} = \frac{1}{r}\left\{\left(\frac{\partial r}{\partial x}\right)^2 + \left(\frac{\partial r}{\partial y}\right)^2\right\}$$

Solution: Since $x = r\cos\theta$, $y = r\sin\theta$

$\therefore$ $r^2 = x^2 + y^2$ ...(1)

differentiating (1) partially w.r.t. x, we get

$$2r\frac{\partial r}{\partial x} = 2x, \quad \text{or} \quad \frac{\partial r}{\partial x} = x/r$$

$$\therefore \quad \frac{\partial^2 r}{\partial x^2} = \frac{\partial}{\partial x}\left(\frac{\partial r}{\partial x}\right) = \frac{\partial}{\partial x}\left(\frac{x}{r}\right)$$

$$= \frac{1}{r} - \frac{x}{r^2}\left(\frac{\partial r}{\partial x}\right)$$

$$= \frac{1}{r} - \frac{x}{r^2}\left(\frac{x}{r}\right) \quad \text{(from above)}$$

or

$$\frac{\partial^2 r}{\partial x^2} = \frac{r^2 - x^2}{r^3} \quad ...(2)$$

Similarly, we can find

$$\frac{\partial r}{\partial y} = \frac{y}{r} \text{ and } \frac{\partial^2 r}{\partial y^2} = \frac{r^2 - y^2}{r^3} \quad ...(3)$$

Adding (2) and (3), we get

$$\frac{\partial^2 r}{\partial x^2} + \frac{\partial^2 r}{\partial y^2} = \frac{r^2 - x^2}{r^3} + \frac{r^2 - y^2}{r^3}$$

$$= \frac{2r^2 - \left(x^2 + y^2\right)}{r^3} = \frac{2r^2 - r^2}{r^3} = \frac{1}{r}$$

Thus, L.H.S., $\frac{\partial^2 r}{\partial x^2}+\frac{\partial^2 r}{\partial y^2}=\frac{1}{r}$ ...(4)

R.H.S. $=\frac{1}{r}\left\{\left(\frac{\partial r}{\partial x}\right)^2+\left(\frac{\partial r}{\partial y}\right)^2\right\}=\frac{1}{r}\left\{\frac{x^2}{r^2}+\frac{y^2}{r^2}\right\}$

$$=\frac{1}{r}\left(\frac{x^2+y^2}{r^2}\right)=\frac{1}{r}\left(\frac{r^2}{r^2}\right)=\frac{1}{r}= \text{L.H.S} \qquad ...(5)$$

Hence, by (4) and (5), we get

$$\frac{\partial^2 r}{\partial x^2}+\frac{\partial^2 r}{\partial y^2}=\frac{1}{r}\left\{\left(\frac{\partial r}{\partial x}\right)^2+\left(\frac{\partial r}{\partial y}\right)^2\right\}$$ Proved.

Example 4 Prove that

$$\frac{\partial^2 u}{\partial x^2}+\frac{\partial^2 u}{\partial y^2}=\frac{\partial^2 u}{\partial \xi^2}+\frac{\partial^2 u}{\partial \eta^2},$$

where $x=\xi\cos\alpha-\eta\sin\alpha$, $y=\xi\sin\alpha+\eta\cos\alpha$.

Solution: Here $x=\xi\cos\alpha-\eta\sin\alpha$ ...(1)

$y=\xi\sin\alpha+\eta\cos\alpha$ ...(2)

from (1) and (2), we have

$$\frac{\partial x}{\partial \xi}=\cos\alpha \text{ and } \frac{\partial x}{\partial \eta}=-\sin\alpha$$

$$\frac{\partial y}{\partial \xi}=\sin\alpha \text{ and } \frac{\partial y}{\partial \eta}=\cos\alpha$$

since u is a function of x and y, where x and y are functions of ξ and η, therefore

$$\frac{\partial u}{\partial \xi}=\frac{\partial u}{\partial x}.\frac{\partial x}{\partial \xi}+\frac{\partial u}{\partial y}.\frac{\partial y}{\partial \xi}$$

or $$\frac{\partial u}{\partial \xi}=\cos\alpha.\frac{\partial u}{\partial x}+\sin\alpha\frac{\partial u}{\partial y} \qquad ...(3)$$

and $$\frac{\partial u}{\partial \eta}=\frac{\partial u}{\partial x}.\frac{\partial x}{\partial \eta}+\frac{\partial u}{\partial y}.\frac{\partial y}{\partial \eta}$$

or $$\frac{\partial u}{\partial \eta}=-\sin\alpha\frac{\partial u}{\partial x}+\cos\alpha\frac{\partial u}{\partial y} \qquad ...(4)$$

from (3) and (4), we have operators

$$\frac{\partial}{\partial \xi} = \cos\alpha \frac{\partial}{\partial x} + \sin\alpha \frac{\partial}{\partial y}$$

$$\frac{\partial}{\partial \eta} = -\sin\alpha \frac{\partial}{\partial x} + \cos\alpha \frac{\partial}{\partial y}$$

$$\therefore \quad \frac{\partial^2 u}{\partial \xi^2} = \frac{\partial}{\partial \xi}\left(\frac{\partial u}{\partial \xi}\right) = \left(\cos\alpha \frac{\partial}{\partial x} + \sin\alpha \frac{\partial}{\partial y}\right).\left(\cos\alpha \frac{\partial u}{\partial x} + \sin\alpha \frac{\partial u}{\partial y}\right)$$

$$= \cos\alpha \frac{\partial}{\partial x}\left(\cos\alpha \frac{\partial u}{\partial x} + \sin\alpha \frac{\partial u}{\partial y}\right) + \sin\alpha \frac{\partial}{\partial y}\left(\cos\alpha \frac{\partial u}{\partial x} + \sin\alpha \frac{\partial u}{\partial y}\right)$$

$$= \cos\alpha\left(\cos\alpha \frac{\partial^2 u}{\partial x^2} + \sin\alpha \frac{\partial^2 u}{\partial x \partial y}\right) + \sin\alpha\left(\cos\alpha \frac{\partial^2 u}{\partial x \partial y} + \sin\alpha \frac{\partial^2 u}{\partial y^2}\right)$$

i.e.,

$$\frac{\partial^2 u}{\partial \xi^2} = \cos^2\alpha \frac{\partial^2 u}{\partial x^2} + 2\cos\alpha \sin\alpha \frac{\partial^2 u}{\partial x \partial y} + \sin^2\alpha \frac{\partial^2 u}{\partial y^2} \quad \text{...(5)}$$

Similarly,

$$\frac{\partial^2 u}{\partial \eta^2} = \sin^2\alpha \frac{\partial^2 u}{\partial x^2} - 2\sin\alpha \cos\alpha \frac{\partial^2 u}{\partial x \partial y} + \cos^2\alpha \frac{\partial^2 u}{\partial y^2} \quad \text{...(6)}$$

Adding (5) and (6), we get

$$\frac{\partial^2 u}{\partial \xi^2} + \frac{\partial^2 u}{\partial \eta^2} = \frac{\partial^2 u}{\partial x^2} + \frac{\partial^2 u}{\partial y^2} \quad \text{Proved.}$$

Note: From the given relations (1) and (2) we can find ξ and η in terms of x and y, then proceeding in the same manner we can show that

$$\frac{\partial^2 u}{\partial x^2} + \frac{\partial^2 u}{\partial y^2} = \frac{\partial^2 u}{\partial \xi^2} + \frac{\partial^2 u}{\partial \eta^2}$$

Example 5 Find $\frac{dy}{dx}$ if $e^x + e^y = 2xy$

Solution: Let $u = e^x + e^y - 2xy = 0$ ...(1)

differentiating (1) partially w.r.t. x and y, we get

$$\frac{\partial u}{\partial x} = e^x - 2y \text{ and } \frac{\partial u}{\partial y} = e^y - 2x$$

$$\therefore \quad \frac{dy}{dx} = -\frac{\partial u \mid \partial x}{\partial u \mid \partial y} = -\frac{e^x - 2y}{e^y - 2x} = \frac{2y - e^x}{e^y - 2x} \quad \text{Answer}$$

Example 6 If $f(x,y) = 0$, $\phi(y,z) = 0$, show that

$$\frac{\partial f}{\partial y} \cdot \frac{\partial \phi}{\partial z} \cdot \frac{dz}{dx} = \frac{\partial f}{\partial x} \cdot \frac{\partial \phi}{\partial y}$$

Solution: If f(x,y) = 0, then

$$\frac{dy}{dx} = \frac{-(\partial f \mid \partial x)}{(\partial f \mid \partial y)} \quad ...(1)$$

If ϕ (y, z) = 0, then $\frac{dz}{dy} = -\frac{(\partial \phi \mid \partial y)}{(\partial \phi \mid \partial z)}$ ...(2)

multiplying (1) and (2), we get

$$\frac{dy}{dx} \cdot \frac{dz}{dy} = \frac{(\partial f \mid \partial x)}{(\partial f \mid \partial y)} \cdot \frac{(\partial \phi \mid \partial y)}{(\partial \phi \mid \partial z)}$$

or $\frac{\partial f}{\partial y} \cdot \frac{\partial \phi}{\partial z} \cdot \frac{dz}{dx} = \frac{\partial f}{\partial x} \cdot \frac{\partial \phi}{\partial y}$. Proved

Example 7 If u = x log xy, where $x^3 + y^3 + 3xy = 1$, find $\frac{du}{dx}$

Solution: Here u is a function of x and y, and y is a function of x.

$$\therefore \quad \frac{du}{dx} = \frac{\partial u}{\partial x} + \frac{\partial u}{\partial y} \cdot \frac{dy}{dx} \quad ...(1)$$

Given u = log x y

$$\therefore \quad \frac{\partial u}{\partial x} = 1. \log xy + x.\frac{1}{xy}.y = 1 + \log x y$$

and $\frac{\partial u}{\partial y} = x.\frac{1}{xy}.x = \frac{x}{y}$

Again we are given that

$$x^3 + y^3 + 3xy = 1 \quad ...(2)$$

differentiating (2) both sides w.r.t. x, we get

$$3x^2 + 3y^2\frac{dy}{dx} + 3y + 3x\frac{dy}{dx} = 0$$

or $$\left(y^2+x\right)\frac{dy}{dx}+\left(x^2+y\right)=0$$

$$\therefore \quad \frac{dy}{dx}=\frac{-\left(x^2+y\right)}{y^2+x}$$

substituting these values in (1), we get

$$\frac{du}{dx}=(1+\log xy)+\frac{x}{y}\left[-\frac{\left(x^2+y\right)}{\left(y^2+x\right)}\right]$$

or $$\frac{du}{dx}=(1+\log xy)-\frac{x}{y}\frac{\left(x^2+y\right)}{\left(y^2+x\right)} \quad \text{Ans.}$$

Example 8 Find $\frac{dy}{dx}$, when $\tan^{-1}\left(\frac{x}{y}\right)+y^3+1=0;\ x>0,\ y>0$

Solution: we have f(x, y) = 0

where $f(x, y) = \tan^{-1}\left(\frac{x}{y}\right)+y^3+1$

Now, we know that, when f(x, y) = 0

$$\Rightarrow \quad \frac{dy}{dx}=\frac{-\partial f \mid \partial x}{\partial f \mid \partial y}$$

Here, $$\frac{\partial f}{\partial x}=\frac{\partial}{\partial x}\left[\tan^{-1}\left(\frac{x}{y}\right)+y^3+1\right]$$

$$=\frac{1}{1+\left(\frac{x}{y}\right)^2}\cdot\frac{1}{y}+0+0$$

$$=\frac{y^2}{x^2+y^2}\cdot\frac{1}{y}=\frac{y}{x^2+y^2}$$

Again, $$\frac{\partial f}{\partial y}=\frac{\partial}{\partial y}\left[\tan^{-1}\left(\frac{x}{y}\right)+y^3+1\right]$$

$$=\frac{1}{1+\left(\frac{x}{y}\right)^2}\cdot\left(-\frac{x}{y^2}\right)+3y^2$$

$$=\frac{y^2}{x^2+y^2}\cdot\left(-\frac{x}{y^2}\right)+3y^2$$

$$=\frac{-x}{x^2+y^2}+3y^2=\frac{-x+3y^4+3x^2y^2}{x^2+y^2}$$

or $$\frac{\partial f}{\partial y}=\frac{3y^4+3y^2x^2-x}{x^2+y^2}$$

$\therefore$ $$\frac{dy}{dx}=\frac{-y\,|\left(x^2+y^2\right)}{\frac{3y^4+3x^2y^2-x}{\left(x^2+y^2\right)}}=\frac{-y}{3y^4+3y^2x^2-x}\quad \text{Ans.}$$

Example 9 If $x + y = 2\,e^{\theta}\cos\phi$ and $x - y = 2\,i\,e^{\theta}\sin\phi$

show that $$\frac{\partial^2 u}{\partial\theta^2}+\frac{\partial^2 u}{\partial\phi^2}=4xy\,\frac{\partial^2 u}{\partial x\partial y}.$$

Solution: we have, $x + y = 2\,e^{\theta}\cos\phi$ ...(1)

and $x - y = 2\,i\,e^{\theta}\sin\phi$ ...(2)

Adding (1) and (2), we get

$2x = 2\,e^{\theta}\,(i\sin\phi + \cos\phi)$

$x = e^{\theta + i\phi}$ $\left(\because\ e^{i\theta}=\cos\theta+i\sin\theta\right)$ ...(3)

subtracting (2) from (1), we get

$2y = 2e^{\theta}\ (\cos\phi - i\sin\phi) = e^{\theta - i\phi}$ ...(4)

clearly, $u = f(x, y)$ and x,y are the functions of θ and ϕ, Hence u is a composite function of θ and ϕ.

Now, $$\frac{\partial u}{\partial\theta}=\frac{\partial u}{\partial x}\cdot\frac{\partial x}{\partial\theta}+\frac{\partial u}{\partial y}\cdot\frac{\partial y}{\partial\theta}$$

$$=x\frac{\partial u}{\partial x}+y\frac{\partial u}{\partial y}\qquad\left[\because\ e^{\theta+i\phi}=x,\ e^{\theta-i\phi}=y\right.$$

and $$\frac{\partial^2 u}{\partial\theta^2}=\frac{\partial}{\partial\theta}\left(\frac{\partial u}{\partial\theta}\right)=\frac{\partial}{\partial x}\left(\frac{\partial u}{\partial\theta}\right)\frac{\partial x}{\partial\theta}+\frac{\partial}{\partial y}\left(\frac{\partial u}{\partial\theta}\right)\frac{\partial y}{\partial\theta}$$

$$= \frac{\partial}{\partial x}\left[x\frac{\partial u}{\partial x} + y\frac{\partial u}{\partial y}\right]x + \frac{\partial}{\partial y}\left[x\frac{\partial u}{\partial x} + y\frac{\partial u}{\partial y}\right]y$$

$$= \left[x\frac{\partial^2 u}{\partial x^2} + \frac{\partial u}{\partial x} + y\frac{\partial^2 u}{\partial x \partial y}\right]x + \left[x\frac{\partial^2 u}{\partial x \partial y} + y\frac{\partial^2 u}{\partial y^2} + y\frac{\partial u}{\partial y}\right]y$$

or $$\frac{\partial^2 u}{\partial \theta^2} = \left[x^2\frac{\partial^2 u}{\partial x^2} + y^2\frac{\partial^2 u}{\partial y^2} + 2xy\frac{\partial^2 u}{\partial x \partial y} + x\frac{\partial u}{\partial x} + y\frac{\partial u}{\partial y}\right] \quad ...(5)$$

Again,

$$\frac{\partial u}{\partial \phi} = \frac{\partial u}{\partial x}.\frac{\partial x}{\partial \phi} + \frac{\partial u}{\partial y}.\frac{\partial y}{\partial \phi}$$

$$= i\left[x\frac{\partial u}{\partial x} - y\frac{\partial u}{\partial y}\right]$$

$\therefore$ $$\frac{\partial^2 u}{\partial \phi^2} = \frac{\partial}{\partial \phi}\left(\frac{\partial u}{\partial \phi}\right) = \frac{\partial}{\partial x}\left(\frac{\partial u}{\partial \phi}\right)\left(\frac{\partial x}{\partial \phi}\right) + \frac{\partial}{\partial y}\left(\frac{\partial u}{\partial \phi}\right)\frac{\partial y}{\partial \phi}$$

$$= i\frac{\partial}{\partial x}\left[x\frac{\partial u}{\partial x} - y\frac{\partial u}{\partial y}\right](ix) + i\frac{\partial}{\partial y}\left[x\frac{\partial u}{\partial x} - y\frac{\partial u}{\partial y}\right].(-iy)$$

$$= -x\frac{\partial}{\partial x}\left[x\frac{\partial u}{\partial x} - y\frac{\partial u}{\partial y}\right] + y\frac{\partial}{\partial y}\left[x\frac{\partial u}{\partial x} - y\frac{\partial u}{\partial y}\right]$$

$$= -x\left[\frac{\partial u}{\partial x} + x\frac{\partial^2 u}{\partial x^2} - y\frac{\partial^2 u}{\partial x \partial y}\right] + y\left[x\frac{\partial^2 u}{\partial x \partial y} - y\frac{\partial^2 u}{\partial y^2} - \frac{\partial u}{\partial y}\right]$$

or $$\frac{\partial^2 u}{\partial \phi^2} = -\left[x\frac{\partial u}{\partial x} + y\frac{\partial u}{\partial y}\right] + 2xy\frac{\partial^2 u}{\partial x \partial y} - \left[x^2\frac{\partial^2 u}{\partial x^2} + y^2\frac{\partial^2 u}{\partial y^2}\right] \quad ...(6)$$

Adding (5) and (6), we get

$$\frac{\partial^2 u}{\partial \theta^2} + \frac{\partial^2 u}{\partial \phi^2} = 4xy\frac{\partial^2 u}{\partial x \partial y}$$ Proved.

Example 10 If $v = v\left(\frac{y-x}{xy}, \frac{z-x}{xz}\right)$, then prove that

$$x^2\frac{\partial v}{\partial x} + y^2\frac{\partial v}{\partial y} + z^2\frac{\partial v}{\partial z} = 0$$

Solution: we have $v = v\left(\dfrac{y-x}{xy}, \dfrac{z-x}{xz}\right)$

Put $r = \dfrac{y-x}{xy}$ and $s = \dfrac{z-x}{xz}$

we hav v = v (r, s) ...(1)

$$r = \frac{1}{x} - \frac{1}{y}, \quad s = \frac{1}{x} - \frac{1}{z}$$

$$\frac{\partial r}{\partial x} = -\frac{1}{x^2}, \qquad \frac{\partial s}{\partial x} = -\frac{1}{x^2}$$

$$\frac{\partial r}{\partial y} = \frac{1}{y^2}, \qquad \frac{\partial s}{\partial z} = \frac{1}{z^2}$$

$$\frac{\partial r}{\partial z} = 0, \qquad \frac{\partial s}{\partial y} = 0$$

we know that

$$\frac{\partial v}{\partial x} = \frac{\partial v}{\partial r}.\frac{\partial r}{\partial x} + \frac{\partial v}{\partial s}.\frac{\partial s}{\partial x} = \frac{\partial v}{\partial r}\left(-\frac{1}{x^2}\right) + \frac{\partial v}{\partial s}\left(-\frac{1}{x^2}\right)$$

$$= -\frac{1}{x^2}\frac{\partial v}{\partial r} - \frac{1}{x^2}\frac{\partial v}{\partial s}$$

$$\Rightarrow \quad x^2\frac{\partial v}{\partial x} = \frac{\partial v}{\partial r} - \frac{\partial v}{\partial s} \qquad ...(2)$$

and

$$\frac{\partial v}{\partial y} = \frac{\partial v}{\partial r}.\frac{\partial r}{\partial y} + \frac{\partial v}{\partial s}.\frac{\partial s}{\partial y} = \frac{1}{y^2}\frac{\partial v}{\partial r} + \frac{\partial v}{\partial s}(0)$$

$$\Rightarrow \quad y^2\frac{\partial v}{\partial y} = \frac{\partial v}{\partial r} \qquad ...(3)$$

$$\frac{\partial v}{\partial z} = \frac{\partial v}{\partial r}.\frac{\partial r}{\partial z} + \frac{\partial v}{\partial s}.\frac{\partial s}{\partial z} = \frac{\partial v}{\partial r} \times 0 + \frac{\partial v}{\partial s}.\frac{1}{z^2} = \frac{1}{z^2}\frac{\partial v}{\partial s}$$

$$\Rightarrow \quad z^2\frac{\partial v}{\partial z} = \frac{\partial v}{\partial s} \qquad ...(4)$$

Adding (2), (3) and (4), we get

$$x^2\frac{\partial v}{\partial x} + y^2\frac{\partial v}{\partial y} + z^2\frac{\partial v}{\partial z} = 0$$

Proved.

Exercise 6.3

1. If $x^y + y^x = c$, find the value of $\frac{dy}{dx}$.

2. If $ax^2 + 2hxy + by^2 = 1$, find $\frac{d^2y}{dx^2}$.

3. If $u = e^{mx}(y - z)$, $y = m \sin x$ and $z = \cos x$, find $\frac{du}{dx}$.

4. If $ax^2 + by^2 + cz^2 = 1$ and $lx + my + nz = 0$, find the values of $\frac{dy}{dx}$ and $\frac{dz}{dx}$.

5. If the curves $f(x,y) = 0$ and $\phi(x,y) = 0$ touch each other, show that at the point of contact,

$$\frac{\partial f}{\partial x} \cdot \frac{\partial \phi}{\partial y} - \frac{\partial f}{\partial y} \cdot \frac{\partial \phi}{\partial x} = 0$$

6. If $x = r\cos\theta$, $y = r\sin\theta$, prove that

(a) $\left(\frac{\partial r}{\partial x}\right)^2 + \left(\frac{\partial r}{\partial y}\right)^2 = 1$ (b) $\frac{\partial^2 r}{\partial x^2} \cdot \frac{\partial^2 r}{\partial y^2} = \left(\frac{\partial^2 r}{\partial x \partial y}\right)^2$

7. If $x = r\cos\theta$, $y = r\sin\theta$, prove that $\frac{\partial r}{\partial x} = \frac{\partial x}{\partial r}$,

$$\frac{1}{r}\frac{\partial x}{\partial \theta} = r\frac{\partial \theta}{\partial x} \text{ and } \frac{\partial^2 \theta}{\partial x^2} + \frac{\partial^2 \theta}{\partial y^2} = 0$$

8. If $u = c \cos hx \cos y$, $v = c \sin hx \sin y$, prove that

$$\frac{\partial^2 v}{\partial x^2} + \frac{\partial^2 u}{\partial y^2} = \frac{c^2}{2}(\cos h\, 2x - \cos 2y)\left(\frac{\partial^2 v}{\partial u^2} + \frac{\partial^2 u}{\partial v^2}\right)$$

9. If $z = z(x, y)$, $x = e^u + e^{-v}$, $y = e^{-u} - e^v$, prove that

$$\frac{\partial z}{\partial u} \cdot \frac{\partial z}{\partial v} = x\frac{\partial z}{\partial x} - y\frac{\partial z}{\partial y}.$$

10. If $x^4 + y^4 = 4a^2xy$, prove that $\left(a^2x - y^3\right)^3 \frac{d^2y}{dx^2} = 2a^2xy\left(3a^4 + x^2y^2\right)$

11. If $u = x^2 + y^2 + z^2 - 2xyz = 1$, Prove that

$$\frac{dx}{\sqrt{1-x^2}} + \frac{dy}{\sqrt{1-y^2}} + \frac{dz}{\sqrt{1-z^2}} = 0$$

Hint: $du = \frac{\partial u}{\partial x}dx + \frac{\partial u}{\partial y}dy + \frac{\partial u}{\partial z}dz = 0$

12. By changing the independent variables x and t to U and V by means of the relationships U = x – at and V = x + at, Prove that

$$a^2\frac{\partial^2 y}{\partial x^2} - \frac{\partial^2 y}{\partial t^2} = 4a^2\frac{\partial^2 y}{\partial U \partial V}$$

13. If z = f(x,y), where x = $e^u \cos v$, y = $e^u \sin v$,

prove that; $\left(\frac{\partial z}{\partial x}\right)^2 + \left(\frac{\partial z}{\partial y}\right)^2 = e^{2u}\left[\left(\frac{\partial z}{\partial u}\right)^2 + \left(\frac{\partial z}{\partial v}\right)^2\right]$

14. If u = f(r) and x = r cos θ, y = r sin θ, then show that

$$\frac{\partial^2 u}{\partial x^2} + \frac{\partial^2 u}{\partial y^2} = f''(r) + \frac{1}{r}f'(r).$$

15. Transform the equation $\frac{\partial^2 u}{\partial x^2} + \frac{\partial^2 u}{\partial y^2} = 0$ into polar co-ordinates.

Answers

1. $-\left[\frac{y^x . \log_e y + yx^{y-1}}{x^y \log_e x + xy^{x-1}}\right]$

2. $-\frac{\left[(hx + by)^2 . a - 2(hx + by)(ax + by)h + (ax + hy)^2 b\right]}{(hx + by)^2}$

3. $e^{mx}(m^2 + 1)\sin x$

4. $\frac{dy}{dx} = \frac{clz - anx}{bny - cmz}, \frac{dz}{dx} = \frac{amx - bly}{bny - cmz}$

15. $\frac{\partial^2 u}{\partial x^2} + \frac{\partial^2 u}{\partial y^2} = \frac{\partial^2 u}{\partial r^2} + \frac{1}{r}\frac{\partial u}{\partial r} + \frac{1}{r^2}\frac{\partial^2 u}{\partial \theta^2}$.

6.9 Taylor's theorem for two variables: Let f(x,y) be a function of two independent variables x and y and it has continuous partial derivatives at all points (x,y), then Taylor's series in two variables is as follows:

$$f(x + h, y + k) = f(x,y) + \left(h\frac{\partial}{\partial x} + k\frac{\partial}{\partial y}\right) f(x, y) + \frac{1}{2!}\left(h\frac{\partial}{\partial x} + k\frac{\partial}{\partial y}\right)^2 f(x,y)$$

$$+ \ldots\ldots\ldots\ldots + \frac{1}{n!}\left(h\frac{\partial}{\partial x} + k\frac{\partial}{\partial y}\right)^n f(x,y) + \ldots\ldots\ldots$$

Proof: Assume that f(x + h, y + k) is a function of single variable, say x, then by Taylor's theorem for functions of single variable, we have

$$f(x+h, y+k) = f(x, y+k) + h\frac{\partial}{\partial x} f(x, y+k) + \frac{h^2}{2!}\frac{\partial^2}{\partial x^2} f(x, y+k) + \ldots\ldots$$

expanding now each term,

$$f(x + h, y + k) = \left[f(x, y) + k\frac{\partial}{\partial y} f(x,y) + \frac{k^2}{2!}\frac{\partial^2}{\partial y^2} f(x,y) + \ldots\ldots\right] +$$

$$h\frac{\partial}{\partial x}\left[f(x,y) + k\frac{\partial}{\partial y} f(x,y) + \ldots\ldots\right] +$$

$$\frac{h^2}{2!}\frac{\partial^2}{\partial x^2}\left[f(x,y) + k\frac{\partial}{\partial y} f(x,y) + \ldots\ldots\right]$$

Hence,

$$f(x+h, y+k) = f(x,y) + \left(h\frac{\partial f}{\partial x} + k\frac{\partial f}{\partial y}\right) + \frac{1}{2!}\left[h^2\frac{\partial^2 f}{\partial x^2} + 2hk\frac{\partial^2 f}{\partial x.\partial y} + k^2\frac{\partial^2 f}{\partial y^2}\right] + \ldots.$$

This is sometimes written symbolically as,

$$f(x+h, y+k) = f(x,y) + \left(h\frac{\partial}{\partial x} + k\frac{\partial}{\partial y}\right) f(x,y) +$$

$$\frac{1}{2!}\left(h\frac{\partial}{\partial x} + k\frac{\partial}{\partial y}\right)^2 f(x,y) + \ldots\ldots$$

$$+ \frac{1}{n!}\left(h\frac{\partial}{\partial x} + k\frac{\partial}{\partial y}\right)^n f(x,y) + \ldots\ldots$$

where $\left(h\frac{\partial}{\partial x} + k\frac{\partial}{\partial y}\right)^2 f = h^2\frac{\partial^2 f}{\partial x^2} + 2hk\frac{\partial^2 f}{\partial x \partial y} + k^2\frac{\partial^2 f}{\partial y^2}$

and $$\left(h\frac{\partial}{\partial x}+k\frac{\partial}{\partial y}\right)^n f = h^n\frac{\partial^2 f}{\partial x^n}+nh^{n-1}k\frac{\partial^n f}{\partial x^{n-1}\partial y}+$$

$$\frac{n(n-1)}{1.2}h^{n-2}k^2\frac{\partial^n f}{\partial x^{n-2}\partial y^2}+\dots\dots\dots+$$

$$\frac{k^n\partial^n f}{\partial y^n}$$

Note: The above theorem can be easily extended to any number of independent variables.

Another form

$$f(x,y)=f(a,b)+\left[(x-a)\left(\frac{\partial f}{\partial x}\right)_{(a,b)}+(y-b)\left(\frac{\partial f}{\partial y}\right)_{(a,b)}\right]+$$

$$\frac{1}{2!}\left[(x-a)^2\left(\frac{\partial^2 f}{\partial x^2}\right)_{(a,b)}+2(x-a)(y-b)\left(\frac{\partial^2 f}{\partial x\partial y}\right)_{(a,b)}+(y-b)^2\left(\frac{\partial^2 f}{\partial y^2}\right)_{(a,b)}\right]+\dots.$$

If a = 0, b = 0, then

$$f(x,y)=f(0,0)+\left[x\left(\frac{\partial f}{\partial x}\right)_{(0,0)}+y\left(\frac{\partial f}{\partial y}\right)_{(0,0)}\right]+$$

$$\frac{1}{2!}\left[x^2\left(\frac{\partial^2 f}{\partial x^2}\right)_{(0,0)}+2xy\left(\frac{\partial^2 f}{\partial x\partial y}\right)_{(0,0)}+y^2\left(\frac{\partial^2 f}{\partial y^2}\right)_{(0,0)}\right]+\dots..$$

This is called Maclaurin's series for two variables.

Illustrative Examples

Example 1 Expand e^{xy} about the point (1,1)

Solution: Let $f(x,y)=e^{xy}\Rightarrow f(1,1)=e^1=e$

$$\frac{\partial f}{\partial x}=y\,e^{xy}\Rightarrow\left(\frac{\partial f}{\partial x}\right)_{(1,1)}=e^1=e$$

$$\frac{\partial f}{\partial y}=x\,e^{xy}\Rightarrow\left(\frac{\partial f}{\partial y}\right)_{\text{at }(1,1)}=e^1=e$$

$$\frac{\partial^2 f}{\partial x^2}=y^2e^{xy}\Rightarrow\left(\frac{\partial^2 f}{\partial x^2}\right)at_{(1,1)}=1^2.e^1=e$$

$$\frac{\partial^2 f}{\partial x \partial y} = xy\,e^{xy} + e^{xy} \Rightarrow \left(\frac{\partial^2 f}{\partial x \partial y}\right)_{at\ (1,1)} = 1.1\,e^1 + e^1 = 2e$$

$$\frac{\partial^2 f}{\partial y^2} = x^2 e^{xy} \Rightarrow \left(\frac{\partial^2 f}{\partial y^2}\right)_{(1,1)} = 1^2.e^1 = e$$

By Taylor's theorem for function of two variables, we have

$$f(x,y) = f(a,b) + \left[(x-a)\frac{\partial}{\partial x} f(a,b) + (y-b)\frac{\partial}{\partial y} f(a,b)\right] +$$

$$\frac{1}{2!}\left[(x-a)^2 \frac{\partial^2 f}{\partial x^2}(a,b) + 2(x-a)(y-b)\frac{\partial^2 f}{\partial x \partial y}(a,b) + (y-b)^2 \frac{\partial^2 f}{\partial y^2}(a,b)\right] + \ldots\ldots$$

$$\Rightarrow \quad e^{xy} = e\left[1 + (x-1) + (y-1)\right] + \frac{1}{2!}\left[(x-1)^2 + 4(x-1)(y-1) + (y-1)^2\right]e + \ldots..$$

Example 2 Expand $e^x \cos y$ near the point $\left(1, \frac{\pi}{4}\right)$ by Taylor's series.

Solution: Let $f(x,y) = e^x \cos y$, then $f\left(1, \frac{\pi}{4}\right) = \frac{e}{\sqrt{2}}$

Now, $\frac{\partial f}{\partial x} = e^x \cos y \Rightarrow \left(\frac{\partial f}{\partial x}\right)_{at\ (1,\ \pi/4)} = \frac{e}{\sqrt{2}}$

$$\frac{\partial f}{\partial y} = -e^x \sin y \Rightarrow \left(\frac{\partial f}{\partial y}\right)_{at\ (1,\ \pi/4)} = \frac{-e}{\sqrt{2}}$$

$$\frac{\partial^2 f}{\partial x^2} = e^x \cos y \Rightarrow \left(\frac{\partial^2 f}{\partial x^2}\right)_{at\ (1,\ \pi/4)} = \frac{e}{\sqrt{2}}$$

$$\frac{\partial^2 f}{\partial y^2} = -e^x \cos y \Rightarrow \left(\frac{\partial^2 f}{\partial y^2}\right)_{at\ (1,\ \pi/4)} = \frac{-e}{\sqrt{2}}$$

and $\frac{\partial^2 f}{\partial x \partial y} = -e^x \sin y \Rightarrow \frac{\partial^2 f}{\partial x \partial y}$ at $\left(1, \frac{\pi}{4}\right) = \frac{-e}{\sqrt{2}}$

By Taylor's series,

$$f(x,y) = f(a,b) + \left[(x-a)\frac{\partial}{\partial x} f(a,b) + (y-b)\frac{\partial}{\partial y} f(a,b)\right]$$

$$+\frac{1}{2!}\left[(x-a)^2\frac{\partial^2}{\partial x^2}f(a,b)+2(x-a)(y-b)\frac{\partial^2 f}{\partial x\partial y}(a,b)+(y-b)^2\frac{\partial^2}{\partial y^2}f(a,b)\right]+.....$$

$$e^x\cos y=\frac{e}{\sqrt{2}}+\left[(x-1)\frac{e}{\sqrt{2}}+\left(y-\frac{\pi}{4}\right)\left(\frac{-e}{\sqrt{2}}\right)\right]+$$

$$\frac{1}{2!}\left[(x-1)^2\frac{e}{\sqrt{2}}+2(x-1)\left(y-\frac{\pi}{4}\right)\left(-\frac{e}{\sqrt{2}}\right)+\left(y-\frac{\pi}{4}\right)^2\left(\frac{-e}{\sqrt{2}}\right)\right]+.....$$

Example 3 Find the expansion of cos x cos y in powers of x, y up to fourth order terms

Solution: Let $f(x,y) = \cos x\cos y \Rightarrow f(0,0) = \cos 0\cos 0 = 1$

$$\frac{\partial f}{\partial x}=-\sin x\cos y \quad\Rightarrow\quad \left(\frac{\partial f}{\partial x}\right)_{(0,0)}=0$$

$$\frac{\partial f}{\partial y}=-\cos x\sin y \quad\Rightarrow\quad \left(\frac{\partial f}{\partial y}\right)_{(0,0)}=0$$

$$\frac{\partial^2 f}{\partial x^2}=-\cos x\cos y \quad\Rightarrow\quad \left(\frac{\partial^2 f}{\partial x^2}\right)_{(0,0)}=-1$$

$$\frac{\partial^2 f}{\partial x\partial y}=\sin x\sin y \quad\Rightarrow\quad \left(\frac{\partial^2 f}{\partial x\partial y}\right)_{(0,0)}=0$$

$$\frac{\partial^2 f}{\partial y^2}=-\cos x\cos y \quad\Rightarrow\quad \left(\frac{\partial^2 f}{\partial y^2}\right)_{(0,0)}=-1$$

$$\frac{\partial^3 f}{\partial x^3}=\sin x\cos y \quad\Rightarrow\quad \left(\frac{\partial^3 f}{\partial x^3}\right)_{(0,0)}=0$$

$$\frac{\partial^3 f}{\partial x^2\partial y}=\cos x\sin y \quad\Rightarrow\quad \left(\frac{\partial^3 f}{\partial x^2\partial y}\right)_{(0,0)}=0$$

$$\frac{\partial^3 f}{\partial x\partial y^2}=\sin x\cos y \quad\Rightarrow\quad \left(\frac{\partial^3 f}{\partial x\partial y^2}\right)_{(0,0)}=0$$

$$\frac{\partial^3 f}{\partial y^3}=\cos x\sin y \quad\Rightarrow\quad \left(\frac{\partial^3 f}{\partial y^3}\right)_{(0,0)}=0$$

$$\frac{\partial^4 f}{\partial x^4} = \cos x \cos y \quad \Rightarrow \quad \left(\frac{\partial^4 f}{\partial x^4}\right)_{(0,0)} = 1$$

$$\frac{\partial^4 f}{\partial x^3 \partial y} = -\sin x \sin y \quad \Rightarrow \quad \left(\frac{\partial^4 f}{\partial x^3 \partial y}\right)_{(0,0)} = 0$$

$$\frac{\partial^4 f}{\partial x^2 \partial y^2} = \cos x \cos y \quad \Rightarrow \quad \left(\frac{\partial^4 f}{\partial x^2 \partial y^2}\right)_{(0,0)} = 1$$

$$\frac{\partial^4 f}{\partial x \partial y^3} = -\sin x \sin y \quad \Rightarrow \quad \left(\frac{\partial^4 f}{\partial x \partial y^3}\right)_{(0,0)} = 0$$

$$\frac{\partial^4 f}{\partial y^4} = \cos x \cos y \quad \Rightarrow \quad \left(\frac{\partial^4 f}{\partial y^4}\right)_{(0,0)} = 1$$

Putting these values in Taylor's theorem, we get

$$f(x,y) = f(0,0) + \left[x\frac{\partial f}{\partial x}(0,0) + y\frac{\partial f}{\partial y}(0,0)\right] +$$

$$\frac{1}{2!}\left[x^2\frac{\partial^2 f}{\partial x^2}(0,0) + 2xy\frac{\partial^2 f}{\partial y \partial x}(0,0) + y^2\frac{\partial^2 f}{\partial y^2}(0,0)\right]$$

$$+\frac{1}{3!}\left[x^3\frac{\partial^3 f}{\partial x^3}(0,0) + 3x^2 y\frac{\partial^3 f}{\partial x^2 \partial y}(0,0) + 3xy^2\frac{\partial^3 f}{\partial x \partial y^2}(0,0) + y^3\frac{\partial^3 f}{\partial y^3}(0,0)\right] +$$

$$\frac{1}{4!}\left[x^4\frac{\partial^4 f}{\partial x^4}(0,0) + 4xy^3\left(\frac{\partial^4 f}{\partial x \partial y^3}(0,0)\right) + 4x^3 y\frac{\partial^4 f}{\partial x^3 \partial y}(0,0) +\right.$$

$$\left.6x^2y^2\frac{\partial^4 f}{\partial x^2 \partial y^2}(0,0) + y^4\frac{\partial^4 f}{\partial y^4}(0,0)\right] + \ldots\ldots$$

$$\Rightarrow \quad \cos x \cos y = 1 + 0 + 0 + \frac{1}{2}(-x^2 + 0 - y^2) + \frac{1}{6}(0 + 0 + 0 + 0) +$$

$$\frac{1}{24}(x^4 + 0 + 6x^2 y^2 + 0 + y^4) + \ldots\ldots\ldots\ldots$$

or $$\cos x \cos y = 1 - \frac{x^2}{2} - \frac{y^2}{2} + \frac{x^4}{24} + \frac{x^2 y^2}{4} + \frac{y^4}{24} + \ldots..$$

Example 4 Obtain Taylor's expansion of $f(x,y) = \tan^{-1} y/x$ about (1,1) upto and including the second degree terms. Hence compute f (1.1, 0.9).

Solution: Let $f(x,y) = \tan^{-1}(y/x) \Rightarrow f(1,1) = -\pi/4$

$$\frac{\partial f}{\partial x} = \frac{1}{1+\frac{y^2}{x^2}}\left(-\frac{y}{x^2}\right) = \frac{-y}{x^2+y^2} \Rightarrow \frac{\partial}{\partial x} f(1,1) = -\frac{1}{2}$$

$$\frac{\partial f}{\partial x} = \frac{1}{1+(y/x)^2}\left(\frac{1}{x}\right) = \frac{x}{x^2+y^2} \Rightarrow \frac{\partial f}{\partial y}(1,1) = \frac{1}{2}$$

$$\frac{\partial^2 f}{\partial x^2} = \frac{2xy}{\left(x^2+y^2\right)^2} \Rightarrow \frac{\partial^2 f}{\partial x^2}(1,1) = \frac{1}{2}$$

$$\frac{\partial^2 f}{\partial y^2} = \frac{-2xy}{\left(x^2+y^2\right)^2} \Rightarrow \frac{\partial^2 f}{\partial y^2}(1,1) = -\frac{1}{2}$$

$$\frac{\partial^2 f}{\partial x \partial y} = \frac{y^2-x^2}{\left(x^2+y^2\right)^2} \Rightarrow \frac{\partial^2 f}{\partial x \partial y}(1,1) = 0$$

Putting these values in Taylor's series, we get

$$f(x,y) = f(a,b) + \left[(x-a)\frac{\partial f}{\partial x} + (y-b)\frac{\partial f}{\partial y}\right] +$$

$$\frac{1}{2!}\left[(x-a)^2\frac{\partial^2 f}{\partial x^2} + 2(x-a)(y-b)\frac{\partial^2 f}{\partial x \partial y} + (y-b)^2\frac{\partial^2 f}{\partial y^2}\right] + \ldots\ldots \quad \ldots(1)$$

since here $a = 1, b = 1$,

Therefore,

$$\tan^{-1}\left(\frac{y}{x}\right) = \frac{\pi}{4} + (x-1)\left(-\frac{1}{2}\right) + (y-1).\frac{1}{2} + \frac{1}{2!}$$

$$\left[(x-1)^2.\frac{1}{2} + 2(x-1)(y-1).0 + (y-1)^2\left(-\frac{1}{2}\right)\right]$$

or

$$\tan^{-1}\left(\frac{y}{x}\right) = \frac{\pi}{4} - \frac{1}{2}(x-1) + \frac{1}{2}(y-1) + \frac{1}{4}(x-1)^2 - \frac{1}{4}(y-1)^2 + \ldots. \quad (2)$$

put $(x - 1) = [1.1 - 1] = 0.1$ and $(y - 1) = [0.9 - 1] = -0.1$ is (2) we get

$$f(1.1, 0.9) = \frac{\pi}{4} - \frac{1}{2}(0.1) + \frac{1}{2}(-0.1) + \frac{1}{4}(0.1)^2 - \frac{1}{4}(-0.1)^2$$

$= 0.78535 - 0.1 = 0.68535$ Proved

Example 5 Find the quadratic Taylor's series expansion of $f(x,y) = 2x^3 + 2y^3 - 4x^2y$ about the point (1,2).

Solution: We have,

$$f(x,y) = 2x^3 + 2y^3 - 4x^2y$$

here, $a = 1, \quad b = 2$

Now, from Taylor's theorem, we have

$$f(x,y) = f(a,b) + \left[(x-a)\frac{\partial f}{\partial x}(a,b) + (y-b)\frac{\partial f}{\partial y}(a,b)\right] +$$

$$\frac{1}{2!}\left[(x-a)^2\frac{\partial^2 f}{\partial x^2}(a,b) + 2(x-a)(y-b)\frac{\partial^2 f}{\partial x \partial y}(a,b) + (y-b)^2\frac{\partial^2 f}{\partial y^2}(a,b)\right] + \ldots \quad (1)$$

now $f(x,y) = 2x^3 + 2y^3 - 4x^2y$

$f(1,2) = 2 \times (1)^3 + 2(2)^3 - 4(1)^2.(2)$

$= 2 + 2 \times 8 - 8 = 10$

$$\frac{\partial f}{\partial x} = 6x^2 - 8xy \Rightarrow \frac{\partial f}{\partial x}(1,2) = 6 - 16 = -10$$

$$\frac{\partial f}{\partial y} = 6y^2 - 4x^2 \Rightarrow \frac{\partial f}{\partial y}(1,2) = 20$$

$$\frac{\partial^2 f}{\partial x^2} = 12x - 8y \Rightarrow \frac{\partial^2 f}{\partial x^2}(1,2) = -4$$

$$\frac{\partial^2 f}{\partial x \partial y} = -8x \Rightarrow \frac{\partial^2 f}{\partial x \partial y}(1,2) = -8$$

$$\frac{\partial^2 f}{\partial y^2} = 12y \Rightarrow \frac{\partial^2 f}{\partial y^2}(1,2) = 24$$

$$\frac{\partial^3 f}{\partial x^3} = 12 \Rightarrow \frac{\partial^3 f}{\partial x^3}(1,2) = 12$$

$$\frac{\partial^3 f}{\partial x^2 \partial y} = -8 \Rightarrow \frac{\partial^3 f}{\partial x^2 \partial y}(1,2) = -8$$

$$\frac{\partial^3 f}{\partial x \partial y^2} = 0 \Rightarrow \frac{\partial^3 f}{\partial x \partial y^2} = 0$$

$$\frac{\partial^3 f}{\partial y^3} = 12 \Rightarrow \frac{\partial^3 f}{\partial y^3} = 12$$

and all other higher order derivatives vanish.
putting these values in (1), we have

$$(2x^3 + 2y^3 - 4x^2y) = 10 + (x-1)(-10) + (y-2)(20) + \frac{1}{2!}$$

$$[(x-1)^2(-4) + 2(x-1)(y-2)(-8) + (y-2)^2.(24)] +$$

$$\frac{1}{3!}[(x-1)^3.(12) + 3(x-1)^2(y-2)(-8) +$$

$$3(x-1)(y-2)^2 \times 0 + 12(y-2)^3] + ...$$

$$= 10 - 10(x-1) + 20(y-2) - 2(x-1)^2 - 8(x-1)(y-2) +$$

$$12(y-2)^2 + 2(x-1)^3 - 4(x-1)^2(y-2) + 2(y-2)^3 + ...$$ which

can be verified by direct expansion also.

Example 6 Find the first three terms of the expansion of the function $e^x \log(1 + y)$ in Taylor's theorem in the neighbourhood of the point (0,0).

Solution: we have

$$f(x,y) = e^x \log(1+y) \Rightarrow f(0,0) = 0$$

$$\frac{\partial f}{\partial x} = e^x \log(1+y) \Rightarrow \left(\frac{\partial f}{\partial x}\right)_{(0,0)} = 0$$

$$\frac{\partial f}{\partial y} = \frac{e^x}{1+y} \Rightarrow \left(\frac{\partial f}{\partial y}\right)_{(0,0)} = 1$$

$$\frac{\partial^2 f}{\partial x^2} = e^x \log(1+y) \Rightarrow \left(\frac{\partial^2 f}{\partial x^2}\right)_{(0,0)} = 0$$

$$\frac{\partial^2 f}{\partial y^2} = \frac{-e^x}{(1+y)^2} \Rightarrow \left(\frac{\partial^2 f}{\partial y^2}\right)_{(0,0)} = -1$$

$$\frac{\partial^2 f}{\partial x \partial y} = \frac{e^x}{1+y} \Rightarrow \left(\frac{\partial^2 f}{\partial x \partial y}\right)_{(0,0)} = 1$$

putting these values in Taylor's theorem, we get

$$f(x,y) = f(0,0) + \left[x\frac{\partial f}{\partial x}(0,0) + y\frac{\partial f}{\partial y}(0,0)\right] +$$

$$\frac{1}{2!}\left[x^2\frac{\partial^2 f}{\partial x^2}(0,0)+2xy\frac{\partial^2 f}{\partial x\partial y}(0,0)+y^2\frac{\partial^2 f}{\partial y^2}(0,0)\right]+\ldots.$$

or $$e^x \log(1+y) = 0 + (x\times 0 + y\times 1) + \frac{1}{2!}[(x^2\times 0 + 2xy\times 1 + y^2(-1)] + \ldots$$

$\Rightarrow$ $$e^x \log(1+y) = y + xy - \frac{y^2}{2}$$

Example 7 Expand $\dfrac{(x+h)(y+k)}{(x+h+y+k)}$ in powers of h and k upto and inclusive of 2nd degree term.

Solution: Here $f(x+h, y+k) = \dfrac{(x+h)(y+k)}{(x+h+y+k)}$, then

$$f(x,y) = \frac{xy}{x+y} \Rightarrow \frac{\partial f}{\partial x} = \frac{(x+y).y-(xy)}{(x+y)^2} = \frac{y^2}{(x+y)^2}$$

similarly, $$\frac{\partial f}{\partial y} = \frac{x^2}{(x+y)^2}, \frac{\partial^2 f}{\partial x^2} = \frac{-2y^2}{(x+y)^3}$$

$$\frac{\partial^2 f}{\partial y^2} = \frac{-2x^2}{(x+y)^3} \text{ and } \frac{\partial^2 f}{\partial x\partial y} = \frac{2xy}{(x+y)^3} \text{ etc.}$$

By Taylor's theorem, we have

$$f(x+h, y+k) = f(x,y) + \left(h\frac{\partial}{\partial x}+k\frac{\partial}{\partial y}\right)f(x,y) +$$

$$\frac{1}{2!}\left(h\frac{\partial}{\partial x}+k\frac{\partial}{\partial y}\right)^2 f(x,y)+\ldots.$$

$\therefore$ $$\frac{(x+h)(y+k)}{(x+h+y+k)} = \frac{xy}{x+y} + \left[h.\frac{y^2}{(x+y)^2} + k.\frac{x^2}{(x+y)^2}\right]$$

$$+\frac{h^2}{2!}\frac{(-2y^2)}{(x+y)^3} + \frac{1}{2!}.2hk.\frac{2xy}{(x+y)^3} + \frac{1}{2!}\frac{k^2(-2x^2)}{(x+y)^3} + \ldots$$

or $$\frac{(x+h)(y+k)}{(x+h+y+k)} = \frac{xy}{(x+y)} + \frac{hy^2}{(x+y)^2} + \frac{kx^2}{(x+y)^2} -$$

$$\frac{h^2y^2}{(x+y)^3}+\frac{2hkxy}{(x+y)^3}-\frac{k^2x^2}{(x+y)^3}+.....$$

Example 8 Expand f(x, y) = $x^2y + 3y - 2$ in powers of (x–1) and (y + 2) by Taylor's theorem.

Solution: We know by Taylor's theorem,

$$f(x,y)=f(a,b)+\left[(x-a)\frac{\partial f}{\partial x}+(y-b)\frac{\partial f}{\partial y}\right]+$$

$$\frac{1}{2!}\left[(x-a)^2\frac{\partial^2 f}{\partial x^2}+2(x-a)(y-b)\frac{\partial^2 f}{\partial x\partial y}+(y-b)^2\frac{\partial^2 f}{\partial y^2}\right]+\ldots \quad \ldots(1)$$

here, a + h = x and h = x – 1 so, a = 1 and b + k = y and k = y + 2 so b = – 2

$f(x,y) = x^2y + 3y - 2 \Rightarrow f(1, -2) = -10$

$$\frac{\partial f}{\partial x}=2xy \Rightarrow \left(\frac{\partial f}{\partial x}\right)_{(1,-2)}=-4$$

$$\frac{\partial f}{\partial y}=x^2+3 \Rightarrow \left(\frac{\partial f}{\partial y}\right)_{(1,-2)}=4$$

$$\frac{\partial^2 f}{\partial x^2}=2y \Rightarrow \left(\frac{\partial^2 f}{\partial x^2}\right)_{(1,-2)}=-4$$

$$\frac{\partial^2 f}{\partial x\partial y}=2x \Rightarrow \left(\frac{\partial^2 f}{\partial x\partial y}\right)_{(1,-2)}=2$$

$$\frac{\partial^2 f}{\partial y^2}=0 \Rightarrow \left(\frac{\partial^2 f}{\partial y^2}\right)_{(1,-2)}=0$$

$$\frac{\partial^3 f}{\partial x^3}=0 \Rightarrow \left(\frac{\partial^3 f}{\partial x^3}\right)_{(1,-2)}=0$$

$$\frac{\partial^3 f}{\partial x^2\partial y}=2 \Rightarrow \left(\frac{\partial^3 f}{\partial x^2\partial y}\right)_{(1,-2)}=2$$

$$\frac{\partial^3 f}{\partial x\partial y^2}=0 \Rightarrow \left(\frac{\partial^3 f}{\partial x\partial y^2}\right)_{(1,-2)}=0$$

$$\frac{\partial^3 f}{\partial y^3} = 0 \quad \Rightarrow \quad \left(\frac{\partial^3 f}{\partial y^3}\right)_{(1,-2)} = 0$$

Putting these values in (1), we get

$$x^2y + 3y - 2 = (-10) + [(x-1)(-4) + (y+2).4] +$$

$$\frac{1}{2!}[(x-1)^2(-4) + 2(x-1)(y+2).2 + (y+2)^2.0] +$$

$$\frac{1}{3!}[(x-1)^3.0 + 3(x-1)^2(y+2).2 + 0 + 0] + ...$$

or $$x^2 + 3y - 2 = -10 - 4(x-1) + 4(y+2) - 2(x-1)^2 + 2(x-1)(y+2) + (x-1)^2(y+2).$$

Example 9 Expand x^y in powers of (x–1) and (y–1) upto third degree term.

Solution: We have $f(x,y) = x^y$ ...(1)

Here, $a + h = x$, and $h = x - 1 \Rightarrow a = 1$

and $b + k = y$ and $k = y - 1 \Rightarrow b = 1$

Also $f(1, 1) = 1$

Now differentiating (1) w.r.t x, we get

$$\frac{\partial f}{\partial x} = yx^{y-1} \quad \Rightarrow \quad \left(\frac{\partial f}{\partial x}\right)_{(1,1)} = 1$$

$$\frac{\partial^2 f}{\partial x^2} = y(y-1)x^{y-2} \quad \Rightarrow \quad \left(\frac{\partial^2 f}{\partial x^2}\right)_{(1,1)} = 0$$

$$\frac{\partial^3 f}{\partial x^3} = y(y-1)(y-2)x^{y-3} \quad \Rightarrow \quad \left(\frac{\partial^3 f}{\partial x^3}\right)_{(1,1)} = 0$$

Again $$\frac{\partial f}{\partial y} = x^y \log x \quad \Rightarrow \quad \left(\frac{\partial f}{\partial y}\right)_{(1,1)} = 0$$

$$\frac{\partial^2 f}{\partial y^2} = x^y (\log x)^2 \Rightarrow \quad \left(\frac{\partial^2 f}{\partial y^2}\right)_{(1,1)} = 0$$

$$\frac{\partial^3 f}{\partial y^3} = x^y (\log x)^3 \Rightarrow \quad \left(\frac{\partial^3 f}{\partial y^3}\right)_{(1,1)} = 0$$

$$\frac{\partial^2 f}{\partial x \partial y} = x^y . \frac{1}{x} + yx^{y-1}\log x = x^{y-1} + yx^{y-1}\log x$$

$$\Rightarrow \quad \left(\frac{\partial^2 f}{\partial x \partial y}\right)_{x=1, y=1} = 1$$

$$\frac{\partial^3 f}{\partial x^2 \partial y} = (y-1)x^{y-2} + y(y-1)x^{y-2}\log x + y\,x^{y-2}$$

$$\left(\frac{\partial^3 f}{\partial x^2 \partial y}\right)_{(1,1)} = 1$$

$$\frac{\partial^3 f}{\partial x \partial y^2} = y\,x^{y-1}(\log x)^2 + 2x^{y-1}\log x \quad \Rightarrow \quad \left(\frac{\partial^3 f}{\partial x \partial y^2}\right)_{(1,1)} = 0$$

Now applying Taylor's theorem, we get

$$f(a+h, b+k) = f(a,b) + \left(h\frac{\partial f}{\partial x} + k\frac{\partial f}{\partial y}\right) +$$

$$\frac{1}{2!}\left[h^2\frac{\partial^2 f}{\partial x^2} + 2hk\frac{\partial^2 f}{\partial x \partial y} + k^2\frac{\partial^2 f}{\partial y^2}\right] +$$

$$\frac{1}{3!}\left[h^3\frac{\partial^3 f}{\partial x^3} + 3h^2k\frac{\partial^3 f}{\partial x^2 \partial y} + 3hk^2\frac{\partial^3 f}{\partial x \partial y^2} + k^3\frac{\partial^3 f}{\partial y^3}\right] + \ldots$$

$$\therefore \quad f(x,y) = x^y = 1 + (x-1) + 0 + \frac{1}{2!}(0 + 2(x-1)(y-1) + 0)$$

$$+ \frac{1}{3!}[0 + 3(x-1)^2(y-1) + 0 + 0] + \ldots\ldots$$

$$\Rightarrow \quad x^y = 1 + (x-1) + (x-1)(y-1) + \frac{1}{2}(x-1)^2(y-1).$$

Example 10 Expand $e^x \sin y$ in powers of x and y as far as term of third degree.

Solution: Let $f(x,y) = e^x \sin y$

Taylor's series in the powers of x and y.

$$f(x,y) = f(0,0) + x\,f_x(0,0) + y\,f_y(0,0) + \frac{x^2}{2!}f_{xx}(0,0) + \frac{y^2}{2!}f_{yy}(0,0) + \frac{2xy}{2!}f_{xy}(0,0) +$$

$$\frac{x^3}{3!}f_{xxx}(0,0) + \frac{y^3}{3!}f_{yyy}(0,0) + \frac{3x^2y}{3!}f_{xxy}(0,0) + \frac{3xy^2}{3!}f_{xyy}(0,0) + \ldots\ldots \quad (A)$$

Now,

$$\begin{aligned}
f(x,y) &= e^x \sin y & \therefore\ & f(0,0) = 0\\
f_x(x,y) &= e^x \sin y & \therefore\ & f_x(0,0) = 0\\
f_y(x,y) &= e^x \cos y & \therefore\ & f_y(0,0) = 1\\
f_{xx}(x,y) &= e^x \sin y & \therefore\ & f_{xx}(0,0) = 0\\
f_{xy}(x,y) &= e^x \cos y & \therefore\ & f_{xy}(0,0) = 1\\
f_{yy}(x,y) &= -e^x \sin y & \therefore\ & f_{yy}(0,0) = -1\\
f_{xxx}(x,y) &= e^x \sin y & \therefore\ & f_{xxx}(0,0) = 0\\
f_{xxy}(x,y) &= e^x \cos y & \therefore\ & f_{xxy}(0,0) = 1\\
f_{xyy}(x,y) &= -e^x \sin y & \therefore\ & f_{xyy}(0,0) = 0\\
f_{yyy}(x,y) &= -e^x \cos y & \therefore\ & f_{yyy}(0,0) = -1
\end{aligned}$$

Putting all these values in equation (A), we get

$$f(x, y) = e^x \sin y = 0 + x(0) + y(1) + \frac{x^2}{2!}(0) + x\,y\,(1) +$$

$$\frac{y^2}{2!}(-1) + \frac{x^3}{3!}(0) + \frac{3x^2y}{3!}(1) + \frac{3xy^2}{3!}(0) + \frac{y^3}{3!}(-1) + \ldots$$

Hence $$e^x \sin y = y + xy - \frac{y^2}{2} + \frac{x^2y}{2} - \frac{y^3}{6} + \ldots\ldots$$

Example 11 If $f(x,y) = \tan^{-1}(xy)$, compute an approximate value of $f(0.9, -1.2)$.

Solution: $f(x,y) = \tan^{-1}(xy) \quad \therefore f(1,-1) = \dfrac{-\pi}{4} \quad \ldots(1)$

Let us expand f(x,y) near the point (1, –1).

$$f(0.9, -1.2) = f(1 - 0.1, -1 - 0.2)$$

$$= f(1, -1) + [(0.1)\, f_x(1, -1) + (-0.2)\, f_y(1, -1)]$$

$$+ \frac{1}{2!}[(-0{:}1)^2 f_{xx}(1, -1) + (-0.2)^2 f_{yy}(1, -1) +$$

$$2(-0.1)(-0.2) f_{xy}(1, -1)] + \ldots \quad \text{(A)}$$

from equation (1)

$$f_x(x,y) = \frac{y}{1+x^2y^2} \qquad \therefore f_x(1,-1) = -\frac{1}{2} \qquad \ldots\text{(ii)}$$

$$f_y(x,y) = \frac{x}{1+x^2y^2} \qquad \therefore f_y(1,-1) = \frac{1}{2} \qquad \ldots\text{(iii)}$$

$$f_{xx}(x,y) = -\frac{2xy^3}{\left(1+x^2y^2\right)^2} \qquad \therefore f_{xx}(1,-1) = \frac{1}{2} \qquad \ldots\text{(iv)}$$

$$f_{yy}(x,y) = -\frac{-2x^3y}{\left(1+x^2y^2\right)^2} \qquad \therefore\ f_{yy}(1,-1) = \frac{1}{2} \qquad \text{...(v)}$$

$$f_{xy}(x,y) = -\frac{1-x^2y^2}{\left(1+x^2y^2\right)^2} \qquad \therefore\ f_{xy}(1,-1) = 0 \qquad \text{...(vi)}$$

putting all these values in equation (A), we get

$$f(0.9, -1.2) = \frac{-\pi}{4} + (-0.1)\left(-\frac{1}{2}\right) + (-0.2)\left(\frac{1}{2}\right) +$$

$$\frac{1}{2}\left[(-0.1)^2\left(\frac{1}{2}\right) + 2(-0.1)(-0.2)(0) + (-0.2)^2\left(\frac{1}{2}\right)\right] + \ldots$$

$$= -\frac{\pi}{4} + 0.05 - 0.1 + \frac{1}{2}(0.005 + 0.04)$$

$$= -\frac{\pi}{4} + 0.05 - 0.1 + 0.0225 = -0.8229 \qquad \text{Answer}$$

6.10 JOCOBIANS

INTRODUCTION

In the study of function of several variables, we find a special function of their derivatives known as functional determinant of the given functions. The functional determinants were first studied by the German mathematician Carl Gustav Jacobi (1804-1851). These functional determinants are called Jacobians, after his name. In this topic we shall deal with these determinants and their properties. An important application of Jacobian is in connection with the change of variables of multiple integrals.

Definition If $u_1, u_2, \ldots\ldots\ldots\ldots u_n$ are functions of n independent variables $x_1, x_2, \ldots\ldots\ldots x_n$,

then determinant $\begin{vmatrix} \frac{\partial u_1}{\partial x_1} & \frac{\partial u_1}{\partial x_2} & \cdots\cdots & \frac{\partial u_1}{\partial x_n} \\ \frac{\partial u_2}{\partial x_1} & \frac{\partial u_2}{\partial x_2} & \cdots\cdots & \frac{\partial u_2}{\partial x_n} \\ \cdots\cdots & \cdots\cdots & \cdots\cdots & \cdots\cdots \\ \cdots\cdots & \cdots\cdots & \cdots\cdots & \cdots\cdots \\ \frac{\partial u_n}{\partial x_1} & \frac{\partial u_n}{\partial x_2} & \cdots\cdots & \frac{\partial u_n}{\partial x_n} \end{vmatrix}$ is called the Jocobian of $u_1, u_2 \ldots\ldots\ldots u_n$ with respect to $x_1, x_2, \ldots\ x_n$ and is denoted by the symbol $\dfrac{\partial(u_1, u_2, \ldots\ldots\ldots u_n)}{\partial(x_1, x_2, \ldots\ldots\ldots x_n)}$. It is also denoted by $J(u_1, u_2, \ldots\ u_n)$ or simply by J when there is no ambiguity with regard to the variables.

6.11 PROPERTIES OF JACOBIANS

1. Let $J_1 = \dfrac{\partial(u, v)}{\partial(x, y)}$ and $J_2 = \dfrac{\partial(x,y)}{\partial(u,v)}$, then $J_1 . J_2 = 1$

Proof: Let $u = f(x, y)$ and $v = g(x, y)$. suppose, on solving for x and y, we get $x = \phi(u, v)$ and $y = \psi(u, v)$.

Then,
$$\left.\begin{aligned} \frac{\partial u}{\partial u} &= 1 = \frac{\partial u}{\partial x}.\frac{\partial x}{\partial u} + \frac{\partial u}{\partial y}.\frac{\partial y}{\partial u} \\ \frac{\partial u}{\partial v} &= 0 = \frac{\partial u}{\partial x}.\frac{\partial x}{\partial v} + \frac{\partial u}{\partial y}.\frac{\partial y}{\partial v} \\ \frac{\partial v}{\partial u} &= 0 = \frac{\partial v}{\partial x}.\frac{\partial x}{\partial u} + \frac{\partial v}{\partial y}.\frac{\partial y}{\partial u} \\ \frac{\partial v}{\partial v} &= 1 = \frac{\partial v}{\partial x}.\frac{\partial x}{\partial v} + \frac{\partial v}{\partial y}.\frac{\partial y}{\partial v} \end{aligned}\right\} \qquad \ldots(1)$$

Now,
$$J_1 J_2 = \frac{\partial(u,v)}{\partial(x,y)}.\frac{\partial(x,y)}{\partial(u,v)} \qquad \ldots(2)$$

$$= \begin{vmatrix} u_x & u_y \\ v_x & v_y \end{vmatrix} . \begin{vmatrix} x_u & x_v \\ y_u & y_v \end{vmatrix} \qquad \left[\text{where } u_x = \frac{\partial u}{\partial x} \text{ etc.}\right.$$

$$= \begin{vmatrix} u_x\, x_u + u_y\, y_u & u_x\, x_v + u_y\, y_v \\ v_x\, x_u + v_y\, y_u & v_x\, x_v + v_y\, y_v \end{vmatrix}$$

$$= \begin{vmatrix} 1 & 0 \\ 0 & 1 \end{vmatrix} = 1 \qquad \text{(from (1)}$$

2. If u, v are functions of s, t, where s, t are functions of x, y then

$$\frac{\partial(u, v)}{\partial(x, y)} = \frac{\partial(u, v)}{\partial(s, t)} \times \frac{\partial(s, t)}{\partial(x, y)}$$

Proof: since u, v are composite function of x, y

$$\therefore \quad \left.\begin{aligned} \frac{\partial u}{\partial x} &= \frac{\partial u}{\partial s}.\frac{\partial s}{\partial x} + \frac{\partial u}{\partial t}.\frac{\partial t}{\partial x} \\ \frac{\partial u}{\partial y} &= \frac{\partial u}{\partial s}.\frac{\partial s}{\partial y} + \frac{\partial u}{\partial t}.\frac{\partial t}{\partial y} \\ \frac{\partial v}{\partial x} &= \frac{\partial v}{\partial s}.\frac{\partial s}{\partial x} + \frac{\partial v}{\partial t}.\frac{\partial t}{\partial x} \\ \frac{\partial v}{\partial y} &= \frac{\partial v}{\partial s}.\frac{\partial s}{\partial y} + \frac{\partial v}{\partial t}.\frac{\partial t}{\partial y} \end{aligned}\right\} \quad (1)$$

Now,
$$\frac{\partial(u,v)}{\partial(s,t)} \times \frac{\partial(s,t)}{\partial(x,y)} = \begin{vmatrix} \frac{\partial u}{\partial s} & \frac{\partial u}{\partial t} \\ \frac{\partial v}{\partial s} & \frac{\partial v}{\partial t} \end{vmatrix} . \begin{vmatrix} \frac{\partial s}{\partial x} & \frac{\partial s}{\partial y} \\ \frac{\partial t}{\partial x} & \frac{\partial t}{\partial y} \end{vmatrix}$$

$$= \begin{vmatrix} \frac{\partial u}{\partial s}.\frac{\partial s}{\partial x} + \frac{\partial u}{\partial t}.\frac{\partial t}{\partial x} & \frac{\partial u}{\partial s}.\frac{\partial s}{\partial y} + \frac{\partial u}{\partial t}.\frac{\partial t}{\partial y} \\ \frac{\partial v}{\partial s}.\frac{\partial s}{\partial x} + \frac{\partial v}{\partial t}.\frac{\partial t}{\partial x} & \frac{\partial v}{\partial s}.\frac{\partial s}{\partial y} + \frac{\partial v}{\partial t}.\frac{\partial t}{\partial y} \end{vmatrix}$$

$$= \begin{vmatrix} \frac{\partial u}{\partial x} & \frac{\partial u}{\partial y} \\ \frac{\partial v}{\partial x} & \frac{\partial v}{\partial y} \end{vmatrix} = \frac{\partial(u,v)}{\partial(x,y)} \qquad \text{(from (i))}$$

6.12 A SPECIAL FORM

If the function $u_1, u_2, \ldots\ldots u_n$ of the variables $x_1, x_2, \ldots\ldots\ldots x_n$ are defined by the relation $u_1 = f_1(x_1)$; $u_2 = f_2(x_1, x_2)$; .... $u_n = f_n(x_1, x_2 \ldots\ldots x_n)$

Then,
$$\frac{\partial(u_1, u_2, \ldots\ldots u_n)}{\partial(x_1, x_2, \ldots\ldots x_n)} = \begin{vmatrix} \frac{\partial u_1}{\partial x_1} & 0 & \ldots\ldots & 0 \\ \frac{\partial u_2}{\partial x_1} & \frac{\partial u_2}{\partial x_2} & \ldots\ldots & 0 \\ \ldots & \ldots & \ldots & \ldots \\ \ldots & \ldots & \ldots & \ldots \\ \frac{\partial u_n}{\partial x_1} & \frac{\partial u_n}{\partial x_2} & \ldots\ldots & \frac{\partial u_n}{\partial x_n} \end{vmatrix} = \frac{\partial u_1}{\partial x_1}.\frac{\partial u_2}{\partial x_2}.\ldots\ldots\frac{\partial u_n}{\partial x_n}.$$

i.e., the jacobians reduce to the product of the leading diagonal elements of the determinants.

3. If u, v, w are functions of three independent variables x, y, z and functionally related, then

$$\frac{\partial(u, v, w)}{\partial(x, y, z)} = 0$$

Proof: Here u, v, w are not independent, then F(u, v, w) = 0 ...(1)

Now differentiating (1) w.r.t. x, we get

$$\frac{\partial F}{\partial u}.\frac{\partial u}{\partial x} + \frac{\partial F}{\partial v}.\frac{\partial v}{\partial x} + \frac{\partial F}{\partial w}.\frac{\partial w}{\partial x} = 0 \qquad ...(2)$$

$$\frac{\partial F}{\partial u}.\frac{\partial u}{\partial y} + \frac{\partial F}{\partial v}.\frac{\partial v}{\partial y} + \frac{\partial F}{\partial w}.\frac{\partial w}{\partial y} = 0 \qquad ...(3)$$

$$\frac{\partial F}{\partial u}.\frac{\partial u}{\partial z} + \frac{\partial F}{\partial v}.\frac{\partial v}{\partial z} + \frac{\partial F}{\partial w}.\frac{\partial w}{\partial z} = 0 \qquad ...(4)$$

eliminating $\frac{\partial F}{\partial u}, \frac{\partial F}{\partial v}, \frac{\partial F}{\partial w}$ from (2), (3) and (4), we have,

$$\begin{vmatrix} \frac{\partial u}{\partial x} & \frac{\partial v}{\partial x} & \frac{\partial w}{\partial x} \\ \frac{\partial u}{\partial y} & \frac{\partial v}{\partial y} & \frac{\partial w}{\partial y} \\ \frac{\partial u}{\partial z} & \frac{\partial v}{\partial y} & \frac{\partial w}{\partial z} \end{vmatrix} = 0$$

interchanging rows and columns, we get

$$\begin{vmatrix} \frac{\partial u}{\partial x} & \frac{\partial u}{\partial y} & \frac{\partial u}{\partial z} \\ \frac{\partial v}{\partial x} & \frac{\partial v}{\partial y} & \frac{\partial v}{\partial z} \\ \frac{\partial w}{\partial x} & \frac{\partial w}{\partial y} & \frac{\partial w}{\partial z} \end{vmatrix} = 0$$

$$\Rightarrow \qquad \frac{\partial(u, v, w)}{\partial(x, y, z)} = 0$$

Illustrative Examples

Example 1 If $u = x + 2y + z$, $v = x - 2y + 3z$ and $w = 2xy - xz + 4yz - 2z^2$, prove that they are not independent. find the relation between u, v and w.

Solution: Here u, v, w. are not independent, if $\dfrac{\partial(u, v, w)}{\partial(x, y, z)} = 0$ ...(1)

Now
$$\frac{\partial(u, v, w)}{\partial(x, y, z)} = \begin{vmatrix} \dfrac{\partial u}{\partial x} & \dfrac{\partial u}{\partial y} & \dfrac{\partial u}{\partial z} \\ \dfrac{\partial v}{\partial x} & \dfrac{\partial v}{\partial y} & \dfrac{\partial v}{\partial z} \\ \dfrac{\partial w}{\partial x} & \dfrac{\partial w}{\partial y} & \dfrac{\partial w}{\partial z} \end{vmatrix} \quad ...(2)$$

Here
$$\frac{\partial u}{\partial x} = 1, \quad \frac{\partial u}{\partial y} = 2, \quad \frac{\partial u}{\partial z} = 1$$
$$\frac{\partial v}{\partial x} = 1, \quad \frac{\partial v}{\partial y} = -2, \quad \frac{\partial v}{\partial z} = 3$$
$$\frac{\partial w}{\partial x} = 2y - z, \quad \frac{\partial w}{\partial y} = 2x + 4z, \quad \frac{\partial w}{\partial z} = -x + 4y - 4z$$

from (2), we have,
$$\frac{\partial(u, v, w)}{\partial(x, y, z)} = \begin{vmatrix} 1 & 2 & 1 \\ 1 & -2 & 3 \\ 2y - z & 2x + 4z & -x + 4y - 4z \end{vmatrix}$$

operating $c_2 \to c_2 - 2c_1, \; c_3 \to c_3 - c_1,$

$$= \begin{vmatrix} 1 & 0 & 0 \\ 1 & -4 & 2 \\ 2y - z & 2x - 4y + 6z & -x + 2y - 3z \end{vmatrix}$$

$$= -4(-x + 2y - 3z) - 2(2x - 4y + 6z)$$
$$= 4x - 8y + 12z - 4x + 8y - 12z = 0$$

Hence, u, v, w are not independent

Now
$$u + v = 2x + 4z \quad ...(3)$$
$$u - v = 4y - 2z \quad ...(4)$$

Multiplying (3) and (4), we get
$$(u + v)(u - v) = (2x + 4z)(4y - 2z)$$
$$u^2 - v^2 = 4(2xy + 4yz - zx - 2z^2)$$

$$\Rightarrow \quad \boxed{u^2 - v^2 = 4w}$$

Example 2 Verify whether the following functions are functionally related, and if so, find the relation between them.

$$u = \frac{x+y}{1-xy}, \; v = \tan^{-1} x + \tan^{-1} y$$

Solution: We know that $\dfrac{\partial(u,v)}{\partial(x,y)} = \begin{vmatrix} \dfrac{\partial u}{\partial x} & \dfrac{\partial u}{\partial y} \\ \dfrac{\partial v}{\partial x} & \dfrac{\partial v}{\partial y} \end{vmatrix}$

$$= \begin{vmatrix} \dfrac{1+y^2}{(1-xy)^2} & \dfrac{(1+x^2)}{(1-xy)^2} \\ \dfrac{1}{(1+x^2)} & \dfrac{1}{(1+y^2)} \end{vmatrix}$$

$$= \frac{1}{(1-xy)^2} - \frac{1}{(1-xy)^2} = 0$$

Hence, u, v are functionally dependent. Now, we know that,

$$\tan^{-1}x + \tan^{-1}y = \tan^{-1}\frac{x+y}{1-xy}$$

$$\Rightarrow \quad v = \tan^{-1} u \text{ (As given)} \quad \therefore \quad \boxed{u = \tan v}$$

Example 3 If $x = r\cos\theta$, $y = r\sin\theta$, find $\dfrac{\partial(x,y)}{\partial(r,\theta)}$ and $\dfrac{\partial(r,\theta)}{\partial(x,y)}$.

Solution: Here, $x = r\cos\theta$, $y = r\sin\theta$

$$\therefore \quad \frac{\partial x}{\partial r} = \cos\theta, \quad \frac{\partial y}{\partial r} = \sin\theta$$

and

$$\frac{\partial x}{\partial \theta} = -r\sin\theta, \quad \frac{\partial y}{\partial \theta} = r\cos\theta$$

we have

$$\frac{\partial(x,y)}{\partial(r,\theta)} = \begin{vmatrix} \dfrac{\partial x}{\partial r} & \dfrac{\partial x}{\partial \theta} \\ \dfrac{\partial y}{\partial r} & \dfrac{\partial y}{\partial \theta} \end{vmatrix} = \begin{vmatrix} \cos\theta & -r\sin\theta \\ \sin\theta & r\cos\theta \end{vmatrix}$$

$$= r\cos^2\theta + r\sin^2\theta = r(\cos^2\theta + \sin^2\theta) = r.$$

Now, $r^2 = x^2 + y^2$ and $\theta = \tan^{-1}(y/x)$

$$\therefore \quad \frac{\partial r}{\partial x} = \frac{x}{r}, \quad \frac{\partial r}{\partial y} = y/r$$

and $$\frac{\partial \theta}{\partial x} = \frac{-y}{x^2+y^2} \quad \Rightarrow \quad \frac{\partial v}{\partial x} = \frac{-y}{r^2}$$

$$\frac{\partial \theta}{\partial y} = \frac{x}{x^2+y^2} \quad \Rightarrow \quad \frac{\partial v}{\partial y} = \frac{x}{r^2}$$

$$\therefore \quad \frac{\partial(r,\theta)}{\partial(r,y)} = \begin{vmatrix} \frac{\partial r}{\partial x} & \frac{\partial r}{\partial y} \\ \frac{\partial \theta}{\partial x} & \frac{\partial \theta}{\partial y} \end{vmatrix} = \begin{vmatrix} x/r & y/r \\ -y/r^2 & \frac{x}{r^2} \end{vmatrix}$$

$$= \frac{x^2}{r^3} + \frac{y^2}{r^3} = \frac{r^2}{r^3} = \frac{1}{r}$$

Remark: $$\frac{\partial(x,y)}{\partial(r,\theta)} \times \frac{\partial(r,\theta)}{\partial(x,y)} = r.\frac{1}{r} = 1$$

Example 4 If $x = r\sin\theta.\cos\phi$, $y = r\sin\theta\sin\phi$,

$z = r\cos\theta$, show that $\dfrac{\partial(x,y,z)}{\partial(r,\theta,\phi)} = r^2\sin\theta$

Solution: Here, we have,

$$\frac{\partial x}{\partial r} = \sin\theta\cos\phi, \qquad \frac{\partial x}{\partial \theta} = r\cos\theta\cos\phi$$

$$\frac{\partial x}{\partial \phi} = -r\sin\theta\sin\phi, \qquad \frac{\partial y}{\partial r} = \sin\theta\sin\phi,$$

$$\frac{\partial y}{\partial \theta} = r\cos\theta\sin\phi, \qquad \frac{\partial y}{\partial \phi} = r\sin\theta\cos\phi$$

and $$\frac{\partial z}{\partial r} = \cos\theta, \qquad \frac{\partial z}{\partial \theta} = -r\sin\theta, \qquad \frac{\partial z}{\partial \phi} = 0$$

Now, we have,

$$\frac{\partial(x,y,z)}{\partial(r,\theta,\phi)} = \begin{vmatrix} \frac{\partial x}{\partial r} & \frac{\partial x}{\partial \theta} & \frac{\partial x}{\partial \phi} \\ \frac{\partial y}{\partial r} & \frac{\partial y}{\partial \theta} & \frac{\partial y}{\partial \phi} \\ \frac{\partial z}{\partial r} & \frac{\partial z}{\partial \theta} & \frac{\partial z}{\partial \phi} \end{vmatrix}$$

$$= \begin{vmatrix} \sin\theta\cos\phi & r\cos\theta\cos\phi & -r\sin\theta\sin\phi \\ \sin\theta\sin\phi & r\cos\theta\sin\phi & r\sin\theta\cos\phi \\ \cos\theta & -r\sin\theta & 0 \end{vmatrix}$$

$$= \cos\theta\,(r^2\sin\theta\cos\theta\cos^2\phi + r^2\sin\theta\cos\theta\sin^2\phi) + r\sin\theta\,(r\sin^2\theta\cos^2\phi + r\sin^2\theta\sin^2\phi)$$

(expanding by 3rd row)

$$= r^2\sin\theta\cos^2\theta + r^2\sin^3\theta$$

$$= r^2\sin\theta$$

Example 5 If $y_1 = \frac{x_2 x_3}{x_1}$, $y_2 = \frac{x_3 x_1}{x_2}$, $y_3 = \frac{x_1 x_2}{x_3}$, prove that Jacobian of y_1, y_2, y_3 with respect to x_1, x_2, x_3 is 4.

Solution: Here $y_1 = \frac{x_2 x_3}{x_1}$, $y_2 = \frac{x_3 x_1}{x_2}$, $y_3 = \frac{x_1 x_2}{x_3}$

we have

$$\frac{\partial(y_1,y_2,y_3)}{\partial(x_1,x_2,x_3)} = \begin{vmatrix} \frac{\partial y_1}{\partial x_1} & \frac{\partial y_1}{\partial x_2} & \frac{\partial y_1}{\partial x_3} \\ \frac{\partial y_2}{\partial x_1} & \frac{\partial y_2}{\partial x_2} & \frac{\partial y_2}{\partial x_3} \\ \frac{\partial y_3}{\partial x_1} & \frac{\partial y_3}{\partial x_2} & \frac{\partial y_3}{\partial x_3} \end{vmatrix}$$

$$= \begin{vmatrix} \frac{-x_2x_3}{x_1^2} & \frac{x_3}{x_1} & \frac{x_2}{x_1} \\ \frac{x_3}{x_2} & \frac{-x_3x_1}{x_2^2} & \frac{x_1}{x_2} \\ \frac{x_2}{x_3} & \frac{x_1}{x_3} & \frac{-x_1x_2}{x_3^2} \end{vmatrix}$$

$$= \frac{1}{x_1^2 x_2^2 x_3^2} \begin{vmatrix} -x_2 x_3 & x_3 x_1 & x_1 x_2 \\ x_2 x_3 & -x_3 x_1 & x_1 x_2 \\ x_2 x_3 & x_3 x_1 & -x_1 x_2 \end{vmatrix}$$

$$= \frac{x_1^2 x_2^2 x_3^2}{x_1^2 x_2^2 x_3^2} \begin{vmatrix} -1 & 1 & 1 \\ 1 & -1 & 1 \\ 1 & 1 & -1 \end{vmatrix}$$

$$= (-1)(1-1) - 1(-1-1) + (1+1) = 0 + 2 + 2 = 4.$$ Answer

Example 6 If $x = r\cos\theta\cos\phi$, $y = r\sin\theta\sqrt{1-m^2\sin^2\phi}$ $z = r\sin\phi\sqrt{1-n^2\sin^2\theta}$, where $m^2 + n^2 = 1$, find the jacobian of $\frac{\partial(x, y, z)}{\partial(r, \theta, \phi)}$

Solution: Squaring and adding the given relation, we have,

$$x^2 + y^2 + z^2 = r^2\cos^2\theta\cos^2\phi + r^2\sin^2\theta(1 - m^2\sin^2\phi) + r^2\sin^2\phi(1 - n^2\sin^2\theta)$$

$$= r^2(\cos^2\theta\cos^2\phi + \sin^2\theta - \sin^2\theta\sin^2\phi + \sin^2\phi) \quad (\because m^2 + n^2 = 1)$$

$$= r^2(\cos^2\theta\cos^2\phi + \sin^2\theta\cos^2\phi + \sin^2\phi) \quad (\because 1 - \sin^2\phi = \cos^2\phi)$$

$$= r^2(\cos^2\phi + \sin^2\phi) = r^2$$

differentiating this partially w.r.t. r, θ, ϕ, we get

$$x\frac{\partial x}{\partial r} + y\frac{\partial y}{\partial r} + z\frac{\partial z}{\partial r} = r \quad \text{...(1)}$$

$$x\frac{\partial x}{\partial \theta} + y\frac{\partial y}{\partial \theta} + z\frac{\partial z}{\partial \theta} = 0 \quad \text{...(2)}$$

and $$x\frac{\partial x}{\partial \phi} + y\frac{\partial y}{\partial \phi} + z\frac{\partial z}{\partial \phi} = 0 \quad \text{...(3)}$$

$$\therefore \quad \frac{\partial(x, y, z)}{\partial(r, \theta, \phi)} = \begin{vmatrix} \frac{\partial x}{\partial r} & \frac{\partial x}{\partial \theta} & \frac{\partial x}{\partial \phi} \\ \frac{\partial y}{\partial r} & \frac{\partial y}{\partial \theta} & \frac{\partial y}{\partial \phi} \\ \frac{\partial z}{\partial r} & \frac{\partial z}{\partial \theta} & \frac{\partial z}{\partial \phi} \end{vmatrix}$$

$$= \frac{1}{x}\begin{vmatrix} r & 0 & 0 \\ \frac{\partial y}{\partial r} & \frac{\partial y}{\partial \theta} & \frac{\partial y}{\partial \phi} \\ \frac{\partial z}{\partial r} & \frac{\partial z}{\partial \theta} & \frac{\partial z}{\partial \phi} \end{vmatrix}$$

on adding ($R_2y + R_3z$) to xR_1 and using (1) to (3)

$$= \frac{r}{x}\begin{vmatrix} \frac{\partial y}{\partial \theta} & \frac{\partial y}{\partial \phi} \\ \frac{\partial z}{\partial \theta} & \frac{\partial z}{\partial \phi} \end{vmatrix}$$

$$= \frac{r}{x}\begin{vmatrix} r\cos\theta\sqrt{1-m^2\sin^2\phi} & \frac{-r\sin\theta.\, m^2\sin\phi\cos\phi}{\sqrt{(1-m^2\sin^2\phi)}} \\ \frac{-r\sin\phi.\, n^2\sin\theta\cos\theta}{\sqrt{1-n^2\sin^2\theta}} & r\cos\phi\sqrt{1-n^2\sin^2\theta} \end{vmatrix}$$

$$= \frac{r}{x}\left[r^2\cos\theta\cos\phi\sqrt{(1-m^2\sin^2\phi)}\sqrt{(1-n^2\sin^2\theta)} - \frac{m^2n^2r^2\sin^2\theta\sin^2\phi\cos\theta\cos\phi}{\sqrt{(1-m^2\sin^2\phi)}\sqrt{(1-n^2\sin^2\theta)}}\right]$$

$$= \frac{r^3\cos\theta\cos\phi}{x}\left[\frac{(1-m^2\sin^2\phi)(1-n^2\sin^2\theta) - m^2n^2\sin^2\theta\sin^2\phi}{\sqrt{(1-m^2\sin^2\phi)}\sqrt{(1-n^2\sin^2\theta)}}\right]$$

$$= \frac{r^2(1-m^2\sin^2\phi - n^2\sin^2\theta)}{\sqrt{(1-m^2\sin^2\phi)(1-n^2\sin^2\theta)}} \qquad (\because m^2+n^2=1)$$

Example 7 If $u = \sin^{-1}x + \sin^{-1}y$ and $v = x\sqrt{1-y^2} + y\sqrt{1-x^2}$ then evaluate the Jacobian $\frac{\partial(u,v)}{\partial(x,y)}$, given that $x^2 < 1, y^2 < 1$.

Also find the functional relation between u and v, if any.

Solution: We have,

$$u = \sin^{-1} x + \sin^{-1} y \qquad ...(1)$$

and

$$v = x\sqrt{1-y^2} + y\sqrt{1-x^2} \qquad ...(2)$$

$$\therefore \quad \frac{\partial u}{\partial x} = \frac{1}{\sqrt{1-x^2}}, \frac{\partial v}{\partial x} = \sqrt{1-y^2} - \frac{xy}{\sqrt{1-x^2}}$$

and

$$\frac{\partial u}{\partial y} = \frac{1}{\sqrt{1-y^2}}, \frac{\partial v}{\partial y} = \sqrt{1-x^2} - \frac{xy}{\sqrt{1-y^2}}$$

Now, we have,

$$\frac{\partial(u,v)}{\partial(x,y)} = \begin{vmatrix} \frac{\partial u}{\partial x} & \frac{\partial u}{\partial y} \\ \frac{\partial v}{\partial x} & \frac{\partial v}{\partial y} \end{vmatrix}$$

$$= \begin{vmatrix} \frac{1}{\sqrt{1-x^2}} & \frac{1}{\sqrt{1-y^2}} \\ \sqrt{1-y^2} - \frac{xy}{\sqrt{1-x^2}} & \sqrt{1-x^2} - \frac{xy}{\sqrt{1-y^2}} \end{vmatrix}$$

$$= \frac{1}{\sqrt{1-x^2}}\left(\sqrt{1-x^2} - \frac{xy}{\sqrt{1-y^2}}\right) - \frac{1}{\sqrt{1-y^2}}\left(\sqrt{1-y^2} - \frac{xy}{\sqrt{1-x^2}}\right)$$

$$= \left(1 - \frac{xy}{\sqrt{1-x^2}\,\sqrt{1-y^2}}\right) - \left(1 - \frac{xy}{\sqrt{1-x^2}\,\sqrt{1-y^2}}\right) = 0$$

Jacobian is also known as functional determinant.

so, the required functional relation between u and v is,

$$\begin{vmatrix} \frac{\partial u}{\partial x} & \frac{\partial u}{\partial y} \\ \frac{\partial v}{\partial x} & \frac{\partial v}{\partial y} \end{vmatrix} = 0 \Rightarrow \quad \frac{\partial u}{\partial x}.\frac{\partial v}{\partial y} - \frac{\partial u}{\partial y}.\frac{\partial v}{\partial x} = 0$$

Now, from (1), $\quad u = \sin^{-1} x + \sin^{-1} y$

$$= \sin^{-1}\left(x\sqrt{1-y^2} + y\sqrt{1-x^2}\right)$$

$= \sin^{-1} v$ (from (2))

or $v = \sin u$, which is the required relationship between u and v.

Example 8 If $y_1 = \cos x_1$, $y_2 = \sin x_1 \cos x_2$ and $y_3 = \sin x_1 \sin x_2 \cos x_3$, then prove that

$$\frac{\partial(y_1, y_2, y_3)}{\partial(x_1, x_2, x_3)} = -\sin^3 x_1 \sin^2 x_2 \sin x_3 .$$

Solution: we have $y_1 = \cos x_1$, $y_2 = \sin x_1 \cos x_2$, $y_3 = \sin x_1 \sin x_2 \cos x_3$

Now
$$\frac{\partial(y_1, y_2, y_3)}{\partial(x_1, x_2, x_3)} = \begin{vmatrix} \frac{\partial y_1}{\partial x_1} & \frac{\partial y_1}{\partial x_2} & \frac{\partial y_1}{\partial x_3} \\ \frac{\partial y_2}{\partial x_1} & \frac{\partial y_2}{\partial x_2} & \frac{\partial y_2}{\partial x_3} \\ \frac{\partial y_3}{\partial x_1} & \frac{\partial y_3}{\partial x_2} & \frac{\partial y_3}{\partial x_3} \end{vmatrix}$$

$$= \begin{vmatrix} -\sin x_1 & 0 & 0 \\ \cos x_1 \cos x_2 & -\sin x_1 \sin x_2 & 0 \\ \cos x_1 \sin x_2 \cos x_3 & \sin x_1 \cos x_2 \cos x_3 & -\sin x_1 \sin x_2 \sin x_3 \end{vmatrix}$$

Expanding R_3, we get

$$= -\sin x_1 \sin x_2 \sin x_3 (\sin^2 x_1 \sin x_2)$$
$$= -\sin^3 x_1 \sin^2 x_2 \sin x_3$$

Example 9 In cylindrical co-ordinates $x = \rho\cos\phi$, $y = \rho\sin\phi$, $z = z$ prove that $\frac{\partial(x, y, z)}{\partial(\rho, \phi, z)} = \rho$

Solution: we have $\frac{\partial x}{\partial \rho} = \cos\phi$, $\frac{\partial x}{\partial \phi} = -\rho\sin\phi$, $\frac{\partial x}{\partial z} = 0$, $\frac{\partial y}{\partial \rho} = \sin\phi$, $\frac{\partial y}{\partial \phi} = \rho\cos\phi$, $\frac{\partial y}{\partial z} = 0$

and $\frac{\partial z}{\partial \rho} = 0$, $\frac{\partial z}{\partial \phi} = 0$, $\frac{\partial z}{\partial z} = 1$

$$\therefore \quad \frac{\partial(x, y, z)}{\partial(\rho, \phi, z)} = \begin{vmatrix} \frac{\partial x}{\partial \rho} & \frac{\partial x}{\partial \phi} & \frac{\partial x}{\partial z} \\ \frac{\partial y}{\partial \rho} & \frac{\partial y}{\partial \phi} & \frac{\partial y}{\partial z} \\ \frac{\partial z}{\partial \rho} & \frac{\partial z}{\partial \phi} & \frac{\partial z}{\partial z} \end{vmatrix} = \begin{vmatrix} \cos\phi & -\rho\sin\phi & 0 \\ \sin\phi & \rho\cos\phi & 0 \\ 0 & 0 & 1 \end{vmatrix}$$

$= \rho.$ Proved

Exercise 6.4

1. If $u = \dfrac{y^2}{2x}$, $v = \dfrac{x^2 + y^2}{2x}$, find $\dfrac{\partial(u, v)}{\partial(x, y)}$.

2. If x = a cos u cos hv, y = a sin u sin h v. prove that

$$\frac{\partial(x, y)}{\partial(u, v)} = \frac{1}{2} a^2 (\cos 2u - \cos h2v)$$

3. If u = 3x + 2y – z, v = x – 2y + z, w = x (x + 2y – z),

Show that $\dfrac{\partial(u, v, w)}{\partial(x, y, z)} = 0$

4. If u = 2xy, v = $x^2 - y^2$, x = r cos θ, y = r sin θ,

find $\dfrac{\partial(u, v)}{\partial(r, \theta)}$

5. If $u = \dfrac{x}{y - z}$, $v = \dfrac{y}{z - x}$, $w = \dfrac{z}{x - y}$, Prove that

$$\frac{\partial(u, v, w)}{\partial(x, y, z)} = 0$$

6. If $x = uv, y = \dfrac{u + v}{u - v}$, find $\dfrac{\partial(u, v)}{\partial(x, y)}$.

7. For the transformation $x = e^u \cos v, y = e^u \sin v$,

show that $\dfrac{\partial(x, y)}{\partial(u, v)} \times \dfrac{\partial(u, v)}{\partial(x, y)} = 1$

8. Verify JJ' = 1, if x = e^v secu, y = e^v tanu.

9. If $u_1 = \dfrac{x_1}{n}$, $u_2 = \dfrac{x_2}{n}$ .... $u_n = \dfrac{x_n}{n}$ and

$x_1^2 + x_2^2 + ... + x_n^2 = 1$, find $\dfrac{\partial(u_1, u_2, ... u_n)}{\partial(x_1, x_2, ... x_n)}$.

10. If $u = \dfrac{y - x}{1 + xy}$ and v = $\tan^{-1}y - \tan^{-1}x$, find $\dfrac{\partial(u, v)}{\partial(x, y)}$.

11. If $x_1 = r\sin\theta_1 \sin\theta_2$, $x_2 = r\sin\theta_1 \cos\theta_2$

$x_3 = r\cos\theta_1 \sin\theta_2$, $x_4 = r\cos\theta_1 \cos\theta_2$

show that $J(x_1, x_2, x_3, x_4) = r^2 \sin\theta_1 \cos\theta_1$

12. Find jacobian of $y_1, y_2, \ldots\ldots y_n$ being given

$y_1 = x_1(1-x_2), y_2 = x_1 x_2 (1-x_3) \ldots\ldots y_{n-1} = x_1 x_2 \ldots\ldots x_{n-1}(1-x_n)$

$y_n = x_1 x_2 \ldots\ldots x_{n-1} x_n$

13. If $x = u(1-v)$, $y = uv$, verify $JJ' = 1$

14. Verify that $\dfrac{\partial(x,y)}{\partial(u,v)} \times \dfrac{\partial(u,v)}{\partial(x,y)} = 1$, if

$x = \sin\theta\cos\phi$, $y = \sin\theta\sin\phi$.

15. If $u = x^2 - y^2$, $v = 2xy$ and $x = r\cos\theta$, $y = r\sin\theta$,

then prove that $\dfrac{\partial(u,v)}{\partial(r,\theta)} = 4r^3$.

Answers

1. $-y/2x$ 4. $-4r^3$ 6. $\dfrac{(u-v)^2}{4uv}$ 9. $\dfrac{1}{x_n^{n-1}}$ 10. 0

12. $x_1^{n-1} x_2^{n-2} \ldots\ldots x_{n-1}$.

6.13 OTHER SPECIAL FORMS

There are functions for which the actual calculation of the Jacobians by evaluating the functional determinant is often found to be a tedious task. But the calculation can be made easier by reducing these functions to the special form which we discussed above or to one of the special forms which we shal! describe here.

1. Form of function of functions: If $v_1, v_2, \ldots\ldots v_n$ are functions of $y_1, y_2, \ldots\ldots y_n$ and $y_1, y_2 \ldots\ldots y_n$ are functions of $x_1, x_2, \ldots\ldots x_n$, then

$$\frac{\partial(v_1, v_2, \ldots\ldots v_n)}{\partial(x_1, x_2 \ldots\ldots x_n)} = \frac{\partial(v_1, v_2, \ldots\ldots v_n)}{\partial(y_1, y_2 \ldots\ldots y_n)} \cdot \frac{\partial(y_1, y_2, \ldots\ldots y_n)}{\partial(x_1, x_2 \ldots\ldots x_n)}$$

Here, we are given that $v_1, v_2, \ldots\ldots v_n$ are functions of $y_1, y_2 \ldots\ldots y_n$ which are themselves functions of $x_1, x_2 \ldots\ldots x_n$, therefore, we have.

$$\left\{\begin{array}{l}\dfrac{\partial v_1}{\partial x_1}=\dfrac{\partial v_1}{\partial y_1}\cdot\dfrac{\partial y_1}{\partial x_1}+\dfrac{\partial v_1}{\partial y_2}\cdot\dfrac{\partial y_2}{\partial x_1}+\ldots\ldots+\dfrac{\partial v_1}{\partial y_n}\cdot\dfrac{\partial y_n}{\partial x_1}=\sum_{r=1}^{n}\dfrac{\partial v_1}{\partial y_r}\cdot\dfrac{\partial y_r}{\partial x_1}\\ \dfrac{\partial v_1}{\partial x_2}=\dfrac{\partial v_1}{\partial y_1}\cdot\dfrac{\partial y_1}{\partial x_2}+\dfrac{\partial v_1}{\partial y_2}\cdot\dfrac{\partial y_2}{\partial x_2}+\ldots\ldots+\dfrac{\partial v_1}{\partial y_n}\cdot\dfrac{\partial y_n}{\partial x_2}=\sum_{r=1}^{n}\dfrac{\partial v_1}{\partial y_r}\cdot\dfrac{\partial y_r}{\partial x_2}\\ -------------------\\ -------------------\\ \dfrac{\partial v_1}{\partial x_n}=\dfrac{\partial v_1}{\partial y_1}\cdot\dfrac{\partial y_1}{\partial x_n}+\dfrac{\partial v_1}{\partial y_2}\cdot\dfrac{\partial y_2}{\partial x_n}+\ldots\ldots+\dfrac{\partial v_1}{\partial y_n}\cdot\dfrac{\partial y_n}{\partial x_n}=\sum_{r=1}^{n}\dfrac{\partial v_1}{\partial y_r}\cdot\dfrac{\partial y_r}{\partial x_n}\end{array}\right.$$

Now, we can find partial derivatives of $y_2, y_3 \ldots\ldots y_n$ w.r.t. $x_1\ x_2 \ldots\ldots x_n$.

Now $\dfrac{\partial(v_1, v_2 \ldots\ldots v_n)}{\partial(y_1, y_2, \ldots\ldots y_n)}\cdot\dfrac{\partial(y_1, y_2 \ldots\ldots y_n)}{\partial(x_1, x_2, \ldots\ldots x_n)}$

$$=\begin{vmatrix}\dfrac{\partial v_1}{\partial y_1} & \dfrac{\partial v_1}{\partial y_2} & \ldots\ldots & \dfrac{\partial v_1}{\partial y_n}\\ \dfrac{\partial v_2}{\partial y_1} & \dfrac{\partial v_2}{\partial y_2} & \ldots\ldots & \dfrac{\partial v_2}{\partial y_n}\\ \dfrac{\partial v_n}{\partial y_1} & \dfrac{\partial v_n}{\partial y_2} & \ldots\ldots & \dfrac{\partial v_n}{\partial y_n}\end{vmatrix}\cdot\begin{vmatrix}\dfrac{\partial y_1}{\partial x_1} & \dfrac{\partial y_1}{\partial x_2} & \ldots\ldots & \dfrac{\partial y_1}{\partial x_n}\\ \dfrac{\partial y_2}{\partial x_1} & \dfrac{\partial y_2}{\partial x_2} & \ldots\ldots & \dfrac{\partial y_2}{\partial x_n}\\ \dfrac{\partial y_n}{\partial x_1} & \dfrac{\partial y_n}{\partial x_2} & \ldots\ldots & \dfrac{\partial y_n}{\partial x_n}\end{vmatrix}$$

$$=\begin{vmatrix}\sum\dfrac{\partial v_1}{\partial y_r}\cdot\dfrac{\partial y_r}{\partial x_1} & \sum\dfrac{\partial v_1}{\partial y_r}\cdot\dfrac{\partial y_r}{\partial x_2} & \ldots\ldots\ldots & \sum\dfrac{\partial v_1}{\partial y_r}\cdot\dfrac{\partial y_r}{\partial x_n}\\ \sum\dfrac{\partial v_2}{\partial y_r}\cdot\dfrac{\partial y_r}{\partial x_1} & \sum\dfrac{\partial v_2}{\partial y_r}\cdot\dfrac{\partial y_r}{\partial x_2} & \ldots\ldots\ldots & \sum\dfrac{\partial v_2}{\partial y_r}\cdot\dfrac{\partial y_r}{\partial x_n}\\ ------ & ------ & ---- & ------\\ \sum\dfrac{\partial v_n}{\partial y_r}\cdot\dfrac{\partial y_r}{\partial x_1} & \sum\dfrac{\partial v_n}{\partial y_r}\cdot\dfrac{\partial y_r}{\partial x_2} & \ldots\ldots\ldots & \sum\dfrac{\partial v_n}{\partial y_r}\cdot\dfrac{\partial y_r}{\partial x_n}\end{vmatrix}$$

(By the rule of determinant's multiplication)

$$=\begin{vmatrix}\dfrac{\partial v_1}{\partial x_1} & \dfrac{\partial v_1}{\partial x_2} & \ldots\ldots & \dfrac{\partial v_1}{\partial x_n}\\ \dfrac{\partial v_2}{\partial x_1} & \dfrac{\partial v_2}{\partial x_2} & \ldots\ldots & \dfrac{\partial v_2}{\partial x_n}\\ \ldots\ldots & \ldots\ldots & \ldots\ldots & \ldots\ldots\\ \dfrac{\partial v_n}{\partial x_1} & \dfrac{\partial v_n}{\partial x_2} & \ldots\ldots & \dfrac{\partial v_n}{\partial x_n}\end{vmatrix} \quad \text{By using relation (1)}$$

$$= \frac{\partial (v_1, v_2 \ldots\ldots\ldots\ldots v_n)}{\partial (x_1, x_2 \ldots\ldots\ldots\ldots x_n)} \text{ which proves the required proposition.}$$

Particular Case: If $y_1, y_2, \ldots\ldots\ldots y_n$ are functions of $x_1, x_2 \ldots\ldots\ldots x_n$ and $x_1, x_2 \ldots\ldots x_n$ are functions of $y_1, y_2, \ldots\ldots\ldots y_n$, then

$$\frac{\partial (y_1, y_2, \ldots\ldots\ldots y_n)}{\partial (x_1, x_2, \ldots\ldots\ldots x_n)} \times \frac{\partial (x_1, x_2, \ldots\ldots\ldots x_n)}{\partial (y_1, y_2, \ldots\ldots\ldots y_n)} = 1$$

This result may be deduced from the previous result by putting $u_1 = x_1$, $u_2 = x_2$ etc.

Note: The above particular case may be regarded as a generalization of the formula $\frac{dy}{dx} \times \frac{dx}{dy} = 1$ of the ordinary differential coefficient.

2. Form of Implicit functions: If $u_1, u_2 \ldots\ldots\ldots u_n$ are the functions of $x_1, x_2 \ldots\ldots\ldots x_n$ given by the implicit relations

$$F_1 (u_1, u_2, \ldots\ldots\ldots\ldots u_n, \; x_1, x_2 \ldots\ldots\ldots\ldots x_n) = 0$$
$$F_2 (u_1, u_2, \ldots\ldots\ldots\ldots u_n, \; x_1, x_2 \ldots\ldots\ldots\ldots x_n) = 0$$

$$F_n (u_1, u_2, \ldots\ldots\ldots\ldots u_n, \; x_1, x_2 \ldots\ldots\ldots\ldots x_n) = 0$$

then
$$\frac{\partial (u_1, u_2 \ldots\ldots\ldots u_n)}{\partial (x_1, x_2 \ldots\ldots\ldots x_n)} = (-1)^n \frac{\dfrac{\partial (F_1, F_2 \ldots\ldots\ldots F_n)}{\partial (x_1, x_2 \ldots\ldots\ldots x_n)}}{\dfrac{\partial (F_1, F_2 \ldots\ldots\ldots F_n)}{\partial (u_1, u_2 \ldots\ldots\ldots u_n)}}$$

Assuming the existence of the concerned partial derivatives, we have on differentiating partially the given relations with respect to $x_1, x_2 \ldots\ldots\ldots x_n$.

$$\frac{\partial F_1}{\partial x_1} + \frac{\partial F_1}{\partial u_1} \cdot \frac{\partial u_1}{\partial x_1} + \frac{\partial F_1}{\partial u_2} \cdot \frac{\partial u_2}{\partial x_1} + \ldots\ldots\ldots\ldots + \frac{\partial F_1}{\partial u_n} \cdot \frac{\partial u_n}{\partial x_1} = 0$$

$$\frac{\partial F_1}{\partial x_2} + \frac{\partial F_1}{\partial u_1} \cdot \frac{\partial u_1}{\partial x_2} + \frac{\partial F_1}{\partial u_2} \cdot \frac{\partial u_2}{\partial x_2} + \ldots\ldots\ldots\ldots + \frac{\partial F_1}{\partial u_n} \cdot \frac{\partial u_n}{\partial x_2} = 0$$

$$\frac{\partial F_1}{\partial x_n} + \frac{\partial F_1}{\partial u_1} \cdot \frac{\partial u_1}{\partial x_n} + \frac{\partial F_1}{\partial u_2} \cdot \frac{\partial u_2}{\partial x_n} + \ldots\ldots\ldots\ldots + \frac{\partial F_1}{\partial u_n} \cdot \frac{\partial u_n}{\partial x_n} = 0$$

Similarly,
$$\frac{\partial F_2}{\partial x_1} + \frac{\partial F_2}{\partial u_1} \cdot \frac{\partial u_1}{\partial x_1} + \frac{\partial F_2}{\partial u_2} \cdot \frac{\partial u_2}{\partial x_1} + \ldots\ldots\ldots\ldots + \frac{\partial F_2}{\partial u_n} \cdot \frac{\partial u_n}{\partial x_1} = 0 \text{ etc}$$

or, $$\sum \frac{\partial F_1}{\partial u_r} \cdot \frac{\partial u_r}{\partial x_1} = -\frac{\partial F_1}{\partial x_1}; \quad \sum \frac{\partial F_1}{\partial u_r} \cdot \frac{\partial u_r}{\partial x_1} = -\frac{\partial F_1}{\partial x_2}; \ldots\ldots \quad \sum \frac{\partial F_1}{\partial u_r} \cdot \frac{\partial u_r}{\partial x_n} = -\frac{\partial F_1}{\partial x_n} \text{ etc}$$

Now, $$\frac{\partial(F_1, F_2, \ldots\ldots F_n)}{\partial(u_1, u_2, \ldots\ldots u_n)} \times \frac{\partial(u_1, u_2, \ldots\ldots u_n)}{\partial(x_1, x_2, \ldots\ldots x_n)}$$

$$\begin{vmatrix} \frac{\partial F_1}{\partial u_1} & \frac{\partial F_1}{\partial u_2} & \ldots\ldots & \frac{\partial F_1}{\partial u_n} \\ \frac{\partial F_2}{\partial u_1} & \frac{\partial F_2}{\partial u_2} & \ldots\ldots & \frac{\partial F_2}{\partial u_n} \\ \ldots\ldots & \ldots\ldots & \ldots\ldots & \ldots\ldots \\ \frac{\partial F_n}{\partial u_1} & \frac{\partial F_n}{\partial u_2} & \ldots\ldots & \frac{\partial F_n}{\partial u_n} \end{vmatrix} \times \begin{vmatrix} \frac{\partial u_1}{\partial x_1} & \frac{\partial u_1}{\partial x_2} & \ldots\ldots & \frac{\partial u_1}{\partial x_n} \\ \frac{\partial u_2}{\partial x_1} & \frac{\partial u_2}{\partial x_2} & \ldots\ldots & \frac{\partial u_2}{\partial x_n} \\ \ldots\ldots & \ldots\ldots & \ldots\ldots & \ldots\ldots \\ \frac{\partial u_n}{\partial x_1} & \frac{\partial u_n}{\partial x_2} & \ldots\ldots & \frac{\partial u_n}{\partial x_n} \end{vmatrix}$$

$$\begin{vmatrix} \sum \frac{\partial F_1}{\partial u_r} \cdot \frac{\partial u_r}{\partial x_1} & \sum \frac{\partial F_1}{\partial u_r} \cdot \frac{\partial u_r}{\partial x_2} & \ldots\ldots & \sum \frac{\partial F_1}{\partial u_r} \cdot \frac{\partial u_r}{\partial x_n} \\ \sum \frac{\partial F_2}{\partial u_r} \cdot \frac{\partial u_r}{\partial x_1} & \sum \frac{\partial F_2}{\partial u_r} \cdot \frac{\partial u_r}{\partial x_2} & \ldots\ldots & \sum \frac{\partial F_2}{\partial u_r} \cdot \frac{\partial u_r}{\partial x_n} \\ \ldots\ldots & \ldots\ldots & \ldots\ldots & \ldots\ldots \\ \sum \frac{\partial F_n}{\partial u_r} \cdot \frac{\partial u_r}{\partial x_1} & \sum \frac{\partial F_n}{\partial u_r} \cdot \frac{\partial u_r}{\partial x_2} & \ldots\ldots & \sum \frac{\partial F_n}{\partial u_r} \cdot \frac{\partial u_r}{\partial x_n} \end{vmatrix}$$

(By the multiplication of two determinants)

$$= \begin{vmatrix} -\frac{\partial F_1}{\partial x_1} & \frac{-\partial F_1}{\partial x_2} & \ldots\ldots & \frac{-\partial F_1}{\partial x_n} \\ -\frac{\partial F_2}{\partial x_1} & \frac{-\partial F_2}{\partial x_2} & \ldots\ldots & \frac{-\partial F_2}{\partial x_n} \\ \ldots\ldots & \ldots\ldots & \ldots\ldots & \ldots\ldots \\ \ldots\ldots & \ldots\ldots & \ldots\ldots & \ldots\ldots \\ -\frac{\partial F_n}{\partial x_1} & \frac{-\partial F_n}{\partial x_2} & \ldots\ldots & \frac{-\partial F_n}{\partial x_n} \end{vmatrix} \quad \text{(using relation (1))}$$

$$= (-1)^n \begin{vmatrix} \frac{\partial F_1}{\partial x_1} & \frac{\partial F_1}{\partial x_2} & \cdots\cdots & \frac{\partial F_1}{\partial x_n} \\ \frac{\partial F_2}{\partial x_1} & \frac{\partial F_2}{\partial x_2} & \cdots\cdots & \frac{\partial F_2}{\partial x_n} \\ \cdots\cdots & \cdots\cdots & \cdots\cdots & \cdots\cdots \\ \cdots\cdots & \cdots\cdots & \cdots\cdots & \cdots\cdots \\ \frac{\partial F_n}{\partial x_1} & \frac{\partial F_n}{\partial x_2} & \cdots\cdots & \frac{\partial F_n}{\partial x_n} \end{vmatrix}.$$

$$= (-1)^n \frac{\partial (F_1, F_2, \ldots\ldots F_n)}{\partial (x_1, x_2, \ldots\ldots x_n)}$$

Hence
$$\frac{\partial (u_1, u_2, \ldots\ldots u_n)}{\partial (x_1, x_2, \ldots\ldots x_n)} = (-1)^n \frac{\dfrac{\partial (F_1, F_2, \ldots\ldots F_n)}{\partial (x_1, x_2, \ldots\ldots x_n)}}{\dfrac{\partial (F_1, F_2, \ldots\ldots F_n)}{\partial (u_1, u_2, \ldots\ldots u_n)}}$$

Note: This result may be regarded as a generalization of the result, $\frac{dy}{dx} = -\frac{\partial f / \partial x}{\partial f / \partial y}$ when we are given an implicit function $F(x, y) = 0$

Particular Case: If the functions $u_1, u_2, \ldots\ldots u_n$, be given implicitly in terms of the variables $x_1, x_2, \ldots\ldots x_n$ by the relations:

$$\phi_1 (x_1, x_2, \ldots\ldots x_n, u_1) = 0$$
$$\phi_1 (x_1, x_2, \ldots\ldots x_{n-1}, u_1, u_2) = 0$$
$$\cdots\cdots\cdots\cdots\cdots\cdots$$
$$\cdots\cdots\cdots\cdots\cdots\cdots$$
$$\phi_n (x_1, u_1, u_2, \ldots\ldots u_n) = 0$$

then we have by one of the above result,

$$\frac{\partial (\phi_1, \phi_2, \phi_3, \ldots\ldots, \phi_n)}{\partial (u_1, u_2, u_3, \ldots\ldots, u_n)} = \frac{\partial \phi_1}{\partial u_1} \cdot \frac{\partial \phi_2}{\partial u_2} \cdots\cdots \frac{\partial \phi_n}{\partial u_n}$$

and
$$\frac{\partial (\phi_1, \phi_2, \phi_3, \ldots\ldots, \phi_n)}{\partial (x_1, x_2, x_3, \ldots\ldots, x_n)} = \frac{\partial \phi_1}{\partial x_1} \cdot \frac{\partial \phi_2}{\partial x_2} \cdots\cdots \frac{\partial \phi_n}{\partial x_n}$$

Hence
$$\frac{\partial (u_1, u_2, \ldots\ldots u_n)}{\partial (x_1, x_2, \ldots\ldots, x_n)} = (-1)^n \frac{\dfrac{\partial \phi_1}{\partial x_1} \cdot \dfrac{\partial \phi_2}{\partial x_2} \cdots\cdots \dfrac{\partial \phi_n}{\partial x_n}}{\dfrac{\partial \phi_1}{\partial u_1} \cdot \dfrac{\partial \phi_2}{\partial u_2} \cdots\cdots \dfrac{\partial \phi_n}{\partial u_n}}$$

Illustrative Examples

Example 1 If $x + y + z = u$, $y + z = uv$, $z = uvw$, show that $\frac{\partial(x, y, z)}{\partial(u, v, w)} = u^2v$

Solution: Let $F_1 = x + y + z - u = 0$

$F_2 = y + z - uv = 0$

and $F_3 = z - uvw = 0$

we have
$$\frac{\partial(x, y, z)}{\partial(u, v, w)} = (-1)^3 \frac{\dfrac{\partial(F_1, F_2, F_3)}{\partial(u, v, w)}}{\dfrac{\partial(F_1, F_2, F_3)}{\partial(x, y, z)}} \quad \ldots(1)$$

Now,
$$\frac{\partial(F_1, F_2, F_3)}{\partial(u, v, w)} = \begin{vmatrix} \frac{\partial F_1}{\partial u} & \frac{\partial F_1}{\partial v} & \frac{\partial F_1}{\partial w} \\ \frac{\partial F_2}{\partial u} & \frac{\partial F_2}{\partial v} & \frac{\partial F_2}{\partial w} \\ \frac{\partial F_3}{\partial u} & \frac{\partial F_3}{\partial v} & \frac{\partial F_3}{\partial w} \end{vmatrix}$$

$$= \begin{vmatrix} -1 & 0 & 0 \\ -v & -u & 0 \\ -vw & -wu & -uv \end{vmatrix} = -u^2v$$

and
$$\frac{\partial(F_1, F_2, F_3)}{\partial(x, y, z)} = \begin{vmatrix} \frac{\partial F_1}{\partial x} & \frac{\partial F_1}{\partial y} & \frac{\partial F_1}{\partial z} \\ \frac{\partial F_2}{\partial x} & \frac{\partial F_2}{\partial y} & \frac{\partial F_2}{\partial z} \\ \frac{\partial F_3}{\partial x} & \frac{\partial F_3}{\partial y} & \frac{\partial F_3}{\partial z} \end{vmatrix}$$

$$= \begin{vmatrix} 1 & 1 & 1 \\ 0 & 1 & 1 \\ 0 & 0 & 1 \end{vmatrix} = 1$$

substituting these values in (1), we get

$$\frac{\partial (x, y, z)}{\partial (u, v, w)} = (-1)^3 \left(-u^2 v\right) = u^2 v.$$

Example 2 If J be the Jacobian of the system u, v with respect to x, y and J' be reciprocal that of J, then prove that JJ' = 1

Solution: Let u = f (x, y) and v = g(x, y), differentiating these w.r.t. u and v, we have

$$1 = \frac{\partial f}{\partial x}.\frac{\partial x}{\partial u} + \frac{\partial f}{\partial y}.\frac{\partial y}{\partial u} = \frac{\partial u}{\partial x}.\frac{\partial x}{\partial u} + \frac{\partial u}{\partial y}.\frac{\partial y}{\partial u};$$

$$0 = \frac{\partial f}{\partial x}.\frac{\partial x}{\partial v} + \frac{\partial f}{\partial y}.\frac{\partial y}{\partial v} = \frac{\partial u}{\partial x}.\frac{\partial x}{\partial v} + \frac{\partial u}{\partial y}.\frac{\partial y}{\partial v}$$

and

$$0 = \frac{\partial g}{\partial x}.\frac{\partial x}{\partial u} + \frac{\partial g}{\partial y}.\frac{\partial y}{\partial u} = \frac{\partial v}{\partial x}.\frac{\partial x}{\partial u} + \frac{\partial v}{\partial y}.\frac{\partial y}{\partial u}$$

$$1 = \frac{\partial g}{\partial x}.\frac{\partial x}{\partial v} + \frac{\partial g}{\partial y}.\frac{\partial y}{\partial v} = \frac{\partial v}{\partial x}.\frac{\partial x}{\partial v} + \frac{\partial v}{\partial y}.\frac{\partial y}{\partial v}$$

Now,

$$J = \frac{\partial(u, u)}{\partial(x, y)} = \begin{vmatrix} \frac{\partial u}{\partial x} & \frac{\partial u}{\partial y} \\ \frac{\partial v}{\partial x} & \frac{\partial v}{\partial y} \end{vmatrix}$$

$$J' = \frac{\partial(x, y)}{\partial(u, v)} = \begin{vmatrix} \frac{\partial x}{\partial u} & \frac{\partial x}{\partial v} \\ \frac{\partial y}{\partial u} & \frac{\partial y}{\partial v} \end{vmatrix}$$

$$JJ' = \begin{vmatrix} \frac{\partial u}{\partial x}.\frac{\partial x}{\partial u} + \frac{\partial u}{\partial y}.\frac{\partial y}{\partial u} & \frac{\partial u}{\partial x}.\frac{\partial x}{\partial v} + \frac{\partial u}{\partial y}.\frac{\partial y}{\partial v} \\ \frac{\partial v}{\partial x}.\frac{\partial x}{\partial u} + \frac{\partial v}{\partial y}.\frac{\partial y}{\partial u} & \frac{\partial v}{\partial x}.\frac{\partial x}{\partial v} + \frac{\partial v}{\partial y}.\frac{\partial y}{\partial v} \end{vmatrix}$$

$$= \begin{vmatrix} 1 & 0 \\ 0 & 1 \end{vmatrix} = 1$$

Example 3 The roots of the equation in λ, $(\lambda - x)^3 + (\lambda - y)^3 + (\lambda - z)^3 = 0$ are u, v, w.

prove that $\dfrac{\partial(u, v, w)}{\partial(x, y, z)} = -2\dfrac{(y-z)(z-x)(x-y)}{(v-w)(w-u)(u-v)}$.

Solution: The equation in λ may be written as,

$$3\lambda^3 - 3\lambda^2 (x + y + z) + 3\lambda (x^2 + y^2 + z^2) - (x^3 + y^3 + z^3) = 0$$

since u, v, w are the roots of this equation,

we have

$$u + v + w = x + y + z \qquad \text{(sum of the roots)}$$

$$uv + vw + wu = x^2 + y^2 + z^2 \qquad \text{(sum of the products of roots taken two at a time)}$$

and $$u\,v\,w = \frac{1}{3}(x^3 + y^3 + z^3) \qquad \text{(product of roots)}$$

Then, we have

$$\frac{\partial(u, v, w)}{\partial(x, y, z)} = (-1)^3 \frac{\dfrac{\partial(F_1, F_2, F_3)}{\partial(x, y, z)}}{\dfrac{\partial(F_1, F_2, F_3)}{\partial(u, v, w)}} \qquad ...(1)$$

Now, $$\frac{\partial(F_1, F_2, F_3)}{\partial(x, y, z)} = \begin{vmatrix} -1 & -1 & -1 \\ -2x & -2y & -2z \\ -x^2 & -y^2 & -z^2 \end{vmatrix} = -2 \begin{vmatrix} 1 & 1 & 1 \\ x & y & z \\ x^2 & y^2 & z^2 \end{vmatrix}$$

$$= -2 \begin{vmatrix} 1 & 0 & 0 \\ x & y-z & z-x \\ x^2 & y^2-z^2 & z^2-x^2 \end{vmatrix} \quad (by\ C_2 \to C_2 - C_3,\ C_3 \to C_3 - C_1)$$

$$= -2 (y - z) (z - x) (z + x - y - z)$$

$$= -2 (y - z) (z - x) (x - y) \qquad ...(2)$$

and $$\frac{\partial(F_1, F_2, F_3)}{\partial(u, v, w)} = \begin{vmatrix} 1 & 1 & 1 \\ v+w & w+u & u+v \\ vw & wu & uv \end{vmatrix}$$

$$= \begin{vmatrix} 1 & 0 & 0 \\ v+w & u-v & v-w \\ v\,w & w(u-v) & u(v-w) \end{vmatrix} \quad (by\ C_2 \to C_2 - C_1,\ C_3 \to C_3 - C_1 \qquad ...(3)$$

$$= (u - v) (v - w) (u - w)$$

Hence, from (1), (2) and (3), we get

$$\frac{\partial(u, v, w)}{\partial(x, y, z)} = (-1)^3 \frac{(-2)(y-z)(z-x)(x-y)}{(u-v)(v-w)(u-w)}$$

$$= -2\frac{(y-z)(z-x)(x-y)}{(v-w)(w-u)(u-v)}$$

Example 4 If $x_1 + x_2 + \ldots\ldots\ldots\ldots x_n = y_1$

$x_2 + x_3 + \ldots\ldots\ldots\ldots x_n = y_1 y_2$

....

....

and $x_n = y_1 y_2 \ldots\ldots\ldots\ldots y_n$; then

show that $\dfrac{\partial(x_1, x_2, \ldots\ldots\ldots x_n)}{\partial(y_1, y_2, \ldots\ldots\ldots y_n)} = y_1^{n-1} . y_2^{n-2} \ldots\ldots\ldots\ldots y_{n-2}^2\, y_{n-1}$.

Solution: Here, let

$F_1 \equiv x_1 + x_2 + \ldots\ldots\ldots\ldots + x_n - y_1 = 0$

$F_2 \equiv x_2 + x_3 + \ldots\ldots\ldots\ldots + x_n - y_1 y_2 = 0$

....

....

$F_n \equiv x_n - y_1 y_2 \ldots\ldots\ldots\ldots y_n = 0$

$$\therefore \quad \frac{\partial(F_1, F_2, F_3 \ldots\ldots\ldots\ldots, F_n)}{\partial(x_1, x_2, x_3, \ldots\ldots\ldots, x_n)} = \begin{vmatrix} 1 & 1 & \ldots\ldots\ldots\ldots & 1 \\ 0 & 1 & \ldots\ldots\ldots\ldots & 1 \\ \ldots & \ldots & \ldots\ldots\ldots\ldots & \ldots \\ 0 & 0 & \ldots\ldots\ldots\ldots & 1 \end{vmatrix} = 1$$

and

$$\frac{\partial(F_1, F_2, \ldots\ldots\ldots\ldots, F_n)}{\partial(y_1, y_2, \ldots\ldots\ldots, y_n)} = \begin{vmatrix} -1 & 0 & \ldots\ldots\ldots\ldots & 0 \\ -y_2 & -y_1 & \ldots\ldots\ldots\ldots & \\ \ldots & \ldots & \ldots\ldots\ldots\ldots & \\ \ldots & \ldots & \ldots\ldots\ldots\ldots & \\ -y_2 y_3 \ldots\ldots y_n & -y_1 y_3 \ldots\ldots y_n & \ldots & -y_1 y_2 \ldots\ldots y_n \end{vmatrix}$$

$$= (-1)^n\, y_1^{n-1}\, y_2^{n-2} \ldots\ldots\ldots\, y_{n-2}^2\, y_{n-1}$$

Hence,
$$\frac{\partial(x_1, x_2 \ldots\ldots\ldots\ldots x_n)}{\partial(y_1, y_2, \ldots\ldots\ldots y_n)} = (-1)^n \frac{\dfrac{\partial(F_1, F_2 \ldots\ldots\ldots F_n)}{\partial(y_1, y_2 \ldots\ldots\ldots y_n)}}{\dfrac{\partial(F_1, F_2 \ldots\ldots\ldots F_n)}{\partial(x_1, x_2 \ldots\ldots\ldots x_n)}}$$

$$= (-1)^n\, y_1^{n-1}\, y_2^{n-2} \ldots\ldots\ldots\ldots y^2_{n-2}\, y_{n-1}$$

$$= y_1^{n-1}\, y_2^{n-2} \ldots\ldots\ldots y^2_{n-2}\, y_{n-1}$$

Example 5 If f(0) = 0, $f'(x) = \dfrac{1}{1+x^2}, f'(y) = \dfrac{1}{1+y^2}$, without using the method of integration, show that $f(x) + f(y) = f\left(\dfrac{x+y}{1-xy}\right)$

Solution: Let $u = f(x) + f(y)$ ...(1)

and $v = \dfrac{x+y}{1-xy}$ ...(2)

we know that

$$J(u,v) = \frac{\partial(u,v)}{\partial(x,y)} = \begin{vmatrix} \dfrac{\partial u}{\partial x} & \dfrac{\partial u}{\partial y} \\ \dfrac{\partial v}{\partial x} & \dfrac{\partial v}{\partial y} \end{vmatrix} \quad ...(3)$$

from (1) and (3)

$$\frac{\partial u}{\partial x} = f'(x) = \frac{1}{1+x^2}, \frac{\partial u}{\partial y} = f'(y) = \frac{1}{1+y^2}, \frac{\partial v}{\partial x} = \frac{1+y^2}{(1-xy)^2}$$

and $$\frac{\partial v}{\partial y} = \frac{1+x^2}{(1-xy)^2}$$

substituting these values in (3), we get

$$\frac{\partial(u,v)}{\partial(x,y)} = \begin{vmatrix} \dfrac{1}{1+x^2} & \dfrac{1}{1+y^2} \\ \dfrac{1+y^2}{(1-xy)^2} & \dfrac{1+x^2}{(1-xy)^2} \end{vmatrix}$$

$$= \frac{1}{(1-xy)^2} \begin{vmatrix} \dfrac{1}{1+x^2} & \dfrac{1}{1+y^2} \\ 1+y^2 & 1+x^2 \end{vmatrix}$$

$$= \frac{1}{(1-xy)^2}\left[\frac{1+x^2}{1+x^2} - \frac{1+y^2}{1+y^2}\right] = 0$$

$$\Rightarrow \quad \frac{\partial(u,v)}{\partial(x,y)} = 0$$

$\Rightarrow$ There is a relationship between u and v,

suppose, $u = \phi(v)$, then

$$f(x)+f(y)=\phi\left[\frac{x+y}{1-xy}\right] \qquad \ldots(4)$$

substituting $y = 0$, we have,

$$f(x) + f(0) = \phi(x) \qquad (\because f(0)=0)$$

Hence $\phi = f$, from (4), we get

$$f(x)+f(y)= f\left[\frac{(x+y)}{(1-xy)}\right].$$

Example 6 If $y_1 = \cos x_1$, $y_2 = \sin x_1 \cos x_2$ and $y_3 = \sin x_1 \sin x_2 \cos x_3$, then prove that

$$\frac{\partial(y_1, y_2, y_3)}{\partial(x_1, x_2, x_3)} = -\sin^3 x_1 \sin^2 x_2 \sin x_3$$

Solution: we have,

$y_1 = \cos x_1$, $y_2 = \sin x_1 \cos x_2$, $y_3 = \sin x_1 \sin x_2 \cos x_3$

we know that,

$$\frac{\partial(y_1, y_2, y_3)}{\partial(x_1, x_2, x_3)} = \begin{vmatrix} \frac{\partial y_1}{\partial x_1} & \frac{\partial y_1}{\partial x_2} & \frac{\partial y_1}{\partial x_3} \\ \frac{\partial y_2}{\partial x_1} & \frac{\partial y_2}{\partial x_2} & \frac{\partial y_2}{\partial x_3} \\ \frac{\partial y_3}{\partial x_1} & \frac{\partial y_3}{\partial x_2} & \frac{\partial y_3}{\partial x_3} \end{vmatrix}$$

$$= \begin{vmatrix} -\sin x_1 & 0 & 0 \\ \cos x_1 \cos x_2 & -\sin x_1 \sin x_2 & 0 \\ \cos x_1 \sin x_2 \cos x_3 & \sin x_1 \cos x_2 \cos x_3 & -\sin x_1 \sin x_2 \sin x_3 \end{vmatrix}$$

expanding about R_3 we get

$$= -\sin x_1 \sin x_2 \sin x_3 \left(\sin^2 x_1 \sin x_2\right)$$

$$= -\sin^3 x_1 \sin^2 x_2 \sin x_3.$$

Example 7. If $u^3 + v^3 + w^3 = x + y + z$, $u + v + w = x^2 + y^2 + z^2$

and $u^2 + v^2 + w^2 = x^3 + y^3 + z^3$, then prove that,

$$\frac{\partial(u, v, w)}{\partial(x, y, z)} = \frac{(x-y)(y-z)(z-x)}{(u-v)(w-u)(v-w)}.$$

Solution: Let $f_1 \equiv u^3 + v^3 + w^3 - x - y - z = 0$

$f_2 \equiv u^2 + v^2 + w^2 - x^3 - y^3 - z^3 = 0$

and $f_3 \equiv u + v + w - x^2 - y^2 - z^2 = 0$

But we know that,

$$\frac{\partial(u,v,w)}{\partial(x,y,z)} = (-1)^3 \frac{\dfrac{\partial(f_1,f_2,f_3)}{\partial(x,y,z)}}{\dfrac{\partial(f_1,f_2,f_3)}{\partial(u,v,w)}} \qquad \text{...(1)}$$

Now, we find

$$\frac{\partial(f_1,f_2,f_3)}{\partial(x,y,z)} = \begin{vmatrix} \dfrac{\partial f_1}{\partial x} & \dfrac{\partial f_1}{\partial y} & \dfrac{\partial f_1}{\partial z} \\ \dfrac{\partial f_2}{\partial x} & \dfrac{\partial f_2}{\partial y} & \dfrac{\partial f_2}{\partial z} \\ \dfrac{\partial f_3}{\partial x} & \dfrac{\partial f_3}{\partial y} & \dfrac{\partial f_3}{\partial z} \end{vmatrix} \qquad \text{...(2)}$$

here,

$$\left.\begin{aligned} &\frac{\partial f_1}{\partial x} = -1, \frac{\partial f_1}{\partial y} = -1, \frac{\partial f_1}{\partial z} = -1 \\ &\frac{\partial f_2}{\partial x} = -3x^2, \frac{\partial f_2}{\partial y} = -3y^2, \frac{\partial f_2}{\partial z} = -3z^2 \\ &\frac{\partial f_3}{\partial x} = -2x, \frac{\partial f_3}{\partial y} = -2y, \frac{\partial f_3}{\partial z} = -2z \end{aligned}\right\} \qquad \text{...(3)}$$

from (2) and (3), we have,

$$\frac{\partial(f_1,f_2,f_3)}{\partial(x,y,z)} = \begin{vmatrix} -1 & -1 & -1 \\ -3x^2 & -3y^2 & -3z^2 \\ -2x & -2y & -2z \end{vmatrix}$$

Operating $c_1 \to c_1 - c_3$ and $c_2 \to c_2 - c_3$, we get

$$= \begin{vmatrix} 0 & 0 & -1 \\ 3(z^2 - x^2) & 3(z^2 - y^2) & 3(z)^2 \\ 2(z - x) & 2(z - y) & -2z \end{vmatrix}$$

$$\Rightarrow \quad 6\begin{vmatrix} 0 & 0 & -1 \\ (z-x)(z+x) & (z-y)(z+y) & z^2 \\ (z-x) & (z-y) & -z \end{vmatrix}$$

$$\Rightarrow \quad 6(z-x)(z-y)\begin{vmatrix} 0 & 0 & -1 \\ (z+x) & (z+y) & z^2 \\ 1 & 1 & -z \end{vmatrix}$$

$$\Rightarrow \quad 6(z-x)(x-y)(-1)\begin{vmatrix} (z+x) & (z+y) \\ 1 & 1 \end{vmatrix}$$

$\Rightarrow \quad -6(z-x)(z-y)(x-y)$

$\Rightarrow \quad 6(x-y)(y-z)(z-x).$

Again,
$$\frac{\partial(f_1, f_2, f_3)}{\partial(u, v, w)} = \begin{vmatrix} \frac{\partial f_1}{\partial u} & \frac{\partial f_1}{\partial v} & \frac{\partial f_1}{\partial w} \\ \frac{\partial f_2}{\partial u} & \frac{\partial f_2}{\partial v} & \frac{\partial f_2}{\partial w} \\ \frac{\partial f_3}{\partial u} & \frac{\partial f_3}{\partial v} & \frac{\partial f_3}{\partial w} \end{vmatrix} \qquad \text{...(4)}$$

Here,
$$\left.\begin{array}{l} \frac{\partial f_1}{\partial u} = 3u^2, \frac{\partial f_1}{\partial v} = 3v^2, \frac{\partial f_1}{\partial w} = 3w^2 \\ \frac{\partial f_2}{\partial u} = 2u, \frac{\partial f_2}{\partial v} = 2v, \frac{\partial f_2}{\partial w} = 2w \\ \frac{\partial f_3}{\partial u} = 1, \frac{\partial f_3}{\partial v} = 1, \frac{\partial f_3}{\partial w} = 1 \end{array}\right\} \qquad \text{...(5)}$$

from (4) and (5), we get

$$\frac{\partial(f_1, f_2, f_3)}{\partial(u, v, w)} = \begin{vmatrix} 3u^2 & 3v^2 & 3w^2 \\ 2u & 2v & 2w \\ 1 & 1 & 1 \end{vmatrix}$$

operating $c_1 \to c_1 - c_3$ and $c_2 \to c_2 - c_3$, we get

$$= 6\begin{vmatrix} u^2 - w^2 & v^2 - w^2 & 3w^2 \\ u - w & v - w & 2w \\ 0 & 0 & 1 \end{vmatrix}$$

$$= 6(u-w)(v-w)\begin{vmatrix} u+w & v+w \\ 1 & 1 \end{vmatrix}$$

$$= 6\,(u-w)\,(v-w)\,(u-v)$$

from (1) $$\frac{\partial(u,v,w)}{\partial(x,y,z)} = \frac{(x-y)(y-z)(z-x)}{(u-v)(v-w)(w-u)}$$

Exercise 6.5

1. If $u^3 + v^3 = x + y, \quad u^2 + v^2 = x^3 + y^3$, prove that

$$\frac{\partial(u,v)}{\partial(x,y)} = \frac{(y-x)(y+x)}{2(uv)(u-v)}$$

2. If $x_1 + x_2 + x_3 + x_4 = y_1, \quad x_2 + x_3 + x_4 = y_1 y_2,$
$x_3 + x_4 = y_1 y_2 y_3,\ x_4 = y_1 y_2 y_3 y_4$, then prove that

$$\frac{\partial(x_1, x_2, x_3, x_4)}{\partial(y_1, y_2, y_3, y_4)} = y_1^3\, y_2^2\, y_3.$$

3. If λ, μ, γ are the roots of the equation $\frac{x}{a+k} + \frac{y}{b+k} + \frac{z}{c+k} = 1$, then prove that,

$$\frac{\partial(x,y,z)}{\partial(\lambda,\mu,\gamma)} = -\frac{(\mu-\gamma)(\gamma-\lambda)(\lambda-\mu)}{(b-c)(c-a)(a-b)}.$$

4. If $u = 2xy$, $v = x^2 - y^2$ and $x = r\cos\theta$, $y = r\sin\theta$, then compute $\frac{\partial(u,v)}{\partial(r,\theta)}$.

5. If $y_1 = 1 - x_1,\ y_2 = x_1(1 - x_2),\ y_3 = x_1 x_2 (1 - x_3); \ldots\ldots\ldots$
$y_r = x_1 x_2 \ldots\ldots\ x_{n-1}(1 - x_n)$; prove that,

$$\frac{\partial(y_1, y_2, \ldots\ldots y_n)}{\partial(x_1, x_2, \ldots\ldots\ x_n)} = (-1)^n\, x_1^{n-2}\, x_3^{n-3} \ldots\ldots x_{n-1}.$$

6. If $u^3 + v + w = x + y^2 + z^2,\ u + v^3 + w = x^2 + y + z^2,$

$u + v + w^3 = x^2 + y^2 + z$, find $\frac{\partial(u,v,w)}{\partial(x,y,z)}$

7. If $u^2x^3 + v^2y^3 + 2a^2(u^3 + v^3) = c_1^5,$
$u^3x^2 + v^3y^2 + 2a^2(x^3 + y^3) = c_2^5,$

find $\frac{\partial(u,v)}{\partial(x,y)}$

8. Prove that $\dfrac{\partial(u,v)}{\partial(x,y)} \cdot \dfrac{\partial(x,y)}{\partial(u,v)} = 1$

9. If $u^3 = xyz$, $\dfrac{1}{v} = \dfrac{1}{x} + \dfrac{1}{y} + \dfrac{1}{z}$ and $w^2 = x^2 + y^2 + z^2$, then show that

$$\frac{\partial(u,v,w)}{\partial(x,y,z)} = \frac{v(x-y)(y-z)(z-x)(x+y+z)}{3u^2w(xy+yz+zx)}$$

10. If $u = \dfrac{x+y}{1-xy}$, $v = \tan^{-1}x + \tan^{-1}y$, Prove that

$$\frac{\partial(u,v)}{\partial(x,y)} = 0,$$

Answers

4. $-4r^3$

6. $\dfrac{1-4(yz+zx+xy)+16\,xyz}{2-3(u^2+v^2+w^2)-27\,u^2v^2w^2}$

7. $\dfrac{xy\,u^2v^2(vx-uy)+3a^2x^2y^2(u^2-v^2)}{uv\,x^2y^2(vx-uy)-3a^2u^2v^2(x^2-y^2)}$.

6.14 DEPENDENCE AND INDEPENDENCE OF FUNCTIONS

The main application of Jacobians is to determine the dependence or independence of functions. If there exists a relationship between the functions, their Jacobian vanishes identically and conversaly, if the Jacobian vanishes, the functions are not all independent. Briefly we can state that the vanishing of Jacobian is the necessary and sufficient condition for the existence of relation between the functions, we now proceed to establish this proposition.

The condition is necessary: Let $u_1, u_2 \ldots\ldots u_n$ be the functions of n independent variables $x_1, x_2, \ldots\ldots x_n$ and let there exists a relation $F(u_1, u_2, \ldots\ldots u_n) = 0$, between them. Then, in general, $\dfrac{\partial(u_1, u_2, \ldots\ldots u_n)}{\partial(x_1, x_2, \ldots\ldots x_n)} = 0$

Differentiating the given relation partially w.r.t. $x_1, x_2 \ldots\ldots x_n$, we get

$$\frac{\partial F}{\partial u_1}\cdot\frac{\partial u_1}{\partial x_1}+\frac{\partial F}{\partial u_2}\cdot\frac{\partial u_2}{\partial x_1}+\text{.........}+\frac{\partial F}{\partial u_n}\cdot\frac{\partial u_n}{\partial x_1}=0$$

$$\frac{\partial F}{\partial u_1}\cdot\frac{\partial u_1}{\partial x_2}+\frac{\partial F}{\partial u_2}\cdot\frac{\partial u_2}{\partial x_2}+\text{.........}+\frac{\partial F}{\partial u_n}\cdot\frac{\partial u_n}{\partial x_2}=0$$

....

....

$$\frac{\partial F}{\partial u_1}\cdot\frac{\partial u_1}{\partial x_n}+\frac{\partial F}{\partial u_2}\cdot\frac{\partial u_2}{\partial x_n}+\text{.........}+\frac{\partial F}{\partial u_n}\cdot\frac{\partial u_n}{\partial x_n}=0$$

eliminating $\frac{\partial F}{\partial u_1}, \frac{\partial F}{\partial u_2}, \ldots\ldots, \frac{\partial F}{\partial u_n}$, from these equations, we obtain

$$\begin{vmatrix} \frac{\partial u_1}{\partial x_1} & \frac{\partial u_2}{\partial x_1} & \ldots\ldots\ldots & \frac{\partial u_n}{\partial x_1} \\ \frac{\partial u_1}{\partial x_2} & \frac{\partial u_2}{\partial x_2} & \ldots\ldots\ldots & \frac{\partial u_n}{\partial x_2} \\ \ldots\ldots & \ldots\ldots & \ldots\ldots & \ldots\ldots \\ \ldots\ldots & \ldots\ldots & \ldots\ldots & \ldots\ldots \\ \frac{\partial u_1}{\partial x_n} & \frac{\partial u_2}{\partial x_n} & \ldots\ldots\ldots & \frac{\partial u_n}{\partial x_n} \end{vmatrix} = 0$$

i.e., $$\frac{\partial(u_1, u_2, \ldots\ldots\ldots u_n)}{\partial(x_1, x_2, \ldots\ldots\ldots x_n)}=0$$

The condition is sufficient: Let u_1, u_2 .... u_n be the functions of the n variables x_1, x_2, ... x_n and let $\frac{\partial(u_1, u_2, \ldots\ldots\ldots, u_n)}{\partial(x_1, x_2, \ldots\ldots, x_n)}=0$, then there must exist a relation between the functions u_1, u_2 ...., u_n. By suitable eliminations, we can always reduce any system of functions to some special form. Here the given functions u_1, u_2, ... u_n are defined by the following set of equations.

$$\phi_1\,(u_1, x_1, x_2, \ldots\ldots\ldots x_n) = 0,$$

$$\phi_2\,(u_1, u_2, x_2, \ldots\ldots, x_n) = 0,$$

....

....

$$\phi_r\,(u_1, u_2, \ldots\ldots\ldots, u_r, x_r, x_{r+1}, \ldots\ldots\, x_n) = 0,$$

$$\phi_n\,(u_1, u_2, \ldots\ldots\ldots, u_n, x_n) = 0,$$

Then the Jacobian of $u_1, u_2, \ldots, u_n$ with respect to $x_1, x_2, \ldots, x_n$ is

$$\frac{\partial(u_1, u_2, \ldots, u_n)}{\partial(x_1, x_2, \ldots, x_n)} = (-1)^n \frac{\dfrac{\partial \phi_1}{\partial x_1} \cdot \dfrac{\partial \phi_2}{\partial x_2} \cdots \dfrac{\partial \phi_n}{\partial x_n}}{\dfrac{\partial \phi_1}{\partial u_1} \cdot \dfrac{\partial \phi_2}{\partial u_2} \cdots \dfrac{\partial \phi_n}{\partial u_n}}$$

But
$$\frac{\partial(u_1, u_2, \ldots, u_n)}{\partial(x_1, x_2, \ldots, x_n)} = 0$$

$$\therefore \qquad \frac{\partial \phi_1}{\partial x_1} \cdot \frac{\partial \phi_2}{\partial x_2} \cdots \frac{\partial \phi_r}{\partial x_r} \cdots \frac{\partial \phi_n}{\partial x_n} = 0$$

i.e., $\dfrac{\partial \phi_r}{\partial x_r} = 0$, for some value of r between 1 and n and this particular function ϕ_r is independent of x_r. Hence, for that particular value of r, the function ϕ_r must not contain x_r, and accordingly the corresponding functional relation is of the form,

$\phi_r(u_1, u_2, \ldots, u_r, x_r, x_{r+1}, x_{r+2}, \ldots, x_n) = 0$. consequently, we can eliminate the (n – r) independent variables, $x_{r+1}, x_{r+2}, \ldots, x_n$, from the n–r + 1 equations given by

$$\phi_r = \phi_{r+1} = , \ldots , = \phi_n = 0;$$

and obtain a functional relation of the form $\phi(u_1, u_2, \ldots, u_n) = 0$ between the functions $u_1, u_2, \ldots, u_n$.

Illustrative Examples

Example 1 If $u = x+2y+z$, $v = x-2y+3z$ and $w = 2xy-xz+4yz-2z^2$; show that J(u, v, w) vanishes and find a relation between u, v and w.

Solution: Here, we have

$$J(u, v, w) = \begin{vmatrix} \dfrac{\partial u}{\partial x} & \dfrac{\partial u}{\partial y} & \dfrac{\partial u}{\partial z} \\ \dfrac{\partial v}{\partial x} & \dfrac{\partial v}{\partial y} & \dfrac{\partial v}{\partial z} \\ \dfrac{\partial w}{\partial x} & \dfrac{\partial w}{\partial y} & \dfrac{\partial w}{\partial z} \end{vmatrix}$$

$$= \begin{vmatrix} 1 & 2 & 1 \\ 1 & -2 & 3 \\ 2y-z & 2x+4z & -x+4y-4z \end{vmatrix}$$

$$= 2\begin{vmatrix} 1 & 1 & 1 \\ 1 & -1 & 3 \\ 2y-z & x+2z & -x+4y-4z \end{vmatrix}$$

$$= 2\begin{vmatrix} 1 & 0 & 0 \\ 1 & -2 & 2 \\ 2y-z & x-2y+3z & -x+2y-3z \end{vmatrix}$$

[on subtracting c_1 from c_2 and c_3]

$$= -2\begin{vmatrix} 1 & 0 & 0 \\ 1 & 2 & 2 \\ 2y-z & -x+2y-3z & -x+2y-3z \end{vmatrix} = 0$$

since c_2 and c_3 are identical.

Vanishing of Jacobian shows that functions u, v, w are not independent.

Now to find the functional relationship, we have,

$$u + v = 2x + 4z, \quad u - v = 4y - 2z,$$

and $$w = x(2y - z) + 2z\,(2y-z)$$

$$= \left(\frac{u+v}{2}\right)\left(\frac{u-v}{2}\right)$$

$\Rightarrow$ $$4w = u^2 - v^2$$

which is the required relation between u, v and w.

Example 2 Show that $ax^2 + 2hxy + by^2$ and $Ax^2 + 2Hxy + By^2$ are independent unless,

$$\frac{a}{A} = \frac{b}{B} = \frac{h}{H}$$

Solution: Let $u = ax^2 + 2hxy + by^2$,

$$v = Ax^2 + 2Hxy + By^2$$

If the functions u and v are not independent, then Jacobian, $\dfrac{\partial(u, v)}{\partial(x, y)} = 0$ i.e.,

$$\begin{vmatrix} \dfrac{\partial u}{\partial x} & \dfrac{\partial u}{\partial y} \\ \dfrac{\partial v}{\partial x} & \dfrac{\partial v}{\partial y} \end{vmatrix} = 0$$

or $$\begin{vmatrix} 2\,(ax + hy) & 2\,(hx + by) \\ 2\,(Ax + Hy) & 2\,(Hx + By) \end{vmatrix} = 0$$

or $$[(ax + hy)\,(Hx + By)] - [(hx + by)\,(Ax + Hy)] = 0$$

or $$(aH - Ah)\,x^2 + (aB - Ab)\,xy + (hB - Hb)\,y^2 = 0$$

since x,y are independent variables, the coefficients of x^2 and y^2 in the above equation must vanish i.e. $aH - Ah = 0$, $hB - Hb = 0$ and it is possible only if

$$\frac{a}{A} = \frac{h}{H} = \frac{b}{B}.$$

Example 3 Show that the functions $u = x + y + z$, $v = xy + yz + zx$ and $w = x^3 + y^3 + z^3 - 3xyz$ are not independent. Also find the relation between u, v and w.

Solution: we have,

$$\frac{\partial(u,v,w)}{\partial(x,y,z)} = \begin{vmatrix} \frac{\partial u}{\partial x} & \frac{\partial u}{\partial y} & \frac{\partial u}{\partial z} \\ \frac{\partial v}{\partial x} & \frac{\partial v}{\partial y} & \frac{\partial v}{\partial z} \\ \frac{\partial w}{\partial x} & \frac{\partial w}{\partial y} & \frac{\partial w}{\partial z} \end{vmatrix}$$

$$= \begin{vmatrix} 1 & 1 & 1 \\ y+z & z+x & x+y \\ 3(x^2-yz) & 3(y^2-zx) & 3(z^2-xy) \end{vmatrix}$$

$$= 3\begin{vmatrix} 1 & 1 & 1 \\ y+z & z+x & x+y \\ (x^2-yz) & (y^2-zx) & (z^2-xy) \end{vmatrix}$$

$$= 3(x-y)(x-z)\begin{vmatrix} 1 & 0 & 0 \\ y+z & 1 & 1 \\ x^2-yz & -(x+y+z) & -(x+y+z) \end{vmatrix} = 0$$

[$\because$ c_2 and c_3 are identical]

Since Jacobian is zero, therefore functions are not independent and so relation can set up between them.

we have $w = x^3 + y^3 + z^3 - 3xyz$

$= (x+y+z)(x^2 + y^2 + z^2 - xy - yz - zx)$

$w = u(u^2 - 3v) = u^3 - 3uv$

$\Rightarrow$ $u^3 = 3uv + w$ is the required relation.

Example 4 If x, y, z are connected by a functional relation $f(x, y, z) = 0$, then prove that

$$\frac{\partial(y,z)}{\partial(x,z)} = \left(\frac{\partial y}{\partial x}\right)_{z=constant}$$

Solution: we have $f(x, y, z) = 0$

$\Rightarrow$ y is a function of x and z. Also from the equation z = z, z may be regarded as a function of x and y.

$$\therefore \quad \frac{\partial(y,z)}{\partial(x,z)} = \begin{vmatrix} \left(\frac{\partial y}{\partial x}\right)_{z=\text{constant}} & \left(\frac{\partial y}{\partial z}\right)_{x=\text{constant}} \\ \frac{\partial z}{\partial x} & \frac{\partial z}{\partial z} \end{vmatrix}$$

$$= \begin{vmatrix} \left(\frac{\partial y}{\partial x}\right)_{z=\text{constant}} & \left(\frac{\partial y}{\partial z}\right)_{x=\text{constant}} \\ 0 & 1 \end{vmatrix}$$

$$\left[\because \frac{\partial z}{\partial x} = 0, \frac{\partial z}{\partial z} = 1\right]$$

$$= \left(\frac{\partial y}{\partial x}\right)_{z=\text{constant}}$$

Example 5 If $f_1 = x^3 = y^3 + z^3 + u^3$,

$f_2 = x^2 + y^2 + z^2 + u^2$, $f_3 = x + y + z + u$, and $f_4 = yzu + zux + uxy + xyz$, prove that a functional relationship exist between f_1, f_2, f_3, f_4.

Solution: Here $f_1 = x^3 + y^3 + z^3 + u^3$,

$f_2 = x^2 + y^2 + z^2 + u^2$

$f_3 = x + y + z + u$ and

$f_4 = yzu + zux + uxy + xyz$

Now, we have,

$$J(f_1, f_2, f_3, f_4) = \frac{\partial(f_1, f_2, f_3, f_4)}{\partial(x, y, z, u)}$$

$$= 6\begin{vmatrix} 1 & 1 & 1 & 1 \\ x & y & z & u \\ x^2 & y^2 & z^2 & u^2 \\ yz+zu+uy & zu+ux+xz & ux+xy+yu & xy+yz+zx \end{vmatrix}$$

operating $(x + y + z + u) R_2 - R_3 + R_4$, we get

$$= 6(xy+xz+xu+yz+yu+zu)\begin{vmatrix} 1 & 1 & 1 & 1 \\ x & y & z & u \\ x^2 & y^2 & z^2 & u^2 \\ 1 & 1 & 1 & 1 \end{vmatrix}$$

Here, the two rows i.e.,

R_1 and R_4 are identical. Hence the determinant vanishes.

$\Rightarrow$ $J(f_1, f_2, f_3, f_4) = 0$

Therefore, these exist a functional relation between the functions $f_1, f_2, f_3 f_4$.

To find the relation, we have

$$\Sigma xy = xy + yz + xu + zx + yu + zu = \frac{1}{2}(f_3 - f_2),$$

and $$\Sigma x \, \Sigma xy = 3\,\Sigma xyz + \Sigma x^2y = 3f_4 + (f_2 f_3 - f_1)$$

Therefore,

$$f_3(f_3^2 - f_2) = 6f_4 + 2(f_2f_3 - f_1)$$

or $$2f_1 - 3f_2f_3 + f_3^2 - 6f_4 = 0$$

Some Additional Examples

Example 1 If $x^2 = u^2 - v^2$, $y = 2uv$; then obtain $\dfrac{\partial(x,y)}{\partial(u,v)}$.

Hint: we have $x = u^2 - v^2$, $y = 2uv$

$\therefore$ $$\frac{\partial x}{\partial u} = 2u, \quad \frac{\partial x}{\partial v} = -2v$$

and $$\frac{\partial y}{\partial u} = 2v, \quad \frac{\partial y}{\partial v} = 2u$$

now, we have,

$$\frac{\partial(x,y)}{\partial(u,v)} = \begin{vmatrix} \frac{\partial x}{\partial u} & \frac{\partial x}{\partial v} \\ \frac{\partial y}{\partial u} & \frac{\partial y}{\partial v} \end{vmatrix} = 4\left(u^2 + v^2\right)$$

Example 2 If $x = a \cos h\,\phi \cos\psi$, $y = a \sin h\phi \sin\psi$, prove that

$$\frac{\partial(x,y)}{\partial(\phi,\psi)} = \frac{1}{2}a^2\left[\cos h2\phi - \cos 2\psi\right]$$

Hint: we have $x = a \cos h\phi \cos\psi$, $y = a \sin h\phi \sin\psi$

Now, $$\frac{\partial x}{\partial \phi} = a \sin h\,\phi \cos\psi, \quad \frac{\partial x}{\partial \psi} = -a \cos h\,\phi \sin\psi$$

and $$\frac{\partial y}{\partial \phi} = a \cos h\,\phi \sin\psi, \quad \frac{\partial y}{\partial \psi} = a \sin h\,\phi \cos\psi$$

then, $$\frac{\partial(x,y)}{\partial(\phi,\psi)} = \begin{vmatrix} \frac{\partial x}{\partial \phi} & \frac{\partial x}{\partial \psi} \\ \frac{\partial y}{\partial \phi} & \frac{\partial y}{\partial \psi} \end{vmatrix}$$

$$= \frac{1}{2} a^2 (\cos h2\phi - \cos 2\psi) .$$

Example 3 If $F_1 = xu + v - y = 0, \quad F_2 = u^2 + vy + w = 0,$

$F_3 = zu - v + vw = 0$, obtain $\frac{\partial(F_1, F_2, F_3)}{\partial(u, v, w)}$

Hint: We have $\frac{\partial F_1}{\partial u} = x, \ \frac{\partial F_1}{\partial v} = 1, \ \frac{\partial F_1}{\partial w} = 0;$

$$\frac{\partial F_2}{\partial u} = 2u, \ \frac{\partial F_2}{\partial v} = y, \ \frac{\partial F_2}{\partial w} = 1,$$

and $\frac{\partial F_3}{\partial u} = z, \frac{\partial F_3}{\partial v} = w - 1, \ \frac{\partial F_3}{\partial w} = v$, then

we have

$$\frac{\partial(F_1, F_2, F_3)}{\partial(u, v, w)} = x(u - 1 - vy) + 2uv - z .$$

Example 4 If $u = xyz, \qquad v = x^2 + y^2 + z^2$ and

$w = x + y + z$, find $\frac{\partial(x,y,z)}{\partial(u, v, w)}$

Hint: Here u, v, w are explicit, so

$$\frac{\partial u}{\partial x} = yz, \frac{\partial u}{\partial y} = zx, \frac{\partial u}{\partial z} = xy$$

$$\frac{\partial v}{\partial x} = 2x, \ \frac{\partial v}{\partial y} = 2y, \ \frac{\partial v}{\partial z} = 2z$$

and $$\frac{\partial w}{\partial x} = 1, \ \frac{\partial w}{\partial y} = 1, \ \frac{\partial w}{\partial z} = 1$$

$$\therefore \qquad J' = \frac{\partial(u,v,w)}{\partial(x,y,z)} = \begin{vmatrix} \frac{\partial u}{\partial x} & \frac{\partial u}{\partial y} & \frac{\partial u}{\partial z} \\ \frac{\partial v}{\partial x} & \frac{\partial v}{\partial y} & \frac{\partial v}{\partial z} \\ \frac{\partial w}{\partial x} & \frac{\partial w}{\partial y} & \frac{\partial w}{\partial z} \end{vmatrix}$$

$$= -2(x-y)(y-z)(z-x)$$

By property, $J.J' = 1$, we have

$$J = \frac{\partial(x,y,z)}{\partial(u,v,w)} = -\frac{1}{2(x-y)(y-z)(z-x)}$$

Example 5 If $x^2 + y^2 + u^2 - v^2 = 0$ and $uv + xy = 0$,

find $\frac{\partial(u,v)}{\partial(x,y)}$.

Hint: Let $f_1 = x^2 + y^2 + u^2 - v^2$, $f_2 = uv + xy$

Now,

$$\frac{\partial f_1}{\partial x} = 2x, \qquad \frac{\partial f_1}{\partial y} = 2y$$

$$\frac{\partial f_2}{\partial x} = y, \qquad \frac{\partial f_2}{\partial y} = x$$

and

$$\frac{\partial f_1}{\partial u} = 2u, \qquad \frac{\partial f_1}{\partial v} = -2v$$

$$\frac{\partial f_2}{\partial u} = v, \qquad \frac{\partial f_2}{\partial v} = u$$

Now, we have,

$$\frac{\partial(u,v)}{\partial(x,y)} = (-1)^2 \frac{\frac{\partial(f_1,f_2)}{\partial(x,y)}}{\frac{\partial(f_1,f_2)}{\partial(u,v)}} = \frac{x^2-y^2}{u^2+v^2}.$$

Example 6 If $x = uv$, $y = \frac{u+v}{u-v}$, prove that

$$\frac{\partial(u,v)}{\partial(x,y)} = \frac{(u-v)^2}{4uv}.$$

Hint: We have, $\frac{\partial x}{\partial u} = v, \frac{\partial x}{\partial v} = u$

and $$\frac{\partial y}{\partial u} = \frac{-2v}{(u-v)^2}, \frac{\partial y}{\partial v} = \frac{2u}{(u-v)^2}$$

Now, $$J' = \frac{\partial(x,y)}{\partial(u,v)} = \begin{vmatrix} \frac{\partial x}{\partial u} & \frac{\partial x}{\partial v} \\ \frac{\partial y}{\partial u} & \frac{\partial y}{\partial v} \end{vmatrix}$$

$$= \frac{4uv}{(u-v)^2}$$

But we know that

$$J.J' = 1 \Rightarrow J = \frac{\partial(u,v)}{\partial(x,y)} = \frac{(u-v)^2}{4uv}$$

Example 7 If $u = x(1 - r^2)^{-1/2}$, $v = y(1 - r^2)^{-1/2}$, $w = z(1 - r^2)^{-1/2}$, where

$r^2 = x^2 + y^2 + z^2$, show that $\frac{\partial(u,v,w)}{\partial(x,y,z)} = (1-r^2)^{-5/2}$

Hint: Here $u = x(1 - r^2)^{-1/2}$

$\Rightarrow$ $x = u(1 - r^2)^{1/2}$

$\Rightarrow$ $x^2 = u^2(1 - r^2) \Rightarrow x^2 = u^2(1 - x^2 - y^2 - z^2)$

$$(\because r^2 = x^2 + y^2 + z^2)$$

similarly,

$y^2 = v^2(1 - r^2) = v^2(1 - x^2 - y^2 - z^2)$

and $z^2 = w^2(1 - r^2) = w^2(1 - x^2 - y^2 - z^2)$

Let $f_1 = x^2 - u^2(1 - x^2 - y^2 - z^2) = 0$

$f_2 = y^2 - v^2(1 - x^2 - y^2 - z^2) = 0$

$f_3 = z^2 - w^2(1 - x^2 - y^2 - z^2) = 0$

then $$\frac{\partial(u,v,w)}{\partial(x,y,z)} = (-1)^3 \frac{\frac{\partial(f_1,f_2,f_3)}{\partial(x,y,z)}}{\frac{\partial(f_1,f_2,f_3)}{\partial(u,v,w)}} \quad ...(1)$$

Now, $$\frac{\partial(f_1, f_2, f_3)}{\partial(x, y, z)} = 8xyz\left(1+u^2+v^2+w^2\right)$$

Also $$\frac{\partial(f_1, f_2, f_3)}{\partial(u, v, w)} = -8uvw\left(1-r^2\right)^3$$

from (1) $$\frac{\partial(u, v, w)}{\partial(x, y, z)} = \left(1-r^2\right)^{-5/2}$$

$$\left(\because u^2+v^2+w^2 = \frac{x^2+y^2+z^2}{1-r^2} = \frac{r^2}{1-r^2}\right)$$

Example 8 If $y_1 = r\cos\theta_1$, $y_2 = r\sin\theta_1\cos\theta_2$, $y_3 = r\sin\theta_1\sin\theta_2\cos\theta_3$, ... $y_n = r\sin\theta_1$, ... $\sin\theta_{n-1}$,

prove that $$\frac{\partial(y_1, y_2, \ldots\ldots\ldots, y_n)}{\partial(r, \theta_1, \theta_2, \ldots\ldots\ldots, \theta_{n-1})} = r^{n-1}\sin^{n-1}\theta_1\sin^{n-3}\theta_2 \ldots\ldots\sin\theta_{n-2}.$$

Hint: squaring and adding, the given relations, we get

$y_1^2 + y_1^2 + \ldots\ldots\ldots\, y_n^2 = r^2$ and let

$f_1 = y_1 - r\cos\theta_1 = 0$, $f_2 = y_2 - r\sin\theta_1\cos\theta_2 = 0$, ...

$f_{n-1} = y_1^2 + y_2^2 + \ldots\ldots + y_n^2 - r^2 = 0$ etc.

Then we put in Jacobian formula.

Example 9 when $x_1^2 + x_2^2 + \ldots\ldots + x_n^2 = r^2$ and

$u_r = x_r(1-r^2)^{-1/2}$, $r = 1, 2, \ldots\ldots\ldots\ldots, n$, show that

$$\frac{\partial(u_1, u_2, \ldots\ldots u_n)}{\partial(x_1, x_2, \ldots\ldots x_n)} = \frac{1}{\left(1-r^2\right)^{\left(\frac{n}{2}+1\right)}}$$

Hint: since $\dfrac{\partial r}{\partial x_r} = \dfrac{x_r}{r}$, we have

$$\frac{\partial u_r}{\partial x_r} = \frac{1}{\left(1-r^2\right)^{1/2}} + \frac{x_r r}{\left(1-r^2\right)^{3/2}};$$

$$\frac{\partial r}{\partial x_r} = \frac{1}{\left(1-r^2\right)^{1/2}} + \frac{x_r^2}{\left(1-r^2\right)^{3/2}}$$

and $$\frac{\partial u_r}{\partial x_s} = \frac{x_r r}{\left(1-r^2\right)^{3/2}};\ \frac{\partial r}{\partial x_r} = \frac{x_r . r}{\left(1-r^2\right)^{3/2}}\ r \neq s$$

Now, we have

$$\frac{\partial(u_1, u_2, ...u_n)}{\partial(x_1, x_2, ...x_n)} = \frac{1}{\left(1-r^2\right)^{3n/2}} \begin{vmatrix} 1-r^2+x_1^2 & x_1x_2 & \cdots\cdots & x_1x_n \\ x_2x_1 & 1-r^2+x_2^2 & \cdots\cdots & x_2x_n \\ \cdots\cdots & \cdots\cdots & \cdots\cdots & \cdots\cdots \\ x_nx_1 & x_nx_2 & \cdots\cdots & 1-r^2+x_n^2 \end{vmatrix}$$

To simplify this determinant, we multiply the first column by x, and add to it, the II and III .... n^{th} column multiplied respectively by x_2, x_3 ...., x_n.

Then we get

$$\frac{\partial(u_1, u_2, ..., u_n)}{\partial(x_1, x_2, ..., x_n)} = \frac{1}{x_1\left(1-r^2\right)^{\frac{3n}{2}}} \begin{vmatrix} x_1 & 0 & \cdots & 0 \\ x_2 & 1-r^2 & \cdots & 0 \\ \cdots & \cdots & \cdots & \cdots \\ \cdots & \cdots & \cdots & \cdots \\ x_n & 0 & \cdots & 1-r^2 \end{vmatrix}$$

$$= \frac{1}{x_1\left(1-r^2\right)^{3n/2}} . x_1\left(1-r^2\right)^{n-1}$$

$$= \frac{1}{\left(1-r^2\right)^{\frac{n}{2}+1}}$$

Example 10 If $u = x + y + z$, $u^2v = y + z$ and

$u^3w = z$, prove that $\dfrac{\partial(u, v, w)}{\partial(x, y, z)} = u^{-5}$

Hint: Let $r = x + y + z$, $s = y + z$, $t = z$, then

$u = r$, $u^2v = s$, $u^3w = t$

Now, we have $$\frac{\partial(u, v, w)}{\partial(x, y, z)} = \frac{\partial(u, v, w)}{\partial(r, s, t)} \times \frac{\partial(r, s, t)}{\partial(x, y, z)}$$

then, we get required result.

Exercise 6.6

1. Investigate whether any relation exists between u, v, w if they are given by $u = x + y + z$, $v = xy + yz + xz$, $w = x^2 + y^2 + z^2$ and find the relation.
2. Show that the functions u, v, w where $u = x + y + z$, $v = xy + xz$, $w = x^2 + y^2 + z^2 + 2yz$ are functionally dependent and find the functional relation.
3. Prove that the functions, $u=x+y-z$, $v=x-y+z$, $w=x^2+y^2+z^2-2yz$ are not independent of one another. Also find the relation between them.
4. Prove that the functions $u = 3x + 2y - z$, $v = x - 2y + z$, $w = x(x + 2y - z)$ are not independent and find the relation between them.
5. Prove that the functions $u_1 = x_1 + x_2 + x_3 + x_4$, $u_2 = x_1^2 + x_2^2 + x_3^2 + x_4^2$, $u_3 = x_1^3 + x_2^3 + x_3^3 + x_4^4$, $u_4 = x_1x_2x_3 + x_2x_3x_4 + x_3x_4x_1 + x_4x_1x_2$ are not independent. find the relation between them.
6. If $X = x + y + z + t$, $Y = x + y - z - t$, $Z = xy - zt$, $T = x^2 + y^2 - z^2 - t^2$, Prove that $J(X, Y, Z, T) = 0$.
7. If u, v, w are functions of four independent variables x_1, x_2, x_3, x_4 and if

$$P = \frac{\partial(u,v,w)}{\partial(x_2,x_3,x_4)},\quad Q = \frac{\partial(u,v,w)}{\partial(x_3,x_4,x_1)},\quad R = \frac{\partial(u,v,w)}{\partial(x_4,x_1,x_2)},\quad S = \frac{\partial(u,v,w)}{\partial(x_1,x_2,x_3)},$$ prove that

$$\frac{\partial P}{\partial x_1} - \frac{\partial Q}{\partial x_2} + \frac{\partial R}{\partial x_3} - \frac{\partial S}{\partial x_4} = 0.$$

8. If $X = \frac{x+y}{z}$, $Y = \frac{y+z}{x}$, $Z = \frac{y(x+y+z)}{xz}$, prove that X, Y, Z, are not independent and find the relation between them.
9. If $u = x + y + z$, $v = x^2 + y^2 + z^2$, $w = x^3 + y^3 + z^3 - 3xyz$. Prove that u, v, w are connected by some functional relation and find it
10. Show that the functions u, v, w defined as $u = \frac{x}{y-z}$, $v = \frac{y}{z-x}$, $w = \frac{z}{x-y}$ are not independent of each other.
11. Show that the functions, $x + 2y + z$, $x - 2y + 3z$, $2xy - xz + 4yz - 2z^2$ are not independent and find the relations between them.
12. Prove that if a functional relation exist between the functions $u = f(x) + f(y) + f(z)$, $y = f(y) f(z) + f(z) f(x) + f(x) f(y)$, $w = f(x) f(y) f(z)$, then 'f' must be a constant.
13. Are the following functions functionally dependent? If so, find the relation between them.

$$u = \frac{r-s}{r+t},\quad v = \frac{r+t}{s+t}.$$

14. Determine functional dependence and find the relation between them.

(i) $X = \frac{x+y}{x-y}$, $Y = \frac{xy}{(x-y)^2}$

(ii) $X = \dfrac{x-y}{x+y}$, $\quad Y = \dfrac{xy}{(x+y)^2}$

15. If $u^{-1} = x^{-1} + y^{-1} + z^{-1}$, $v^2 = x^2 + y^2 + z^2$, $w^3 = xyz$, prove that

$$\frac{\partial(u, v, w)}{\partial(x, y, z)} = \frac{u(y-z)(z-x)(x-y)(x+y+z)}{2v\,w^2\,(yz+zx+xy)}$$

16. If $u_r = x_r(1 - x_1^2 - x_2^2 \ldots\ldots\ldots - x_n^2)$, $r = 1, 2, \ldots\ldots\ldots n$, prove that

$$\frac{\partial(u_1, u_2, \ldots\ldots, u_n)}{\partial(x_1, x_2, \ldots\ldots, x_n)} = \left(1 - x_1^2 - x_2^2 \ldots\ldots - x_n^2\right)^{-\frac{n+2}{2}}$$

Answers

1. $u^2 = 2v + w$ 2. $u^2 = 2v + w$ 3. $u^2 + v^2 = 2w$ 4. $u^2 - v^2 = 8w$

5. $XY - 2Z = T$ 8. $XY = Z + 1$ 9. $u^3 - 3uv + 2w = 0$ 11. $u^2 - v^2 = 4w$

13. $u = \dfrac{1}{1-v}$ 14. (i) $X^2 - 4X = 1$

(ii) $4Y = 1 - X^2$

6.15 EXTREMA OF FUNCTIONS OF SEVERAL VARIABLES AND LAGRANGE'S METHOD OF MULTIPLIERS.

Introduction: In this exercise, we shall determine the values of a function which are greatest or least in their immediate neighbourhood, technically they are termed as maximum and minimum values of the function. A knowledge of these values is helpful in tracing of curves and in determining the greatest and least values of a function in any given finite interval.

- **Maxima and minima of a function:**

A function f(x) is said to have a maximum and minimum at x = a, if there exists a small nbd. of 'a' i.e., if a positive number $\in$ exists, such that $f(x) < f(a)$ or $f(x) > f(a)$ for all values of x for which $0 < |(x-a)| < \in$.

Note: 1 The term extreme values is used for maximum as well as for a minimum value and the points where these values occur are called extreme points.

Note: 2 It should also be noted that a maximum or a minimum value f(a) of f(x) is the greatest or the least value of the function in only a small nbd. of the point x = a and not necessarily in the entire interval of definition of f(x). Infact, a function may have several maximum and minimum values.

- **Stationary value:** A function f(x) is said to be stationary at x = a, if the derivative f'(x) vanishes at x = a i.e., if f'(a) = 0. The value f(a) is said to be a stationary value or a turning value of f(x). The term 'stationary' arises from the fact that the rate of change, f'(x) of the function f(x) w.r.t x is zero for a value of x for which f(x) is stationary.

It should be noted that a maximum or a minimum value of a function is also a stationary value but a stationary value may neither be a maximum nor a minimum value. For example, the function

$f(x) = (x - 1)^3 + 2$ has neither a maximum nor a minimum at $x = 1$, despite the fact that f(x) has stationary value at this point.

6.16 MAXIMA AND MINIMA OF A FUNCTION OF SEVERAL INDEPENDENT VARIABLES.

The definitionof maximum or minimum value of a function of several independent variables can be given in the same way as is given for the function of a single variable. Thus, the function f(x, y, z, ...) is said to have a maximum or minimum value at the point (a, b, c ....) if f(x, y, z, ...) < f(a, b, ...) or

$$f(x, y, z, \ldots\ldots) > f(a, b, \ldots\ldots)$$

for all points (x, y, z, ...) in the nbd. of the point (a, b, c, ...).

6.17 MAXIMA AND MINIMA OF A FUNCTION OF TWO INDEPENDENT VARIABLES:

A function f(x, y) of two independent variables is said to have maximum (local maximum) at the point (a, b) if there exists a small positive number δ such that $f(a + h, b + k) < f(a, b)$ for all values of h, k such that $|h| < \delta$, $|k| < \delta$.

A function f(x, y) is said to have minimum (local minimum) at (a, b) if there exists a small positive number δ such that $f(a+h, b+k) > f(a, b)$ for all h, k such that $|h| < \delta$ and $|k| < \delta$.

We now establish conditions for existence of maximum and minimum of a function of two independent variables.

NECESSARY CONDITIONS FOR THE EXISTENCE OF MAXIMA AND MINIMA:

Let f(a, b) be an extreme (a maximum or a minimum) value of the function f(x, y) at the point (a, b). Then, by definition $f(a+h, b+k) - f(a, b)$ should preserve a fixed sign (positive for minimum and negative for maximum) for all sufficiently small values of h and k.

Now, from Taylor's series expansion, we have,

$$f(a + h, b + k) - f(a, b) = \left(h\frac{\partial f}{\partial x} + k\frac{\partial f}{\partial y}\right)_{\substack{x=a\\y=b}} + \frac{1}{2!}\left(h^2\frac{\partial^2 f}{\partial x^2} + 2hk\frac{\partial^2 f}{\partial x \partial y} + k^2\frac{\partial^2 f}{\partial y^2}\right)_{\substack{x=a\\y=b}} + \ldots\ldots \quad (1)$$

By taking h, k sufficiently small, the first degree terms in h, k can be made to govern the sign of the R.H.S. and therefore of the L.H.S. of (1). But the first degree terms change their sign when the signs of h and k are reversed. Hence in order that L.H.S. may preserve a fixed sign, it is necessary that

$$\left(h\frac{\partial f}{\partial x} + k\frac{\partial f}{\partial y}\right)_{\substack{x=a\\y=b}} = 0 \quad \ldots(2)$$

since h and k are arbitrary and independent of each other, we must have

$$\left(\frac{\partial f}{\partial x}\right)_{\substack{x=a\\y=b}} = 0 \text{ and } \left(\frac{\partial f}{\partial y}\right)_{\substack{x=a\\y=b}} = 0$$

Hence, the necessary condition for f(x, y) to have a maximum or minimum value at (a, b) are that the first order partial derivatives $\frac{\partial f}{\partial x}$ and $\frac{\partial f}{\partial y}$ must vanish for x = a and y = b.

SUFFICIENT CONDITIONS FOR THE EXISTENCE OF MAXIMA AND MINIMA:

If for the function f(x, y), $\frac{\partial f}{\partial x}$ and $\frac{\partial f}{\partial y}$ both vanish at x = a and y = b, then from (1)

$$f(a+h, b+k) - f(a,b) = \frac{1}{2!}\left(h^2 \frac{\partial^2 f}{\partial x^2} + 2hk\frac{\partial^2 f}{\partial x \partial y} + k^2 \frac{\partial^2 f}{\partial y^2}\right)_{\substack{x=a \\ y=b}} + \ldots\ldots$$

$$= \frac{1}{2!}\left(rh^2 + 2shk + tk^2\right) + \ldots.. \qquad \ldots(3)$$

where, for brevity, we have put

$$r = \left(\frac{\partial^2 f}{\partial x^2}\right)_{\substack{x=a \\ y=b}},\ s = \left(\frac{\partial^2 f}{\partial x \partial y}\right)_{\substack{x=a \\ y=b}},\ t = \left(\frac{\partial^2 f}{\partial y^2}\right)_{\substack{x=a \\ y=b}}$$

for the sufficiently small values of h and k.

The sign of R.H.S. in (3) is governed by the second degree terms in h, k and therefore of the L.H.S. Thus, if for all possible values of h, k, the sign of $(rh^2 + 2shk + tk^2)$ is positive, we have a minima of f(x, y) and if it is negative, we have a maxima. To identify the sign of expression $(rh^2 + 2hks + tk^2)$, we write this as, $rh^2 + 2hks + tk^2 = \frac{1}{r}(r^2h^2 + 2rshk + tk^2)$

$$= \frac{1}{r}[(rh + sk)^2 + (t - s^2)k^2] \qquad \ldots(4)$$

we have now following cases:

1. If $r > 0$ and $rt - s^2 > 0$, the sign of (4) is positive.
2. If $r < 0$ and $(rt - s^2) > 0$, the sign of (4) is negative.
3. If $rt - s^2 < 0$, the sign of (4) cannot be identified.
4. If $rt - s^2 = 0$ and $rh + sk \neq 0$, the sign of (4) is positive and negative according as r is positive or negative, however, if $rt - s^2 = 0$ and $rh + sk = 0$, further investigations are required.

Hence, if $r > 0$ and $rt - s^2 > 0$, the function f(x, y) has a minimum at (a, b). If $r < 0$ and $rt - s^2 > 0$, the function f(x,y) has a maximum at (a, b). However, if $rt - s^2 < 0$, the function has neither a maximum or nor minimum at (a, b).

Again, if $rt - s^2 = 0$ and $rh + sk \neq 0$, the sign of r only will decide the maxima and minima of the function.

If $rt - s^2 = 0$ and $rh + sk = 0$, then second degree terms in (4) vanish and we must consider the terms of higher order in h and k,

The above conditions are sufficient conditions (also called Lagrange's conditions) to determine a maximum and a minimum of a function.

Working Rule: In order to determine a maximum or a minimum of a function f(x, y), we proceed as follows:

1. Find $\frac{\partial f}{\partial x}$ and $\frac{\partial f}{\partial y}$, equate them to zero. Solve these simultaneous equations for x and y. Let the roots be (a_i, b_i), i = 1, 2, 3, ... These (a_i, b_i) are called stationary points.
2. Find $\frac{\partial^2 f}{\partial x^2}, \frac{\partial^2 f}{\partial x \partial y}$, and $\frac{\partial^2 f}{\partial y^2}$, substitute $x = a_i; y = b_i$ in turn. calculate the value $rt - s^2$ for each pair of values.
3. If $r > 0$ and $rt - s^2 > 0$ for a pair of roots, f(x, y) is a minimum for this pair. If $r < 0$ and $rt - s^2 > 0$, it is a maximum. If $rt - s^2 < 0$, the function has neither a maximum nor a minimum and the point is called saddle point. If $rt - s^2 = 0$, the case is undecided and further investigations are necessary to decide it.

Illustrative Examples

Example 1 Find the maximum and minimum values of the function (x – 1) (x – 2) (x – 3).

Solution: Let

$$f(x) = (x-1)(x-2)(x-3)$$
$$= x^3 - 6x^2 + 11x - 6$$

then for maximum or minimum, f'(x) = 0, i.e.,

$$\Rightarrow \quad 3x^2 - 12x + 11 = 0$$

$$\Rightarrow \quad x = 2 + \frac{1}{\sqrt{3}}, \quad 2 - \frac{1}{\sqrt{3}}$$

Also $f''(x) = 6x + 12$

clearly $f''\left(2 + \frac{1}{\sqrt{3}}\right)$ is positive and

$f''\left(2 - \frac{1}{\sqrt{3}}\right)$ is negative.

Hence, the given function has maximum at $2 - \frac{1}{\sqrt{3}}$ and a minimum at $2 + \frac{1}{\sqrt{3}}$. The maximum value of f(x) is, therefore

$$\left(1 - \frac{1}{\sqrt{3}}\right)\left(-\frac{1}{\sqrt{3}}\right)\left(-1 - \frac{1}{\sqrt{3}}\right) = \frac{2}{3\sqrt{3}}$$

and the maximum value is

$$\left(1+\frac{1}{\sqrt{3}}\right)\left(\frac{1}{\sqrt{3}}\right)\left(-1+\frac{1}{\sqrt{3}}\right)=-\frac{2}{3\sqrt{3}}$$

Example 2 Investigate $f(x) = x^5 - 5x^4$ for maximum and minimum.

Solution: Here $f(x) = x^5 - 5x^4$

$$\therefore \quad f'(x) = 5x^4 - 20x^3 \Rightarrow f'(x) = 5x^3(x-4)$$

For maximum and minimum, $f'(x) = 0$

$$\Rightarrow \quad x = 0 \text{ or } 4$$

Now $f''(x) = 20x^3 - 60x^2$

Now $f''(0) = 0$, which indicates that we should proceed further to identify maximum or minimum at $x = 0$. for this obtain $f'''(x)$. If $f'''(0) = 0$, then $f^{iv}(0) > 0$ gives minimum and $f^{iv}(0) < 0$ gives maximum.

If $f^{iv}(x)$ is also zero, then for maximum or minimum $f^{v}(x)$ must vanish, then $f^{vi}(x) > 0$ is for minimum and $f^{vi}(x) < 0$ is for maximum and so on. Thus for maximum and minimum odd order derivatives must vanish whereas sign of even order derivative identifies the maximum or minimum.

Now from above

$f^{ii}(x) = 20x^3 - 60x^2$

$\therefore$ $f^{iii}(x) = 60x^2 - 120x = 0$ for $x = 0$ and

$f^{iv}(x) = 120x - 120$

i.e., $f^{iv}(0) = -120 < 0$ $\therefore$ at $x = 0$, there is maximum.

Again $f^{ii}(4) = 20(4)^3 - 60(4)^2$

$= 320 > 0$ which implies $f(x)$ is minimum at $x = 4$.

Hence at $x = 0$, $f(x)$ is maximum and at $x = 4$, it is minimum.

Example 3 A gas holder is a cylindrical vessel closed at the top and open at the bottom (which dips into water), what should be the ratio of the height to the diameter in order that for a given value its construction may require the least amount of material?

Solution: Let R be the radius of cylindrical vessel and d be the depth of tank.

Let V be the given capacity of the tank and A be the area of the sheet required to construct it, then

$$V = \pi R^2 d \qquad \ldots(1)$$

$$\therefore \quad d = \frac{V}{\pi R^2}$$

and A = Area of base + Area of cylindrical wall

$$A = \pi R^2 + 2\pi R d$$

$$A = \pi R^2 + 2\pi R.\frac{V}{\pi R^2} \qquad \text{[from (1)]}$$

$$\Rightarrow \quad A = \pi R^2 + \frac{2V}{R} \Rightarrow \frac{dA}{dR} = 2\pi R - \frac{2V}{R^2}$$

Again $$\frac{d^2A}{dR^2} = 2\pi + \frac{4V}{R^3}$$

For maximum or minimum, $\frac{dA}{dR} = 0$

$$\Rightarrow \quad 2\pi R - \frac{2V}{R^2} = 0 \Rightarrow \boxed{V = \pi R^3}$$

$$\left(\frac{d^2A}{dR^2}\right)_{V=\pi R^3} = 6\pi > 0, \text{ (minimum)}$$

Hence, $$d = \frac{V}{\pi R^2} \Rightarrow d = \frac{\pi R^3}{\pi R^2} = R$$

$$\Rightarrow \quad \frac{d}{2R} = \frac{1}{2} \Rightarrow \frac{Height}{Diameter} = \frac{d}{D} = \frac{1}{2}.$$

Example 4 Assuming, that the petrol burnt (per hour) in driving a motor boat varies as the cube of its velocity. Find the most economical speed when going against a current of c km/hour.

Solution: Let the velocity of the motor boat be v km./hour. The velocity of the boat relative to the current (v - c) km./hour. Let the total distance to be covered be d km. Then time taken $\frac{d}{v-c}$ hours.

Also, we are given that the petrol burnt in one hour = αv^3, where α is constant of proportionality.

$\therefore$ If P be the total amount of petrol burnt in covering d kms., then

$$P = \frac{\alpha v^3 d}{v-c} \quad \text{...(1)}$$

$$\therefore \quad \frac{dP}{dv} = \alpha d.\left[\frac{(v-c).3v^2 - v^3.1}{(v-c)^2}\right] = \frac{\alpha d\left(2v^3 - 3cv^2\right)}{(v-c)^2}$$

and differentiating, we have

$$\frac{d^2P}{dv^2} = \frac{\alpha d}{(v-c)^3}\left[(v-c)^2\left(6v^2 - 6vc\right) - \left(2v^3 - 3cv^2\right).\,2(v-c)\right]$$

$$= \frac{\alpha d}{(v-c)^3}\left[6v.(v-c)^2 - 2v^2\,(2v - 3c)\right]$$

for maximum or minimum, $\frac{dP}{dv} = 0$

$$2v^3 - 3cv^2 = 0 \Rightarrow v = 0 \text{ or } \frac{3}{2}c$$

If $v = 0$, $P = 0$, i.e., no petrol is burnt.

If $v = \frac{3c}{2}$, $\frac{d^2P}{dv^2} = +ve$, i.e., P is minimum.

$\therefore$ Petrol burnt is minimum when $v = \frac{3c}{2}$, i.e., the most economical speed is $\left(\frac{3c}{2}\right)$ km/hour.

Example 5 Obtain the semi-vertical angle of a right circular cone of given total surface including the area of the base) and having maximum value.

Solution: Let h be the height and α be the semi-vertical angle of the cone. Then its slant height = h sec α and radius of its base = h tan α

Let S be the given total surface and V be the volume of the cone. Then

$$S = (\text{area of curved surface} + \text{area of base})$$

$$\Rightarrow \quad S = \pi rl + \pi r^2 = \pi h^2 \sec\alpha \tan\alpha + \pi h^2 \tan^2\alpha$$

or $$h^2 = \frac{S}{\pi \tan\alpha\,(\sec\alpha + \tan\alpha)}$$

$$\text{Volume } V = \frac{1}{3}\pi r^2 h = \frac{1}{3}\pi h^3 \tan^2\alpha.$$

or $$V = \frac{\pi}{3}\left(\frac{S}{\pi \tan\alpha\,(\sec\alpha + \tan\alpha)}\right)\tan^2\theta \qquad \text{(from (1)}$$

or $$V = \frac{S^{3/2}\sqrt{\tan\alpha}}{3\sqrt{\pi}\,(\sec\alpha + \tan\alpha)^{3/2}}$$

or $$V^2 = \frac{S^3 \tan\alpha}{9\pi(\sec\alpha + \tan\alpha)^3} = u(say)$$

and if u is minimum, then V^2 is maximum.

$$\therefore \quad \frac{du}{d\alpha} = \frac{S^3}{9\pi}\left[\frac{(\sec\alpha + \tan\alpha)^3 \sec^2\alpha - \tan\alpha\, 3(\sec\alpha + \tan\alpha)^2 (\sec\alpha \tan\alpha + \sec^2\alpha)}{(\sec\alpha + \tan\alpha)^6}\right]$$

or $$\frac{du}{d\alpha} = \frac{S^3}{9\pi}\left[\frac{(\sec\alpha + \tan\alpha)^3 (\sec^2\alpha - 3\sec\alpha \tan\alpha)}{(\sec\alpha + \tan\alpha)^6}\right]$$

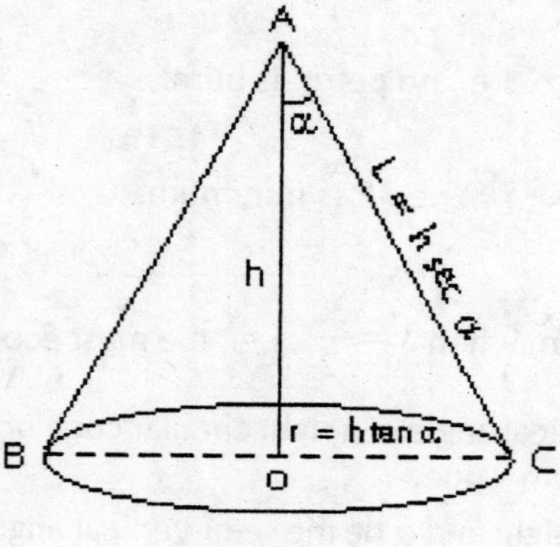

Fig. (1)

or $$\frac{du}{d\alpha} = \frac{S^3}{9\pi}\left[\frac{\sec\alpha\,(\sec\alpha - 3\tan\alpha)}{(\sec\alpha + \tan\alpha)^3}\right]$$

For maximum or minimum $\frac{du}{d\alpha} = 0$, we get $\sec\alpha(\sec\alpha - 3\tan\alpha) = 0$. But $\sec\alpha \neq 0$

$\therefore$ $\cos\alpha \neq \infty$ for any value of α.

Hence $\sec\alpha - 3\tan\alpha = 0$

i.e., $1 - 3\sin\alpha = 0 \Rightarrow \sin\alpha = 1/3 \Rightarrow \alpha = \sin^{-1}(1/3)$

Also for $\alpha = \sin^{-1}(1/3)$, we can show that

$\frac{d^2u}{d\alpha^2}$ is negative.

u or V^2 is maximum

Hence, for maximum value of V,

$$\alpha = \sin^{-1}(1/3).$$

Example 6 Find the semi vertical angle of the cone of maximum volume and of a given slant height.

Solution: Let h be the slant height and θ be the semi-vertical angle of cone. Then radius of the base r = OC = h sin θ and height of the cone H = OA = h cos θ

Let V be the volume of the cone, then

$$V = \frac{1}{3}\pi r^2 H = \frac{1}{3}\pi\,(h\sin\theta)^2\,(h\cos\theta)$$

$$= \frac{1}{3}\pi h^3 \sin^2\theta\cos\theta \qquad \text{....(1)}$$

differentiating (1) w.r.t.θ, we get

$$\frac{dV}{d\theta} = \frac{\pi}{3} h^3 \left[\sin^2\theta\,(-\sin\theta) + \cos\theta.\, 2\sin\theta\cos\theta\right]$$

$$= \frac{\pi}{3} h^3 \sin\theta \left(2\cos^2\theta - \sin^2\theta\right)$$

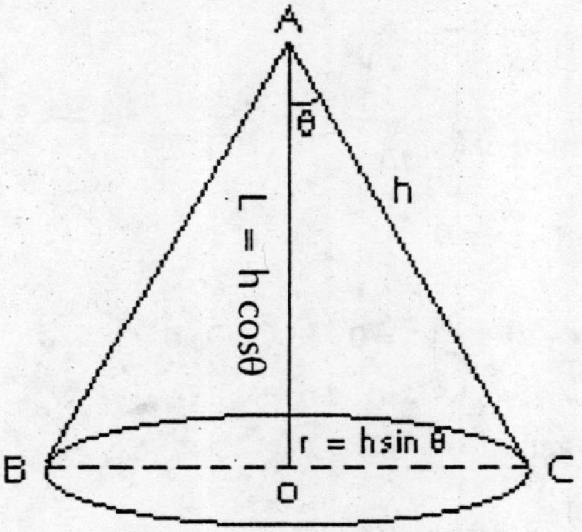

Fig. (2)

Again,

$$\frac{d^2V}{d\theta^2} = \frac{\pi}{3} h^3 \left[\sin\theta\,(-4\cos\theta\sin\theta - 2\sin\theta\cos\theta) + \left(2\cos^2\theta - \sin^2\theta\right).\cos\theta\right]$$

$$= \frac{\pi}{3} h^3 \left[-6\sin^2\theta\cos\theta + \cos\theta\left(2\cos^2\theta - \sin^2\theta\right)\right]$$

For maximum or minimum, $\frac{dV}{d\theta} = 0$

$\Rightarrow$ $\sin\theta = 0$ or $\tan^2\theta = 2$

$\Rightarrow$ $\theta = 0$ or $\theta = \tan^{-1}\sqrt{2}$

when $\theta = 0$, volume becomes zero and cone becomes a straight line which is not the case.

when $\theta = \tan^{-1}\sqrt{2}$, we have $\frac{d^2V}{d\theta^2} = \frac{\pi}{3} h^3 \left[-6.\frac{2}{3}.\frac{1}{\sqrt{3}} + 0\right]$

$= -$ve (sign)

Hence the volume of the cone is maximum when $\theta = \tan^{-1}\sqrt{2}$. when $\theta = -\tan^{-1}\sqrt{2}$, volume of core becomes negative which is meaningless, hence is not the case.

More Illustrative Examples

Example 1 Find the maximum or minimum values of the function $x^3y^2(1-x-y)$.

Solution: Let $f(x,y) = x^3y^2(1-x-y)$

$$\therefore \quad \frac{\partial f}{\partial x} = 3x^2y^2(1-x-y) - x^3y^2$$

and $$\frac{\partial f}{\partial y} = 2yx^3(1-x-y) - x^3y^2$$

for maximum or minimum of f(x, y),

$$\frac{\partial f}{\partial x} = 0 \text{ and } \frac{\partial f}{\partial y} = 0$$

$$\therefore \quad 3x^2y^2(1-x-y) - x^3y^2 = 0 \qquad ...(1)$$

and $$2x^3y(1-x-y) - x^3y^2 = 0 \qquad ...(2)$$

solving (1) and (2), we get

$$x = \frac{1}{2}, y = \frac{1}{3}, x = 0, y = 0$$

The stationary points are:

$$(0, 0), \left(\frac{1}{2}, \frac{1}{3}\right)$$

and also $r = \dfrac{\partial^2 f}{\partial x^2} = 6xy^2 - 12x^2y^2 - 6xy^3;$

$$s = \frac{\partial^2 f}{\partial x \partial y} = 6x^2y - 8x^3y - 9x^2y^2;$$

$$t = \frac{\partial^2 f}{\partial y^2} = 2x^3 - 2x^4 - 6x^3y$$

Now $rt - s^2$ at $\left(\frac{1}{2}, \frac{1}{3}\right) = \dfrac{1}{144} > 0$ and also

$$r \text{ at } \left(\frac{1}{2}, \frac{1}{3}\right) = -\frac{1}{9} < 0$$

Hence f has a maximum at $\left(\frac{1}{2}, \frac{1}{3}\right)$ and the maximum value is $\dfrac{1}{432}$. **Answer**

Example 2 Find a point within triangle such that the sum of the squares of its distances from the three vertices, is a minimum.

Solution: Let (x_n, y_n); n = 1, 2, 3 be the vertices of the triangle and let (x, y) be any point inside the triangle.

Then let $$v = \sum_{n=1}^{3}\left[(x-x_n)^2 + (y-y_n)^2\right] \quad ...(1)$$

$$\therefore \quad \frac{\partial v}{\partial x} = \sum_{n=1}^{3} 2(x-x_n) = 2[(x-x_1)+(x-x_2)+(x-x_3)]$$

and $$\frac{\partial v}{\partial y} = \sum_{n=1}^{3} 2(y-y_n) = 2[(y-y_1)+(y-y_2)+(y-y_3)]$$

and also $r = \dfrac{\partial^2 v}{\partial x^2} = 6$, $s = \dfrac{\partial^2 v}{\partial x \partial y} = 0$ and $t = \dfrac{\partial^2 v}{\partial y^2} = 6$

Fro a maximum or minimum of v, we have

$$\frac{\partial v}{\partial x} = 0 \Rightarrow 2[(x-x_1)+(x-x_2)+(x-x_3)] = 0$$

$$\Rightarrow \quad \boxed{x = \frac{x_1+x_2+x_3}{3}}$$

and $$\frac{\partial v}{\partial y} = 0 \Rightarrow 2[(y-y_1)+(y-y_2)+(y-y_3)]$$

$$\Rightarrow \quad \boxed{y = \frac{y_1+y_2+y_3}{3}}$$

Thus $\left(\dfrac{x_1+x_2+x_3}{3}, \dfrac{y_1+y_2+y_3}{3}\right)$ is the only point at which v may have a maximum or minimum.

At this point r = 6, s = 0, t = 6, so that $rt - s^2 = 36$ (+ve)

Hence the stationary value of v at the point $\left(\dfrac{x_1+x_2+x_3}{3}, \dfrac{y_1+y_2+y_3}{3}\right)$ is an extreme value.

But r > 0, then this extreme value of v is a minimum.

Thus the required point at which v is minimum is $\left(\dfrac{x_1+x_2+x_3}{3}, \dfrac{y_1+y_2+y_3}{3}\right)$. This is known as centroid of the triangle.

Example 3 Find the stationary point of the function $z = x^3y^2(12 - x - y)$ satisfying the condition x > 0, y > 0 and examine their nature.

Solution: Let $z = x^3y^2(12 - x - y)$ ...(1)

$$\therefore \quad \frac{\partial z}{\partial x} = x^3y^2(-1) + (12 - x - y)\, y^2 . 3x^2$$

$$= 3x^2y^2(12 - x - y) - x^3y^2$$

$$= 36x^2y^2 - 4x^3y^2 - 3x^2y^3$$

$$\therefore \quad \frac{\partial z}{\partial x} = 36x^2y^2 - 4x^3y^2 - 3x^2y^3 \quad \text{...(2)}$$

Also $$\frac{\partial z}{\partial y} = x^3y^2(-1) + (12 - x - y)\, x^3 . 2y$$

$$= 24x^3y - 2x^4y - 2x^3y^2 - x^3y^2$$

or $$\frac{\partial z}{\partial y} = 24x^3y - 2x^4y - 3x^3y^2 \quad \text{...(3)}$$

For maximum or minimum value of z, we have

$$\frac{\partial z}{\partial x} = 0, \quad \frac{\partial z}{\partial y} = 0$$

$$\therefore \quad \frac{\partial z}{\partial x} = 0 \Rightarrow 36\, x^2y^2 - 4x^3y^2 - 3x^2y^3 = 0$$

$$\Rightarrow \quad x^2y^2(36 - 4x - 3y) = 0$$

$$\Rightarrow \quad x = 0,\ y = 0,\ 4x + 3y = 36 \quad \text{...(4)}$$

Also $$\frac{\partial z}{\partial y} = 24x^3y - 2x^4y - 3x^3y^2 = 0$$

or $$x^3y(24 - 2x - 3y) = 0$$

$$\Rightarrow \quad x = 0,\ y = 0,\ 2x + 3y = 24 \quad \text{...(5)}$$

solving equations (4) and (5), we get

$$x = 0,\ y = 0,\ x = 6 \text{ and } y = 4$$

Now $$r = \frac{\partial^2 z}{\partial x^2} = 72xy^2 - 12x^2y^2 - 6xy^3 \quad \text{...(6)}$$

$$t = \frac{\partial^2 z}{\partial y^2} = 24x^3 - 2x^4 - 6x^3y \quad \text{...(7)}$$

Also $$s = \frac{\partial^2 z}{\partial x \partial y} = \frac{\partial}{\partial x}\left(24x^3y - 2x^4y - 3x^3y^2\right)$$

$\Rightarrow \quad s = 72x^2y - 8x^3y - 9x^2y^2 \quad ...(8)$

Thus at x = 6 and y = 4

$r = 72 \times 6 \times 16 - 12 \times (36) \times (16) - 6 \times 6 \times 64$

$= -2054 = (-ve)$

$t = 24\,(216) - 2(1216) - 6(216 \times 4) = -2592.$

$\therefore \quad s = 72\,(36)\,(4) - 8(216)\,(4) - 9(36) \times (16) = -1728$

$\therefore \quad rt - s^2 = (-2504) \times (-2592) - (-1728) > 0$

Hence $\quad rt - s^2 > 0,\ r < 0$

we have maximum at x = 6, y = 4.

Example 4 Locate the stationary points of the function $x^3 + y^3 - 3axy$ and determine their nature.

Solution: Let $f(x,y) = x^3 + y^3 - 3axy$

The stationary points are given by

$$\frac{\partial f}{\partial x} = 3x^2 - 3ay = 0 \quad ...(1)$$

and

$$\frac{\partial f}{\partial y} = 3y^2 - 3ax = 0 \quad ...(2)$$

from (1) $y = \frac{x^2}{a}$, substituting is (2), we get

$x^4 - a^3x = 0$, or $x(x - a)\,(x^2 + ax + a^2) = 0$

$\Rightarrow \quad x = 0, a$ (discarding the imaginary roots)

$\therefore \quad y = 0$, when $x = 0$; $\quad y = a$ when $x = a$

Hence the stationary points are (0,0) and (a, a)

Now, $\quad \frac{\partial^2 f}{\partial x^2} = 6x, \ \frac{\partial^2 f}{\partial x \partial y} = -3a, \ \frac{\partial^2 f}{\partial y^2} = 6y$

i.e., $\quad r = 6x,\ s = -3a$ and $t = 6y$.

At (0,0), $rt - s^2 = 0 - (-3a)^2 = -9a^2$ and $r = 0$

Hence at this point, the function has neither a maxima nor a minima, i.e., this point is a saddle point.

At (a,a) $\Rightarrow r = 6a,\ rt - s^2 = 6a.\,6a - (-3a)^2 = 27a^2$

Hence at this point, the function has a minimum if $a > 0$, and has a maximum if $a < 0$.

Example 5 Discuss the maxima and minima of the function, $f(x, y) = \sin x \sin y \sin(x + y)$.

Solution: The points of maxima and minima are given by,

$$\frac{\partial f}{\partial x} = \sin y\left[\cos x \sin(x+y) + \sin x \cos(x+y)\right] = 0$$

$\Rightarrow \quad \sin x\,[\sin(2x + y)] = 0 \quad ...(1)$

and $$\frac{\partial f}{\partial y} = \sin x\left[\cos y \sin(x+y) + \sin y \cos(x+y)\right] = 0$$

$\Rightarrow$ $$\sin x \sin(x + 2y) = 0 \quad ...(2)$$

Solving these equations, the stationary points are (0,0) and $\left(\frac{\pi}{3}, \frac{\pi}{3}\right)$.

Now, $$r = \frac{\partial^2 f}{\partial x^2} = 2 \sin y \cos(2x + y),$$

$$s = \frac{\partial^2 f}{\partial x \partial y} = \sin y \cos(2x + y) + \cos y \sin(2x + y)$$

$$= \sin(2x + 2y),$$

and $$t = \frac{\partial^2 f}{\partial y^2} = 2 \sin x \cos(x + 2y),$$

At $$\left(\frac{\pi}{3}, \frac{\pi}{3}\right), r = 2 \sin\frac{\pi}{3} \cos \pi = -\sqrt{3},$$

$$s = \sin\frac{4\pi}{3} = \frac{-\sqrt{3}}{2}$$

and $$t = 2 \sin\frac{\pi}{3} \cos \pi = -\sqrt{3}.$$

clearly $r < 0$ and $rt - s^2 = 3 - \frac{3}{4} = \frac{9}{4} > 0$. Hence f(x,y) is maximum at $\left(\frac{\pi}{3}, \frac{\pi}{3}\right)$ and this maximum value is given by $f_{max} = \sin\frac{\pi}{3} \sin\frac{\pi}{3} \sin\frac{2\pi}{3} = \frac{3\sqrt{3}}{8}$.

At (0,0), we get

$r = 1, s = t = 0$ such that $rt - s^2 = 0$. Thus no conclusion can be made at this stage and higher order terms must be considered.

However, at this point f(x,y) = 0, which may be minimum of the given function.

Example 6 Discuss the maximum and minimum of $x^2 + y^2 + 6x + 12$

Solution: Let $f(x,y) = x^2 + y^2 + 6x + 12$

For maximum and minimum

$$\frac{\partial f}{\partial x} = 2x + 6 = 0 \quad ...(1)$$

or $2x + 6 = 0 \Rightarrow \boxed{x = -3}$

and $\dfrac{\partial f}{\partial y} = 2y = 0$ ...(2)

$\Rightarrow \boxed{y = 0}$.

Now, $r = \dfrac{\partial^2 f}{\partial x^2} = 2, \quad s = \dfrac{\partial^2 f}{\partial x \partial y} = 0, \quad t = \dfrac{\partial^2 f}{\partial y^2} = 2$

At $(-3, 0)$, $rt - s^2 = 2 \times 2 - 0 = 4 > 0$

$r = 2 > 0$

Hence f(x,y) is minimum when

$x = -3$ and $y = 0$

Minimum value is $f(-3, 0) = 9 + 0 - 18 + 12 = 3$

Example 7 Discuss the maximum or minimum of the function $u = xy + \dfrac{a^3}{x} + \dfrac{a^3}{y}$.

Solution: we have $u = xy + \dfrac{a^3}{x} + \dfrac{a^3}{y}$

for maximum or minimum of u, we have

$$\frac{\partial u}{\partial x} = y - \frac{a^3}{x^2} = 0 \Rightarrow a^3 = x^2 y$$

and

$$\frac{\partial u}{\partial y} = x - \frac{a^3}{y^2} = 0 \Rightarrow a^3 = xy^2$$

These relations give

$$a^3 = x^2 y = xy^2 \Rightarrow x = y = a$$

Now

$$r = \frac{\partial^2 u}{\partial x^2} = \frac{2a^3}{x^3} = 2 \text{ at } (a, a)$$

$$s = \frac{\partial^2 u}{\partial x \partial y} = 1 \text{ at } (a, a)$$

and

$$t = \frac{\partial^2 u}{\partial y^2} = \frac{2a^3}{y^3} = 2 \text{ at } (a, a)$$

so that $rt - s^2 = (2)(2) - (1)^2 = 3$ (+ve)

Also r = 2 = positive. Hence the given function has a minimum at x = y = a. Also the minimum value at x = a, y = a is

$$u = a^2 + a^2 + a^2 = 3a^2.$$

Example 8 Test the function $f(x, y) = (x^2 + y^2)\,e^{-(x^2+y^2)}$ for maximum and minimum for points not on the circle $x^2 + y^2 = 1$.

Solution: Let $f(x,y) = (x^2 + y^2)\,e^{-(x^2+y^2)}$

$$\therefore \quad \frac{\partial f}{\partial x} = (x^2 + y^2)\,e^{-(x^2+y^2)}.(-2x) + 2xe^{-(x^2+y^2)}$$

$$= e^{-(x^2+y^2)}\left[1-(x^2+y^2)\right].2x$$

and
$$\frac{\partial f}{\partial y} = e^{-(x^2+y^2)}\left[1-(x^2+y^2)\right]2y$$

For maximum or minimum

$$\frac{\partial f}{\partial x} = 0 \text{ and } \frac{\partial f}{\partial y} = 0$$

$$\Rightarrow \quad 2x\left[1-(x^2+y^2)\right]e^{-(x^2+y^2)} = 0$$

and
$$2y\left[1-(x^2+y^2)\right]e^{-(x^2+y^2)} = 0$$

$$\Rightarrow \quad x = 0,\ y = 0 \text{ and } x^2 + y^2 = 1$$

leaving $x^2 + y^2 = 1$ as given, we take $x = 0, y = 0$.

Hence the point (0,0) is only stationary point.

Again,

$$\frac{\partial^2 f}{\partial x^2} = r = (2 - 6x^2 - 2y^2)\,e^{-(x^2+y^2)} + (2x - 2x^3 - 2xy^2)\,e^{-(x^2+y^2)}.(-2x)$$

$$= e^{-(x^2+y^2)}\left[4x^4 - 10x^2 + 4x^2y^2 - 2y^2 + 2\right]$$

and
$$\frac{\partial^2 f}{\partial x \partial y} = s = (-4xy)\,e^{-(x^2+y^2)} + (2x - 2x^3 - 2xy^2)\,e^{-(x^2+y^2)}.(-2y)$$

$$= e^{-(x^2+y^2)}\left[-8xy + 4x^3y + 4xy^3\right]$$

and $$\frac{\partial^2 f}{\partial y^2} = t = e^{-(x^2+y^2)}\left[2-2x^2-10y^2-4x^2y^2+4y^4\right]$$

At point (0,0), $r = 2$, $s = 0$, $t = 2$

$$rt - s^2 = 4 > 0, \text{ Also } r > 0$$

$\therefore$ f(x,y) has a minimum value at (0,0). Minimum value is $(0+0)e^{-0} = 0$.

Example 9 Find the maximum and minimum values of $x^4 + y^4 - x^2 + xy - y^2$

Solution: Let $f(x, y) = x^4 + y^4 - x^2 + xy - y^2$

for maximum or minimum,

$$\frac{\partial f}{\partial x} = 0 \Rightarrow 4x^3 - 2x + y = 0 \qquad \text{...(1)}$$

and $$\frac{\partial f}{\partial y} = 0 \Rightarrow 4y^3 + x - 2y = 0 \qquad \text{...(2)}$$

we have by subtraction

$$4\left(x^3 - y^3\right) - 3(x - y) = 0$$

$$\Rightarrow \quad x - y = 0$$

Hence from equation (1), $x = y = 0$ or $\pm\frac{1}{2}$, Again by adding (1) and (2), we get

$$4\left(x^3 + y^3\right) - (x + y) = 0$$

$$\Rightarrow \quad x + y = 0$$

from equation (2), we get $x = -y = \pm\frac{\sqrt{3}}{2}$

further, we h:ve,

$$r = \frac{\partial^2 f}{\partial x^2} = 12x^2 - 2, \; t = \frac{\partial^2 f}{\partial y^2} = 12y^2 - 2 \text{ and } s = \frac{\partial^2 f}{\partial x \partial y} = 1$$

Therefore $rt - s^2 = 4(6x^2 - 1)(6y^2 - 1) - 1$

$= 3 > 0$, when $x = 0, y = 0$

$= 0$, when $x = y = \pm\frac{1}{2}$

$= 48 > 0$, when $x = -y = \pm\frac{\sqrt{3}}{2}$

Also $\quad r = -2 < 0$, when $x = 0, y = 0$

$= 1 > 0$, when $x = y = \pm\frac{1}{2}$

$= 7 > 0$, when $x = -y = \pm\frac{\sqrt{3}}{2}$.

Hence there is maximum when $x = 0, y = 0$ and a minimum when $x = -y = \pm\frac{\sqrt{3}}{2}$, the values of $f(x, y)$ being zero and $-\frac{9}{8}$ respectively.

when $x = y = \pm\frac{1}{2}$, the test fails. Proceeding to higher degree terms of the Taylor's, expansion in this case, since

$$f\left(\pm\frac{1}{2}, \pm\frac{1}{2}\right) = -\frac{1}{8}, \ \frac{\partial f}{\partial x} = \frac{\partial f}{\partial y} = 0$$

At $\left(\pm\frac{1}{2}, \pm\frac{1}{2}\right)$

$$\frac{\partial^2 f}{\partial x^2} = 12x^2 - 2 = 1, \quad \frac{\partial^2 f}{\partial y^2} = 12y^2 - 2 = 1, \quad \frac{\partial^2 f}{\partial y^2} = 1$$

$$\frac{\partial^3 f}{\partial x^3} = 24x = \pm 12, \quad \frac{\partial^3 f}{\partial y^3} = 24y = \pm 12$$

and
$$\frac{\partial^4 f}{\partial x^4} = \frac{\partial^4 f}{\partial y^4} = 24, \quad \frac{\partial^4 f}{\partial x^3 \partial y} = \frac{\partial^4 f}{\partial x^2 \partial y^2} = \frac{\partial^4 f}{\partial x \partial y^3} = 0$$

we have
$$f\left(\pm\frac{1}{2} + h, \ \pm\frac{1}{2} + k\right)$$

$$= f\left(\pm\frac{1}{2}, \pm\frac{1}{2}\right) + \left(h\frac{\partial}{\partial x} + k\frac{\partial}{\partial y}\right) f + \frac{1}{2!}\left(h\frac{\partial}{\partial x} + k\frac{\partial}{\partial y}\right)^2 f$$

$$+ \frac{1}{3!}\left(h\frac{\partial}{\partial x} + k\frac{\partial}{\partial y}\right)^3 f + \frac{1}{4!}\left(h\frac{\partial}{\partial x} + k\frac{\partial}{\partial y}\right)^2 f +$$

$$= -\frac{1}{8} + \frac{1}{2}(h + k)^2 + 2\left(h^3 + k^3\right) + \left(h^4 + k^4\right).$$

(we may also obtain this expansion algebraically from the function f(x,y)). Therefore when h + k = 0, i.e., h = –k, we find $f\left(\pm\frac{1}{2}+h, \pm\frac{1}{2}+k\right)-\left(-\frac{1}{8}\right)=2k^4>0.$

Hence there is a minimum when x = y = $\pm\frac{1}{2}$, the minimum value of 'f' being $\left(-\frac{1}{8}\right)$.

6.18 MAXIMA AND MINIMA OF A FUNCTION OF THREE INDEPENDENT VARIABLES:

Let f(x, y, z) be a function of three independent variables and let f(a, b, c) be an extreme value of the function at the point (a, b, c). Then by definition f(a + h, b+k, c+l) – f(a, b, c) should preserve a fixed sign (positive for minimum and negative for a maximum) for all sufficiently small values of h, k and l.

Now from Taylor's expansion, we have f(a+h, b+k, c+l) – f(a, b, c)

$$=\frac{1}{2!}\left[A h^2+B k^2+C l^2+2Fkl+2Glh+2Hhk\right]+\text{ higher order terms} \quad ...(1)$$

when $\frac{\partial f}{\partial x}=\frac{\partial f}{\partial y}=\frac{\partial f}{\partial z}=0$ at x = a, y = b and z = c,

where $A=\frac{\partial^2 f}{\partial x^2}, \quad B=\frac{\partial^2 f}{\partial y^2}, \quad C=\frac{\partial^2 f}{\partial z^2}, \quad F=\frac{\partial^2 z}{\partial y \partial z},$

$G=\frac{\partial^2 f}{\partial z \partial x}, \quad H=\frac{\partial^2 f}{\partial x \partial y}$ at the point (a, b, c).

Now by taking h, k and l sufficiently small the term within the square brackets in (1) can be made to govern the sign of the R.H.S. and therefore of L.H.S. If for all such values of h, k and l, this expansion is positive, then the function f(x, y, z) has minimum at (a, b, c) and if the sign is negative then f(x, y, z) has maximum at (a, b, c).

To identify the sign, let us write

$$Ah^2+Bk^2+Cl^2+2Fkl+2Glh+2Hhk$$

$$=\frac{1}{A}\left[A^2h^2+ABk^2+ACl^2+2AFkl+2AGlh+2AHhk\right]$$

$$=\frac{1}{A}\left[(Ah+Hk+Gl)^2+\left(AB-H^2\right)k^2+2(AF-GH)\,kl+\left(CA-G^2\right)l^2\right]$$

This will have the same sign as A, if AB - $H^2>0$, and $(AB-H^2)(CA-h^2)-(AF-GH)^2>0$ i.e., if $AB-H^2>0$, and $A(ABC+2FGH-AF^2-BG^2-CH^2)>0$

we may also write these as

$$AB - H^2 = \begin{vmatrix} A & H \\ H & B \end{vmatrix} = D_1 \text{ (say) and}$$

$$ABC + 2FGH - AF^2 - BG^2 - CH^2 = \begin{vmatrix} A & H & G \\ H & B & F \\ G & F & C \end{vmatrix} = D_2 \text{ (say)}$$

hence the above expression will be positive if A, D_1, D_2 are all positive and will be negative, if these ar alternately negative and positive. Hence the function f(x, y, z) will have minimum, if $A > 0$, $D_1 > 0$ and $D_2 > 0$; and will have maximum, if $A < 0$, $D_1 > 0$, $D_2 < 0$.

Working Rule

Let the function u = f(x, y, z) have three independent variables x, y and z.

1. Find $\frac{\partial u}{\partial x}, \frac{\partial u}{\partial y}, \frac{\partial u}{\partial z}$ and equate to zero to obtain the values of x, y, z.

2. Find $A = \frac{\partial^2 u}{\partial x^2}$, $B = \frac{\partial^2 u}{\partial y^2}$, $C = \frac{\partial^2 u}{\partial z^2}$, $F = \frac{\partial^2 u}{\partial y \partial z}$, $G = \frac{\partial^2 u}{\partial z \partial x}$, $H = \frac{\partial^2 u}{\partial x \partial y}$.

Now, we have the following cases:

(a) The function u = f(x, y, z) will be minimum at (a, b, c) if the expressions

$$A, \begin{vmatrix} A & H \\ H & B \end{vmatrix}, \begin{vmatrix} A & H & G \\ H & B & F \\ G & F & C \end{vmatrix} \text{ be all positive at (a, b, c)}$$

(b) The function u = f(x, y, z) will be maximum at (a, b, c) if the expressions

$$A, \begin{vmatrix} A & H \\ H & B \end{vmatrix}, \begin{vmatrix} A & H & G \\ H & B & F \\ G & F & C \end{vmatrix} \text{ be alternately negative and positive,}$$

(c) If cases (a) and (b) are not satisfied, we have neither maximum nor minimum.

Illustrative Examples

Example 1 Discuss the maximum and minimum of the function, $f(x, y, z) = x^2 + y^2 + z^2 - xy + x - 2z$.

Solution: Here for maximum and minimum of the function, we have,

$$\frac{\partial f}{\partial x} = 2x - y + 1 = 0$$

$$\frac{\partial f}{\partial y} = 2y - x = 0$$

and $$\frac{\partial f}{\partial z} = 2z - 2 = 0$$

These equations, on solving, give

$$x = -\frac{2}{3},\ y = -\frac{1}{3},\ z = 1$$

Now, $$A = \frac{\partial^2 f}{\partial x^2} = 2,\quad B = \frac{\partial^2 f}{\partial y^2} = 2,\quad C = \frac{\partial^2 f}{\partial z^2} = 2,$$

$$F = \frac{\partial^2 f}{\partial y \partial z} = 0,\quad G = \frac{\partial^2 f}{\partial z \partial x} = 0, \text{ and } H = \frac{\partial^2 f}{\partial x \partial y} = -1$$

Hence at $\left(-\frac{2}{3}, -\frac{1}{3}, 1\right)$, we have

$$A = 2, D_1 = \begin{vmatrix} 2 & -1 \\ -1 & 2 \end{vmatrix} = 3, D_2 = \begin{vmatrix} +2 & -1 & 0 \\ -1 & 2 & 0 \\ 0 & 0 & 2 \end{vmatrix} = 6$$

since $A > 0, D_1 > 0, D_2 > 0$, we conclude that the function has a minimum at $\left(-\frac{2}{3}, -\frac{1}{3}, 1\right)$ and the minimum value is $-\frac{4}{3}$.

Example 2 Show that the function

$u = (x + y + z)^3 - 3(x + y + z) - 24xyz + a^3$ has minimum at (1, 1, 1) and maximum at (–1, –1, –1).

Solution: We have $u = (x+y+z)^3 - 3(x+y+z) - 24xyz + a^3$ ...(1)

$$\frac{\partial u}{\partial x} = 3(x+y+z)^2 - 3 - 24\,yz \qquad ...(2)$$

$$\frac{\partial u}{\partial y} = 3(x+y+z)^2 - 3 - 24zx \qquad ...(3)$$

and $$\frac{\partial u}{\partial z} = 3(x+y+z)^2 - 3 - 24\,xy \qquad ...(4)$$

for the maximum or minimum of u, we have

$$\frac{\partial u}{\partial x} = \frac{\partial u}{\partial y} = \frac{\partial u}{\partial z} = 0$$

from equation (2), (3) and (4), we get

$$x = y = z$$

substituting $y = x$ and $z = x$ in (2), we get

$$27x^2 - 3 - 24x^2 = 0$$

$\Rightarrow$ $x = \pm 1$

$\Rightarrow$ $x = y = z = 1$ and $x = y = z = -1$ are solution of (2), (3) and (4).

Hence the stationary points are

$$(1, 1, 1) \text{ and } (-1, -1, -1)$$

Now $A = \dfrac{\partial^2 u}{\partial x^2} = 6(x + y + z)$, $B = \dfrac{\partial^2 u}{\partial y^2} = 6(x + y + z)$

$$C = \frac{\partial^2 u}{\partial z^2} = 6(x + y + z), \quad F = \frac{\partial^2 u}{\partial y \partial z} = 6(x + y + z) - 24x$$

$$G = \frac{\partial^2 u}{\partial z \partial x} = 6(x + y + z) - 24y$$

$$H = \frac{\partial^2 u}{\partial x \partial y} = 6(x + y + z) - 24z.$$

for $(1, 1, 1)$

$A = 18$, $B = 18$, $C = 18$, $F = -6$, $G = -6$ and $H = -6$ at the point $(1, 1, 1)$,

we have $A = 18 > 0$

$$D_1 = \begin{vmatrix} A & H \\ H & B \end{vmatrix} = \begin{vmatrix} 18 & -6 \\ -6 & 18 \end{vmatrix} = 288 > 0$$

and $$D_2 = \begin{vmatrix} A & H & G \\ H & B & F \\ G & F & C \end{vmatrix} = \begin{vmatrix} 18 & -6 & -6 \\ -6 & 18 & -6 \\ -6 & -6 & 18 \end{vmatrix}$$

$$= 3426 > 0$$

since all these three expressions are positive therefore we have u is minimum at the point (1, 1, 1).

Again for $(-1, -1, -1)$

$A = -18$, $B = -18$, $C = -18$, $F = 6$, $G = 6$ and $H = 6$

$\therefore$ at the point $(-1, -1, -1)$ we have

$$A = -18 < 0$$

$$D_1 = \begin{vmatrix} A & H \\ H & B \end{vmatrix} = \begin{vmatrix} -18 & 6 \\ 6 & -18 \end{vmatrix} = 288 > 0$$

and $$D_2 = \begin{vmatrix} A & H & G \\ H & B & F \\ G & F & C \end{vmatrix} = \begin{vmatrix} -18 & 6 & 6 \\ 6 & -18 & 6 \\ 6 & 6 & -18 \end{vmatrix}$$

$$= -3426 < 0$$

Hence, the above three expressions are alternately negative and positive.

Hence u is maximum at (–1, –1, –1).

Example 3 Prove that the rectangular solid of maximum volume that can be inscribed in a given sphere is a cube.

Solution: Let the equation of the sphere be $x^2 + y^2 + z^2 = a^2$ and let (x, y, z) be one corner of the rectangular solid inscribed in it. Then the sides of the solid are parallel to the co-ordinate planes and the other corners are $(\pm x, \pm y, \pm z)$. Therefore the edges of the cuboid are 2x, 2y, 2z and its volume V is given by

$$V = 8xyz$$

or $$V^2 = 64\, x^2y^2\left(a^2 - x^2 - y^2\right) \qquad \left[\because z^2 = a^2 - x^2 - y^2\right] \qquad ...(1)$$

since V is maximum and V^2 is also maximum, we shall determine the maxima of (1).

$$\frac{\partial V^2}{\partial x} = 64y^2\left(2a^2x - 4x^3 - 2xy^2\right) = 0,$$

and $$\frac{\partial V^2}{\partial y} = 64x^2\left(2a^2y - 2x^2y - 4y^3\right) = 0$$

omitting the factors xy^2 and x^2y (which do not give a maximum), we get

$$2x^2 + y^2 = a^2, \quad x^2 + 2y^2 = a^2$$

By symmetry (or by subtraction),

we get x = y, therefore

$$x = y = z = \frac{a}{\sqrt{3}} \qquad ...(2)$$

Now, $$r = \frac{\partial^2\left(V^2\right)}{\partial x^2} = 64\, y^2\left(2a^2 - 12x^2 - 2y^2\right)$$

$$= \frac{64.a^2}{3}\left(2a^2 - 4a^2 - \frac{2a^2}{3}\right)$$

$$t = -64 \times \frac{8a^4}{3}, \text{ (by symmetry)}$$

$$r = \frac{\partial^2 (V^2)}{\partial x \partial y} = 64\,xy\left(4a^2 - 8x^2 - 8y^2\right)$$

$$= \frac{64}{3}a^2.\left(4a^2 - \frac{8a^2}{3} - \frac{8a^2}{3}\right)$$

$$= \frac{-64}{9} \times 4a^4$$

Therefore, $rt - s^2 > 0$, also r is –ve, so that the value (2) gives a maximum. Hence the rectangular solid of maximum volume inscribed in a sphere is a cube.

Example 4 Find the relative maximum/minimum value of the function $f(x, y) = 4x^2 + 9y^2 - 8x - 12y + 4$ within the rectangular region bounded by the lines x = 2, y = 3 and the co-ordinate axes.

Solution: Let $f(x, y) = 4x^2 + 9y^2 - 8x - 12y + 4$

Now $\frac{\partial f}{\partial x} = 8x - 8,\ \frac{\partial f}{\partial y} = 18y - 12,$

$$\frac{\partial^2 f}{\partial x^2} = 8,\quad \frac{\partial^2 f}{\partial y^2} = 18,\quad \frac{\partial^2 f}{\partial x \partial y} = 0$$

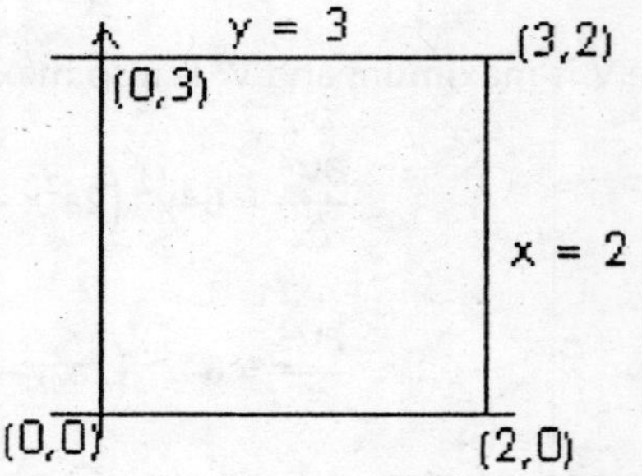

Fig. (3)

For extremum, we have

$$\frac{\partial f}{\partial x} = \frac{\partial f}{\partial y} = 0$$

we have $\frac{\partial f}{\partial x} = 0 \Rightarrow 8x - 8 = 0 \Rightarrow x = 1$

$$\frac{\partial f}{\partial y} = 0 \Rightarrow 18y - 12 = 0 \Rightarrow y = 2/3$$

Thus $\left(1, \frac{2}{3}\right)$ is a stationary point. Also from the given rectangular region, we have four other stationary points, (0,0), (0, 3), (3, 2) and (2, 0).

Now in order to find relative maximum or minimum value, we find f(x, y) at each point.

$$f(0,0) = 4,\ f(0, 3) = 4 \times 0 + 9 \times (3)^2 - 8 \times 0 - 12 \times 3 + 4$$

$$= (81) - 36 + 4 = 49$$

$$f(2, 0) = 0, \qquad f(3, 2) = 28$$

and $f\left(1, \frac{2}{3}\right) = -12.$

Thus, the function has relative minimum value = −12 at point $\left(1, \frac{2}{3}\right)$ in the given rectangular region.

Example 5 A rectangular box, open at the top, is to have a volume of 32 cc. Find the dimensions of the box requiring minimum material for its construction.

Solution: Suppose l be the length 'b' be the breath and h be the height of the box and S be the surface area and V its volume.

Given that $V = 32$ cc

volume of the box = l. b. h = 32.

or $$b = \frac{32}{l.h}$$

and $$S = 2(l + b)h + lb \qquad ...(1)$$

substituting the value of 'b' in equation (1), we get

$$S = 2\left(l + \frac{32}{lh}\right)h + l\left(\frac{32}{lh}\right)$$

$$\Rightarrow \quad S = 2lh + \frac{64}{l} + \frac{32}{h} \qquad ...(2)$$

Now differentiating (2) partially w.r.t. 'l', we get

$$\frac{\partial s}{\partial l} = 2h - \frac{64}{l^2} \qquad ...(3)$$

Again differentiating (2) partially w.r.t. 'h', we get

$$\frac{\partial s}{\partial h} = 2l - \frac{32}{h^2} \qquad ...(4)$$

For maximum or minimum

$$\frac{\partial s}{\partial l} = 0 \Rightarrow 2h - \frac{64}{l^2} = 0$$

$$\Rightarrow \quad h = \frac{32}{l^2} \qquad ...(5)$$

and $$\frac{\partial s}{\partial h} = 0 \Rightarrow 2l - \frac{32}{h^2} = 0 \Rightarrow l = \frac{16}{h^2} \qquad ...(6)$$

from equations (5) and (6), l = 4, h = 2 and b = 4

Again $$r = \frac{\partial^2 s}{\partial l^2} = \frac{128}{l^2} = \frac{128}{64} = 2.$$

$$s = \frac{\partial^2 s}{\partial l \partial h} = 2,\ t = \frac{\partial^2 s}{\partial h^2} = \frac{64}{h^2} = \frac{64}{8} = 8$$

$$\therefore \quad rt - s^2 = \frac{\partial^2 s}{\partial l^2} \cdot \frac{\partial^2 s}{\partial h^2} - \left(\frac{\partial^2 s}{\partial l \partial h} \right)^2$$

$$\therefore \quad = 2 \times 8 - (2)^2 = 12$$

so $\quad r = 2(+\text{ve})$, so S is minimum for $l = 4$, $b = 4$, $h = 2$.

Example 6 Divide 24 into three parts such that the continued product of the first, square of the second and cube of the third, is maximum.

Solution: Let 24 be divided in the parts x, y and z.

Then $\quad x + y + z = 24 \quad$ ...(1)

Given that $\quad f(x, y, z) = x^3y^2z$

$\Rightarrow \quad f(x, y) = x^3y^2(24 - x - y) \quad$ (from (1))

or $\quad f(x, y) = 24x^3y^2 - x^4y^2 - x^3y^3$

Now,
$$\frac{\partial f}{\partial x} = 72x^2y^2 - 4x^3y^2 - 3x^2y^3,$$

$$\frac{\partial^2 f}{\partial x^2} = r = 144xy^2 - 12x^2y^2 - 6xy^3$$

$$\frac{\partial f}{\partial y} = 48x^3y - 2x^4y - 3x^3y^2,$$

$$\frac{\partial^2 f}{\partial y^2} = t = 48x^3 - 2x^4 - 6x^3y \text{ and}$$

$$s = \frac{\partial^2 f}{\partial x \partial y} = 144x^2y - 8x^3y - 9x^2y^2$$

for maximum or minimum

$$\frac{\partial f}{\partial x} = 0 = 72x^2y^2 - 4x^3y^2 - 3x^2y^3$$

$$= (72 - 4x - 3y)\, x^2y^2 = 0 \quad \text{...(2)}$$

and
$$\frac{\partial f}{\partial y} = 0 = x^3y\,(48 - 2x - 3y)$$

$$\Rightarrow \quad 48 - 2x - 3y = 0,\ x = 0,\ y = 0 \quad \text{...(3)}$$

solving (2) and (3), we get

$x = 12, \quad y = 8$ etc.

At (12, 8), we have $rt - s^2 = +ve$ and $r < 0$, therefore f(x, y) is maximum at (12, 8).

Put x = 12 and y = 8 in (1), we get z = 4, The value of x = 12, y = 8 and z = 4.

This is division of 24 for maximum f(x, y, z)

Example 7 A rectangular box open at the top is to have given capacity, find the dimensions of the box requiring minimum material for its construction.

Solution: Let x, y, z be the dimensions of the box. Then its volume V and surfaces S is given by

$$V = xyz \qquad ...(1)$$

and

$$S = xy + 2(yz + zx) \qquad ...(2)$$

For the box requiring least material, S must be minimum under the condition(1).

Let us define a function *f* by the relation $f = xy + 2(yz + zx) + \lambda(xyz - V)$.

for maximum or minimum of f, we have

$$\frac{\partial f}{\partial x} = y + 2z + \lambda yz = 0 \qquad ...(3)$$

$$\frac{\partial f}{\partial y} = x + 2z + \lambda zx = 0 \qquad ...(4)$$

and

$$\frac{\partial f}{\partial z} = 2(x + y) + \lambda xy = 0 \qquad ...(5)$$

Multiplying (3) by x, (4) by y and subtracting, we get

$2zx - 2yz = 0 \Rightarrow x = y$ [z = 0 is discarded as it does not satisfy (1)]

similarly, from (4) and (5), we get y = 2z

Hence the dimensions of the box are

$$x = y = 2z = (2V)^{1/3} \quad \text{[by (1)]} \qquad ...(6)$$

Now, as x and y are independent variables, we have on differentiating (1) partially w.r.t. x

$$yz + xy\frac{\partial z}{\partial x} = 0 \Rightarrow \frac{\partial z}{\partial x} = -\frac{z}{x}$$

similarly,

$$\frac{\partial z}{\partial y} = -\frac{z}{y}$$

$$\therefore \quad \frac{\partial S}{\partial x} = y + 2z + 2(x + y)\frac{\partial z}{\partial x} = y + 2z - \frac{2z}{x}(x + y) = y - \frac{2yz}{x}$$

$$\therefore \quad r = \frac{\partial^2 S}{\partial x^2} = \frac{2yz}{x^2} - \frac{2y}{x}\cdot\frac{\partial z}{\partial x} = \frac{2yz}{x^2} + \frac{2yz}{x^2} = \frac{4yz}{x^2}$$

$$= \frac{2(2V)^{1/3}(2V)^{1/3}}{(2V)^{2/3}} = 2 \qquad \text{(by (6))}$$

Similarly, $t = 2$,

and $$s = \frac{\partial^2 S}{\partial y \partial x} = 1 - \frac{2y}{x}\frac{\partial z}{\partial y} - \frac{2z}{x} = 1 - \frac{2z}{x} + \frac{2z}{x} = 1$$

$\therefore \quad rt - s^2 = 4 - 1 = 3 > 0$

since r is +ve and $rt - s^2 > 0$, therefore when $x = y = 2z = (2V)^{1/3}$, the surface S is minimum.

Exercise 6.7

1. Find the stationary points of the function $z = x^4 + y^4 - 2x^2 + 4xy - 2y^2$ and examine their nature.
2. Examine $f(x, y) = x^3 + 3x^2y - 15x^2 - 15y^2 + 72x$ for extreme values.
3. Find the maximum value of $f(x,y) = xye^{-(2x+3y)}$.
4. Discuss the maximum and minimum of the following:

   (i) $u(x, y) = x^2 + y^2 + \frac{2}{x} + \frac{2}{y}$.

   (ii) $u(x, y) = x^2 + xy + y^2 + 3x + 4y$

   (iii) $f(x,y) = 2 \sin \frac{1}{2}(x + y) \cos \frac{1}{2}(x - y) + \cos (x + y)$

   (iv) $f(x, y) = x^2 + y^2 + 6x + 2$.

5. Show that $\frac{a}{7}, \frac{2a}{7}, \frac{3a}{7}$ is stationary point of the function, $u = axy^2z^3 - x^2y^2z^3 - xy^3z^3 - 3xy^2z^4$ and determine if it is also an extreme point.
6. Find the maximum and minimum values of $u = y^2 + 2z^2 - 5x^4 + 4x^5$.
7. The sum of the three positive numbers is constant. Show that their product is maximum when they are equal.
8. Find a plane triangle ABC such that $u = \sin^m A \sin^n B \sin^p C$ has maximum value.
9. Prove that the height of the cylinder of maximum volume that can be inscribed in a sphere of radius r is $\frac{2r}{\sqrt{3}}$.
10. Find the absolute maximum and minimum values of $u(x, y) = 2 + 2x + 2y - x^2 - y^2$ on triangular plate in the first quadrant bounded by the lines $x = 0$, $y = 0$ and $y = ax$.
11. Find the values of x and y, for which $x^2 + y^2 + 6x - 12 = 0$, has a minimum value and also find this minimum value
12. A tree trunk of length l metres has the shape of a frustum of a circular cone with radii of its ends r_1 and r_2 metres, where $r_1 > r_2$. Find the length of a beam of uniform square cross-section which can be cut from the tree trunk so that the beam has the maximum volume.
13. Show that the function

    $u(x, y) = x^3 + y^3 - 63(x + y) + 12xy$ is maximum at $(-7, -7)$ and minimum at $(3,3)$.

14. The electric time constant of cylindrical coil of wire can be expressed approximately by $\alpha = \dfrac{mxyz}{ax + by + cz}$, where z is the axial length of the coil, y is the difference between the external and internal radii and x is the mean radius; a, b, c and m denote positive constants. If the volume of the coil is fixed, find the value of x and y which make the time constant α as large as possible.
15. ABCD is a quadrilateral having no re-entrant angle and P is a point in its plane. Find the position of P for which the sum of the distances from the vertices is minimum.
16. In a plane ΔABC, find the maximum value of cos A cos B cos C.
    hint: $A + B + C = \pi$
17. Examine the function $u(x, y) = x^3y^2(6 - x - y)$ for the maximum or minimum for points not at origin.
18. Discuss the extremum values of the following:
    (i) $x^3 + y^3 + 3xy$ (ii) $x^3 + y^3 - 63(x + y) + 12xy$
    (iii) $x^3y^2(1 - x - y)$ (iv) $x^2 + 2xy + 2y^2 + 2x + y$
    (v) $\sin x + \sin y + \sin(x+y)$ (vi) $x^3 - y^2 - 7x^2 + 4y + 15x - 13$
    (vii) $x^2y^2 - 5x^2 - 8xy - 5y^2$.
19. Discuss the maximum and minimum of $u = \left(ax^2 + by^2\right)^{-x^2-y^2}$, $a \neq b$.
    what happens when a = b = 1?.
20. Show the maximum value of $(ax + by + cz)e^{-(\alpha^2x^2 + \beta^2y^2 + \gamma^2z^2)}$ is
    $$\left[\frac{1}{2e}\left(\frac{a^2}{\alpha^2} + \frac{b^2}{\beta^2} + \frac{c^2}{\gamma^2}\right)\right]^{1/2}.$$

Answers

1. $(0,0), \left(\sqrt{2}, -\sqrt{2}\right), \left(-\sqrt{2}, \sqrt{2}\right)$; z has maximum at $\left(\sqrt{2}, -\sqrt{2}\right)$.
2. (4, 0); 1/2 and (6, 0), 108.
3. $\left(\frac{1}{2}, \frac{1}{3}\right); \frac{1}{6e^2}; (0,0)$, the function does not have any extreme value.
4. (i) Minimum at (1, 1)
   (ii) Minimum at $\left(\sqrt{2}, -\sqrt{2}\right), \left(-\sqrt{2}, \sqrt{2}\right)$
   (iii) At (0, 0) the case is doubtful; at $\left(\frac{2\pi}{3}, \frac{2\pi}{3}\right)$, f(x,y) is minimum; at $\left(\frac{\pi}{3}, \frac{\pi}{3}\right)$, f(x, y) is maximum.
   (iv) At (– 3, 0), minimum value = 3

5. maximum value $= \dfrac{108a^7}{7^7}$

6. minimum at (1, 0, 0); neither maximum nor minimum at (0, 0, 0).

8. u is maximum when $\dfrac{\tan A}{m} = \dfrac{\tan B}{n} = \dfrac{\tan C}{p}$.

10. At (1, 1) maximum value = 4.

11. (–3, 0), minimum value = 3.

12. $\dfrac{8r_1^3 l}{27(r_1 - r_2)}$

15. P is the point of intersection of the diagonals of the quadrilateral.

16. At $\left(\dfrac{\pi}{3}, \dfrac{\pi}{3}\right)$, maximum value $= \dfrac{1}{8}$.

17. Maximum at (3, 2)

18. (i) Maximum value = 1 at (–1, –1).

(ii) Maximum value = +784 at (–7, –7)

Minimum value = – 216 at (3,3)

(iii) No extreme value at (0, 0)

Maximum value $= \dfrac{1}{432}$ at $\left(\dfrac{1}{2}, \dfrac{1}{3}\right)$

(iv) Minimum value $= -\dfrac{5}{4}$ at $\left(-\dfrac{3}{2}, \dfrac{1}{2}\right)$

(v) Maximum at $x = y = \dfrac{\pi}{3}$.

(vi) Neither maximum nor minimum at (3, 2).

Maximum at $x = \dfrac{5}{3}, y = 2$

(vii) Maximum at x = 0, y = 0

Neither maximum nor minimum at (3, 3), (–3, –3) and (1, –1).

19. Maximum or Minimum at (0, 0) according as a and b both are negative or positive; Maximum or minimum at (0, ± 1) according as b > 0 and a < b or b < 0 and a > b. Maximum or minimum at (± 1, 0) according as a > 0 and b < a or a < 0 and b > a. when a = b = 1, maximum at all points where $x^2 + y^2 = 1$ and minimum at (0, 0).

6.19 MAXIMUM AND MINIMUM OF FUNCTIONS SUBJECTED TO CONSTRAINTS:

In this section, we shall consider the problems of determining the maxima and minima of functions of several variables in a different form.

Suppose

$$u = f(x_1, x_2, \ldots\ldots\ldots x_n) \qquad \ldots(1)$$

is a function of n variables $x_1, x_2, \ldots\ldots\ldots x_n$ which are not all independent, but are connected by means of m ($m < n$) equations (constraints)

$$\left.\begin{aligned} &\phi_1 (x_1, x_2, \ldots\ldots\ldots, x_n) = 0 \\ &\phi_2 (x_1, x_2, \ldots\ldots\ldots, x_n) = 0 \\ &\ldots\ldots\ldots\ldots\ldots\ldots\ldots\ldots\ldots\ldots \\ &\ldots\ldots\ldots\ldots\ldots\ldots\ldots\ldots\ldots\ldots \\ &\phi_m (x_1, x_2, \ldots\ldots\ldots\ldots x_n) = 0 \end{aligned}\right\} \qquad \ldots(2)$$

So that only n-m of the variables are independent and we can eliminate m dependent variables between (1) and the m equations (2). Thus u becomes a function of (n-m) variables without any constraints and the problem is now equivalent to the evaluation of maxima and minima of a function of (n-m) independent variables. The method of elimination of the variables is quite inconvenient and sometimes it is rather impossible. We, therefore, explain below the method known as Lagrange's method of undetermined multipliers.

6.20 LAGRANGE'S METHOD OF UNDETERMINED MULTIPLIERS.

For a maximum or minimum of u, we have

$$du = \frac{\partial f}{\partial x_1} dx_1 + \frac{\partial f}{\partial x_2} dx_2 + \ldots\ldots\ldots\ldots + \frac{\partial f}{\partial x_n} dx_n = 0 \qquad \ldots(3)$$

where the differentials $dx_1, dx_2, \ldots\ldots\ldots\ldots, dx_n$ not all independent, but are connected by the equations,

$$\left.\begin{aligned} d\phi_1 &= \frac{\partial \phi_1}{\partial x_1} dx_1 + \frac{\partial \phi_1}{\partial x_2} dx_2 + \ldots\ldots\ldots\ldots + \frac{\partial \phi_1}{\partial x_n} dx_n = 0 \\ d\phi_2 &= \frac{\partial \phi_2}{\partial x_1} dx_1 + \frac{\partial \phi_2}{\partial x_2} dx_2 + \ldots\ldots\ldots\ldots + \frac{\partial \phi_2}{\partial x_n} dx_n = 0 \\ &\ldots\ldots\ldots\ldots\ldots\ldots\ldots\ldots\ldots\ldots\ldots\ldots\ldots\ldots\ldots\ldots \\ &\ldots\ldots\ldots\ldots\ldots\ldots\ldots\ldots\ldots\ldots\ldots\ldots\ldots\ldots\ldots\ldots \\ d\phi_m &= \frac{\partial \phi_m}{\partial x_1} dx_1 + \frac{\partial \phi_m}{\partial x_2} dx_2 + \ldots\ldots\ldots\ldots + \frac{\partial \phi_m}{\partial x_n} dx_n = 0 \end{aligned}\right\} \qquad \ldots(4)$$

obtained by differentiating the given m equations (2).

Multiplying equations (4) by some constraints $\lambda_1, \lambda_2, \ldots\ldots \lambda_m$ respectively and adding them to (3), we get

$$P_1\,dx_1 + P_2dx_2 + \ldots\ldots\ldots + P_ndx_n = 0 \qquad \ldots(5)$$

where $$P_r = \frac{\partial f}{\partial x_r} + \lambda_1 \frac{\partial \phi_1}{\partial x_r} + \lambda_2 \frac{\partial \phi_2}{\partial x_r} + \ldots\ldots\ldots + \lambda_m \frac{\partial \phi_m}{\partial x_r}$$

since $\lambda_1, \lambda_2, \ldots\ldots\ldots \lambda_m$ are arbitrary, these can be so chosen that they satisfy the m equations

$$P_1 = P_2 = \ldots = P_m = 0 \qquad \ldots(6)$$

Equation (5) then reduces to

$$P_{m+1}\,dx_{m+1} + P_{m+2}\,dx_{m+2} + \ldots\ldots\ldots + P_ndx_n = 0 \qquad \ldots(7)$$

As already explained, (n-m) of the n variables are independent (no matter which of them are). we regard $x_{m+1}, x_{m+2}, \ldots\ldots\ldots x_n$ as independent variables, consequently the coefficients of $dx_{m+1}, dx_{m+2}, \ldots\ldots$ dx_n in equation (7) must be separately zero. Hence we must have

$$P_{m+1} = P_{m+2} = \ldots\ldots\ldots = P_n = 0$$

In this way, we obtain (n+m) equations

$$P_1 = P_2 = \ldots\ldots\ldots = P_n = 0$$

and $$\phi_1 = \phi_2 = \ldots\ldots\ldots = \phi_m = 0$$

which determine the m multipliers $\lambda_1, \lambda_2, \ldots\ldots\ldots \lambda_m$ and the values of n variables $x_1, x_2, \ldots\ldots x_n$ for which the function $u = f(x_1, x_2, \ldots\ldots\ldots x_n)$ possesses a maximum or minimum under the constraints (2).

Note: The lagrange's method does not determine the nature of the stationary point.

Remark: The above equations which determine the values of the variables for which the function u has a maximum or minimum may be conveniently written as follows:

Multiplying equations (4) by $\lambda_1, \lambda_2, \ldots\ldots \lambda_m$ respectively and add them to (3), we get

$$\left[\frac{\partial f}{\partial x_1} + \lambda_1 \frac{\partial \phi_1}{\partial x_1} + \lambda_2 \frac{\partial \phi_2}{\partial x_1} + \ldots\ldots + \lambda_m \frac{\partial \phi_m}{\partial x_1}\right] dx_1 +$$

$$\left[\frac{\partial f}{\partial x_2} + \lambda_1 \frac{\partial \phi_1}{\partial x_2} + \lambda_2 \frac{\partial \phi_2}{\partial x_2} + \ldots\ldots + \lambda_m \frac{\partial \phi_m}{\partial x_2}\right] dx_2 +$$

$$\ldots\ldots + \left[\frac{\partial f}{\partial x_n} + \lambda_1 \frac{\partial \phi_1}{\partial x_n} + \lambda_2 \frac{\partial \phi_2}{\partial x_n} + \ldots\ldots + \lambda_m \frac{\partial \phi_m}{\partial x_n}\right] dx_n = 0$$

or $$\frac{\partial f}{\partial x_1} dx_1 + \frac{\partial f}{\partial x_2} dx_2 + \ldots\ldots + \frac{\partial f}{\partial x_n} dx_n = 0 \qquad \ldots(A)$$

where we have defined the function F by the relation

$$F = f + \lambda_1\phi_1 + \lambda_2\phi_2 + \ldots\ldots + \lambda_m\phi_m \qquad \ldots(B)$$

Evidently, F is the function of n variables $x_1, x_2, \ldots\ldots\ldots x_n$.

The constants $\lambda_1, \lambda_2, \ldots\ldots\ldots, \lambda_m$ are at our choice, choose them to satisfy the m equations

$$\frac{\partial f}{\partial x_1} = \frac{\partial f}{\partial x_2} = \frac{\partial f}{\partial x_3} = \ldots\ldots = \frac{\partial f}{\partial x_m} = 0 \qquad \ldots(C)$$

Equation (A) then reduces to

$$\frac{\partial f}{\partial x_{m+1}} dx_{m+1} + \frac{\partial f}{\partial x_{m+2}} dx_{m+2} + \ldots\ldots + \frac{\partial f}{\partial x_n} dx_n = 0 \qquad \ldots(D)$$

since (n-m) of n variables are independent, let them be $x_{m+1}, x_{m+2}, \ldots\ldots, x_n$. Therefore, the coefficients of $dx_{m+1}, dx_{m+2}, \ldots\ldots, dx_n$ in (D) must be separately zero. Thus we must have

$$\frac{\partial f}{\partial x_{m+1}} = \frac{\partial f}{\partial x_{m+2}} = \ldots\ldots = \frac{\partial f}{\partial x_n} = 0.$$

Hence the conditions, that the function has a maximum or minimum are,

$$\frac{\partial f}{\partial x_1} = \frac{\partial f}{\partial x_2} = \frac{\partial f}{\partial x_3} = \ldots\ldots = \frac{\partial f}{\partial x_n} = 0 \qquad \ldots(E)$$

which together with m equations (2) determine the required values of the variables.

But equations (E) are also the conditions that the function F regarded as a function of n independent variables $x_1, x_2, x_3, \ldots\ldots, x_n$ should have a maximum or minimum. It follows that the problem of maximizing or minimizing $u = f(x_1, x_2, \ldots\ldots, x_n)$ subject to m constraints is (2) is equivalent to finding of a maximum or minimum of F.

Working Rule: Let $u = F(x, y, z)$ be a function of three variables x, y, z which are connected by the relation $\phi(x, y, z) = 0$

(i) construct a new function $U = F(x, y, z) + \lambda\phi(x, y, z)$

(ii) Obtain the equations $\frac{\partial U}{\partial x} = 0, \frac{\partial U}{\partial y} = 0$ and $\frac{\partial U}{\partial z} = 0$

(iii) Solve the above equations together with $\phi(x, y, z) = 0$. The values of x, y, z so obtained will give the stationary values of f(x, y, z).

Remarks: Although the lagrange's method is often very useful in applications yet the drawback is that we cannot determine the nature of the stationary point. This can sometimes be determined from physical consideration of the problems.

Illustrative Examples

Example 1 Discuss maximum and minimum of the function $u(x, y, z) = \sin x \sin y \sin z$, where x, y, z are the angles of the triangle.

Solution: Here we have,

$$u(x, y, z) = \sin x \sin y \sin z \qquad \ldots(1)$$

where the variables are connected by the relation $x + y + z = \pi$ ...(2)

For maximum or minimum of u, we have

$$du = \cos x \sin y \sin z\, dx + \sin x \cos y \sin z\, dy + \sin x \sin y \cos z\, dz = 0 \qquad \ldots(3)$$

Also from (2)

$$dx + dy + dz = 0 \qquad ...(4)$$

Multiplying (4) by λ and adding it to (3), then equating the coefficients of dx, dy, dz to zero, we get

$$\cos x \sin y \sin z + \lambda = 0$$

$$\sin x \cos y \sin z + \lambda = 0$$

$$\sin x \sin y \cos z + \lambda = 0$$

from these equations, we get

$$-\lambda = \cos x \sin y \sin z = \sin x \cos y \sin z = \sin x \sin y \cos z$$

i.e., $\cot x = \cot y = \cot z$ (on dividing $\sin x \sin y \sin z$)

i.e., $x = y = z = \frac{\pi}{3}$, by using (2)

since only relation (2) is given among x, y, z therefore we may regard two variables say, x, y as independent and z as dependent. Now to determine a maximum or minimum, we have to find r, s, t from (2), on differentiation, we get

$$1+\frac{\partial z}{\partial x} = 0, \; and \; 1+\frac{\partial z}{\partial y} = 0$$

Now from (1)

$$\frac{\partial u}{\partial x} = \cos x \sin y \sin z + \sin x \sin y \cos z \frac{\partial z}{\partial x}.$$

$$= \cos x \sin y \sin z - \sin x \sin y \cos z \left[\because \frac{\partial z}{\partial x} = -1\right.$$

$$= \sin y\,(\cos x \sin z - \sin x \cos z)$$

$$= \sin y \sin(z-x).$$

and $\frac{\partial^2 u}{\partial x^2} = \sin y \cos(z-x)\left(\frac{\partial z}{\partial x} - 1\right) = -2\sin y \cos(z-x).$

similarly, $\frac{\partial^2 u}{\partial y^2} = -2\sin x \cos(z-y)$

Now, $\frac{\partial^2 u}{\partial x \partial y} = \cos y \sin(z-x) + \sin y \cos(z-x)\left(\frac{\partial z}{\partial y}\right)$

$$= \cos y \sin(z-x) - \sin y \cos(z-x)$$

$$= \sin(z-x-y)$$

Hence putting $x = y = z = \frac{\pi}{3}$, we get

$$r = \frac{\partial^2 u}{\partial x^2} = -2\sin\frac{\pi}{3}, \cos 0 = -\sqrt{3}.$$

and $$t = -\sqrt{3}, s = \frac{\partial^2 u}{\partial y \partial x} = \sin\left(\frac{\pi}{3} - \frac{\pi}{3} - \frac{\pi}{3}\right) = -\frac{\sqrt{3}}{2}$$

$\therefore$ $$rt - s^2 = \left(-\sqrt{3}\right)^2 - \left(\frac{-\sqrt{3}}{2}\right)^2 = 3 - \frac{3}{4} > 0$$

since $r < 0$ and $rt - s^2 > 0$, $\therefore$ u is maximum at $x = y = z = \frac{\pi}{3}$.

Example 2 Find the maximum or minimum value of $u = x^2 + y^2 + z^2$, subject to the condition $xy + yz + zx = 3a^2$.

Solution: we have,

$$u = x^2 + y^2 + z^2 \qquad ...(1)$$

and $$\phi = xy + yz + zx - 3a^2 \qquad ...(2)$$

Consider the lagrange's function,

$$F(x, y, z) = (x^2 + y^2 + z^2) + \lambda(xy + yz + zx - 3a^2)$$

for stationary values, $dF = 0$

$$[2x + \lambda(y + z)] + [2y + \lambda(z + x)]\,dy + [2z + \lambda(x + y)]\,dz = 0$$

$\Rightarrow$ $$2x + \lambda(y + z) = 0 \qquad ...(3)$$

$$2y + \lambda(z + x) = 0 \qquad ...(4)$$

$$2z + \lambda(x + y) = 0 \qquad ...(5)$$

Multiplying (3) by y, (4) by z and (5) by x, and adding, we get

$$2(xy + yz + zx) + \lambda(y^2 + yz + zx + z^2 + x^2 + yz) = 0$$

$\Rightarrow$ $$2(3a^2) + \lambda(u + 3a^2) = 0 \qquad ...(A)$$

$\because$ $xy + yz + zx = 3a^2$ and $u = x^2 + y^2 + z^2$

Now, after multiplying (3) by x, (4) by y and (5) by z and adding, we get

$$u = -3a^2\lambda$$

Now (A) $\Rightarrow$ $$6a^2 + \lambda(-3a^2\lambda + 3a^2) = 0$$

$$6a^2 - 3a^2\lambda(\lambda - 1) = 0$$

$\Rightarrow$ $2 - \lambda^2 + \lambda = 0$ $\Rightarrow$ $\lambda = -1, 2$

for $\lambda = -1$,

$$\lambda = \frac{-u}{3a^2} = \frac{x^2 + y^2 + z^2}{3a^2};\ u = 3a^2 \text{ maximum value}$$

for $\lambda = 2$; $u = -6a^2$ is the minimum value.

Example 3 Find the maximum value of $x^2 + y^2 + z^2$, given that $ax + by + cz = P$.

Solution: Let $u(x, y, z) = x^2 + y^2 + z^2$,

and given that $ax + by + cz = P$ ...(1)

For maximum or minimum

$$du = 2xdx + 2ydy + 2zdz = 0$$

and $adx + bdy + cdz = 0$

Therefore, using lagrange's method of undetermined multipliers,

$$2x + \lambda a = 0, \quad 2y + \lambda b = 0, \quad 2z + \lambda c = 0 \quad ...(2)$$

Multiplying, these equations by x, y, z respectively and adding, we get,

$$2(x^2 + y^2 + z^2) + \lambda(ax + by + cz) = 0$$

$$\Rightarrow \quad 2u_m + \lambda p = 0 \quad ...(3)$$

where u_m indicates the minimum value of u. Also from (1) and (2), we get

$$a\left(-\frac{\lambda a}{2}\right) + b\left(-\frac{\lambda b}{2}\right) + c\left(-\frac{\lambda c}{2}\right) = P$$

$$\Rightarrow \quad \lambda = \frac{-2P}{a^2 + b^2 + c^2}$$

substituting these values in equation (3), we get

$$u_m = \frac{P^2}{(a^2 + b^2 + c^2)}$$

This is the minimum value (and not maximum) can be seen that from the geometrical consideration, u represents the square of the distance of a point on the plane (1) from the origin. This distance decreases as the point approaches from infinity towards the origin and after achieving a minimum again starts increasing as the point moves away

Example 4 Find the maxima and minima of $x^2 + y^2 + z^2$ subject to the conditions $ax^2 + by^2 + cz^2 = 1$ and $lx + my + nz = 0$

Solution: Here $u = x^2 + y^2 + z^2$ ...(1)

where the relations between the variables x, y and z are given by

$$ax^2 + by^2 + cz^2 = 1 \quad ...(2)$$

and $lx + my + nz = 0$ ...(3)

for the maxima and minima, we have $du = 0$

$$\Rightarrow \quad 2x\,dx + 2ydy + 2zdz = 0$$

$$\Rightarrow \quad xdx + ydy + zdz = 0 \quad ...(4)$$

from (2) and (3), we get

$$axdx + bydy + czdz = 0 \quad ...(5)$$

and $\quad ldx + mdy + ndz = 0 \quad ...(6)$

Now multiplying (4) by (l), (5) by l_1 and (6) by l_2 and adding, we get

$$(xdx + ydy + zdz) + l_1 (axdx + bydy + czdz) + l_2 (ldx + mdy + ndz) = 0$$

or
$$(x + al_1x + l\,l_2)dx + (y + bl_1y + ml_2)dy + (z + c\,l_1z + nl_2)\,dz = 0$$

Now equating the coefficients of dx, dy, dz to zero, we get

$$x + l_1ax + l_2l = 0 \quad ...(7)$$

$$y + l_1by + l_2m = 0 \quad ...(8)$$

and
$$z + l_1cy + l_2n = 0 \quad ...(9)$$

multiplying the equations (7), (8) and (9) by x, y and z respectively, then adding, we get

$$(x^2 + y^2 + z^2) + l_1 (ax^2 + by^2 + cz^2) + l_2 (lx + my + nz) = 0$$

$$\Rightarrow \quad u + l_1.1 + l_2.(0) = 0 \qquad \text{(from (1), (2) and (3))}$$

$$\Rightarrow \quad u = -l_1$$

Putting for l_1 in the equations (7), (8) and (9) we get

$$x = \frac{l_2l}{au-1}, \quad y = \frac{l_2m}{bu-1}, \quad z = \frac{l_2n}{cu-1} \quad ...(10)$$

from (3) and (10), we get

$$\frac{l_2l^2}{au-1} + \frac{l_2m^2}{bu-1} + \frac{l_2n^2}{cu-1} = 0$$

$$\Rightarrow \quad \frac{l^2}{au-1} + \frac{m^2}{bu-1} + \frac{n^2}{cu-1} = 0 \quad ...(11)$$

which gives the maximum and minimum of

$$u = x^2 + y^2 + z^2$$

Note: (i) The equation (11) is a quadratic in u. So it gives two stationary values of u. (ii) Geometrically, the surface $ax^2 + by^2 + cz^2 = 1$ denotes an ellipsoid whose center is origin and $lx + my + nz = 0$ denotes a plane passing through the origin. The points (x, y, z) satisfying both the conditions (2) and (3) lie on the conic in which (2) and (3) intersect. The maximum value of this distance is the major axis of this conic and the minimum value of this distance is the minor axis of this conic. Hence, equation (11) gives the squares of the lengths of the semi-axes of the conic of intersection.

Example 5 Find the maximum and minimum value of $u = \frac{x^2}{a^4} + \frac{y^2}{b^4} + \frac{z^2}{c^4}$ subject to the conditions $lx + my + nz = 0$ and $\frac{x^2}{a^2} + \frac{y^2}{b^2} + \frac{z^2}{c^2} = 1$

Solution: Let us define $F = \frac{x^2}{a^4} + \frac{y^2}{b^4} + \frac{z^2}{c^4} + \lambda_1(lx + my + nz) + \lambda_2\left(\frac{x^2}{a^2} + \frac{y^2}{b^2} + \frac{z^2}{c^2} - 1\right)$

for maxima or minima of F, we have

$$\frac{\partial F}{\partial x} = \frac{2x}{a^4} + \lambda_1 . l + \lambda_2 \frac{2x}{a^2} = 0 \quad ...(1)$$

$$\frac{\partial F}{\partial y} = \frac{2y}{b^4} + \lambda_1 . m + \lambda_2 \frac{2y}{b^2} = 0 \quad ...(2)$$

and $$\frac{\partial F}{\partial z} = \frac{2z}{c^4} + \lambda_1 . n + \lambda_2 \frac{2z}{c^2} = 0 \quad ...(3)$$

multiplying (1) by x, (2) by y and (3) by z, and adding, we get

$$2u + \lambda_1 (0) + 2\lambda_2 = 0 \Rightarrow \lambda_2 = -u.$$

$\therefore$ from (1), $$\frac{2x}{a^4} + \lambda_1 . l - \frac{2ux}{a^2} = 0$$

$$\Rightarrow \quad x = -\frac{\lambda_1 \, l \, a^4}{2\left(1 - ua^2\right)}$$

similarly, $$y = -\frac{\lambda_1 \, m \, b^4}{2\left(1 - ub^2\right)}$$

and $$z = -\frac{\lambda_1 \, n \, c^4}{2\left(1 - uc^2\right)}$$

substituting these values of x, y, z in

$$lx + my + nz = 0, \text{ we get}$$

$$\frac{l^2 a^4}{1 - a^2 u} + \frac{m^2 b^4}{1 - b^2 u} + \frac{n^2 c^4}{1 - c^2 u} = 0 \quad ...(4)$$

This equation gives the required maximum or minimum value of u.

Geometrical Interpretation; $\frac{x^2}{a^4} + \frac{y^2}{b^4} + \frac{z^2}{c^4}$ is the reciprocal of the square of the distance of the tangent plane $\frac{Xx}{a^2} + \frac{Yy}{b^2} + \frac{Zz}{c^2} = 1$ to the ellipsoid $\frac{x^2}{a^2} + \frac{y^2}{b^2} + \frac{z^2}{c^2} = 1$ from the origin. If the point (x, y, z) lies on the plane $lx + my + nz = 0$, then the problem is to find the maximum or minimum of the reciprocal of the square of the distance of the tangent plane to the ellipsoid from the origin at the points common to the plane $lx + my + nz = 0$ and the ellipsoid.

Example 6 Using the lagrange's method (of undetermined multipliers). Find shortest distance from the point (1, 2, 2) to the sphere $x^2 + y^2 + z^2 = 36$.

Solution : Let a point (x, y, z) be on the sphere, then its distance from (1, 2, 2) is

$$F = \sqrt{(x-1)^2 + (y-2)^2 + (z-2)^2}$$

Obviously, the distance and the square of the distance have extremum values for the same values of x, y, z.

Consider $F(x, y, z) = \{(x-1)^2 + (y-2)^2 + (z-2)^2\} + \lambda(x^2 + y^2 + z^2 - 36)$

where λ is a multiplier. For the stationary

value of F, $\frac{\partial F}{\partial x} = 0$, $\frac{\partial F}{\partial y} = 0$, $\frac{\partial F}{\partial z} = 0$

$$\therefore \qquad \frac{\partial F}{\partial x} = 0 = 2(x-1) + 2\lambda x$$

$$\frac{\partial F}{\partial y} = 0 = 2(y-2) + 2\lambda y$$

and
$$\frac{\partial F}{\partial z} = 0 = 2(z-2) + 2\lambda z$$

from these, we obtain, $x = \frac{1}{1+\lambda}$, $y = \frac{2}{1+\lambda}$, $z = \frac{2}{1+\lambda}$ putting these in equation

$x^2 + y^2 + z^2 = 36$, we obtain

$$\frac{1}{(1+\lambda)^2} + \frac{4}{(1+\lambda)^2} + \frac{4}{(1+\lambda)^2} = 36$$

$$\Rightarrow \qquad (1+\lambda) = \pm\frac{1}{2} \text{ giving } \lambda = -\frac{1}{2}, -\frac{3}{2}.$$

Hence, the points are (2, 4, 4) and (−2, −4, −4), thus the minimum distance

$$= \sqrt{(1-2)^2 + (2-4)^2 + (2-4)^2}$$

$$= \sqrt{(-1)^2 + (-2)^2 + (-2)^2}$$

$$= \sqrt{9} = 3.$$

and the maximum value is

$$= \sqrt{(1+2)^2 + (2+4)^2 + (2+4)^2}$$

$$= \sqrt{9 + 36 + 36}$$

$$= \sqrt{81} = 9.$$

Example 7 If two variables x and y are connected by the relation $ax^2 + by^2 = ab$, show that the maximum and minimum values of the function $x^2 + y^2 + xy$ will be the values of u given by the relation

$$4(u - a)(u - b) = ab$$

Solution: we have $u = x^2 + y^2 + xy$

where $ax^2 + by^2 = ab$

consider $f(x, y) = x^2 + y^2 + xy + \lambda(ax^2 + by^2 - ab)$, where λ is a multiplier. The condition for maximum or minimum are

$$\frac{\partial f}{\partial x} = 2x + y + 2\lambda ax = 0$$

and
$$\frac{\partial f}{\partial y} = 2y + x + 2\lambda by = 0$$

multiplying these equations by x and y respectively, and adding, we get by the relations,

$$u + \lambda ab = 0$$

Also writing the equation as

$$2(1 + \lambda a)x = -y \text{ and } x = -2(1 + \lambda b)y.$$

since x and y cannot be zero, therefore we find

$$4(1 + \lambda a)(1 + \lambda b) = 1 \quad \ldots(1)$$

Therefore $4(1 - 4/b)(1 - 4/a) = 1$

or $4(u - a)(u - b) = ab \quad \ldots(2)$

which gives the stationary values of u,

To find whether these are the extremum values of u, we consider

$$d^2f = 2(dx^2 + dxdy + dy^2) + 2\lambda(adx^2 + bdy^2)$$
$$= 2(1 + \lambda a)dx^2 + 2dxdy + 2(1 + \lambda b)dy^2$$
$$= 2(1 + \lambda a)\{dx + 2(1 + \lambda b)dy\}^2 \quad \text{by} \quad (1)$$

Hence d^2f has a fixed sign (the same as that of $(1 + \lambda a)$ or $(1 + \lambda b)$) and the values of u is given by (2) are the true maxima or minima.

from (2), we obtain

$$4u^2 - 4u(a + b) + 3ab = 0$$

or $2u = (a + b) \pm \{a - b\}^2 + ab\}^{1/2}$

Therefore, supposing a and b to be positive, we see that with the upper sign,

$$2u > (a + b)\,|a - b|$$

i.e., $u >$ a or b whichever is greater and with lower sign,

$$2u < (a - b) - |a - b|,$$

i.e., $u <$ a or be whichever is less.

Again, since $u + \lambda ab = 0$, we find that when $u > a\ (a > b)$, $1 + \lambda b < 0$, and when $u < b\ (b < a)$, $1 + \lambda a > 0$.

It follows that the value of u with upper sign is the maximum while the value with the lower sign is minimum.

Further, since $2(1 + \lambda a) x = -y$, we observe that when $1 + \lambda a < 0$, x and y have the same sign, and there are two points in the first and the third quadrants at which u is maximum.

Similarly, at two points, lying in the second and the fourth quadrants, u is minimum.

Example 8 Find the maximum or minimum values of $x^p y^q z^r$ subject to the condition

$$ax + by + cz = p + q + r.$$

Solution: Here we have $V = x^p y^q z^r$ ...(1)

The variables x, y, z are connected by the relation $ax + by + cz = p + q + r$ ...(2)

from (1) taking logarithm on both sides, we get

$$\log V = p \log x + q \log y + r \log z.$$

$$\Rightarrow \quad \frac{1}{V} dV = \frac{p}{x} dx + \frac{q}{y} dy + \frac{r}{z} dz$$

for the maxima and minima of V, $dV = 0$

$$\Rightarrow \quad \frac{p}{x} dx + \frac{q}{y} dy + \frac{r}{z} dz = 0 \quad ...(3)$$

from (2), we get $a\,dx + b\,dy + c\,dz = 0$ ...(4)

multiplying (3) by –1 and (4) by l and equating the coefficients of dx, dy and dz to zero,

we get, $$x = \frac{-p}{la}, \quad y = \frac{-q}{lb}, \quad z = \frac{-r}{lc}$$

substituting the values of x, y and z in (2), we get $l = -1$.

Therefore u is stationary when

$$x = \frac{p}{a}, \quad y = \frac{q}{b} \text{ and } z = \frac{r}{c}$$

Now, we find the nature of this stationary value of V.

Let us regard x and y as independent variable and z as the function of x and y from (1), we get

$$\log V = p \log x + q \log y + r \log z$$

$$\frac{1}{V} \frac{\partial V}{\partial x} = \frac{p}{x} + \frac{r}{z} \frac{\partial z}{\partial x}$$

Again differentiating (2) partially w.r.t. x, we get

$$\frac{\partial z}{\partial x} = \frac{-a}{c} \Rightarrow \frac{1}{V} \frac{\partial V}{\partial x} = \frac{p}{x} - \frac{r}{z} \cdot \frac{a}{c},$$

$$\Rightarrow \quad \frac{1}{V} \frac{\partial^2 V}{\partial x^2} - \frac{1}{V^2} \left(\frac{\partial V}{\partial x}\right)^2 = \frac{-p}{x^2} + \frac{a}{c} \cdot \frac{r}{z^2} \frac{\partial z}{\partial x} = \frac{-p}{x^2} - \frac{a^2}{c^2} \frac{r}{z^2}$$

At stationary point $\frac{\partial V}{\partial x} = 0$ which gives $\frac{1}{V^2}\frac{\partial^2 V}{\partial x^2} = \frac{-p}{x^2} - \frac{a^2 r}{c^2 z^2}$ which is negative at stationary point.

$\Rightarrow$ V is maximum at the stationary point.

$\therefore$ The maximum value is given by

$$V = \left(\frac{p}{a}\right)^p \left(\frac{q}{b}\right)^q \left(\frac{r}{c}\right)^r$$

Example 9 Let t be the temperature at any point (x, y, z) in space s = 400 xyz^2. Find the maximum temperature at the surface of a unit sphere $x^2 + y^2 + z^2 = 1$.

Solution: Let $s = 400\,xyz^2$ and $x^2 + y^2 + z^2 = 1$

let $F(x, y, z) = x^2 + y^2 + z^2 - 1$

Consider the lagrange's method, we have

$$\frac{\partial s}{\partial x} + \lambda\frac{\partial F}{\partial x} = 0$$

$$\Rightarrow \quad 400\,yz^2 + \lambda(2x) = 0 \quad ...(1)$$

$$\frac{\partial s}{\partial y} + \lambda\frac{\partial F}{\partial y} = 0$$

$$\Rightarrow \quad 400\,xz^2 + (2y)\,\lambda = 0 \quad ...(2)$$

and

$$\frac{\partial s}{\partial z} + \lambda\frac{\partial F}{\partial z} = 0$$

$$\Rightarrow \quad 800\,xyz + \lambda(2z) = 0 \quad ...(3)$$

multiplying equation (1) by x, (2) by y and (3) by z and adding, we have

$$1600\,xyz^2 + 2\lambda\,(x^2 + y^2 + z^2) = 0$$

$$\Rightarrow \quad 1600\,xyz^2 + 2\lambda(1) = 0$$

$$\Rightarrow \quad \lambda = -800\,xyz^2$$

substituting the value of λ in (1), we get

$$400\,yz^2 + 2x\,(-800\,xy\,z^2) = 0$$

$$1 - 4x^2 = 0 \Rightarrow \quad x = \pm\frac{1}{2}$$

similarly $\quad y = \pm\frac{1}{2}$

substituting the value of λ in equation (3), we get

$$800\,xyz - 1600\,xy\,z^3 = 0$$

$$1 - 2z^2 = 0 \quad \Rightarrow \quad z = \pm\frac{1}{\sqrt{2}}.$$

substituting the value of x, y and z in s, we get

$$s = 400 \times \frac{1}{2} \times \frac{1}{2} \times \frac{1}{2} = 50$$

Example 10 Find the volume of the largest rectangular parallelopiped that can be inscribed in the ellipsoid

$$\frac{x^2}{a^2} + \frac{y^2}{b^2} + \frac{z^2}{c^2} = 1$$

Solution: If (x, y, z) are the coordinates of one vertex of the parallelopiped, the lengths of three coterminous edges are 2x, 2y, 2z so that volume is given by

$$V = (2x)(2y)(2z) = 8xyz \qquad \text{...(1)}$$

subject to the condition $\frac{x^2}{a^2} + \frac{y^2}{b^2} + \frac{z^2}{c^2} = 1$ ...(2)

To find the maximum value of V

Taking log on both sides of (1), we get

$$\log V = \log s + \log x + \log y + \log z$$

$$\Rightarrow \quad \frac{1}{V} dV = \frac{dx}{x} + \frac{dy}{y} + \frac{dz}{z}$$

for maximum V, $\frac{dx}{x} + \frac{dy}{y} + \frac{dz}{z} = 0$ ...(3)

Also, $\frac{2x}{a^2} dx + \frac{2y}{b^2} dy + \frac{2z}{c^2} dz + 0$ ...(4)

multiplying (4) by λ and adding to (3) and then equating the coefficients of dx, dy, dz to zero, we get

$$\frac{1}{x} + \frac{\lambda x}{a^2} = 0, \quad \frac{1}{y} + \frac{\lambda y}{b^2} = 0, \quad \frac{1}{z} + \frac{\lambda z}{c^2} = 0$$

multiplying these equations by x, y, z respectively and adding, we have

$$3 + \lambda\left(\frac{x^2}{a^2} + \frac{y^2}{b^2} + \frac{z^2}{c^2}\right) = 0$$

$$3 + \lambda(1) = 0 \qquad \text{[from (2)]}$$

$$\Rightarrow \quad \boxed{\lambda = -3}$$

Also since, $\frac{1}{x} + \frac{\lambda x}{a^2} = 0$

$$\Rightarrow \quad 1 + (-3)\frac{x^2}{a^2} = 0 \Rightarrow x^2 = \frac{a^2}{3}$$

$$\Rightarrow \qquad x = \pm \frac{a}{\sqrt{3}}$$

Taking +ve sign, $\quad x = \dfrac{a}{\sqrt{3}}$

similarly, $\quad y = \dfrac{b}{\sqrt{3}}$ and $z = \dfrac{c}{\sqrt{3}}$

Thus $\quad V = 8xyz = \dfrac{8abc}{3\sqrt{3}}$ ...(5)

To determine (5) is minimum or maximum. Let us consider z as the dependent variable of x and y. Differentiating (2) partially w.r.t. x, we get,

$$\frac{2x}{a^2} + \frac{2z}{c^2}\cdot\frac{\partial z}{\partial x} = 0 \Rightarrow \frac{\partial z}{\partial x} = -\frac{c^2 x}{a^2 z}$$

$$\therefore \quad \frac{\partial V}{\partial x} = 8yz + 8xy\frac{\partial z}{\partial x} = 8y\left[z + x\left(\frac{-c^2 x}{a^2 z}\right)\right]$$

$$\therefore \quad r = \frac{\partial^2 V}{\partial x^2} = 8y\left(\frac{\partial z}{\partial x} - \frac{2xc^2}{a^2 z} + \frac{c^2 x^2}{a^2 z^2}\cdot\frac{\partial z}{\partial x}\right)$$

$$= -8y\left(\frac{c^2 x}{a^2 z} + \frac{2xc^2}{a^2 z} + \frac{c^2 x^2}{a^2}\cdot\frac{c^2 x}{a^2 z^3}\right) < 0$$

$$\therefore \quad V \text{ is maximum at } \left(\frac{a}{\sqrt{3}}, \frac{b}{\sqrt{3}}, \frac{c}{\sqrt{3}}\right).$$

Example 11 Divide a number n into three parts x, y, z such that u = ayz + bzx + cxy shall have maximum or minimum and determine which it is.

Solution: Let n has been divided into three parts x, y, z.

$x + y + z = n$ ...(1)

and $u = ayz + bzx + cxy$ ...(2)

For u to be maximum or minimum $du = 0$

$\Rightarrow$ $(bz + cy)\,dx + (cx + az)\,dy + (ay + bx)\,dz = 0$ ...(3)

from (1), on differentiation, $dx + dy + dz + 0$ ...(4)

multiplying (3) and (4) by '1' and 'λ' respectively, adding and equating to zero the coefficients. of dx, dy, dz, we get

$bz + cy + \lambda = 0$ ...(5)

$cx + az + \lambda = 0$ ...(6)

$$ay + bx + \lambda = 0 \qquad ...(7)$$

multiplying (5) by x, (6) by y and (7) by z, and adding, we get

$$2u + n\lambda = 0 \;\Rightarrow\; \lambda = -\frac{2u}{n}$$

Now, from (5), (6) and (7), we have

$$0.x + cy + bz - \frac{2u}{n} = 0 \qquad ...(8)$$

$$cx + 0.y + az - \frac{2u}{n} = 0 \qquad ...(9)$$

and $$bx + ay + 0.z - \frac{2u}{n} = 0 \qquad ...(10)$$

Also from (1) $$x + y + z - n = 0 \qquad ...(11)$$

eliminating x, y, z from (8), (9), (10) and (11), we get

$$\begin{vmatrix} o & c & b & \frac{2u}{n} \\ c & o & a & \frac{2u}{n} \\ b & a & o & \frac{2u}{n} \\ 1 & 1 & 1 & n \end{vmatrix} = 0 \qquad ...(12)$$

which gives maximum or minimum values of u. Now to discuss maximum or minimum from (1), let us assume z as a function of x and y, we get

$$\therefore \qquad 1 + \frac{\partial z}{\partial x} = 0 \Rightarrow \frac{\partial z}{\partial x} = -1$$

similarly $$\frac{\partial z}{\partial y} = -1$$

Now $$\frac{\partial u}{\partial x} = bz + cy + (bx + ay)\frac{\partial z}{\partial x}$$

$$= bz + cy - (bx + ay)$$

$$\therefore \qquad r = \frac{\partial^2 u}{\partial x^2} = b\frac{\partial z}{\partial x} - b = -2b$$

$$s = \frac{\partial^2 u}{\partial x \partial y} = b\frac{\partial z}{\partial y} + c - a = -b + c - a$$

$= c - a - b.$

and $t = \dfrac{\partial^2 u}{\partial y^2} = -2a.$

we have $rt - s^2 = (-2b)(-2a) - (-c - a - b)^2$

$= 2ab + 2bc + 2ac - a^2 - b^2 - c^2$

$= a(b + c - a) + b(c + a - b) + c(a + b - c)$

Let us assume that a, b, c form the three sides of a triangle, then

$b + c - a,\ c + a - b,\ a + b - c$ are positive

$\therefore$ $rt - s^2 = +ve$

and $r = -2b = -ve$

Hence the values of u given by (12) is maximum under the sides of the triangle.

Example 12 Obtain the extreme values of $x^3 + y^3 + z^3 + 3kxyz$, $k \neq 2$, when x, y, z are subject to condition $x + y + z = 1$ and prove that the symmetrical extreme value is maximum or minimum according as $k > 2$ or $k < 2$.

Solution: We have $u = x^3 + y^3 + z^3 + 3kxyz$,

where $x + y + z = 1$

Consider the Lagrange's function,

$f(x, y, z) = x^3 + y^3 + z^3 + 3kxyz + \lambda(x + y + z - 1)$

for extreme values of u, $df = 0$

$$\Rightarrow \quad \frac{\partial f}{\partial x} = 3\left(x^2 + kyz\right) + \lambda = 0 \qquad \text{...(1)}$$

$$\frac{\partial f}{\partial y} = 3\left(y^2 + kzx\right) + \lambda = 0 \qquad \text{...(2)}$$

$$\text{and} \quad \frac{\partial f}{\partial z} = 3\left(z^2 + kxy\right) + \lambda = 0 \qquad \text{...(3)}$$

subtracting (2) from (1), we get

$(x - y)(x + y - kz) = 0$

and subtracting (3) from (1), we get

$(x - z)(x + z - ky) = 0$

Therefore either (i) $x = y = z = \dfrac{1}{3}$ or

(ii) $x = y = \dfrac{z}{k-1} = \dfrac{1}{k+1}$

(iii) $x = z = \dfrac{y}{k-1} = \dfrac{1}{k+1}$ or

(iv) $y = z = \frac{x}{k-1} = \frac{1}{k+1}$, The solution (i) is symmetrical in x, y, z giving

$u = \frac{1}{9}(1+k)$; the solutions (ii); (iii) and (iv) are non symmetrical but make $u = \frac{k^3+1}{(1+k)^3}$ in each case. It will be notice that if k = 2, there is only one solution, viz. (i).

further, we have

$$d^2f = 6(xdx^2 + ydy^2) + 6k\,(x\,dydz + ydzdx + zdxdy)$$
$$= 2(dx^2 + dy^2 + dz^2) + 2x\,(dydz + dzdx + dxdy) \qquad \text{(from (1))}$$

But $\quad dx + dy + dz = 0$, for all $x + y + z = 1$

therefore $\quad dx^2 + dy^2 + dz^2 = -2\,(dydz + dzdx + dxdy)$

Now, it is clear that d^2f is negative when $k > 2$ and +ve when $k < 2$.

Hence the value of u in the symmetrical case is maximum or minimum according as $k > 2$ or $k < 2$.

Some Additional Examples

Example 1 Find the stationary points of $f(x,y) = y^2 + 4xy + 3x^2 + x^3$. Examine them for the extreme values of the function.

Hint: Let $f(x,y) = y^2 + 4xy + 3x^2 + x^3$

$$\therefore \quad \frac{\partial f}{\partial x} = 4y + 6x + 3x^2$$

and $$\frac{\partial f}{\partial y} = 2y + 4x$$

Now, for maximum or minimum $\frac{\partial f}{\partial x} = 0$, $\frac{\partial f}{\partial y} = 0$, gives $4y + 6x + 3x^2 = 0$ and $2y + 4x = 0$.

Solving these equations, we get $x = 0, \frac{2}{3}$ and $y = 0, \frac{-4}{3}$.

Hence the points (0,0) and $\left(\frac{2}{3}, \frac{-4}{3}\right)$ are stationary points.

Now, $\quad r = \frac{\partial^2 f}{\partial x^2} = 6 + 6x, \quad s = \frac{\partial^2 f}{\partial x \partial y} = 4, \quad t = \frac{\partial^2 f}{\partial y^2} = 2$

At (0,0), r = 6, s = 4 and t = 2 ⇒ there is no extremum value at (0,0)

At $\quad \left(\frac{2}{3}, \frac{-4}{3}\right)$, r = 10, s = 4 and t = 2.

$\Rightarrow$ there is a minimum value at point $\left(\frac{2}{3}, \frac{-4}{3}\right)$.

Hence the minimum values $= \frac{-4}{27}$.

Example 2 When travelling x km/h a truck burns diesel oil at the rate of $\frac{1}{300}\left(\frac{900}{x} + x\right)$ l/km. If diesel oil costs 40 paise per litre and driver is paid Rs. 1.50 per hour, find the steady speed that will minimize the total cost of trip of 500 km.

Hints: Velocity of truck = x km/h

Velocity of burning diesel oil $= \frac{1}{300}\left(\frac{900}{x} + x\right) l/km.$

cost of diesel oil = Rs. 0.40 per litre and

Wages of driver are Rs. 1.50 per hour and

total distance of trip = 500 km.

$\therefore$ Total diesel burn in 500 km $= \frac{500}{300}\left(\frac{900}{x} + x\right)$ *liter* and its cost = $\frac{5}{3} \times \left(\frac{900}{x} + x\right) \times 0.40$ rupees.

Time taken in running 500 km $= \frac{900}{x}$ hours.

wages of driver $= \frac{500}{x} \times 1.50 = \frac{750}{x}$ rupees.

Total cost of trip $(c) = \frac{2}{3}\left(\frac{900}{x} + x\right) + \frac{750}{x}$

or $c = \frac{1350}{x} + \frac{2}{3}x$

$\therefore$ $\frac{dc}{dx} = \frac{-1350}{x^2} + \frac{2}{3} = 0$ (for maximum or minimum)

$\Rightarrow$ $x^2 = 2025 \Rightarrow x = 45$ km/h

Again $\frac{d^2c}{dx^2} = +ve$. Hence c is minimum when x = 45 km/h.

Example 3 Find the minimum value of $x^2 + y^2$ subject to the condition $ax + by - c = 0$.

Hint: Consider lagrange's function

$$f(x, y) = x^2 + y^2 + \lambda(ax + by - c)$$

Now $\frac{\partial f}{\partial x} = 2x + \lambda a = 0$ and $\frac{\partial f}{\partial y} = 2y + \lambda b = 0$

solving $x = -\frac{1}{2}a\lambda, \quad y = \frac{-1}{2}b\lambda.$

putting these values in $ax + by = c$, we find $\lambda = \frac{-2c}{a^2 + b^2}$.

It gives the minimum value of $(x^2 + y^2)$ at $\left(-\frac{a}{2}\lambda, \frac{-b}{2}\lambda\right)$, where $\lambda = \frac{-2c}{a^2 + b^2}$

Example 4 Find the minimum value of the function $f(x, y) = x^2 + y^2 + z^2 + xy + ax + by$, where a, b are constants.

Hints: $\frac{\partial f}{\partial x} = 2x + y + a = 0$ (for maximum or minimum)

and $\frac{\partial f}{\partial y} = 2y + x + b = 0$

solving these equations, we get

$$x = \frac{b - 2a}{3}, \; y = \frac{a - 2b}{3}.$$

Again, $r = \frac{\partial^2 f}{\partial x^2} = 2, \quad s = \frac{\partial^2 f}{\partial x \partial y} = 1, \; t = \frac{\partial^2 f}{\partial y^2} = 2$

$$rt - s^2 > 0$$

Hence at the point $\left(\frac{b - 2a}{3}, \frac{a - 2b}{3}\right)$ the given function has a minimum and the minimum value is $\frac{-(3a^2 + 3b^2 - 2ab)}{9}$.

Example 5 Find the maximum value of u, where

$$u = \frac{xyz}{(x + a)(x + y)(y + z)(z + b)}$$

Hint: Taking log on both sides, we have $\log u = \log x + \log y + \log z - \log(x + a) - \log(x + y) - \log(y + z) - \log(z + b)$.

Now, on differentiaing $\frac{1}{u}\frac{\partial u}{\partial x} = \frac{ay^2 - x^2}{x(x+a)(x+y)}$

$$\frac{1}{u}\frac{\partial u}{\partial y} = \frac{xz - y^2}{y(x+y)(y+z)}$$

and $$\frac{1}{u}\frac{\partial u}{\partial z} = \frac{by - z^2}{z(y+z)(z+b)}$$

For maximum or minimum,

$$\frac{\partial u}{\partial x} = 0, \quad \frac{\partial u}{\partial y} = 0, \quad \frac{\partial u}{\partial z} = 0$$

$\Rightarrow$ $x^2 = ay,\ y^2 = zx,\ z^2 = by$

$\Rightarrow$ a, x, y; x, y, z and y, z, b are all in geometrical progresion.

$\therefore$ a, x, y, z, b are in geometrical progression.

Let r be the common ratio.

Then,

$$b = ar^4,\ x = a\,r,\ y = ar^2,\ z = ar^3,$$

To find A, B, C, F, G, H, we put

$$\begin{vmatrix} A & H & G \\ H & B & F \\ G & F & C \end{vmatrix} = -ve, \quad A < 0,\ AB - H^2 > 0.$$

Hence u is maximum and minimum value of $u = \frac{1}{\left(a^{1/4} + b^{1/4}\right)^4}$.

Example 6 If $u = ax^2 + by^2 + cz^2$, where $x^2 + y^2 + z^2 = 1$ and $lx + my + nz = 0$, show that stationary values of u satisfy the equation $\frac{l^2}{a-u} + \frac{m^2}{b-u} + \frac{n^2}{c-u} = 0$

Hint: Let $u = ax^2 + by^2 + cz^2$, $f = x^2 + y^2 + z^2 - 1$, $g = lx + my + nz$

$\Rightarrow$ $$\frac{\partial u}{\partial x} = 2ax, \quad \frac{\partial u}{\partial y} = 2by, \quad \frac{\partial u}{\partial z} = 2zc$$

$$\frac{\partial f}{\partial x} = 2x, \quad \frac{\partial f}{\partial y} = 2y, \quad \frac{\partial f}{\partial z} = 2c$$

and $$\frac{\partial g}{\partial x} = l, \quad \frac{\partial g}{\partial y} = m, \quad \frac{\partial g}{\partial z} = h$$

Then by lagrange's method of multiplier, we have

$$\frac{\partial u}{\partial x}+\lambda_1\frac{\partial f}{\partial x}+\lambda_2\frac{\partial g}{\partial x}=0$$

$\Rightarrow$ $\quad 2ax + 2x\lambda_1 + \lambda_2 l = 0$ ...(1)

similarly $\quad 2by + 2y\lambda_1 + \lambda_2 m = 0$ ...(2)

and $\quad 2cz + 2z\lambda_1 + \lambda_3 n = 0$ ...(3)

multiply (1), (2) (3) by x, y, and z respectively and adding, we get

$$\boxed{\lambda_1=-u}$$

putting the value of λ_1 in (1), (2) and (3), we get

$$x=-\frac{\lambda_2 l}{2(a-u)},\qquad y=-\frac{\lambda_2 m}{2(b-u)} \text{ and}$$

$$z=-\frac{\lambda_2 n}{2(c-u)}$$

Putting these values of x, y, z is

$lx + my + nz = 0$, we get

$$-\frac{\lambda_2 l^2}{2(a-u)}-\frac{\lambda_2 m^2}{2(b-u)}-\frac{\lambda_2 n^2}{2(c-u)}=0$$

$$\Rightarrow \qquad \frac{l^2}{a-u}+\frac{m^2}{b-u}+\frac{n^2}{c-u}=0.$$

Example 7 Find the points on the surface $z^2 = xy + 1$ nearest to origin.

Hint: Let (x, y, z) be a point on the surface $z^2 = xy + 1$.

Let s be the distance of this point from the origin, $s=\sqrt{x^2+y^2+z^2}$

$\Rightarrow$ $\quad f(x, y, z) = s^2 = x^2 + y^2 + z^2,$

then $\quad f(x, y, z) = x^2 + y^2 + xy + 1,\quad$ from (1)

Now, $\quad \dfrac{\partial f}{\partial x}=2x+y,\qquad r=\dfrac{\partial^2 f}{\partial x^2}=2$

$$\frac{\partial f}{\partial y}=2y+x,\qquad t=\frac{\partial^2 f}{\partial y^2}=2$$

and $\quad s=\dfrac{\partial^2 f}{\partial x\partial y}=1$

for maximum or minimum

$$\frac{\partial f}{\partial x} = 0 \quad \text{and} \quad \frac{\partial f}{\partial y} = 0 \qquad ...(2)$$

$\Rightarrow$ $2x + y = 0$ and $2y + x = 0$ ...(3)

solving (2) and (3), we get $x = 0, y = 0$.

Here $rt - s^2 = 3$ (+ve) and also r is positive.

$\therefore$ f(x, y) is minimum at (0,0). Put $x = 0, y = 0$ in (1), we get $z = \pm 1$.

Hence the points on $z^2 - xy - 1 = 0$ nearest to origin are $(0, 0, \pm 1)$.

Example 8 Find the maximum value of f, where $f = \sin x \sin y \sin (x + y)$.

Hint: For maximum or minimum of f, we have

$$\frac{\partial f}{\partial x} = 0 = \sin y \left\{\sin x \cos (x + y) + \cos x \sin (x + y)\right\} = 0$$

$$= \sin y \sin (2x + y) = 0$$

$$\frac{\partial f}{\partial y} = \sin x \left\{\sin y \cos (x + y) + \cos y \sin (x + y)\right\} = 0$$

$$= \sin x \sin (x + 2y) = 0$$

These relations give either $x = 0 = y$ or

$$\tan x = \tan y = -\tan (x + y)$$

$$\Rightarrow \quad x = 0 = y \quad \text{or} \quad x = y = \frac{\pi}{3}$$

when $x = y = \dfrac{\pi}{3}$, $r = \dfrac{\partial^2 f}{\partial x^2} = -\sqrt{3}$

$$s = \frac{\partial^2 f}{\partial x \partial y} = -\frac{\sqrt{3}}{2} \text{ and } t = \frac{\partial^2 f}{\partial y^2} = -\sqrt{3}$$

Therefore, $rt - s^2 = \dfrac{9}{4}$ (+ve)

Also r is (–ve). Hence f(x, y) has a maximum at $x = y = \dfrac{\pi}{3}$, and maximum value $= 3\sqrt{\dfrac{3}{8}}$.

Example 9 Find the maximum and minimum of $x^2 + y^2$, subject to the condition

$ax^2 + by^2 + 2hxy = 1$.

Hints: we have $u = x^2 + y^2$ ...(1)

where x, y connected by the relation

$ax^2 + by^2 + 2hxy = 1$ ...(2)

For maximum or minimum of u,

$$du = 0$$

$\Rightarrow$ $2xdx + 2ydy = 0 \Rightarrow xdx + ydy = 0$ ...(3)

and $(ax + hy)\,dx + (hx + by)\,dy = 0$ ...(4)

Then by using lagrange's multipliers, we have

$x + \lambda\,(ax + hy) = 0$ ...(5)

and $y + \lambda\,(hx + by) = 0$ ...(6)

multiplying (5) by x and (6) by y and adding, we get,

$$x^2 + y^2 + \left\{\lambda\left(ax^2 + 2hxy + by^2\right)\right\} = 0$$

$\Rightarrow$ $u + \lambda = 0 \Rightarrow \lambda = -u$

from (5), $x - u\,(ax + hy) = 0$

$\Rightarrow$ $x(1 - au) - huy = 0 \Rightarrow \left(a - \dfrac{1}{u}\right)x + hy = 0$ ...(7)

and $hx + \left(b - \dfrac{1}{u}\right)y = 0$ ...(8)

eliminating x and y from (7) and (8), we get

$$\begin{vmatrix} a - \dfrac{1}{u} & h \\ h & b - \dfrac{1}{u} \end{vmatrix} = 0 \Rightarrow \left(a - \frac{1}{u}\right)\left(b - \frac{1}{u}\right) = h^2 \qquad ...(9)$$

Hence the required maximum or minimum values of $u = x^2 + y^2$ are the roots of (9).

Example 10 To find the maximum value of the product of the cosines of the angle of triangle.

Hint: we have $u = \cos A \cos B \cos C$ ...(1)

where the variables A, B and C are connected by the relations $A + B + C = \pi$ ...(2)

since we can conveniently eliminate any one of the variables, we find

$$u = -\cos A \cos B \cos (A + B).$$

where A and B are two independent variables.

For maximum or minimum,

$$\frac{\partial u}{\partial A} = \sin(2A + B)\cos B = 0$$

and

$$\frac{\partial u}{\partial B} = \sin(A + 2B)\cos A + 0$$

Now cos A or cos B cannot vanish, for that will make u = 0, which is not obviously its maximum value. Also 2 A + B or A + 2B cannot be zero for A and B are +ve.

$\therefore$ we take, $2A + B = \pi = A + 2B$

or $\quad A = B = \frac{\pi}{3}$ which makes $u = \frac{1}{8}$.

Now, $\quad r = \frac{\partial^2 u}{\partial A^2} = -1, \quad t = \frac{\partial^2 u}{\partial B^2} = -1, \quad s = \frac{\partial^2 u}{\partial A \partial B} = \frac{1}{2}$

so $\quad rt - s^2 = \frac{3}{4} > 0$ and $r < 0$

hence u has maximum value for $A = B = C = \frac{\pi}{3}$.

Exercise 6.8

1. Find the minimum value of $x + y + z$, subject to the condition $\left(\frac{a}{x}\right) + \left(\frac{b}{y}\right) + \left(\frac{c}{z}\right) = 1$.
2. Given $F = \frac{5xyz}{x + 2y + 4z}$. Find the values of x, y, z for which F is maximum subject to the condition $xyz = 8$.
3. Find the minimum value of the function $x^2 + y^2 + z^2$ subject to the condition $ax + by + cz = d$.
4. Find the stationary values of $u = x^2 + y^2 + z^2$ subject to the condition $ax^2 + by^2 + cz^2 = 1$ and $lx + my + nz = 0$
5. Using the method of multipliers, find the maximum product of the numbers, x, y and z when $x^2 + y^2 + z^2 = 16$.
6. Find the minimum and maximum distances from the point (1, 2, –1) to the sphere $x^2 + y^2 + z^2 = 24$.
7. Discuss the maximum and minimum of $x^2 + y^2 + z^2$ subject to the following conditions

   (i) $ax^2 + by^2 + cz^2 = 1$ (ii) $yz + zx + xy = 3a^2$

   (iii) $x + y + z = 3a$ (iv) $xyz = a^2$
8. Prove that of all rectangular parallelopipeds of the same volume, the cube has the least surface.
9. Show that if the perimeter of a triangle is constant, its area is maximum when it is equilateral.
10. Use the lagrange's method of multipliers to find the minimum value of $x^2 + y^2 + z^2$ subject to the conditions $x + y + z = 1$, $xyz = -1$.
11. Find the shortest and largest distances from the origin to the curve $5x^2 + 6xy + 5y^2 - 8 = 0$
12. Find the maximum and minimum values of $x^2 + y^2 + z^2$, subject to condition

    $$ax^2 + by^2 + cz^2 + 2fyz + 2gzx + 2hxy = 1.$$
13. Prove that the maximum and minimum of the radii vectors of the sections of the surface $\left(x^2 + y^2 + z^2\right)^2 = \frac{x^2}{a^2} + \frac{y^2}{b^2} + \frac{z^2}{c^2}$ by the plane $\lambda x + \mu y + \gamma z = 0$ are given by the equation

$$\frac{a^2\lambda^2}{1-a^2r^2}+\frac{b^2\mu^2}{1-b^2r^2}+\frac{c^2\gamma^2}{1-c^2r^2}=0.$$

14. If $\frac{x_1^2}{a_1^2}+\frac{x_2^2}{a_2^2}+\dots\dots\dots\dots+\frac{x_n^2}{a_n^2}=1$, where $a_1^2 > a_2^2 > \dots\dots > a_n^2$, Prove that the stationary value of $x_1^2 + x_2^2 + \dots\dots + x_n^2$ are $a_1^2\, a_2^2 \dots\dots\dots, a_n^2$, out of these value, prove that only a_1^2 is true maximum and a_n^2 is true minimum.
15. Find the maximum value of $x^a y^b z^c$ subject to $x + y + z = \lambda$, $x > 0$, $y > 0$, $z > 0$ where a, b, c are positive constants.
16. Find the triangular pyramid of given base and altitude which has the least surface.
17. Given $a^x b^y c^z = A$, Find the maximum value of $(x+1)(y+1)(z+1)$. Interpret the result.
18. Find the maximum and minimum value of ρ, where, $\frac{k}{\rho} = r x^2 + 2sxy + t y^2$, being given that $(1 + p^2) x^2 + 2pq\, xy + (1 + q^2)y^2 = 1$.
19. A closed rectangular box is to be made of 300 sq. cm. of sheet metal in such a way that the perimeter of its base is 25 times the height of the box, calculate the dimension that would give the maximum capacity.
20. If $u = a_1^3x^2 + a_2^3y^2 + a_3^3z^2$, where $x^{-1} + y^{-1} + z^{-1} = 1$, prove that stationary value of u is given by $x=\frac{\Sigma a}{a_1}$, $y=\frac{\Sigma a}{a_2}$, $z=\frac{\Sigma a}{a_3}$.
21. Find the minimum value of $\sum_{r=1}^{n} X_r$, with the condition $\sum_{r=1}^{n} \alpha_r x_r = k$
22. Find a point within a tetrahedron such that the sum of the squares of its distances from the faces of the tetrahedron shall be minimum.

Answers

1. $\left(\sqrt{a}+\sqrt{b}+\sqrt{c}\right)^2$
2. Maximum at $x = 4$, $y = 2$, $z = 1$
3. Minimum value at $\frac{x}{a}=\frac{y}{b}=\frac{z}{c}=\frac{d}{a^2+b^2+c^2}$
4. Stationary values are: $\frac{l^2}{au-1}+\frac{m^2}{bu-1}+\frac{n^2}{cu-1}=0$
5. $3\sqrt{3}$
6. Minimum distance $= \sqrt{6}$ and maximum distance $= 3\sqrt{6}$.

7. (i) Maximum and minimum are the roots of $\left(\frac{1}{a}-u\right)\left(\frac{1}{b}-u\right)\left(\frac{1}{c}-u\right)=0$

(ii) Minimum value = $3a^2$

(iii) Minimum at (a, a, a)

(iv) Minimum at (a, a, a); (–a, –a, +a); (–a, a, –a); (a, –a, –a)

10. Minimum value = 3

11. Shortest distance = 1, greatest distance = 4

12. Maximum or minimum value of u are given by

$$\begin{vmatrix} a-u^{-1} & h & g \\ h & b-u^{-1} & f \\ g & f & c-u^{-1} \end{vmatrix} = 0$$

14. Stationary points are given by

$$\begin{vmatrix} o & c & b & \frac{2u}{n} \\ c & o & a & \frac{2u}{n} \\ b & a & o & \frac{2u}{n} \\ 1 & 1 & 1 & n \end{vmatrix}, \text{ value is maximum}$$

15. $a^a b^b c^c \left(\frac{\lambda}{a+b+c}\right)^{a+b+c}$

16. The foot of the altitude is the incentre of the base, and the minimum surface (excluding the base) of the triangular pyramid is $\left(p^2 s^2 + \Delta^2\right)^{1/2}$, where p is the altitude, s is the semi-perimeter of the base and Δ is area.

17. $\{\log(A\,a\,b\,c)\}^3 / (27 \log a \log b \log c)$

18. One value of p given by the equation

$k^2\left(1+p^2+q^2\right) - kp\{\left(1+q^2\right)r - 2pq\,s + \left(1+p^2\right)t\} + \rho^2\left(rt - s^2\right) = 0$ is a maximum and the other is a minimum.

19. $25\sqrt{\frac{2}{11}} \times 25\sqrt{\frac{2}{11}} \times 4\sqrt{\frac{2}{11}}$.

21. $\dfrac{k^2}{\left(\alpha_1^2 + \alpha_2^2 + \ldots\ldots + \alpha_n^2\right)}$

22. The point where distances from the faces of the tetra hedron are AK, BK, CK, DK, where A, B, C, D are the areas of the faces and $K = 3\left(\dfrac{volume\ ABCD}{\Sigma A^2}\right)$.

6.21 DIFFERENTIATION UNDER INTEGRAL SIGN:

Suppose F(α) is the value of a definite integral $\int_a^b f(x,\alpha)\,dx$, where α is a parameter. To find F(α), first we have to evaluate the integral $\int_a^b f(x,\alpha)\,dx$, and then differentiate F(α) w.r.t. α. however it is not always possible to evaluate the integral and then to find the derivative, such problems are solved by reversing the order of the integration and differentiation i.e., first differentiate F(x, α) partially w.r.t. 'α' and then integrate it, thus in general $\int_a^b f(x,\alpha)\,dx = F(\alpha)$.

6.22 LEIBNITZ'S RULE FOR DIFFERENTIATION UNDER THE INTEGRAL SIGN:

(a) When the limits of integration are constant i.e. independent of parameter.

Statement: If $f(x,\alpha)$ and $\dfrac{\partial}{\partial\alpha} f(x,\alpha)$ be continuous function of x and α, then

$$\frac{d}{d\alpha}\left[\int_a^b f(x,\alpha)\,dx\right] = \int_a^b \frac{\partial}{\partial\alpha} f(x,\alpha)\,dx$$

Proof: Let $\int_a^b \dfrac{\partial}{\partial\alpha} f(x,\alpha)\,dx = F(\alpha)$, ...(i)

where α is a parameter

Suppose δα is the increment in α then

(i) becomes

$$F(\alpha + \delta\alpha) = \int_a^b f(x, \alpha + \delta\alpha)\,dx \qquad \ldots\text{(ii)}$$

subtracting (i) from (ii), then, we get

$$F(\alpha + \delta\alpha) - F(\alpha) = \int_a^b f(x, \alpha + \delta\alpha)\,dx - \int_a^b f(x,\alpha)\,dx$$

$$F(\alpha+\delta\alpha)-F(\alpha)=\int_a^b\left[f(x,\alpha+\delta\alpha)-f(x,\alpha)\right]dx$$

dividing by $\delta\alpha$ on both sides, we get

$$\frac{F(\alpha+\delta\alpha)-F(\alpha)}{\delta\alpha}=\int_a^b\frac{f(x,\alpha+\delta\alpha)-f(x,\alpha)}{\delta\alpha}dx$$

Taking limits on both sides as $\delta\alpha \to 0$, then

$$\lim_{\delta\alpha\to 0}\frac{F(\alpha+\delta\alpha)-F(\alpha)}{\delta\alpha}=\lim_{\delta\alpha\to 0}\int_a^b\frac{f(x,\alpha+\delta\alpha)-f(x,\alpha)}{\delta\alpha}dx$$

$$\therefore \quad \frac{dF}{d\alpha}=\int_a^b\frac{\partial}{\partial\alpha}f(x,\alpha)\,dx$$

This formula is useful for evaluating definite integrals which are otherwise impossible to evaluate.

Illustrative Examples

Example 1 Prove that

$$\int_0^\infty\frac{\tan^{-1}ax}{x\left(1+x^2\right)}dx=\frac{\pi}{2}\log(1+a),\ a\ge 0$$

Solution: Let $F(a)=\int_0^\infty\frac{\tan^{-1}ax}{x\left(1+x^2\right)}dx$

Here a is parameter ...(A)

Differentiating (A) w.r.t. 'a' on both sides, we get,

$$F'(a)=\int_0^\infty\frac{x\,dx}{x\left(1+x^2\right)\left(1+a^2x^2\right)}$$

$$=\int_0^\infty\frac{dx}{\left(1+x^2\right)\left(1+a^2x^2\right)} \qquad \text{...(B)}$$

Now, $$\frac{1}{\left(1+x^2\right)\left(1+a^2x^2\right)}=\frac{Ax+B}{1+x^2}+\frac{(Cx+D)}{1+a^2x^2} \qquad \text{...(C)}$$

$\therefore$ $1=(Ax+B)(1+a^2x^2)+(Cx+D)(1+x^2)$ comparing the coefficient of various like power on both sides

Coefficient of x^3,

$$0=a^2A+C \qquad \text{...(i)}$$

Coefficient of x^2,

$$0 = a^2 B + D \quad ...(ii)$$

Coefficient of x,

$$0 = A + C \quad ...(iii)$$

Constant term

$$1 = B + D \quad ...(iv)$$

from (iii) $A = -C$

from (i) $a^2 A + C = 0$

$\therefore$ $-a^2 C + C = 0$

or $C(1 - a^2) = 0$

or $C = 0$ and $A = 0$

from (iv) $B = 1 - D$

from (ii) $B a^2 + D = 0$

or $(1 - D) a^2 + D = 0$

or $-a^2 = D(1 - a^2)$

$$\therefore \quad D = \frac{a^2}{a^2 - 1}$$

and $$B = 1 - D = 1 - \frac{a^2}{a^2 - 1} = \frac{-1}{a^2 - 1}$$

$$\therefore \quad B = \frac{1}{1 - a^2}$$

Putting these values of partial fraction in (C), we get

$$F'(a) = \int_0^\infty \left[\frac{Ax + B}{1 + x^2} + \frac{(Cx + D)}{1 + a^2 x^2} \right] dx$$

$$\therefore \quad F'(a) = \int_0^\infty \frac{dx}{(1 - a^2)(1 + x^2)} + \int_0^\infty \frac{a^2}{a^2 - 1} \frac{dx}{(1 + a^2 x^2)}$$

$$= \frac{1}{(1 - a^2)} \left[\tan^{-1} x \right]_0^\infty + \frac{a^2}{a^2 - 1} \int_0^\infty \frac{dx}{a^2 \left(x^2 + \frac{1}{a^2} \right)}$$

$$= \frac{\pi}{2(1 - a^2)} + \frac{a}{a^2 - 1} \cdot \frac{\pi}{2}$$

$$\therefore \quad F'(a) = \frac{\pi}{2(1-a^2)} + \frac{\pi a}{2(a^2-1)} \qquad \text{...(D)}$$

Integrating (D) w.r.t. 'a' on both sides, we get

$$F(a) = \frac{\pi}{2} \int \frac{da}{1-a^2} + \frac{\pi}{2} \int \frac{a}{a^2-1} da$$

$$F(a) = \frac{\pi}{2} \left[\int \left\{ \frac{1}{(1+a)} + \frac{1}{1-a} \right\} da \right] + \frac{\pi}{2} \times \frac{1}{2} \int \frac{-2a}{1-a^2} da + C$$

$$\therefore \quad F(a) = \frac{\pi}{2} \left[\frac{1}{2} \left\{ \int \frac{da}{1+a} + \frac{da}{1-a} \right\} \right] + \frac{\pi}{2} \times \frac{1}{2} \log(1-a^2) + C$$

$$\therefore \quad F(a) = \frac{\pi}{2} \left[\log(1+a)^{\frac{1}{2}} - \log(1-a)^{1/2} \right] + \frac{\pi}{2} \log(1-a^2)^{1/2} + C \qquad \text{...(E)}$$

when a = 0, Then from equation (A)

$$F(0) = 0$$

Again when a = 0 then from (E)

$$F(0) = 0 = \frac{\pi}{2} \times 0 + \frac{\pi}{2} \times 0 + C$$

$$\therefore \quad C = 0$$

putting C = 0 in (E), then

$$F(a) = \frac{\pi}{2} \left[\log\left(\frac{1+a}{1-a}\right)^{1/2} + \log(1-a^2)^{1/2} \right] + 0$$

$$F(a) = \frac{\pi}{2} \log(1+a).$$

Example 2 Evaluate $\int_0^1 \frac{x^\alpha - 1}{\log x} dx, \ \alpha \geq 0$

Solution: Let $F(\alpha) = \int_0^1 \frac{x^\alpha - 1}{\log x} dx$ ...(1)

Here α is parameter. Differentiate (1) w.r.t. α on both sides

$$F'(\alpha) = \frac{d}{d\alpha} \int_0^1 \frac{x^\alpha - 1}{\log x} dx$$

$$= \int_0^1 \frac{\partial}{\partial \alpha} \frac{x^\alpha - 1}{\log x} dx = \int_0^1 \frac{x^\alpha \log x}{\log x} dx$$

$$F'(\alpha) = \int_0^1 x^\alpha \, dx$$

Integration w.r.t. x on R.H.S., we get

$$F'(\alpha) = \left[\frac{x^{\alpha+1}}{\alpha+1}\right]_0^1 = \frac{1}{\alpha+1}$$

$$F'(\alpha) = \frac{1}{\alpha+1} \quad \text{...(ii)}$$

on integrating (ii) w.r.t. α, then we get

$$F(\alpha) = \int \frac{d\alpha}{\alpha+1}$$

$$\therefore \quad F(\alpha) = \log(1+\alpha) + c \quad \text{...(iii)}$$

when $\alpha = 0$ from (i) $F(0) = 0$

similarly, when $\alpha = 0$, from (iii)

$$F(0) = \log(1+0) + C$$

$$0 = C$$

putting the value of C is (iii), we get

$$F(\alpha) = \log(1+\alpha) + 0$$

$$\therefore \quad F(\alpha) = \log(1+\alpha).$$

Example 3 Evaluate $\int_0^\pi \log(1 + a\cos x)\, dx$, using the method of differentiation under the sign of integration.

Solution: Let $F(a) = \int_0^\pi \log(1 + a\cos x)dx$ ...(1)

where a is parameter

differentiating (i) w.r.t. a on both sides, we get

$$F'(a) = \frac{d}{da}\int_0^\pi \log(1 + a\cos x)\, dx$$

$$F'(a) = \int_0^\pi \frac{\partial}{\partial a}\log(1 + a\cos x)\, dx$$

$$F'(a) = \int_0^\pi \frac{\cos x}{1 + a\cos x}\, dx \qquad \text{...(ii)}$$

$$= \frac{1}{a} \int_0^\pi \frac{1 + a\cos x - 1}{1 + a\cos x}\, dx$$

or $$F'(a) = \int_0^\pi \left\{ \frac{1}{a} + \frac{-\frac{1}{a}}{1 + a\cos x} \right\} dx$$

on integrating R.H.S. w.r.t. x, we get

$$F'(a) = \frac{\pi}{a} - \frac{1}{a} \int_0^\pi \frac{dx}{1 + a\cos x}$$

From Trigonometry, $\cos x = \dfrac{1 - \tan^2 x/2}{1 + \tan^2 x/2}$

$$\therefore \quad F'(a) = \frac{\pi}{a} - \frac{1}{a} \int_0^\pi \frac{\sec^2 x/2}{\left(1 + \tan^2 \frac{x}{2}\right) + a\left(1 - \tan^2 \frac{x}{2}\right)}\, dx$$

put $$\tan \frac{x}{2} = t$$

$$\therefore \quad \frac{1}{2} \sec^2 \frac{x}{2}\, dx = dt$$

when x = 0, t = 0, when x = π, t = ∞

$$\therefore \quad F'(a) = \frac{\pi}{a} - \frac{1}{a} \int_0^\infty \frac{2dt}{\left(1+t^2\right) + a\left(1-t^2\right)}$$

$$= \frac{\pi}{a} - \frac{1}{a} \int_0^\infty \frac{2dt}{(1+a) + (1-a)\, t^2}$$

$$\therefore \quad F'(a) = \frac{\pi}{a} - \frac{2}{a(1-a)} \int_0^\infty \frac{dt}{\left\{\sqrt{\frac{1+a}{1-a}}\right\}^2 + t^2}$$

$$= \frac{\pi}{a} - \frac{2}{a(1-a)}\sqrt{\frac{1-a}{1+a}}\left\{\tan^{-1}\frac{t}{\sqrt{\frac{1+a}{1-a}}}\right\}_0^{\infty}$$

$$= \frac{\pi}{a} - \frac{2}{a\sqrt{1-a^2}}\cdot\frac{\pi}{2}$$

$$F'(a) = \frac{\pi}{a} - \frac{\pi}{a\sqrt{1-a^2}} \qquad \text{...(iii)}$$

or integrating (iii) w.r.t. a, we get

$$F(a) = \pi\int\frac{da}{a} - \pi\int\frac{da}{a\sqrt{1-a^2}} + C$$

$$= \pi\log a - \pi\int\frac{da}{a\sqrt{1-a^2}} + C$$

Put a = cost

da = – sin t dt

$$= \pi\log a - \pi\int\frac{-\sin t\,dt}{\cos t\sqrt{1-\cos^2 t}} + C$$

$$F(a) = \pi\log a + \pi\int\frac{\sin t}{\cos t\sin t}dt + C$$

$$F(a) = \pi\log a + \pi\int\sec t\,dt + C$$

$$F(a) = \pi\log a + \pi\log(\sec t + \tan t) + C$$

$$F(a) = \pi\log a + \pi\log\left\{\frac{1}{a} + \frac{\sqrt{1-a^2}}{a}\right\} + C$$

$$F(a) = \pi\log a + \pi\log\left\{1 + \sqrt{1-a^2}\right\} - \pi\log a + C$$

$$F(a) = \pi\log\left\{1 + \sqrt{1-a^2}\right\} + C \qquad \text{...(iv)}$$

init ally when a = 0, then from (i)

F(0) = 0

Again from (iv), when a = 0, then

$$F(0) = \pi \log \left\{1 + \sqrt{1-0}\right\} + C$$

or $$0 = \pi \log 2 + C$$

$\therefore$ $$C = -\pi \log 2$$

Putting the value of c in equation (iv), then we get

$$F(a) = \pi \log \left[1 + \sqrt{1-a^2}\right] - \pi \log 2$$

$$F(a) = \pi \log \frac{1 + \sqrt{1-a^2}}{2}$$

Example 4 (a) Differentiating $\int_0^x \frac{dx}{x^2 + a^2} = \frac{1}{a} \tan^{-1} \frac{x}{a}$ under the integral sign. Find the value of

$$\int_0^x \frac{dx}{\left(x^2 + a^2\right)^2}.$$

Solution: Let $$\int_0^x \frac{dx}{\left(x^2 + a^2\right)^2} = \frac{1}{a} \tan^{-1} \frac{x}{a} \quad \text{...(i)}$$

where a is a parameter

Differentiating (i) w.r.t. 'a' on both sides, then we get

$$\frac{d}{da} \int_0^x \frac{dx}{\left(x^2 + a^2\right)} = \frac{d}{da}\left[\frac{1}{a} \tan^{-1}\left(\frac{x}{a}\right)\right]$$

$$\int_0^x \frac{-2a}{\left(x^2 + a^2\right)} dx = \frac{-1}{a^2} \tan^{-1} \frac{x}{a} + \frac{1}{a} \frac{1}{\left(1 + \frac{x^2}{a^2}\right)} \cdot \left(-\frac{x}{a^2}\right)$$

on simplification,

$$\int_0^x \frac{dx}{\left(a^2 + x^2\right)^2} = \frac{1}{2a^3} \tan^{-1}\left(\frac{x}{a}\right) + \frac{x}{2a^2\left(a^2 + x^2\right)^2}$$

Example 4 (b) By successive differentiation $\int_0^1 x^m \, dx = \frac{1}{m+1}$ w.r.t. m. Evaluate $\int_0^1 x^m (\log x)^n \, dx.$.

Solution: Let $$\int_0^1 x^m \, dx = \frac{1}{m+1} \quad \text{...(1)}$$

differentiating (i) w.r.t 'm' on both sides $\frac{d}{dm}\int_0^1 x^m dx = \frac{d}{dm}\left(\frac{1}{m+1}\right)$

or $$\int_0^1 x^m \log x \, dx = \frac{(-1)}{(m+1)^2} \qquad \text{...(ii)}$$

Again differentiating (ii) w.r.t. 'm' on both sides, we get

$$\int_0^1 x^m (\log x)^2 \, dx = \frac{(-1)(-2)}{(m+1)^3}$$

$$\int_0^1 x^m (\log x)^n \, dx = \left(\frac{(-1)(-2)\,......................\,(-n)}{(m+1)^{n+1}}\right)$$

Hence $$\int_0^1 x^m (\log x)^n \, dx = \frac{(-1)^n \, n!}{(m+1)^{n+1}}.$$

Example 5 Prove that

$$\int_0^{\pi/2} \frac{\log\left(1+y\sin^2 x\right)}{\sin^2 x} dx = \pi\left[\sqrt{1+y}-1\right]$$

where y is a parameter.

Solution: Let $$F(y) = \int_0^{\pi/2} \frac{\log\left(1+y\sin^2 x\right)}{\sin^2 x} dx \qquad \text{...(i)}$$

Differentating (i) w.r.t. y on both sides

$$F'(y) = \int_0^{\pi/2} \frac{\sin^2 x}{\sin^2 x\left(1+y\sin^2 x\right)} dx$$

$$= \int_0^{\pi/2} \frac{dx}{1+y\sin^2 x}$$

$$= \int_0^{\pi/2} \frac{dx}{\cos^2 x+\sin^2 x+y\sin^2 x} \qquad \left[\because 1=\cos^2 x+\sin^2 x\right]$$

or $$F'(y) = \int_0^{\pi/2} \frac{dx}{\cos^2 x+(1+y)\sin^2 x}$$

$$= \int_0^{\pi/2} \frac{dx \sec^2 x}{1+(1+y)\tan^2 x}$$

Put $\sqrt{1+y}\tan x = t$

$$\sqrt{1+y}\sec^2 x\, dx = dt$$

$$\sec^2 x\, dx = \frac{dt}{\sqrt{1+y}}$$

when $x = \frac{\pi}{2}$, then $t = \infty$

when $x = 0$, then $t = 0$

$$\therefore \quad F'(y) = \int_0^{\infty} \frac{dt}{\sqrt{1+y}\left[1+t^2\right]}$$

$$F'(y) = \frac{1}{\sqrt{1+y}}\left[\tan^{-1} t\right]_0^{\infty}$$

$$F'(y) = \frac{\pi}{2\sqrt{1+y}} \qquad \text{...(ii)}$$

Integrating (ii) w.r.t. 'y' on both sides, we get

$$F(y) = \frac{\pi}{2}\int (1+y)^{-\frac{1}{2}}\, dy = \frac{\pi}{2}\left[\frac{(1+y)^{1/2}}{1/2} + C\right]$$

$$F(y) = \pi\sqrt{1+y} + C$$

Initially when $y = 0$, from (i)

$$F(0) = 0 \qquad \text{...(iii)}$$

Again from (iii)

$$F(0) = \pi\sqrt{1+0} + C$$

$$\therefore \quad C = -\pi$$

Hence $$\int_0^{\pi} \log\frac{\left(1+y\sin^2 x\right)}{\sin^2 x}\, dx = \pi\left[\sqrt{1+y} - 1\right].$$

Example 6 (a) Prove that

$$\int_0^{\infty} e^{-x^2}\cos\alpha x\, dx = \frac{\sqrt{\pi}}{2} e^{-\frac{\alpha^2}{4}}, \text{ where } \alpha \text{ is parameter.}$$

Solution: Let $F(\alpha) = \int_0^\infty e^{-x^2} \cos \alpha x \, dx$ ...(i)

Differentiating (i) w.r.t α on both sides, we get

$$F'(\alpha) = \int_0^\infty e^{-x^2} (-x \sin \alpha x) \, dx$$

$$= \frac{1}{2} \int_0^\infty \sin \alpha x \left(-2x \, e^{-x^2}\right) dx$$

on integrating by parts

$$= \frac{1}{2}\left[\left(\sin \alpha x \, e^{-x^2}\right)_0^\infty - \int_0^\infty \alpha \cos \alpha \, x \, e^{-x^2} \, dx\right]$$

$$= \frac{1}{2}\left[(0-0) - \int_0^\infty \alpha \cos \alpha \, x \, e^{-x^2} \, dx\right]$$

$$= -\frac{\alpha}{2} \int_0^\infty e^{-x^2} \cos \alpha \, x \, d\, x$$

$$F'(\alpha) = \frac{-\alpha}{2} F(\alpha) \qquad \text{...(ii)}$$

Separating the variable,

$$\frac{F'(\alpha)}{F(\alpha)} = -\frac{\alpha}{2}$$

Integrating w.r.t. α, then, we get

$$\log F(\alpha) = \frac{-\alpha^2}{4} + C \qquad \text{...(iii)}$$

Initially when $\alpha = 0$ from (i)

$$F(0) = \int_0^\infty e^{-x^2} \, dx$$

Put $x^2 = z$

$\therefore$ $2x dx = dz$

when $x = 0,$ then $z = 0$

when $x = \infty,$ then $z = \infty$

$$dx = \frac{dz}{2\sqrt{z}}$$

$$\therefore \quad F(0) = \int_0^\infty e^{-z} \times \frac{dz}{2} \times z^{-1/2}$$

$$= \frac{1}{2} \int_0^\infty e^{-z} z^{-1/2} dz$$

$$= \frac{1}{2} \int_0^\infty e^{-z} z^{\frac{1}{2}-1} dz$$

$$= \frac{1}{2} \Gamma \frac{1}{2}$$ (By gamma function

$$\therefore \quad \Gamma(\tfrac{1}{2}) = \sqrt{\pi} \qquad \because \Gamma\left(\frac{1}{2}\right) = \sqrt{\pi}$$

$$= \frac{1}{2} \frac{\pi}{2}$$

$$\therefore \quad F(0) = \frac{\sqrt{\pi}}{2}$$

Again when $\alpha = 0$, from (iii)

$$\log F(0) = \frac{-0}{4} + C$$

$$\log \frac{\sqrt{\pi}}{2} = + C$$

$$\therefore \quad C = \log \frac{\sqrt{\pi}}{2}$$

Putting the value of C in (iii), then

$$\log F(\alpha) = \frac{-\alpha^2}{4} + \log \frac{\sqrt{\pi}}{2}$$

or
$$\log F(\alpha) - \log \frac{\sqrt{\pi}}{2} = -\frac{\alpha^2}{4}$$

or
$$\log \frac{F(\alpha)}{\frac{\sqrt{\pi}}{2}} = e^{-\alpha^2/4}$$

or
$$F(\alpha) = \frac{\sqrt{\pi}}{2} \times e^{-\frac{\alpha^2}{4}}$$

Hence $$\int_0^\infty e^{-x^2} \cos \alpha x dx = \frac{\sqrt{\pi}}{2} e^{-\alpha^2/4}$$

Example 6 (b) Prove that

$$\int_0^\infty xe^{-x^2} \sin \alpha x\, dx = \frac{\sqrt{\pi}}{4} \alpha e^{-\alpha^2/4}$$

Solution: From the result of above example, i.e.,

$$\int_0^\infty xe^{-x^2} \cos \alpha x \, dx = \frac{\sqrt{\pi}}{2} e^{-\alpha^2/4}, \alpha \text{ is a parameter} \quad ...(1)$$

differentiation (1) w.r.t. α on both sides, we get

$$\int_0^\infty -xe^{-x^2} \sin \alpha x \, dx = \frac{\sqrt{\pi}}{2} e^{-\alpha^2/4} \left(\frac{-2\alpha}{4}\right)$$

$$\therefore \quad \int_0^\infty xe^{-x^2} \sin \alpha x \, dx = \frac{\sqrt{\pi}}{4} \alpha \, e^{-\alpha^2/4}$$

Hence

$$\int_0^\infty x\, e^{-x^2} \sin \alpha x \, dx = \frac{\sqrt{\pi}}{4} \alpha \, e^{-\alpha^2/4}.$$

Example 7 Evaluate the Integral,

$$\int_0^\infty \frac{e^{-x} \sin bx}{x} dx$$

Solution: Let $I(b) = \int_0^\infty \frac{e^{-x} \sin bx}{x} dx$, where b is a parameter ...(i)

Differentiating (i) w.r.t. 'b' on both sides, we get

$$\frac{dI}{db} = \int_0^\infty \frac{\partial}{\partial b}\left(\frac{e^{-x} \sin bx}{x}\right) dx$$

$$= \int_0^\infty x \frac{e^{-x} \cos bx}{x} dx$$

$$= \int_0^\infty e^{-x} \cos bx \, dx$$

$$\therefore \quad \frac{dI}{db} = \left[\frac{e^{-x}}{1+b^2}(-\cos bx + b\sin bx)\right]_0^{\infty}$$

$$\left\{\because \int e^{ax}\cos bx\,dx = \frac{e^{ax}}{a^2+b^2}(a\cos bx + b\sin bx)\right\}$$

$$\frac{dI}{db} = \frac{1}{1+b^2} \qquad \text{...(ii)}$$

Integrating both sides of (ii) w.r.t b, we have

$$I(b) = \tan^{-1} b + c \qquad \text{...(iii)}$$

At initially, when b = 0, from (i)

$$I(0) = 0$$

similarly when b = 0 form (iii)

$$I(0) = \tan^{-1}0 + c$$

$$0 = c$$

Hence (iii) becomes

$$I = \tan^{-1}b$$

or

$$\int_0^{\infty} \frac{e^{-x}\sin bx}{x}\,dx = \tan^{-1} b$$

6.23 WHEN THE LIMIT OF INTEGRATION ARE GIVEN IN THE FORM OF A PARAMETER.

If f(x, α), $\dfrac{\partial f(x,\alpha)}{\partial \alpha}$ be continuous function of x and α, then

$$\frac{d}{d\alpha}\left[\int_{\phi(\alpha)}^{\varphi(\alpha)} f(x,\alpha)\,dx\right] = \int_{\phi(\alpha)}^{\varphi(\alpha)} \frac{\partial}{\partial\alpha} f(x,\alpha)\,dx + \frac{d\varphi}{d\alpha} f[\varphi(\alpha),\alpha] - \frac{d\phi}{d\alpha} f[\phi(\alpha),\alpha]$$

provided $\phi(\alpha)$ and $\varphi(\alpha)$ possesses continuous first order derivatives w.r.t. α.

Example 8 Verify the rule of differentiation under the sign of Integration for

$$\int_0^{a^2} \tan^{-1}\left(\frac{x}{a}\right) dx, \text{ where a is a parameter.}$$

Solution: Let $I(a) = \int_0^{a^2} \tan^{-1}\left(\frac{x}{a}\right) dx,$ ...(i)

Differentiating (i) w.r.t. 'a' on both sides we get

$$\frac{dI(a)}{da} = \int_0^{a^2} \frac{\partial}{\partial a}\left(\tan^{-1}\frac{x}{a}\right) dx + 2a\tan^{-1}\frac{a^2}{a} - 0$$

$$= \int_0^{a^2} \frac{1}{1+\frac{x^2}{a^2}}\left(-\frac{x}{a^2}\right) dx + 2a\tan^{-1} a$$

$$= \int_0^{a^2} \frac{-x}{a^2+x^2} dx + 2a\tan^{-1} a$$

$$= \left[-\frac{1}{2}\log\left(a^2+x^2\right)\right]_0^{a^2} + 2a\tan^{-1} a$$

$$= -\frac{1}{2}\log\left(a^2+a^4\right) + \frac{1}{2}\log a^2 + 2a\tan^{-1} a$$

$$= -\frac{1}{2}\log\frac{a^2+a^4}{a^2} + 2a\tan^{-1} a$$

$$\frac{dI(a)}{da} = -\frac{1}{2}\log\left(1+a^2\right) + 2a\tan^{-1} a \qquad \text{(ii)}$$

Now on integrating (i) by parts w.r.t. x we have,

$$I(a) = \int_0^{a^2} 1.\tan^{-1}\frac{x}{a} dx$$

$$= \left\{\tan^{-1}\left(\frac{x}{a}\right)x\right\}_0^{a^2} - \int_0^{a^2} \frac{1}{1+\frac{x^2}{a^2}}.\frac{1}{a}x\, dx$$

$$= \left(a^2\tan^{-1}\frac{a^2}{a} - 0\right) - \int_0^{a^2} \frac{ax}{a^2+x^2} dx$$

$$= a^2\tan^{-1} a - \frac{a}{2}\left[\log\left(x^2+a^2\right)\right]_0^{a^2}$$

$$\therefore \quad I(a) = a^2 \tan^{-1} a - \frac{a}{2}\log\left(a^2+1\right) \qquad \text{(A)}$$

Differentiating (A) w.r.t. 'a' on both sides

$$\frac{d}{da} I(a) = a^2 \frac{1}{1+a^2} + 2a\tan^{-1} a - \frac{a}{2}\frac{2a}{1+a^2} - \frac{1}{2}\log\left(a^2+1\right)$$

$$= \frac{a^2}{1+a^2} + 2a\tan^{-1} a - \frac{a^2}{1+a^2} - \frac{1}{2}\log\left(a^2+1\right) \qquad \text{(iii)}$$

$$\therefore \quad \frac{dI(a)}{da} = -\frac{1}{2}\log\left(a^2+1\right) + 2a\tan^{-1} a$$

Hence from equation (ii) and (iii), the rule is verified,

Example 9 Evaluate

$$\int_0^{\alpha} \frac{\log(1+\alpha x)}{1+x^2}\,dx \text{ and hence show that}$$

$$\int_0^{1} \frac{\log(1+x)}{1+x^2}\,dx = \log 2$$

Solution: Let $F(\alpha) = \int_0^{\alpha} \frac{\log(1+\alpha x)}{1+x^2}\,dx$ here α is a parameter. ...(i)

Differentiating (i) w.r.t. α on both sides, we get

$$F'(\alpha) = \int_0^{\alpha} \frac{\partial}{\partial\alpha}\frac{\log(1+\alpha x)}{1+x^2}\,dx - \frac{d(0)}{d\alpha} + \frac{d\alpha}{d\alpha}\left[\frac{\log\left(1+\alpha^2\right)}{1+\alpha^2}\right]$$

$$= \int_0^{\alpha} \frac{x}{(1+\alpha x)\left(1+x^2\right)}\,dx + \frac{\log\left(1+\alpha^2\right)}{1+\alpha^2} \qquad \text{...(ii)}$$

Taking $\frac{x}{(1+\alpha x)\left(1+x^2\right)} = \frac{A}{(1+\alpha x)} + \frac{Bx+C}{1+x^2}$

$$x = A\left(1+x^2\right) + (Bx+C)(1+\alpha x) \qquad \text{...(A)}$$

Put $1+\alpha x = 0 \qquad \therefore \quad x = -\frac{1}{\alpha}$

$$-\frac{1}{\alpha} = A\left(1+\frac{1}{\alpha^2}\right) + \left[\left(B\times -\frac{1}{\alpha^2}\right)+C\right](1-1)$$

$$-\frac{1}{\alpha} = A\left(\frac{\alpha^2+1}{\alpha^2}\right) \qquad \therefore \quad A = \frac{-\alpha^2}{\alpha^2+1}$$

Again, comparing the coefficient of like power on both sides of (A)

Coefficient of x^2 $\quad 0 = A + B\alpha$ ...(B)

Coefficient of x $\quad 1 = C\alpha + B$ ...(C)

Constant term $\quad 0 = A + C$ ...(D)

$\therefore \quad A = -C$

from equation (B), (C) and (D), on simplification

$$B = \frac{1}{1+\alpha^2}$$

and
$$C = \frac{\alpha}{1+\alpha^2}$$

from (ii)

$$\therefore \quad F'(\alpha) = \frac{\alpha}{\alpha^2+1}\int_0^\alpha \frac{dx}{1+\alpha x} + \frac{1}{1+\alpha^2}\int_0^\alpha \frac{x}{1+x^2}dx$$

$$+\frac{\alpha}{1+\alpha^2}\int_0^\alpha \frac{1}{1+x^2}dx + \frac{\log\left(1+\alpha^2\right)}{1+\alpha^2}$$

or
$$F'(\alpha) = \frac{-\alpha}{1+\alpha}\int_0^\alpha \frac{dx}{1+\alpha x} + \frac{1}{2\left(1+\alpha^2\right)}\int_0^\alpha \frac{2x}{1+x^2}dx +$$

$$\frac{\alpha}{1+\alpha^2}\int_0^\alpha \frac{dx}{1+x^2} + \frac{\log\left(1+\alpha^2\right)}{1+\alpha^2}$$

$$= -\frac{\alpha}{1+\alpha^2}\left\{\frac{1}{\alpha}\log\left(1+\alpha x\right)\right\}_0^\alpha + \frac{1}{2\left(1+\alpha^2\right)}\cdot\left\{\log\left(1+x^2\right)\right\}_0^\alpha$$

$$+\frac{\alpha}{1+\alpha^2}\left[\tan^{-1}x\right]_0^{\alpha}+\frac{\log\left(1+\alpha^2\right)}{\left(1+\alpha^2\right)}$$

$$F'(\alpha)=\frac{1}{1+\alpha^2}\log\left(1+\alpha^2\right)+\frac{\log\left(1+\alpha^2\right)}{2\left(1+\alpha^2\right)}+$$

$$\frac{\alpha\tan^{-1}\alpha}{1+\alpha^2}+\frac{\log\left(1+\alpha^2\right)}{\left(1+\alpha^2\right)}$$

$\therefore$ $$F'(\alpha)=\frac{\log\left(1+\alpha^2\right)}{2\left(1+\alpha^2\right)}+\frac{\alpha\tan^{-1}\alpha}{1+\alpha^2} \qquad \text{...(iii)}$$

on integrating (iii) both sides w.r.t. α, then

$$F(\alpha)=\frac{1}{2}\int\frac{\log\left(1+\alpha^2\right)}{\left(1+\alpha^2\right)}d\alpha+\int\frac{\alpha\tan^{-1}\alpha}{\left(1+\alpha^2\right)}d\alpha$$

Integrating by parts, taking $\log(1+\alpha^2)$ as first function.

$$=\frac{1}{2}\log\left(1+\alpha^2\right)\tan^{-1}\alpha-\frac{1}{2}\int\frac{2\alpha}{\left(1+\alpha^2\right)}\tan^{-1}\alpha dx$$

$$+\int\frac{\alpha\tan^{-1}\alpha}{1+\alpha^2}+c$$

$$F(\alpha)=\frac{1}{2}\log\left(1+\alpha^2\right)\tan^{-1}\alpha+c \qquad \text{...(iv)}$$

Initially, when $\alpha = 0$, from (i)

$F(0) = 0$

Again, when $\alpha = 0$ from (iv)

$$F(0)=\frac{1}{2}\log(1+0)\tan^{-1}0+c$$

$$0=\frac{1}{2}\times 0+c$$

$\therefore$ $c = 0$

Hence $$F(\alpha) = \frac{1}{2}\log(1+\alpha)\tan^{-1}\alpha$$

$$= \int_0^{\alpha} \frac{\log(1+\alpha x)}{1+x^2} = \frac{1}{2}\log(1+\alpha)\tan^{-1}\alpha \qquad \text{...(v)}$$

To show that $$\int_0^1 \frac{\log(1+x)}{1+x^2}\,dx = \frac{\pi}{8}\log_e 2$$

Put $\alpha = 1$ in (iv), then

$$F(\alpha) = \int_0^1 \frac{\log(1+x)}{1+x^2}\,dx = \frac{1}{2}\log 2\tan^{-1}(1)$$

$$= \frac{1}{2}\log 2 \times \frac{\pi}{4} = \frac{\pi}{8}\log_e 2$$

Example 10 If $y = \int_0^x f(t)\sin[k(x-t)]\,dt$. Prove that y satisfy the differential equation

$$\frac{d^2y}{dx^2} + k^2 y = k f(x).$$

Solution: Let $y(x) = \int_0^x f(t)\sin[k(x-t)]dt$, Here x is a parameter

Differentiating (i) w.r.t x on both sides, we get

$$\frac{dy}{dx} = \frac{d}{dx}\int_0^x f(t)\sin k(x-t)\,dt + \frac{dx}{dx}\times f(x)\sin k(x-x) - 0$$

$$= \int_0^x f(t)\frac{\partial}{\partial x}\{\sin k(t-x)\}\,dt$$

(∴ leibnitz rule for differentiation under the sign of integration

$$= \int_0^x f(t)\,k\cos k(x-t)\,dt \qquad \text{...(ii)}$$

$$\frac{dy}{dx} = k\int_0^x f(t)\cos k(x-t)\,dt$$

Again, differentiating (ii) w.r.t. x on both sides, we get

$$\frac{d^2y}{dx^2} = k\int_0^x f(t)\,k\{-\sin k(x-t)\}\,dt +$$

$$k\frac{dx}{dx}f(x)\cos k(x-x)-0$$

$$\frac{d^2y}{dx^2}=-k^2\int_0^x f(t)\sin k(x-t)\,dt+k f(x)$$

Hence $$\frac{d^2y}{dx^2}+k^2y(x)=kf(x).$$

Exercise 6.9

1. Prove that

$$\int_0^\infty \frac{e^{-ax}\sin x}{x}\,dx=\cot^{-1}a \text{ and hence deduce that } \int_0^\infty \frac{\sin x}{x}\,dx=\frac{\pi}{2}$$

2. $$\int_0^\infty \frac{1-e^{-ax}}{x}e^{-x}\,dx=\log(1+a),\ (a>-1)$$

3. (a) Prove that

$$\int_0^{\pi/2}\frac{\log(1+\sin\alpha\cos x)}{\cos x}\,dx=\pi\alpha, \text{ where } \alpha \text{ is a parameter}$$

(b) $$\int_0^{\pi/2}\frac{\log(1+\cos\alpha\cos x)}{\cos x}\,dx=\frac{1}{2}\left\{\frac{\pi^2}{4}-\alpha^2\right\}$$

4. If $|a|<1$, Evaluate $\int_0^\pi \frac{\log(1+a\cos x)}{\cos x}\,dx$

5. Show that $\int_{\frac{\pi}{2a}}^{\pi/2a}\frac{\sin ax}{x}\,dx=constant$

Answers

4. $\pi\sin^{-1}a$.

CHAPTER 7

Double and Triple Integrations

7.1 Introduction

In this chapter, we shall study how the process of integration can be extended to functions of more than one variable. It will give us two generalisations of the definite integral, namely multiple integrals and repeated integrals.

Definition

The process of integration of a function of one variable can be extended to a function of more than one variable, when the function is continuous or piece wise continuous in the given region, such an integral is known as multiple integral. The multiple integral is usually called improper when the range of integration extends to infinity or when the integrand becomes infinite at a point inside the region or on the boundary of region.

7.2 Double Integrals

Let $f(x, y)$ be function of two independent variables x and y defined at every point of a finite region A of xy – plane. Now divide A into n sub-regions $R_1, R_2 \ldots R_n$ of elementary areas $\Delta A_1, \Delta A_2 \ldots \Delta A_n$ respectively.

Let $P_k(x_k, y_k)$ be a point inside the sub-region R_k. Then from the sum

$$\sum_{k=1}^{n} f(P_k)\,\Delta A_k \qquad \ldots(1)$$

Double integral of $f(x, y)$ over the region A is the limit of above sum as $n \to \infty$ and $\Delta A_k \to 0$. Further it is denoted by I_A. Thus,

$$I_A = \iint_A f(P)dA = \iint_A f(x,y)dxdy = \lim_{\substack{n\to\infty \\ \Delta A_k \to 0}} \sum_{k=1}^{n} f(P_k)\Delta A_k \quad ...(2)$$

The area A is called the region or field of integration and dA or $dxdy$ is called elementary area.

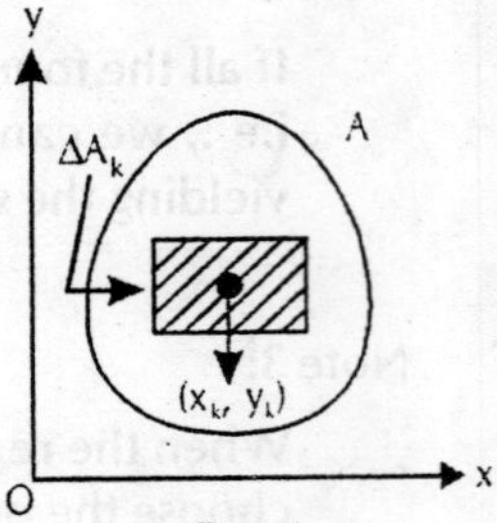

Fig. 1

Properties of Double Integral

1. $\iint_A af(x,y)dA = a\iint_A f(x,y)dA$ where a is constant

2. $\iint_A [f(x,y)+g(x,y)]dA = \iint_A f(x,y)dA + \iint_A g(x,y)dA$

3. $\iint_A f(x,y)dA = \iint_{A_1} f(x,y)dA + \iint_{A_2} f(x,y)dA$

Where A is the union of disjoint regions A_1 and A_2

7.3 Evaluation of Double Integrals

The process of integration of a function of one variable can be adopted as an alternative and effective method for evaluating integrals of two or more variables. We thus change the double integral into two successive single integrals as shown in fig. 2.

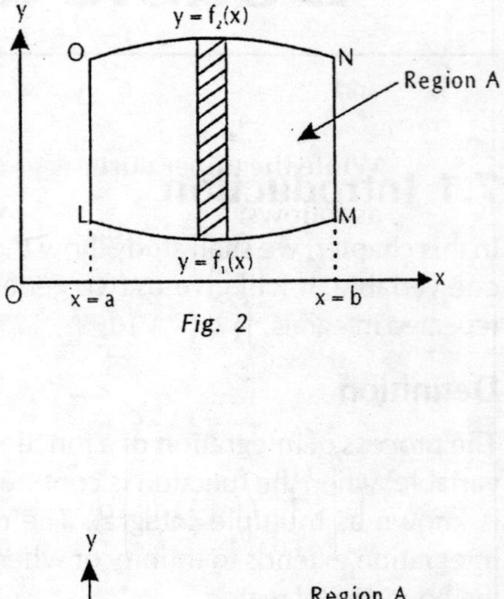

Fig. 2

Let the region A is bounded by the curves

$y = f_1(x)$, $y = f_2(x)$, $x = a$ and $x = b$, then

$$\iint_A f(x,y)dA = \int_{x=a}^{b} \left\{ \int_{y=f_1(x)}^{f_2(x)} f(x,y)dy \right\} dx \quad ...(1)$$

where the integration w.r.t y is performed treating x as constant.

On the other hand, if the region A is bounded by the curves $x = f_1(y)$, $x = f_2(y)$, $y = c$ and $y = d$ as shown in fig. 3, then

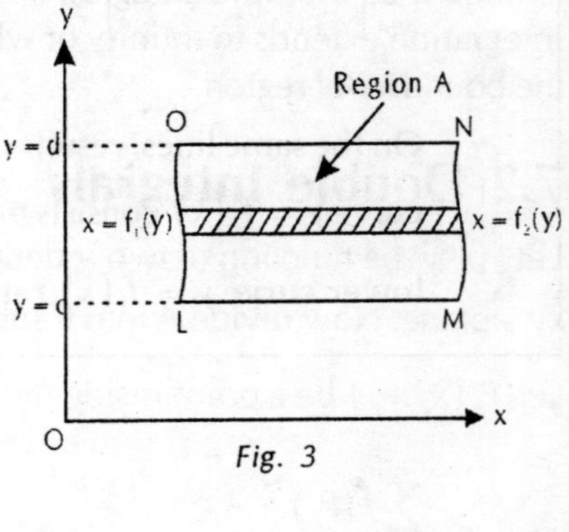

Fig. 3

$$\iint_A f(x,y)dA = \int_{y=c}^{d} \left\{ \int_{x=f_1(y)}^{f_2(y)} f(x,y)dx \right\} dy \quad ...(2)$$

where the integration w.r.t. x is performed treating y as constant.

Note 1:

The braces in (1) and (2) can be omitted, since it is conventional to integrate first w.r.t. a variable whose differential appears first (for example in (1) the order of integration is first *y* and later *x*).

Note 2:

If all the four limits are constants then the order of integration can be done in either way i.e ., we can integrate first w.r.t *x* and later w.r.t. *y* or first w.r.t.*y* and later w.r.t. *x*, yielding the same results.

Note 3:

When the region *A* is regular (in both *x* and *y* directions), then draw a rough sketch of *A*, choose the order of integration and use (1) or (2) whichever is easier for integration.

Note 4:

In fig – 2, suppose lower curve $y = f_1(x)$ consists of two different curves as

$$y_1 = \phi_1(x) \qquad a \le x \le e$$

and

$$y_2 = \phi_2(x) \qquad e \le x \le b$$

While the upper curve is $y = f_2(x)$. Then the integral along the region *A* can be obtained as follows:

$$\iint_A f(x,y)\,dydx = \int_{x=a}^{b}\left\{\int_{y=f_1(x)}^{f_2(x)} f(x,y)\,dy\right\}dx$$

$$= \int_{x=a}^{e}\left\{\int_{y=\phi_1(x)}^{f_2(x)} f(x,y)\,dy\right\}dx + \int_{x=e}^{b}\left\{\int_{y=\phi_2(x)}^{f_2(x)} f(x,y)\,dy\right\}dx$$

...(3)

On the same lines, results can be obtained if the upper curve $y = f_2(x)$ consists of two different curves while lower curve $y = f_1(x)$ remains same.

Illustrative Examples

Example 1

Evaluate $\int_1^a \int_1^b \frac{1}{xy} dxdy$

Solution:

Given integral can be written as

$$\int_{x=1}^{a} \int_{y=1}^{b} \frac{1}{xy} dxdy$$

$$= \int_{x=1}^{a} \left\{ \int_{y=1}^{b} \frac{1}{x} \cdot \frac{1}{y} dy \right\} dx$$

$$= \int_{x=1}^{a} \left[\frac{1}{x} \log y \right]_{y=1}^{b} dx \quad \text{(treating x as constant)}$$

$$= \int_{x=1}^{a} \frac{1}{x} \log b \, dx = [\log b \cdot \log x]_{x=1}^{a}$$

$$= \log b \cdot \log a \qquad \text{Answer}$$

Example 2

Evaluate $\iint_A xydxdy$ over the positive quadrant of the circle $x^2 + y^2 = a^2$.

Solution :

Here the region of integration is the positive quadrant of circle $x^2 + y^2 = a^2$, centered at (0,0) and radius is a. This has been shown in fig – 4. We choose the elementary strip parallel to y axis in the given region. This strip moves from $y = 0$ to the circle $x^2 + y^2 = a^2$ at which $y = \sqrt{a^2 - x^2}$. Also, along x – axis region moves from x = 0 to x = a.

Therefore , the given integral can be written as

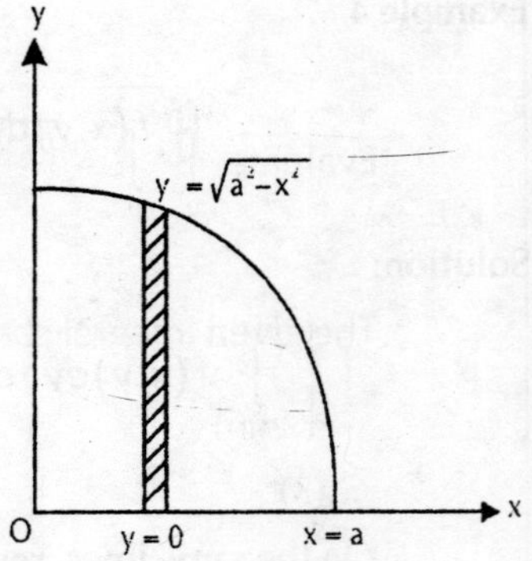

Fig. 4

$$\iint_A xydxdy = \int_{x=0}^{a} \left\{ \int_{y=0}^{\sqrt{a^2-x^2}} xydy \right\} dx$$

$$= \int_{x=0}^{a} \left(\frac{xy^2}{2} \right)_{y=0}^{\sqrt{a^2-x^2}} dx = \int_0^a \frac{x}{2}(a^2 - x^2) dx$$

$$= \frac{1}{2} \int_0^a (a^2x - x^3) dx = \frac{1}{2} \left[a^2 \frac{x^2}{2} - \frac{x^4}{4} \right]_0^a = \frac{1}{2} \cdot \frac{a^4}{4}$$

$$= \frac{a^4}{8} \qquad \text{Answer}$$

Example 3

Evaluate $\int_0^1 \int_0^{\sqrt{1+x^2}} \frac{dxdy}{1+x^2+y^2}$

Solution:

Let $$I = \int_0^1 \int_0^{\sqrt{1+x^2}} \frac{dxdy}{1+x^2+y^2}$$

$$= \int_{x=0}^{1} \left\{ \int_{y=0}^{\sqrt{1+x^2}} \frac{1}{(1+x^2)+y^2} dy \right\} dx$$

$$= \int_{x=0}^{1} \left[\frac{1}{\sqrt{1+x^2}} \tan^{-1} \frac{y}{1+x^2} \right]_{y=0}^{\sqrt{1+x^2}} dx$$

$$= \int_0^1 \frac{1}{\sqrt{1+x^2}} \cdot \frac{\pi}{4} dx$$

$$= \frac{\pi}{4} \left[\log\left(x+\sqrt{x^2+1}\right) \right]_{x=0}^{1} = \frac{\pi}{4} \log\left(1+\sqrt{2}\right) \qquad \text{Answer}$$

Example 4

Evaluate $\int_0^a \int_0^{\sqrt{a^2-y^2}} \sqrt{(a^2-x^2-y^2)}dydx$

Solution:

The given integral can be written as

$$I = \int_{y=0}^{a} \left\{ \int_{x=0}^{\sqrt{a^2-y^2}} \sqrt{(a^2-x^2-y^2)} dx \right\} dy$$

$$= \int_{y=0}^{a} \left[\frac{x\sqrt{(a^2-y^2-x^2)}}{2} + \frac{a^2-y^2}{2} \sin^{-1} \frac{x}{\sqrt{a^2-y^2}} \right]_{x=0}^{\sqrt{a^2-y^2}} dy$$

$$\left(\because \int \sqrt{a^2-x^2} dx = \frac{x\sqrt{a^2-x^2}}{2} + \frac{a^2}{2} \sin^{-1} \frac{x}{a} \right)$$

$$= \int_{y=0}^{a} \frac{a^2 - y^2}{2} \cdot \frac{\pi}{2} dy = \frac{\pi}{4} \int_0^a (a^2 - y^2) dy$$

$$= \frac{\pi}{4}\left(a^2 y - \frac{y^3}{3}\right)_0^a = \frac{\pi a^3}{6} \quad \text{Answer}$$

Example 5

Prove that $\iint_A \sqrt{4x^2 - y^2}\, dxdy = \frac{1}{3}\left(\frac{\pi}{3} + \frac{\sqrt{3}}{2}\right)$ where the region A of integration is the triangle given by $y = 0$, $y = x$ and $x = 1$.

Solution:

The region of integration has been shown through fig – 5. The given triangle is OLM. We choose the elementary strip parallel to y – axis and *it moves from* $y = 0$ to $y = x$ and x varies from $x = 0$ to $x = 1$. Thus the given integral can be written as

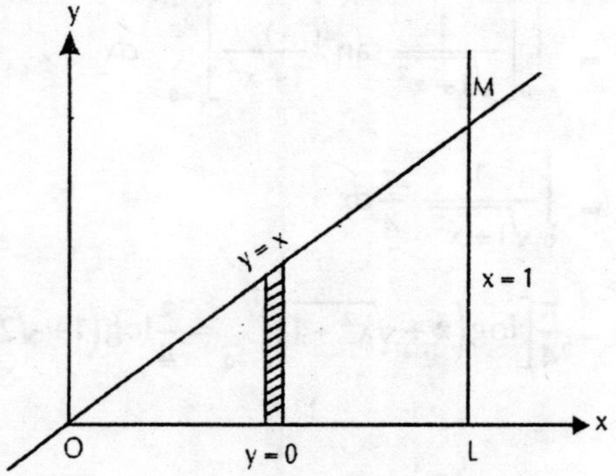

Fig. 5

$$\iint_A \sqrt{4x^2 - y^2}\, dxdy = \int_{x=0}^{1} \left\{ \int_{y=0}^{x} \sqrt{4x^2 - y^2}\, dy \right\} dx$$

$$= \int_{x=0}^{1} \left[\frac{1}{2} y \sqrt{4x^2 - y^2} + \frac{1}{2} \cdot 4x^2 \sin^{-1}\left(\frac{y}{2x}\right) \right]_{y=0}^{x} dx$$

$$= \int_{x=0}^{1} \frac{1}{2}\left(\sqrt{3} \cdot x^2 + 4x^2 \cdot \frac{\pi}{6}\right) dx$$

$$= \frac{1}{2}\left[\sqrt{3} \cdot \frac{x^3}{3} + \frac{2\pi}{9} x^3\right]_{x=0}^{1} = \frac{1}{2}\left[\frac{\sqrt{3}}{3} + \frac{2\pi}{9}\right] = \frac{1}{3}\left[\frac{\sqrt{3}}{2} + \frac{\pi}{3}\right] \quad \text{Proved}$$

Example 6

Evaluate $I_A = \iint_A (x^2 + y^2)\,dxdy$ where A is bounded by $y = x$ and $y^2 = 4x$.

Solution:

The given region A has been shown in fig – 6. The points of intersections of curves $y = x$ and $y^2 = 4x$. are (0,0) and (4,4). Let us choose the elementary strip parallel to y – axis and it moves from $y = x$ to $2\sqrt{x}$ while x varies from $x = 0$ to 4. Thus

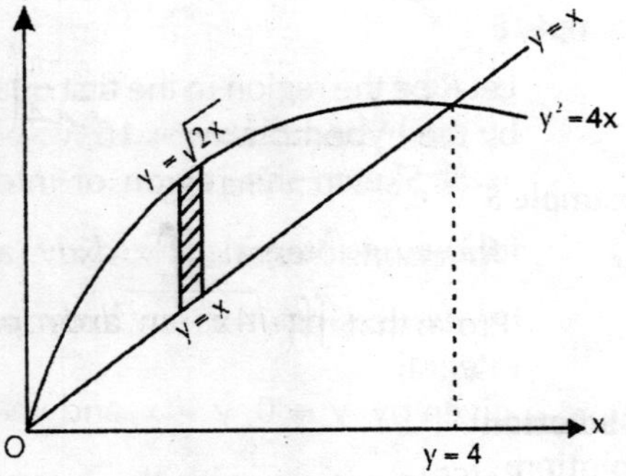

Fig. 6

$$I_A = \int_{x=0}^{4}\left\{\int_{y=x}^{2\sqrt{x}} (x^2 + y^2)\,dy\right\}dx$$

$$= \int_0^4 \left(x^2 y + \frac{y^3}{3}\right)_{y=x}^{2\sqrt{x}} dx = \int_0^4 \left(2x^{\frac{5}{2}} + \frac{8}{3}x^{\frac{3}{2}} - \frac{4}{3}x^3\right)dx$$

$$= \frac{768}{35}$$

Example 7

Evaluate $\iint_A x^3 y\,dxdy$ where A is region enclosed by the ellipse $\dfrac{x^2}{a^2} + \dfrac{y^2}{b^2} = 1$ in first quadrant.

Solution :

The given region has been shown in fig – 7. The given integral can be written as

$$I = \iint_A x^3 y\,dxdy$$

$$= \int_0^a \int_{y=0}^{\frac{b}{a}\sqrt{a^2-x^2}} x^3 y\,dydx$$

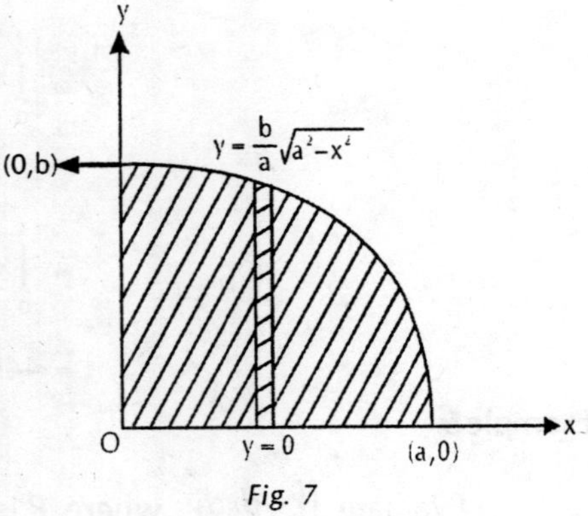

Fig. 7

$$= \int_{x=0}^{a}\left[\frac{x^3 y^2}{2}\right]_{y=0}^{\frac{b}{a}\sqrt{a^2-x^2}} dx = \frac{b^2}{2a^2}\int_0^a (a^2x^3 - x^5)\,dx$$

$$= \frac{b^2}{2a^2}\left[\frac{a^2x^4}{4} - \frac{x^6}{6}\right]_0^a = \frac{b^2a^4}{24} \quad \text{Answer}$$

Example 8

Let R be the region in the first quadrant bounded by the hyperbola $xy = 16$, $y = x$, $y = 0$ and $x = 8$. Sketch the region of integration of the following integral $\iint_R x^2 dxdy$ and evaluate it by expressing it as an appropriate repeated integral.

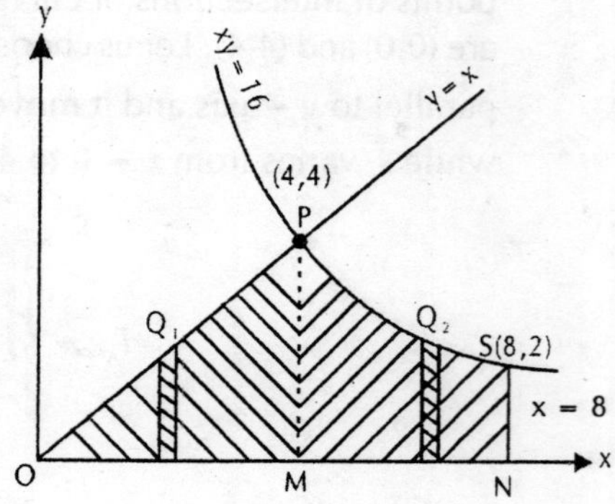

Fig. 8

Solution:

Solving $x = y$ with the curve $xy = 16$, we obtain the co – ordinate of $P(4,4)$. The shaded portion consists of two regions:

(i) region OMP (ii) region PMNS where the line NS is $x = 8$.

For the first region, y varies from 0 to x and x varies from 0 to 4. For the second region, y varies from 0 to $16/x$ and x varies from 4 to 8.

Thus,

$$\iint_R x^2 dxdy = \int_{x=0}^{4}\int_{y=0}^{x} x^2 dxdy + \int_{x=4}^{8}\int_{y=0}^{\frac{16}{x}} x^2 dxdy$$

$$= \int_0^4 x^2\left\{\int_0^x dy\right\}dx + \int_4^8 x^2\left\{\int_0^{\frac{16}{x}} dy\right\}dx$$

$$= \int_0^4 x^3 dx + \int_4^8 16x dx = \left[\frac{x^4}{4}\right]_0^4 + 16\left[\frac{x^2}{2}\right]_4^8$$

$= 448.$ Answer

Example 9

Evaluate $\iint_R y dxdy$, where R is the region bounded by the parabolas $y^2 = 4x$ and $x^2 = 4y$.

Solution:

The points of intersections of the curves $y^2 = 4x$ and $x^2 = 4y$ are (0,0) and (4,4). The given region R has been shown in the Fig. 9 through shaded area .

Thus in the region, y varies from $\frac{x^2}{4}$ to $2\sqrt{x}$ and x varies from 0 to 4.

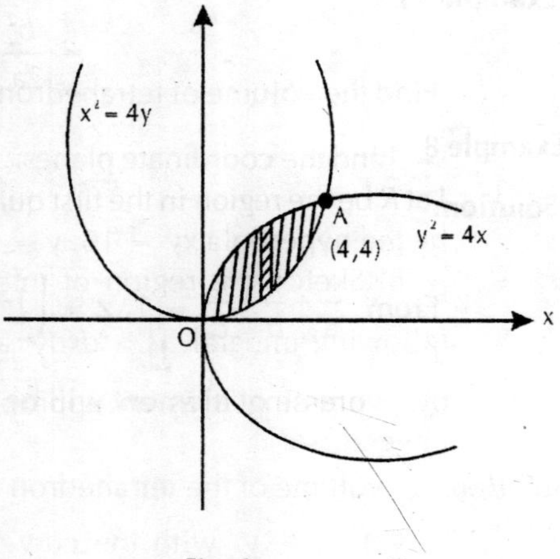

Fig. 9

Hence $\iint_R y\,dx\,dy$

$$= \int_{x=0}^{4}\int_{y=\frac{x^2}{4}}^{2\sqrt{x}} y\,dy\,dx$$

$$= \int_0^4 \left[\frac{y^2}{2}\right]_{x^2/4}^{2\sqrt{x}} dx$$

$$= \int_0^4 \left(2x - \frac{x^4}{32}\right)dx = \left[x^2 - \frac{x^5}{160}\right]_0^4 = 4^2 - \frac{4^5}{160}$$

$$= \frac{48}{5} \qquad \text{Answer}$$

Example 10

Evaluate $\iint_R \sqrt{xy - y^2}\,dy\,dx$, where R is triangle with vertices (0,0), (10,1) and (1,1).

Solution :

Given triangle has been represented by OBA in the Fig. 10. Here the elementary strip has been choosen parallel to x – axis. The line OA is $x = y$ while the line OB is $x = 10y$. So, for the given region x varies from y to $10y$ while y varies from 0 to 1.

Thus, the given integral may be written as

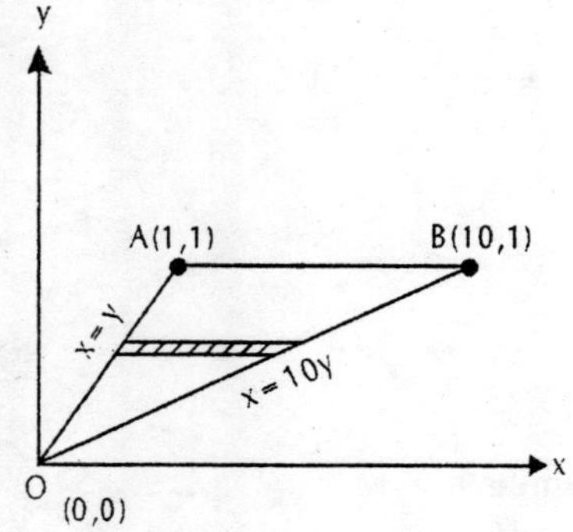

Fig. 10

$$\iint_R \sqrt{xy - y^2}\,dy\,dx = \int_{y=0}^{1}\left\{\int_{x=y}^{10y}\sqrt{xy - y^2}\,dx\right\}dy$$

$$= \int_{y=0}^{1}\left[\frac{2}{3}\frac{1}{y}\left(xy - y^2\right)^{\frac{3}{2}}\right]_y^{10y} dy$$

$$= \int_0^1 \frac{2}{3}\frac{1}{y}\left(9y^2\right)^{\frac{3}{2}} dy = 18\int_0^1 y^2 dy = 6 \qquad \text{Answer}$$

Example 11

Find the volume of tetrahedron bounded by $\frac{x}{a}+\frac{y}{b}+\frac{z}{c} = 1$ and the coordinate planes.

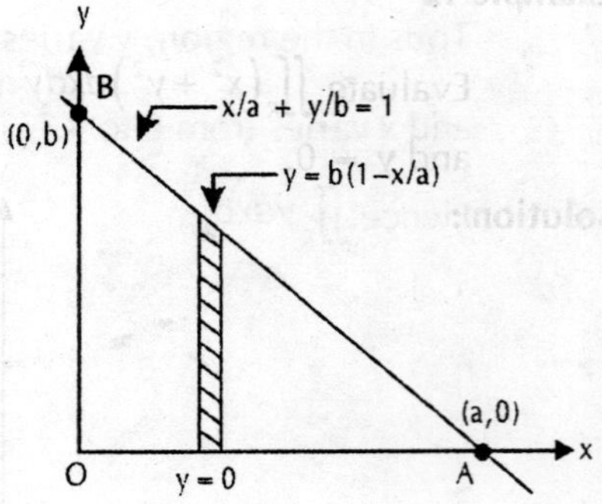

Fig. 11

Solution:

From $\frac{x}{a}+\frac{y}{b}+\frac{z}{c} = 1, z = \left(1-\frac{x}{a}-\frac{y}{b}\right)c$ ...(1)

$\therefore$ volume of element will be $z \cdot dx \cdot dy$

$\therefore$ volume of the tetrahedron will be $V = \iint_R zdxdy$

$$= \iint_R c\left(1-\frac{x}{a}-\frac{y}{b}\right)dxdy \quad \text{by (1)}$$

where R represents the area enclosed by straight line $\frac{x}{a}+\frac{y}{b} = 1$ and the axis in the plane $z = 0$. It has been shown in fig – 11as triangle OBA.

Thus,

$$V = \int_{x=0}^{a}\int_{y=0}^{b\left(1-\frac{x}{a}\right)} c\left(1-\frac{x}{a}-\frac{y}{b}\right)dydx$$

$$= \int_0^a c\left[\left(1-\frac{x}{a}\right)y-\frac{y^2}{2b}\right]_0^{b\left(1-\frac{x}{a}\right)} dx$$

$$= c\int_0^a\left[b\left(1-\frac{x}{a}\right)^2-\frac{b}{2}\left(1-\frac{x}{a}\right)^2\right]dx$$

$$= \frac{bc}{2}\int_0^a\left(1-\frac{x}{a}\right)^2 dx = \frac{bc}{2}\left[-\frac{a}{3}\left(1-\frac{x}{a}\right)^3\right]_0^a$$

$$= \frac{abc}{6} \quad \text{Answer}$$

Example 12

Evaluate $\iint_R \left(x^2+y^2\right)dxdy$ throughout the area enclosed by the curves $y=4x$, $x+y=3$, $y=2$ and $y=0$.

Solution :

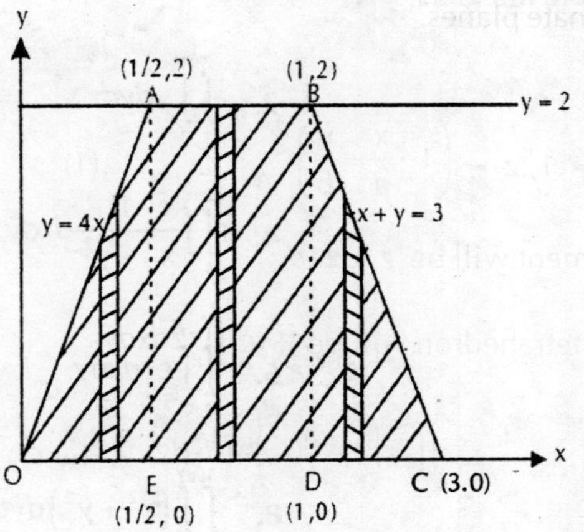

Fig. 12

The given region R has been shown in fig – 12 and it is OABC. We subdivide this region in three portions as OAE, EABD and DBC. The co – ordinates of A,B and C are : $\left(\frac{1}{2},2\right)$, (1,2) and (3,0) respectively.

$$\therefore \quad \iint_R \left(x^2+y^2\right)dxdy$$

$$= \iint_{OAE}\left(x^2+y^2\right)dxdy+\iint_{EABD}\left(x^2+y^2\right)dxdy+\iint_{DBC}\left(x^2+y^2\right)dxdy$$

$$= \int_{x=0}^{\frac{1}{2}}\int_{y=0}^{4x}\left(x^2+y^2\right)dydx+\int_{x=\frac{1}{2}}^{1}\int_{y=0}^{2}\left(x^2+y^2\right)dydx+\int_{x=1}^{3}\int_{y=0}^{(3-x)}\left(x^2+y^2\right)dy\cdot dx$$

$$= \int_0^{\frac{1}{2}}\left[x^2y+\frac{y^3}{3}\right]_0^{4x}dx+\int_{\frac{1}{2}}^{1}\left[x^2y+\frac{y^3}{3}\right]_0^{2}dx+\int_1^3\left[x^2y+\frac{y^3}{3}\right]_0^{(3-x)}dx$$

$$= \int_0^{\frac{1}{2}}\frac{76}{3}x^3dx+\int_{\frac{1}{2}}^{1}\left(2x^2+\frac{8}{3}\right)dx+\int_1^3\left(3x^2-x^3+\frac{(3-x)^3}{3}\right)dx$$

$$= \frac{19}{48}+\frac{23}{12}+\frac{22}{3}=9\frac{31}{48}$$ Answer

Exercise 7.1

Evaluate the following double integrals:

1. $\int_0^1\int_0^2 (x+y)\,dxdy$

2. $\int_1^2\int_0^{3y} y\,dydx$

3. $\int_0^a\int_0^b (x^2+y^2)\,dxdy$

4. $\int_1^2\int_0^x \frac{1}{x^2+y^2}\,dxdy$

5. $\int_0^1\int_0^{\sqrt{1+x^2}} \frac{1}{1+x^2}\,dxdy$

6. $\int_0^1\int_0^{x^2} e^{\frac{y}{x}}\,dxdy$

7. $\int_0^2\int_0^{\sqrt{2x-x^2}} x\,dxdy$

8. $\int_0^1\int_y^{\sqrt{y}} (x^2+y^2)\,dydx$

9. $\int_0^{\pi/2}\int_{\pi/2}^{\pi} \cos(x+y)\,dy\,dx$

10. $\int_0^{\pi}\int_0^{x} x\sin y\,dy\,dx$

11. Prove that $\int_0^1\left[\int_0^1 \frac{x-y}{(x+y)^3}dy\right]dx \neq \int_0^1\left[\int_0^1 \frac{x-y}{(x+y)^3}dx\right]dy$

12. Evaluate $\iint x^2y^2\,dxdy$ over the region $x^2+y^2 \le 1$

13. Evaluate $\iint_A (x^2+y^2)\,dxdy$ over the region bounded by $x = 0$, $y = 0$, $x + y = 1$

Answers

1. 3	**2.** 7	**3.** $\left(\frac{1}{3}\right)ab(a^2+b^2)$
4. $\left(\frac{1}{4}\right)\log 2$	**5.** $\log(1+\sqrt{2})$	**6.** $\frac{1}{2}$
7. $\frac{\pi}{2}$	**8.** $\frac{3}{35}$	**9.** -2
10. $2+(\pi^2)/2$	**12.** $\frac{\pi}{24}$	**13.** $\frac{1}{6}$

7.4 Double Integral in Polar Co-ordinates

Let $f(r,\theta)$ be a function of two independent variable r and θ (the polar co – ordinates), defined at every point of a finite region A of the r, θ – plane . Subdivide this region into n elementary areas δA_1, $\delta A_2 \ldots \delta A_n$. Let $P_k(r_k,\theta_k)$ be any point inside the kth elementary area δA_k . We now form the sum

$$\sum_{k=1}^{n} f(r_k,\theta_k)\,\delta A_k \qquad \ldots(1)$$

Take the limit of (1) as the number of elementary areas increases indefinitely while its area decreases, Thus, we define the double integral of $f(r,\theta)$ over the region A if the limit of (1) exists as $n \to \infty$ and $\delta A_k \to 0$, and is denoted by

$$\iint_A f(r,\theta)\,dA = \lim_{\substack{n\to\infty \\ \delta A_k \to 0}} \sum_{k=1}^{n} f(r_k,\theta_k)\,\delta A_k \qquad \ldots(2)$$

The method of evaluating integrals in polar co – ordinates is almost the same as in cartesian co-ord nates. However, usually integration w.r.t r is performed first and w.r.t. θ afterwards.

Illustrative Examples

Example 1

$$\int_0^{\pi}\int_0^{a(1+\cos\theta)} r\,d\theta\,dr$$

Solution:

We have

$$I = \int_{\theta=0}^{\pi}\int_{r=0}^{a(1+\cos\theta)} r\,d\theta\,dr$$

$$= \int_{\theta=0}^{\pi}\left\{\int_{r=0}^{a(1+\cos\theta)} r\,dr\right\} d\theta$$

$$= \int_0^{\pi}\left[\frac{r^2}{2}\right]_0^{a(1+\cos\theta)} d\theta$$

$$= \frac{a^2}{2}\int_0^{\pi}(1+\cos\theta)^2\,d\theta$$

$$= 2a^2\int_0^{\pi}\cos^4\frac{\theta}{2}\,d\theta$$

Put $\frac{\theta}{2} = \phi \quad \therefore \; d\theta = 2d\phi$ and when $\theta = 0$, $\phi = 0$, when $\theta = \pi$, $\phi = \frac{\pi}{2}$

$$\therefore \; I = 2a^2\int_0^{\pi/2}\cos^4\phi \cdot 2d\phi = 4a^2\,\frac{3}{4}\cdot\frac{1}{2}\cdot\frac{\pi}{2} = \frac{3}{4}\pi a^2$$

Examples 2

Transform the integral $\int_0^a \int_0^{\sqrt{a^2-x^2}} (1-x^2-y^2)^{\frac{1}{2}}\, dx\, dy$ by changing into polar co-ordinates and hence evaluate it.

Solution :

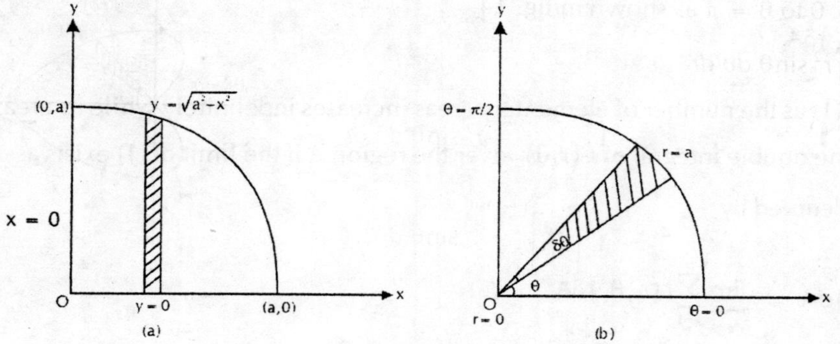

fig. 13

Here the limits of integration shows that the region of integration lies between $y = 0$, $y = \sqrt{a^2 - x^2}$, $x = 0$, $x = a$ which represent the circle $x^2 + y^2 = a^2$ lies in first quadrant (see fig.13(a)). Now we take the limits for polar co-ordinates system as shown in fig. 13(b).

Thus, for the given region r varies from 0 to a and θ varies from 0 to $\frac{\pi}{2}$.

$$\therefore \quad \int_0^a \int_0^{\sqrt{a^2-x^2}} (1-x^2-y^2)^{\frac{1}{2}}\, dx\, dy$$

$$= \int_{\theta=0}^{\pi/2} \int_{r=0}^{a} \sqrt{1-a^2\cos^2\theta - a^2\sin^2\theta}\; r d\theta dr$$

[Putting $x = a\cos\theta$, $y = a\sin\theta$]

$$= \sqrt{1-a^2} \int_0^{\pi/2} \left[\frac{r^2}{2}\right]_0^a d\theta \;=\; \sqrt{1-a^2} \cdot \frac{a^2}{2} [\theta]_0^{\pi/2}$$

$$= \frac{\pi}{4} a^2 \sqrt{(1-a^2)}$$

Example 3

Evaluate $\iint r^2 \sin\theta \, d\theta \, dr$ over the area of cardioid $r = a(1 + \cos\theta)$ above the initial line.

Solution :

The region of integration can be covered by the radial strip whose ends are at $r = 0$, $r = a(1 + \cos\theta)$ and the region lies from $\theta = 0$ to $\theta = \pi$ as shown in fig. 14.

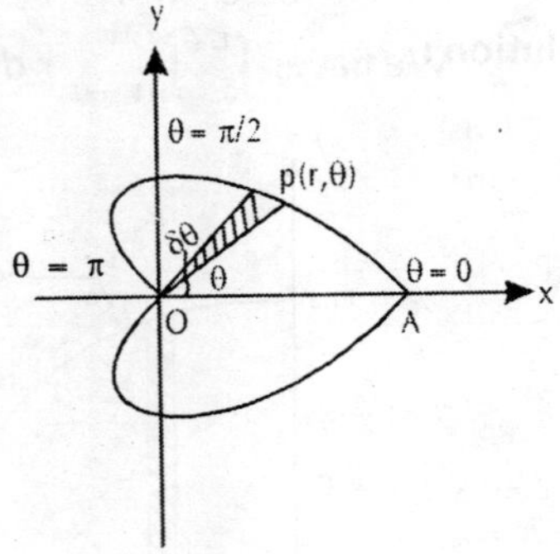

Fig. 14

Thus, $\iint r^2 \sin\theta \, d\theta \, dr$

$$= \int_{\theta=0}^{\pi} \int_{r=0}^{a(1+\cos\theta)} r^2 \sin\theta \, d\theta \, dr$$

$$= \int_0^{\pi} \left[\frac{r^3}{3}\right]_0^{a(1+\cos\theta)} \sin\theta \, d\theta$$

$$= \frac{16}{3} a^3 \int_0^{\pi} \cos^7 \frac{\theta}{2} \sin\frac{\theta}{2} \, d\theta$$

$$= \frac{16}{3} a^3 \int_0^{\pi/2} 2 \sin\phi \cos^7 \phi \, d\phi \qquad \text{where } \frac{\theta}{2} = \phi \text{ and } d\theta = 2d\phi$$

$$= \frac{32}{3} a^3 \left[-\frac{\cos^8 \phi}{8}\right]_0^{\pi/2}$$

$$= \frac{4}{3} a^3 \qquad \text{Answer}$$

7.5 Double Integration by Transforming Cartesian to Polar Co-ordinates

Sometimes it is convenient to solve a double integral of cartesian co-ordinates by changing into polar co-ordinates. In such cases, we use the substitution :

$x = r\cos\theta$, $y = r\sin\theta$ and $r \, d\theta \, dr$ as the elementary area in place of $dx \, dy$ in cartesian co-ordinates.

Note that while we made the above changes, the limits of integration should also be suitably changed so as to cover the whole region.

Thus $$\iint_A f(x,y) dA = \iint_A f(x,y) dx \, dy$$

$$= \iint_A f(r\cos\theta, r\sin\theta) r \, d\theta \, dr \qquad .....(1)$$

Also, it is important to note that, in the polar co-ordinates, we choose the elementary area made up by the circular arcs centred at origin and the series of radial lines through the origin.

Example 1

Evaluate $\int_0^{\pi/2}\int_{a(1-\cos\theta)}^{a\sin\theta} r\,d\theta\,dr$

Solution :

We have, $\int_0^{\pi/2}\int_{a(1-\cos\theta)}^{a\sin\theta} r\,d\theta\,dr$

$$= \int_0^{\pi/2}\left[\frac{r^2}{2}\right]_{a(1-\cos\theta)}^{a\sin\theta} d\theta$$

$$= \frac{1}{2}a^2\int_0^{\pi/2}\left[\sin^2\theta-(1-\cos\theta)^2\right]d\theta$$

$$= \frac{1}{2}a^2\int_0^{\pi/2}\left[\sin^2\theta-1-\cos^2\theta+2\cos\theta\right]d\theta$$

$$= \frac{1}{2}a^2\int_0^{\pi/2}\left[-(1-\sin^2\theta)-\cos^2\theta+2\cos\theta\right]d\theta$$

$$= \frac{1}{2}a^2\int_0^{\pi/2}\left[2\cos\theta-2\cos^2\theta\right]d\theta$$

$$= a^2\int_0^{\pi/2}\left[\cos\theta-\frac{1}{2}(1+\cos 2\theta)\right]d\theta$$

$$= a^2\left[\sin\theta-\frac{\theta}{2}-\frac{1}{4}\sin 2\theta\right]_0^{\pi/2}$$

$$= a^2\left[1-\frac{\pi}{4}\right] = \frac{a^2}{4}(4-\pi)$$

Example 2

Evaluate $\int_0^{\infty}\int_0^{\infty} e^{-(x^2+y^2)}dx\,dy$

Solution :

The region of integration is obtained by the limits $x = 0$ to $x = \infty$ and $y = 0$ to $y = \infty$. So, the region of integration is the entire xy-plane lies in first quadrant.

Changing the given integral to polar co-ordinates by putting $x = r\cos\theta$, $y = r\sin\theta$ and the limits for the given region will be $r = 0$ to ∞ and $\theta = 0$ to $\frac{\pi}{2}$

$\therefore$ given integral $= \int_{\theta=0}^{\pi/2}\int_{r=0}^{\infty} e^{-r^2}\cdot r\,d\theta\,dr$

$$= \int_{\theta=0}^{\pi/2}\left[\frac{-e^{-r^2}}{2}\right]_0^{\infty} d\theta$$

$$= \frac{1}{2}\int_0^{\pi/2} d\theta \quad = \frac{\pi}{4}$$

Example 3

Evaluate $\iint \frac{\left(1-\frac{x^2}{a^2}-\frac{y^2}{b^2}\right)^{\frac{1}{2}}}{\left(1+\frac{x^2}{a^2}+\frac{y^2}{b^2}\right)^{\frac{1}{2}}} dxdy$, over the positive quadrant of the curve $\frac{x^2}{a^2}+\frac{y^2}{b^2} = 1.$

Solution :

Let $\quad x = aX, y = bY$

$\Rightarrow \quad dx = adX, \; dy = bdY$ then the given integral becomes

$$I = \iint \sqrt{\left(\frac{1-X^2-Y^2}{1+X^2+Y^2}\right)}\, ab\, dX\, dY$$

Over the positive quadrant of the circle $X^2 + Y^2 = 1$

Changing to polar coordinates

$X = r\cos\theta$; $Y = r\sin\theta$, $dx\,dy = r\,dr\,d\theta$

This limit becomes $r = 0$ to 1 and $\theta = 0$ to $\pi/2$

$$I = ab\int_0^{\frac{\pi}{2}}\int_0^1 \sqrt{\left(\frac{1-r^2}{1+r^2}\right)}\, r\, dr\, d\theta$$

$$= ab\int_0^{\pi/2} I_1 \, d\theta \qquad \text{...(i)}$$

where
$$I_1 = \int_0^1 r\sqrt{\frac{1-r^2}{1+r^2}}\,dr$$

put
$$r^2 = \sin t,\ 2rdr = \cos t \, dt$$

$\Rightarrow$
$$rdr = \frac{1}{2}\cos t \, dt$$

$$I_1 = \int_0^{\pi/2}\left[\frac{1-\sin t}{1+\sin t}\right]^{\frac{1}{2}}\frac{\cos t}{2}dt$$

But
$$\frac{1-\sin t}{1+\sin t} = \frac{(1-\sin t)^2}{1-\sin^2 t} = \frac{(1-\sin t)^2}{\cos^2 t}$$

$\therefore$
$$I_1 = \int_0^{\pi/2}\frac{1-\sin t}{\cos t}\cdot\frac{\cos t}{2}dt$$

$$= \frac{1}{2}[t+\cos t]_0^{\pi/2} = \frac{1}{2}\left(\frac{\pi}{2}-1\right) = \frac{1}{4}(\pi-2)$$

using this eqn. (1) becomes

$$I = ab\int_0^{\pi/2}\frac{1}{4}(\pi-2)\,d\theta = \frac{ab}{4}(\pi-2)[\theta]_0^{\pi/2}$$

$$= \frac{ab}{8}\pi(\pi-2) \qquad \text{Answer}$$

Example 4

Evaluate $\iint\sqrt{4-x^2-y^2}\,dxdy$ over the region bounded by the semi-circle $x^2+y^2-2x = 0$ lying in first quadrant.

Solution:

Changing the given integral into polar co-ordinates by substituting $x = r\cos\theta$, $y = r\sin\theta \quad \Rightarrow dxdy = rd\theta dr$.

Now the given circle $x^2+y^2-2x = 0$ transforms to

$$r^2 - 2r\cos\theta = 0 \quad \Rightarrow \quad r(r-2\cos\theta) = 0 \quad \Rightarrow r = 0,\ 2\cos\theta$$

So, the limits of r are from 0 to $2\cos\theta$, while θ varies from 0 to $\frac{\pi}{2}$ in first quadrant.

$$\therefore \iint \left(4-x^2-y^2\right)^{\frac{1}{2}} dxdy = \int_{\theta=0}^{\frac{\pi}{2}} \int_{r=0}^{2\cos\theta} \left(4-r^2\right)^{\frac{1}{2}} \cdot rd\theta dr$$

$$= \int_0^{\frac{\pi}{2}} \{ \int_0^{2\cos\theta} \left(4-r^2\right)^{\frac{1}{2}} \cdot rdr\} d\theta = -\frac{1}{2}\int_0^{\frac{\pi}{2}} \left[\frac{\left(4-r^2\right)^{\frac{3}{2}}}{\frac{3}{2}}\right]_0^{2\cos\theta} d\theta$$

$$= -\frac{1}{3}\int_0^{\frac{\pi}{2}} \left(8\sin^3\theta - 8\right) d\theta = \frac{4}{3}\left(\pi - \frac{4}{3}\right) \quad \text{Answer}$$

Example 5

Evaluate $\iint_R r\sin\theta drd\theta$ over the area of cardioid $r = a(1+\cos\theta)$ about the initial line.

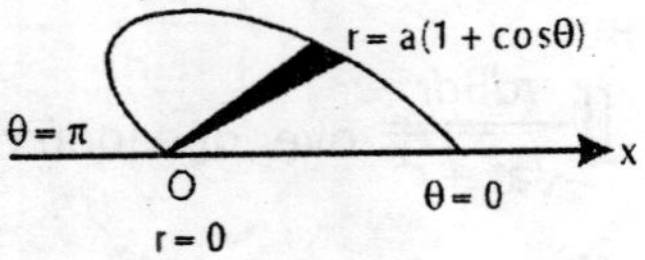

Fig. 15

Solution :

The given region has shown in fig – 15. Here, r varies from 0 to $a(1+\cos\theta)$ and θ from 0 to π.

$$\text{Thus,} \quad \iint_R r\sin\theta drd\theta = \int_{\theta=0}^{\pi} \left\{ \int_{r=0}^{a(1+\cos\theta)} r\sin\theta dr \right\} d\theta$$

$$= \int_0^{\pi} \sin\theta \left[\frac{r^2}{2}\right]_0^{a(1+\cos\theta)} d\theta = \frac{a^2}{2}\int_0^{\pi} \sin\theta(1+\cos\theta)^2 d\theta$$

$$= \frac{a^2}{2}\int_0^{\pi} 2\sin\frac{\theta}{2}\cos\frac{\theta}{2}\left(2\cos^2\frac{\theta}{2}\right)^2 d\theta = 4a^2\int_0^{\pi} \sin\frac{\theta}{2}\cos^5\frac{\theta}{2} d\theta$$

Put $\frac{\theta}{2} = \phi \quad \Rightarrow d\theta = 2d\phi$, we get

$$= 4a^2\int_0^{\frac{\pi}{2}} 2\sin\phi\cos^5\phi d\phi = -8a^2\int_0^{\frac{\pi}{2}} \cos^5\phi(-\sin\phi)d\phi$$

$$= -8a^2\left\{\frac{\cos^6\phi}{6}\right\}_0^{\frac{\pi}{2}} = -\frac{4a^2}{3}(0-1) = \frac{4a^2}{3} \quad \text{Answer}$$

Exercise 7.2

Evaluate the following integrals:

1. $\int_0^{\pi}\int_0^{a\sin\theta} r\,d\theta\,dr$

2. $\int_0^{\frac{\pi}{2}}\int_0^{a\cos\theta} r\sin\theta\,d\theta\,dr$

3. $\int_0^{\pi}\int_0^{a(1+\cos\theta)} r^3\sin\theta\cos\theta\,d\theta\,dr$

4. $\int_0^{\frac{\pi}{2}}\int_0^{a} r^n\sin^n\theta\cos\theta\,dr$, for $n+1>0$

5. $\iint \frac{r\,d\theta\,dr}{\sqrt{a^2+r^2}}$ over one loop of $r^2 = a^2\cos 2\theta$

6. $\iint r^2\,d\theta\,dr$ over the area of circle $r = a\cos\theta$

Evaluate the following integral by changing into polar co – ordinates:

7. $\int_0^{a}\int_0^{x} \frac{x\,dx\,dy}{x^2+y^2}$

8. $\int_0^{2}\int_0^{\sqrt{2x-x^2}} \frac{x\,dx\,dy}{\sqrt{x^2+y^2}}$

9. $\int_0^{1}\int_0^{\sqrt{2x-x^2}} (x^2+y^2)\,dx\,dy$

Answers

1. $\frac{\pi a^2}{4}$
2. $\frac{a^2}{6}$
3. $\left(\frac{16}{15}\right)a^4$
4. $\frac{a^{n+1}}{(n+1)^2}$
5. $\frac{a}{2}(4-\pi)$
6. $\frac{\pi a^3}{9}$
7. $\frac{\pi a}{4}$
8. $\frac{4}{3}$
9. $\left(\frac{3\pi}{8}\right)-1$

7.6 Change of Variables from One System to the Other

In several cases, the evaluation of double integrals becomes easy when there is a change of variables.

Let D be the domain in xy – plane and let x, y be the rectangular cartesian co – ordinates of any point P in D. Let u, v be new variables in domain D^* such that x,y and u,v are connected through the continuous functions (transformations).

i.e $$x = g(u, v) \text{ and } y = h(u,v) \quad \text{...(1)}$$

Then the points u, v are called the curvilinear co – ordinates of the point P* in domain D* which uniquely corresponds to P in domain D. Solving (1) for u and v, we obtain

$$u = g^*(x,y) \text{ and } v = h^*(x,y) \quad \text{...(2)}$$

Then a given double integral in cartesian co – ordinates x, y can be transformed to a double integral in terms of new variables u, v as follows:

$$\iint_D f(x,y)\,dxdy = \iint_{D^*} F(u,v)|J|\,dudv \quad \text{...(3)}$$

where J is the Jacobian

$$J = J\left(\frac{x,y}{u,v}\right) = \frac{\partial(x,y)}{\partial(u,v)} = \begin{vmatrix} \frac{\partial x}{\partial u} & \frac{\partial x}{\partial v} \\ \frac{\partial y}{\partial u} & \frac{\partial y}{\partial v} \end{vmatrix}$$

and $$f(x,y) = f(x(u,v), y(u,v)) = F(u,v)$$

(3) is known as formula for transformation of co – ordinates in double integral. This result remains valid for triple integrals.

Remark (i)

To change the cartesian co –ordinates to polar co– ordinates, we have $x = r\cos\theta$, $y = r\sin\theta$

$$\therefore \quad J = \frac{\partial(x,y)}{\partial(r,\theta)} = \begin{vmatrix} \frac{\partial x}{\partial r} & \frac{\partial x}{\partial \theta} \\ \frac{\partial y}{\partial r} & \frac{\partial y}{r\theta} \end{vmatrix} = \begin{vmatrix} \cos\theta & -r\sin\theta \\ \sin\theta & r\cos\theta \end{vmatrix}$$

$$= r(\cos^2\theta + \sin^2\theta) = r$$

$$\therefore \quad \iint_R f(x,y)\,dxdy = \iint_{R^*} F(r\cos\theta, r\sin\theta)\,r\,drd\theta$$

Remark (ii):

To change cartesian co – ordinates (x, y, z) to cylindrical co – ordinates (r, θ, z), we have

$x = r\cos\theta, \quad y = r\sin\theta, \quad z = z$, see fig. 16

$$J = \frac{\partial(x,y,z)}{\partial(r,\theta,z)} = \begin{vmatrix} \frac{\partial x}{\partial r} & \frac{\partial x}{\partial \theta} & \frac{\partial x}{\partial z} \\ \frac{\partial y}{\partial r} & \frac{\partial y}{\partial \theta} & \frac{\partial y}{\partial z} \\ \frac{\partial z}{\partial r} & \frac{\partial z}{\partial \theta} & \frac{\partial z}{\partial z} \end{vmatrix}$$

$$= \begin{vmatrix} \cos\theta & -r\sin\theta & 0 \\ \sin\theta & r\cos\theta & 0 \\ 0 & 0 & 1 \end{vmatrix} = r$$

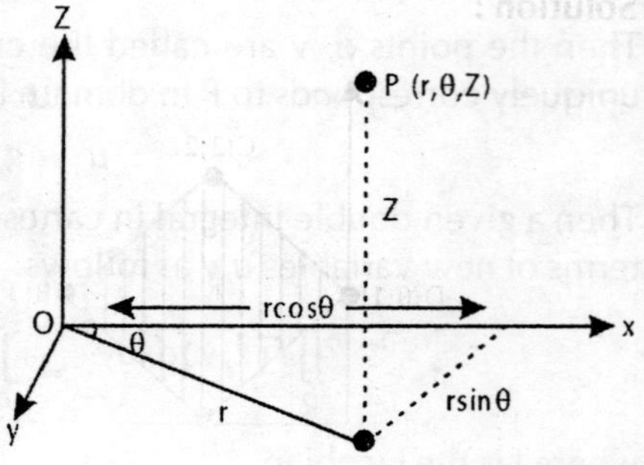

Fig. 16

$$\therefore \quad \iiint_R f(x,y,z)\,dxdydz = \iiint_{R^*} F(r,\theta,z)\,rdrd\theta dz$$

Remark (iii):

To change cartesian co – ordinates (x, y, z) to spherical polar co – ordinates (r, θ, ϕ), we have from fig – 17.

$$x = r\sin\theta\cos\phi,\ y = r\sin\theta\sin\phi,\ z = r\cos\theta$$

$$\therefore \quad J = \frac{\partial(x,y,z)}{\partial(r,\theta,\phi)} = \begin{vmatrix} \frac{\partial x}{\partial r} & \frac{\partial x}{\partial \theta} & \frac{\partial x}{\partial \phi} \\ \frac{\partial y}{\partial r} & \frac{\partial y}{\partial \theta} & \frac{\partial y}{\partial \phi} \\ \frac{\partial z}{\partial r} & \frac{\partial z}{\partial \theta} & \frac{\partial z}{\partial \phi} \end{vmatrix}$$

$$= r^2\sin\theta$$

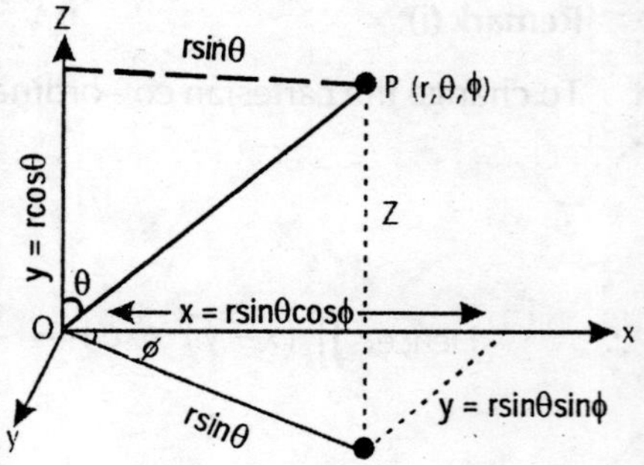

Fig. 17

$$\therefore \quad \iiint_v f(x,y,z)\,dxdydz = \iiint_{v^*} F(r,\theta,\phi)\,r^2\sin\theta\,drd\theta d\phi$$

Illustrative Examples

Example 1 :

Evaluate $\iint_R (x+y)^2\, dxdy$, where R is the parallelogram in xy – plane with vertices (1,0), (3,1), (2, 2) (0,1) using the transformation $u = x + y$, $v = x - 2y$.

Solution :

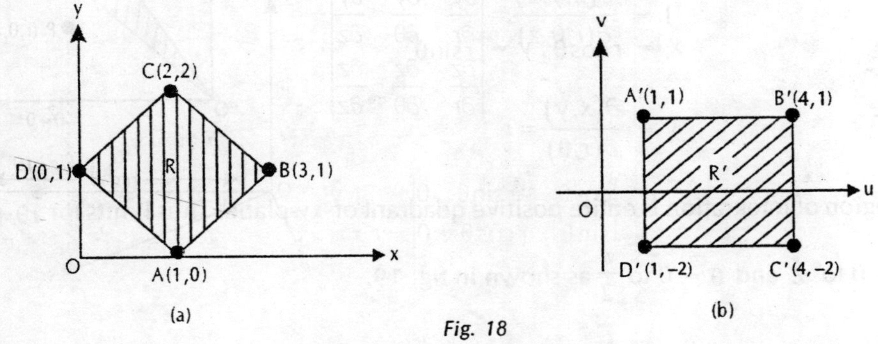

Fig. 18

The Region R i.e the parallelogram ABCD in the xy – plane transforms into region R′ (rectangle A′B′C′D′) in uv – plane as shown in fig – 18 (a) and (b) respectively.

By taking the transformation $u = x+y$, $v = x - 2y$ ...(1)

We obtain $x = \frac{1}{3}(2u+v)$, $y = \frac{1}{3}(u-v)$

$$\therefore \qquad J = \frac{\partial(x,y)}{\partial(u,v)} = \begin{vmatrix} \frac{\partial x}{\partial u} & \frac{\partial x}{\partial v} \\ \frac{\partial y}{\partial u} & \frac{\partial y}{\partial v} \end{vmatrix} = \begin{vmatrix} \frac{2}{3} & \frac{1}{3} \\ \frac{1}{3} & \frac{-1}{3} \end{vmatrix} = -\frac{1}{3}$$

$$\therefore \qquad |J| = \frac{1}{3}$$

$$\text{Hence } \iint_R (x+y)^2\, dxdy = \iint_{R'} u^2 \cdot \frac{1}{3} dudv$$

$$= \frac{1}{3}\int_{u=1}^{4}\int_{v=-2}^{1} u^2 dudv$$

$$= \frac{1}{3}\left(\frac{u^3}{3}\right)_1^4 (v)_{-2}^{1} = 21$$

Example 2

Evaluate $\int_0^\infty\int_0^\infty e^{-(x^2+y^2)}dydx$ and hence show that

$$\int_0^\infty e^{-x^2}dx = \frac{\sqrt{\pi}}{2}$$

Solution :

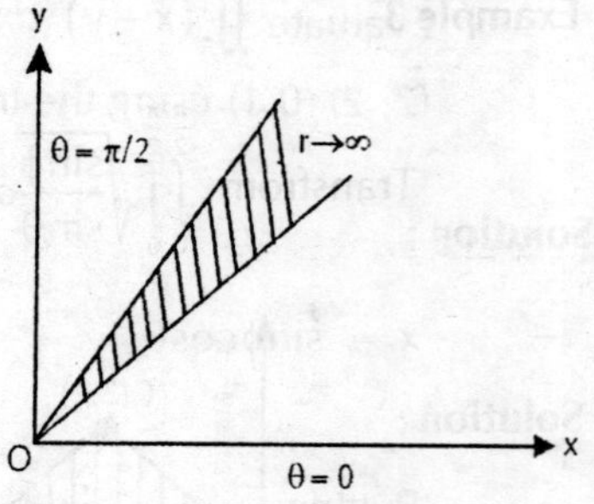

Fig. 19

Putting $x = r\cos\theta,\ y = r\sin\theta$

$$J = \frac{\partial(x,y)}{\partial(r,\theta)} = r$$

The region of integration is entire positive quadrant of xy-plane. Thus limits for $r\theta$-plane will be $r = 0$ to ∞ and $\theta = 0$ to $\frac{\pi}{2}$ as shown in fig. 19.

$$\therefore \quad \int_0^\infty\int_0^\infty e^{-(x^2+y^2)}dxdy = \int_{\theta=0}^{\frac{\pi}{2}}\int_{r=0}^{\infty} e^{-r^2}rd\theta dr$$

$$= [\theta]_0^{\frac{\pi}{2}}\left[\frac{-e^{-r^2}}{2}\right]_0^\infty$$

$$= \frac{\pi}{4} \qquad \ldots(1)$$

Again let

$$I = \int_0^\infty e^{-x^2}dx \qquad \ldots(2)$$

also, $I = \int_0^\infty e^{-y^2}dy$ (by the property of definite integral) ...(3)

multiplying (2) and (3), we get

$$I^2 = \int_0^\infty\int_0^\infty e^{-(x^2+y^2)}dxdy$$

$$\Rightarrow \quad I^2 = \frac{\pi}{4} \qquad \text{[from (1)]}$$

$$\Rightarrow \qquad I = \frac{\sqrt{\pi}}{2} \quad \text{Hence proved.}$$

Example 3

Transfrom $\int_0^{\frac{\pi}{2}}\int_0^{\frac{\pi}{2}} \sqrt{\frac{\sin\phi}{\sin\theta}}\, d\phi d\theta$ by the substitution

$x = \sin\phi\cos\theta,\ y = \sin\phi\sin\theta$.

Solution :

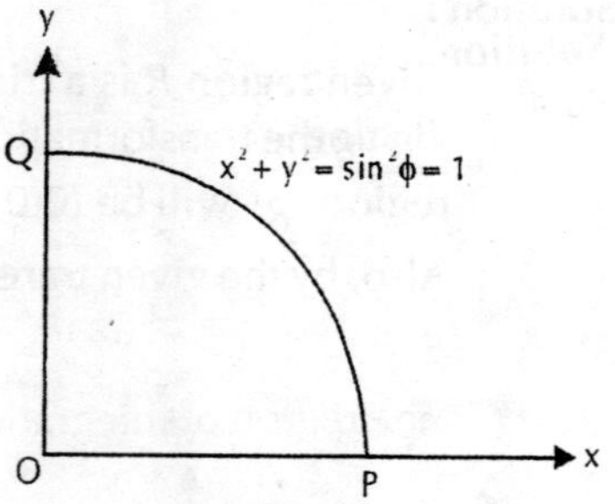

Fig. 20

Putting $\qquad x = \sin\phi\cos\theta,\ y = \sin\phi\sin\theta$, then

$$x^2 + y^2 = \sin^2\phi \text{ and } \theta = \tan^{-1}\left(\frac{y}{x}\right)$$

Here, θ and ϕ both vary from 0 to $\frac{\pi}{2}$ so limits of y are 0 to 1 and limits of x are 0 to $\sqrt{1-y^2}$ (see Fig. 20).

Now $\qquad J = \frac{\partial(x,y)}{\partial(\theta,\phi)} = -\sin\phi\cos\phi$

Therefore, $\qquad \int_0^{\frac{\pi}{2}}\int_0^{\frac{\pi}{2}} \sqrt{\frac{\sin\phi}{\sin\theta}}\, d\phi d\theta$

$$= \int_{y=0}^{1}\int_{x=0}^{\sqrt{1-y^2}} \sqrt{\frac{\sin\phi}{\sin\theta}}\, \frac{dydx}{\sin\phi\cos\phi}$$

$$= \int_0^1\int_0^{\sqrt{1-y^2}} \frac{1}{\cos\phi}\frac{dydx}{\sqrt{\sin\phi\sin\theta}} = \int_0^1\int_0^{\sqrt{1-y^2}} \frac{1}{\sqrt{1-y^2-x^2}}\frac{dydx}{\sqrt{y}}$$

$$\left[\because x^2 + y^2 = \sin^2\phi\right]$$

$$= \int_0^1 \left[\sin^{-1}\frac{x}{\sqrt{1-y^2}}\right]_0^{\sqrt{1-y^2}} \frac{dy}{\sqrt{y}} = \frac{\pi}{2}\int_0^1 \frac{dy}{\sqrt{y}} = \frac{\pi}{2}\left(2\sqrt{y}\right)_0^1 = \pi \quad \text{Answer}$$

Example 4

By using the transformation $x - y = u$, $x + y = v$, prove that

$\iint_R \sin\left(\frac{x-y}{x+y}\right) dxdy = 0$, where R is bounded by the co–ordinates axes and the plane $x + y = 1$ in positive quadrant.

Solution :

Given region R is a triangle having vertices at (0,0), (1,0) and (0,1) as shown in fig – 21(a). Using the transformation $x - y = u$ and $x + y = v$. The corresponding points in the transformed region R^1 will be (0,0), (1,1) and (–1,1) respectively as shown in fig 21(b).

Also, by the given transformation, we obtain

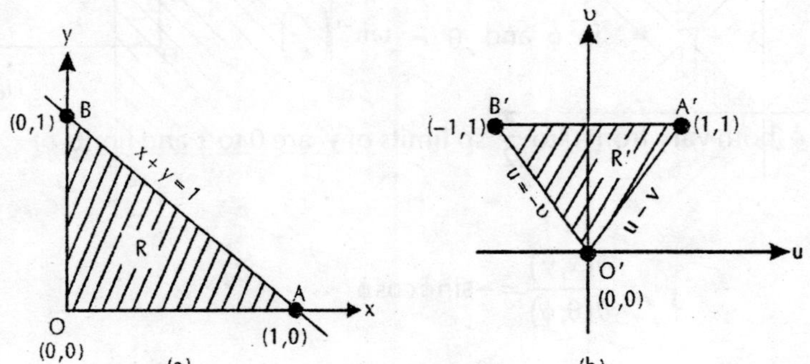

Fig. 21

$$x = \frac{1}{2}(u+v),\ y = \frac{1}{2}(v-u)$$

Again
$$J = \frac{\partial(x,y)}{\partial(u,v)} = \begin{vmatrix} \frac{1}{2} & \frac{1}{2} \\ -\frac{1}{2} & \frac{1}{2} \end{vmatrix} = \frac{1}{2}$$

$$\therefore \quad \iint_R \sin\left(\frac{x-y}{x+y}\right) dxdy = \iint_{R^1} \sin\left(\frac{u}{v}\right) \cdot \frac{1}{2} dudv$$

$$= \frac{1}{2}\int_{v=0}^{1} \left[\frac{-\cos\left(\frac{u}{v}\right)}{\frac{1}{v}}\right]_{u=-v}^{v} dv = \frac{1}{2}\int_0^1 v\left[-\cos 1 + \cos(-1)\right] dv$$

$$= 0$$

Answer

Example 5

Evaluate $\iint_R (x+y)^2\, dxdy$, where R is region bounded by the parallelogram $x + y = 0$, $x + y = 2$, $3x - 2y = 3$, $3x - 2y = 0$.

Solution :

By changing the variables x, y to the new variables u, v by the transformation $u = x + y$, $v = 3x - 2y$, the given parallelogram R reduces to a rectangle R^1 as shown in fig – 22.

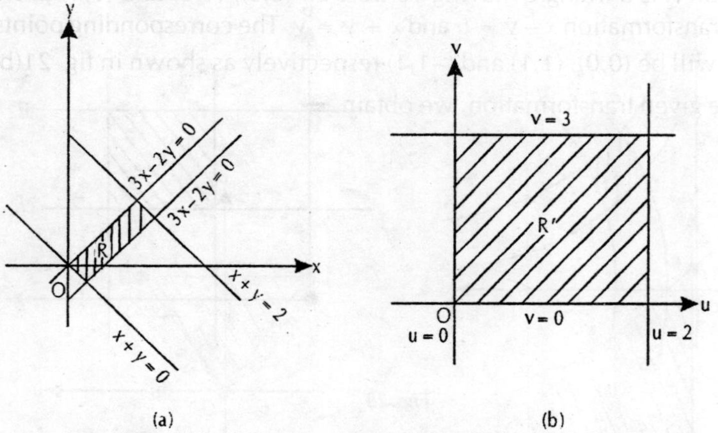

Fig. 22

Also,
$$J = \frac{\partial(x,y)}{\partial(u,v)} = \begin{vmatrix} \frac{\partial x}{\partial u} & \frac{\partial x}{\partial v} \\ \frac{\partial y}{\partial u} & \frac{\partial y}{\partial v} \end{vmatrix} = -\frac{1}{5} \qquad \left\{ \begin{array}{l} \text{as } x = \frac{1}{5}(2u+v) \\ y = \frac{1}{5}(3u-v) \end{array} \right\}$$

$$\therefore \quad |J| = \frac{1}{5}$$

Now
$$\iint_R (x+y)^2\, dxdy = \iint_{R^1} u^2 \cdot \frac{1}{5} dudv = \int_{v=0}^{3} \int_{u=0}^{2} \frac{1}{5} u^2 dudv$$

$$= \frac{1}{5}\int_0^3 \left[\frac{u^3}{3}\right]_0^2 dv = \frac{8}{15}[v]_0^3 = \frac{8}{5} \qquad \text{Answer}$$

Example 6

Evaluate $\iint_R (x^2+y^2)\,dxdy$ where R is the region in the first quadrant bounded by $x^2 - y^2 = a$, $x^2 - y^2 = b$, $2xy = c$, $2xy = d$ and $0 < a < b$, $0 < c < d$.

Solution :

By changing the variables x, y to the new variables u, v by the transformation $x^2 - y^2 = u$, $xy = v$, the given region R reduces to a rectangle R' in uv – plane as shown in fig – 23.

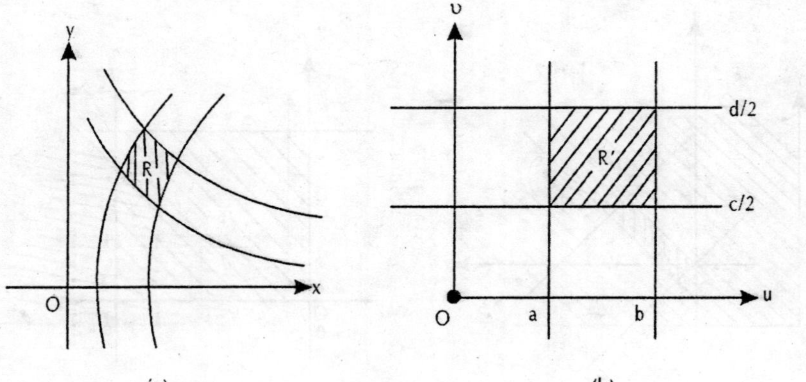

Fig. 23

Now
$$\frac{\partial(u,v)}{\partial(x,y)} = \begin{vmatrix} \frac{\partial u}{\partial x} & \frac{\partial u}{\partial y} \\ \frac{\partial v}{\partial x} & \frac{\partial v}{\partial y} \end{vmatrix} = \begin{vmatrix} 2x & -2y \\ y & x \end{vmatrix} = 2(x^2+y^2)$$

$$\therefore \quad J = \frac{\partial(x,y)}{\partial(u,v)} = \frac{1}{2(x^2+y^2)}.$$ Thus,

$$\iint_R (x^2+y^2)\,dxdy = \iint_{R'} (x^2+y^2)\frac{1}{2(x^2+y^2)}\,dudv$$

$$= \int_{v=\frac{c}{2}}^{\frac{d}{2}} \int_{u=a}^{b} \frac{1}{2}\,dudv = \frac{1}{2}(b-a)\cdot\left(\frac{d}{2}-\frac{c}{2}\right)$$

$$= \frac{(b-a)(d-c)}{4}$$ Answer

Example 7

Evaluate $\iint_R e^{-(x+y)} \sin\left(\frac{\pi y}{x+y}\right) dxdy$, where R is the entire first quadrant in xy – plane.

Solution :

In the first quadrant both x and y vary from 0 to ∞ . Now choose the transformation $u = x + y$, $v = y$ so that $x = u - v$ and $y = v$. New region R' in uv – plane has been shown in fig-24.

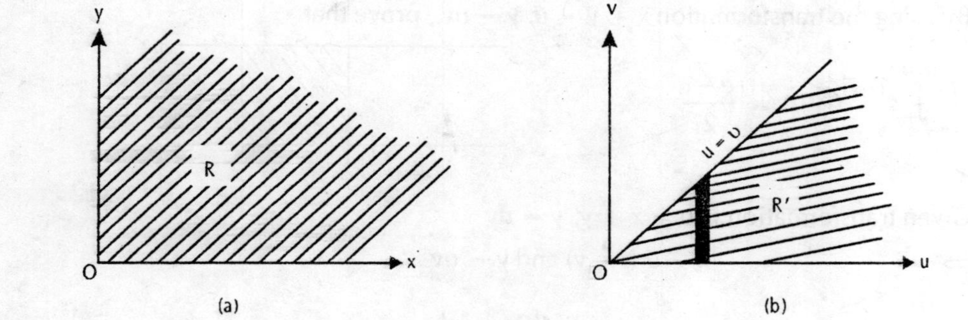

Fig. 24

We observe the following :

(i) Origin (0, 0) in xy – plane: $0 = u - v$, $v = 0$, $u = 0$. Thus (0, 0) in xy – plane corresponds to (0, 0) in uv – plane.

(ii) x axis : $y = 0$ i.e $v = 0$ so $x = u - v = u - 0 = u$ since $x \geq 0$ so $u \geq 0$ ($\because x = u$). Thus x – axis get transformed to u –axis .

(iii) y – axis : $x = 0$ so $u - v = 0 \Rightarrow u = v$ and since $y \geq 0$, $y = v$

$\therefore$ $v \geq 0$. Thus y –axis get transformed to line $u = v$.

(iv) Any point $x > 0, y > 0$: since $v = y > 0$ and $u - v = x > 0$

$\Rightarrow$ $u > v$. Thus any point $(x > 0, y > 0)$ corresponds to (u, v) where $u > v > 0$.

Hence the entire first quadrant in xy –plane get transformed to the region R′ between u – axis and line $u = v$.

Now $$J = \frac{\partial(x,y)}{\partial(u,v)} = \begin{vmatrix} \frac{\partial x}{\partial u} & \frac{\partial x}{\partial v} \\ \frac{\partial y}{\partial u} & \frac{\partial y}{\partial v} \end{vmatrix} = \begin{vmatrix} 1 & -1 \\ 0 & 1 \end{vmatrix} = 1$$

$$\therefore \iint_R e^{-(x+y)} \sin\left(\frac{\pi y}{x+y}\right) dxdy = \iint_{R'} e^{-u} \sin\left(\frac{\pi v}{u}\right) dudv$$

$$= \int_{u=0}^{\infty}\int_{v=0}^{u} e^{-u}\sin\left(\frac{\pi v}{u}\right)dudv = \int_0^{\infty} e^{-u}\left[\frac{u}{\pi}\left(-\cos\frac{\pi v}{u}\right)\right]_0^u du$$

$$= \frac{2}{\pi}\int_0^{\infty} ue^{-u}du = \frac{2}{\pi}\left[\frac{ue^{-u}}{-1} - e^{-u}\right]_0^{\infty} = \frac{2}{\pi}$$

Example 8

By using the transformation $x + y = u, y = uv,$ prove that

$$\int_{x=0}^{1}\int_{y=0}^{(1-x)} e^{\frac{y}{(x+y)}}dxdy = \frac{e-1}{2}$$

Solution :

Given transformation is $u = x + y, y = uv$

$\Rightarrow$ $x = u(1 - v)$ and $y = uv$

Now $$J = \frac{\partial(x,y)}{\partial(u,v)} = \begin{vmatrix} \frac{\partial x}{\partial u} & \frac{\partial x}{\partial v} \\ \frac{\partial y}{\partial u} & \frac{\partial y}{\partial v} \end{vmatrix} = \begin{vmatrix} 1-v & -u \\ v & u \end{vmatrix} = u$$

$\therefore$ Required integral $$= \int_{u=0}^{1}\int_{v=1}^{1} e^{\frac{uv}{u}} ududv = \int_0^1\int_0^1 e^v ududv$$

$$= \left[e^v\right]_0^1 \left[\frac{u^2}{2}\right]_0^1 = \frac{1}{2}(e-1)$$ Hence proved

Exercise 7.3

1. Evaluate $\iint \sqrt{\frac{1-x^2-y^2}{1+x^2+y^2}}dxdy$, over the positive quadrant of the circle $x^2 + y^2 = 1$

2. Using the transformation $x + y = u, y = v$, evaluate $\int_0^{\pi}\int_0^{\pi} |\cos(x+y)|dxdy$.

3. Evaluate the following by changing into polar co–ordinates:

(i) $\int_{-a}^{a}\int_{-\sqrt{a^2-x^2}}^{\sqrt{a^2-x^2}} dydx$ (ii) $\int_0^1\int_y^1 \frac{xdydx}{x^2+y^2}$

(iii) $\int_0^a \int_0^{\sqrt{a^2-y^2}} y^2\sqrt{(x^2+y^2)}\,dydx$ (iv) $\int_{-\infty}^{\infty}\int_{-\infty}^{\infty} \dfrac{a\,dxdy}{(x^2+y^2+a^2)^{\frac{3}{2}}(x^2+y^2+b^2)^{\frac{1}{2}}}$

4. Using the transformation $u = x - y,\ v = x + y$, show that

$\iint_A \cos\left(\dfrac{x-y}{x+y}\right)dxdy = \dfrac{1}{2}\sin 1$, where A is bounded by $x = 0,\ y = 0,\ x + y = 1$

5. Evaluate $\int_0^{4a}\int_{\frac{y^2}{4a}}^{y}\left(\dfrac{x^2-y^2}{x^2+y^2}\right)dydx$ by changing into polar co-ordinates.

6. If $x = u\sin\alpha + v\cos\alpha,\ y = u\cos\alpha - v\sin\alpha$, prove that

$\iint \dfrac{f(x,y)}{\sqrt{1-x^2-y^2}}dxdy = \iint \dfrac{\phi(u,v)}{\sqrt{1-u^2-v^2}}dudv$

7. By using the transformation $u = x + y,\ v = y$, prove that

$\int_0^{\infty}\int_0^{\infty} e^{-(x+y)}\sin\left(\dfrac{\pi y}{x+y}\right)dxdy = \dfrac{1}{\pi}$

8. Evaluate $\iint \sqrt{a^2 - x^2 - y^2}\,dxdy$ over the semi – circle $x^2 + y^2 = ax$ in the positive quadrant.

9. Calculate $\int_0^1\int_x^{\sqrt{2x-x^2}} \sqrt{(x^2+y^2)}\,dxdy$, by changing into polar co – ordinates.

10. Evaluate $\int_0^1\int_0^x \dfrac{x^3\,dxdy}{\sqrt{(x^2+y^2)}}$ by changing into polar co – ordinates.

11. Show that if $x = a\sin\theta\cos\phi,\ y = b\sin\theta\sin\phi$; the integral

$\iint_A \left(\dfrac{x^2}{a^2}+\dfrac{y^2}{b^2}\right)dxdy = \iint_{A'} ab\sin^3\theta\cos\theta\,d\theta d\phi$

12. Change the following integrals into polar co- ordinates and prove that

(i) $\int_0^a\int_y^a \dfrac{1}{x^2+y^2}dydx = \dfrac{\pi a}{4}$ (ii) $\int_0^1\int_x^{\sqrt{(2x-x^2)}} (x^2+y^2)\,dxdy = \dfrac{3\pi}{8} - 1$

(iii) $\int_0^a\int_0^{\sqrt{(a^2-x^2)}} y^2\sqrt{(x^2+y^2)}\,dxdy = \dfrac{\pi a^5}{20}$

13. Evaluate the integral $\iint xy\left(\frac{x^2}{a^2}+\frac{y^2}{b^2}\right)dxdy$ over the positive quadrant of the ellipse

$$\frac{x^2}{a^2}+\frac{y^2}{b^2} = 1.$$

Hint: Put $\frac{x}{a} = r\cos\theta,\ \frac{y}{b} = r\sin\theta$

14. Transform the integral

$\int_{-1}^{1}\int_{\infty}^{1/x} v\, dx\, dy$ by the substitution $u = 1 + x, v = xy$.

Answers

1. $\frac{\pi^2}{8}-\frac{\pi}{4}$ **2.** 2π **3.** (i) πa^2 (ii) $\frac{\pi}{4}$ (iii) $\frac{\pi a^5}{20}$ (iv) $\frac{2\pi}{a+b}$

5. $\frac{4}{3}a^2(3\pi-10)$ **8.** $\frac{a^3}{3}\left(\frac{\pi}{2}-\frac{2}{3}\right)$ **9.** $\frac{2}{9}(8-5\sqrt{2})$

10. $\frac{1}{4}\log(1+\sqrt{2})$ **13.** $\frac{a^{n+4}}{2(n+4)}$ **14.** $\int_0^2\int_{(a-1)^2}^{1}\frac{v}{u-1}du\,dv$

7.7 Change of Order of Integration

As already discussed for the double integral with variable limits

$$I_A = \iint_A f(x,y)\,dA \qquad \text{...(1)}$$

The limits of integration can be fixed from a rough sketch of the region of integration. Then (1) can be evaluated as a two fold iterated integral using either

$$I_A = \int_{x=a}^{b}\int_{y=y_1(x)}^{y_2(x)} f(x,y)\,dydx \qquad \text{...(2)}$$

or

$$I_A = \int_{y=c}^{d}\int_{x=x_1(y)}^{x_2(y)} f(x,y)\,dxdy \qquad \text{...(3)}$$

In each specific problem, depending upon the type of region A and or the nature of integrand , choose either of the form (2) or (3) whichever is easier to evaluate. Thus, in several problems, the evaluation of double integral becomes easier with the change of order of integration, which of course, changes the limits of integration also.

Illustrative Examples

Example 1

Change the order of integration in the double integral

$\int_0^a \int_0^x f(x,y)\,dxdy$.

Solution:

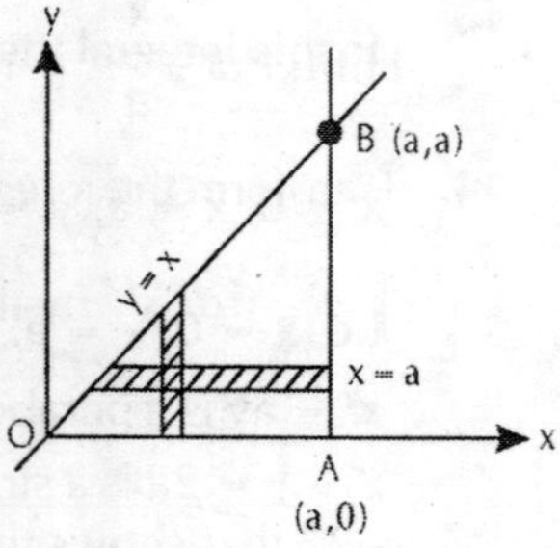

Fig. 25

The given limits show that region is bounded by the curves $x = 0, x = a, y = 0$ and $y = x$. Thus the region of integration is a triangle OAB as shown in fig. 25. In order to change the order of integration, draw a strip parallel to x - axis and choose the limits accordingly.

Thus, $\int_0^a \int_0^x f(x,y)\,dxdy = \int_0^a \int_y^a f(x,y)\,dydx$ Answer

Example 2

Show that

$$\int_0^{4a} \int_{\frac{x^2}{4a}}^{2\sqrt{ax}} f(x,y)\,dxdy = \int_0^{4a} \int_{\frac{y^2}{4a}}^{2\sqrt{ay}} f(x,y)\,dydx$$

Solution:

Limits of L.H.S. integral show that the given region is

bounded by the curves $x = 0, x = 4a,$ $y = \dfrac{x^2}{4a}$

and $y = 2\sqrt{ax}$.

$\Rightarrow$ $x = 0, x = 4a, x^2 = 4ay$ and $y^2 = 4ax$.

Given region has been shown in fig - 26. Thus, to change the order of integration, choose the elementary strip parallel to x -axis instead of y - axis and take the limits accordingly.

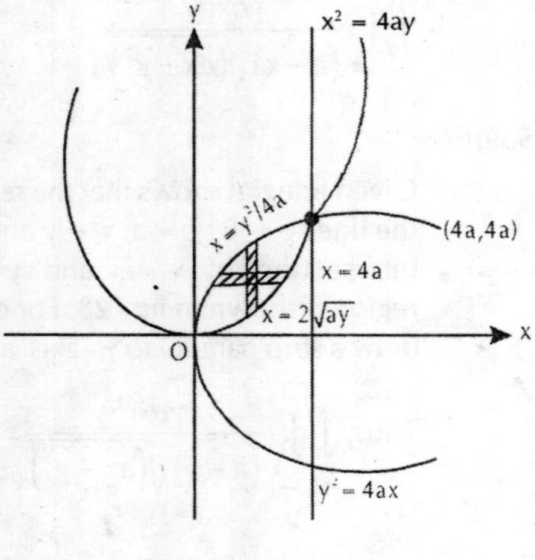

Fig. 26

Thus, $\int_0^{4a} \int_{\frac{x^2}{4a}}^{2\sqrt{ax}} f(x,y)\,dxdy$

$= \int_0^{4a} \int_{\frac{y^2}{4a}}^{2\sqrt{ay}} f(x,y)\,dydx$ Prroved

Example 3

Change the order of integration of

$$\int_0^a \int_{\frac{x^2}{a}}^{2a-x} f(x,y)\,dxdy$$

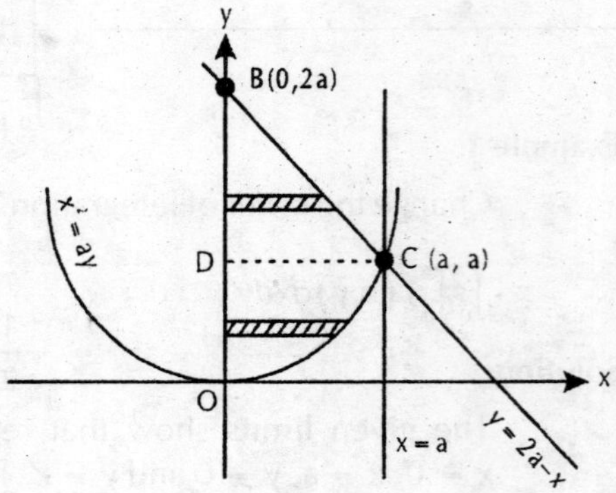

Fig. 27

Solution:

In this integral the region is bounded by

$$x = 0,\ x = a,\ y = \frac{x^2}{a} \text{ and } y = 2a - x$$

i.e $x = 0$, $x = a$, $x^2 = ay$ and $x + y = 2a$

$x^2 = ay$ is a parabola with vertex at origin and about y-axis.

$x + y = 2a$ is a straight line. The point of intersection of these two curves is $C(a, a)$. See the fig - 27 that shows the given region OCB. Now divide the whole region OBC into two parts as OCD andDCB. Draw the elementary strips parallel to, x - axis and choose the limits accordingly as

$$\int_0^a \int_{\frac{x^2}{a}}^{2a-x} f(x,y)\,dxdy = \int_0^a \int_0^{\sqrt{ay}} f(x,y)\,dydx + \int_a^{2a} \int_0^{2a-y} f(x,y)\,dydx \qquad \text{Proved}$$

Example 4

By changing the order of integration, evaluate

$$\int_0^a \int_{\frac{y^2}{a}}^{y} \frac{ydxdy}{(a-x)\sqrt{(ax-y^2)}}$$

Solution:

Given integral shows that the region of integration is bounded by the lines $y = 0$, $y = a$, $x = y$ and parabola $y^2 = ax$. The points of intersections of $x = y$ and $y^2 = ax$ are (0, 0) and (a, a). The region is shown in fig - 28. For changing the order of integration, draw a strip parallel to y - axis and choose the limits accordingly.

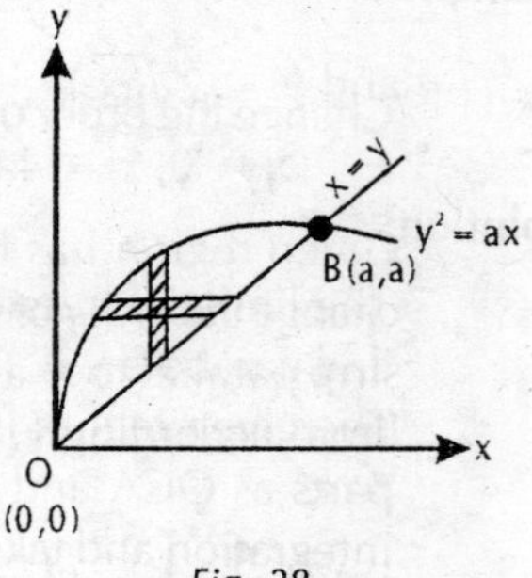

Fig. 28

$$\text{Thus, } \int_0^a \int_{\frac{y^2}{a}}^{y} \frac{ydxdy}{(a-x)\sqrt{(ax-y^2)}}$$

$$= \int_0^a \int_x^{\sqrt{ax}} \frac{ydydx}{(a-x)\sqrt{(ax-y^2)}}$$

$$= \int_0^a \frac{1}{(a-x)} dx \left[\int_x^{\sqrt{ax}} \frac{ydy}{(ax-y^2)^{\frac{1}{2}}} \right] = -\frac{1}{2}\int_0^a \frac{1}{(a-x)} dx \left[\int_x^{\sqrt{ax}} \frac{-2ydy}{(ax-y^2)^{\frac{1}{2}}} \right]$$

$$= -\frac{1}{2}\int_0^a \frac{1}{(a-x)} dx \left[\frac{(ax-y^2)^{1/2}}{\frac{1}{2}} \right]_x^{\sqrt{ax}}$$

$$= \int_0^a \frac{(ax-x^2)^{\frac{1}{2}}}{a-x} dx = \int_0^a \sqrt{x} \cdot \frac{\sqrt{a-x}}{a-x} dx = \int_0^a \frac{\sqrt{x}}{\sqrt{a-x}} dx$$

Now put $x = a\sin^2\theta$ and the limits are : $a = a\sin^2\theta$

$$\Rightarrow \quad \sin\theta = 1,\ \theta = \frac{\pi}{2} \text{ and } 0 = a\sin^2\theta \Rightarrow \sin\theta = 0 \Rightarrow \theta = 0$$

$$\therefore \quad I = \int_0^{\frac{\pi}{2}} \sqrt{\frac{a\sin^2\theta}{a-a\sin^2\theta}}\, 2a\sin\theta\cos\theta d\theta = \int_0^{\frac{\pi}{2}} \frac{\sin\theta}{\cos\theta} \cdot 2a\sin\theta\cos\theta d\theta$$

$$= 2a\int_0^{\frac{\pi}{2}} \sin^2\theta d\theta = 2a \cdot \frac{1}{2} \cdot \frac{\pi}{2} = \frac{\pi a}{2}$$

Example 5

Change the order of integration of $\int_0^1 \int_{x^2}^{(2-x)} xydxdy$ and hence evaluate.

Solution:

The limits of integration show that the region is bounded by the line $x = 0$, $x = 1$, $x + y = 2$ and parabola $y = x^2$. The points of intersection of line $x + y = 2$ and parabola $y = x^2$ is (1, 1). The given region is shown in fig - 29 below by OBA. Now , let us divide the region into two parts as OCA and ABC. Choose the elementary strips parallel to x -axis to change order of integration and take the limits accordingly.

Thus, $$\int_0^1 \int_{x^2}^{(2-x)} xydxdy = \int_0^1 \int_0^{\sqrt{y}} xydydx + \int_1^2 \int_0^{2-y} xydydx$$

$$= \int_0^1 y\left[\frac{x^2}{2}\right]_0^{\sqrt{y}} dy + \int_1^2 y\left[\frac{x^2}{2}\right]_0^{2-y} dy$$

$$= \frac{1}{2}\int_0^1 y^2 dy + \frac{1}{2}\int_1^2 y(2-y)^2\, dy$$

$$= \frac{1}{2}\left[\frac{y^3}{3}\right]_0^1 + \frac{1}{2}\left[2y^2 - \frac{4y^3}{3} + \frac{y^4}{4}\right]_1^2$$

$$= \frac{1}{6} + \frac{1}{2}\left[\left(8 - \frac{32}{3} + 4\right) - \left(2 - \frac{4}{3} + \frac{1}{4}\right)\right] = \frac{8}{3}$$

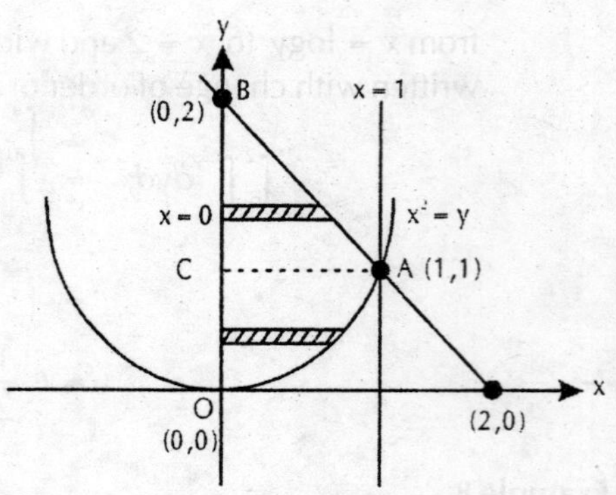

Fig. 29

Example 6

Change the order of integration and evaluate $\int_0^1 \int_{e^x}^{e} \frac{1}{\log y} dx dy$.

Solution:

The given limits show that the integration and region is bounded by $y = e^x$, $y = e$, $x = 0$ and $x = 1$. It is shaded portion in fig - 30.

Thus, $\int_0^1 \int_{e^x}^{e} \frac{1}{\log y} dx dy = \int_1^e \int_0^{\log y} \frac{dy dx}{\log y}$

$$= \int_1^e \left[\frac{x}{\log y}\right]_0^{\log y} dy$$

$$= \int_1^e dy$$

$$= e - 1$$

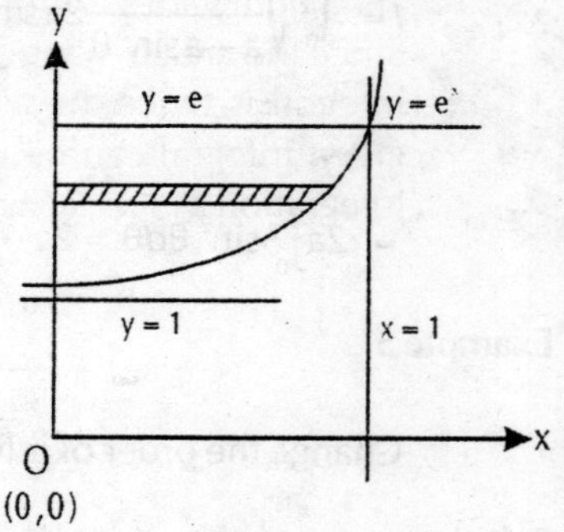

Fig. 30

Example 7:

Change the order of integration and evaluate $\int_0^2 \int_1^{e^x} dy dx$

Solution :

- Here integration is done w.r.t. y first and later w.r.t x. So, the region of integration ABE is shaded as shown in fig - 31, and is bounded by the curve $y = e^x$ and the straight lines $y = 1$, $x = 0$ and $x = 2$. Here, A (0, 1), B (2, 1), E (2, e^2). When the order of integration is changed, consider a horizontal strip whose length varies

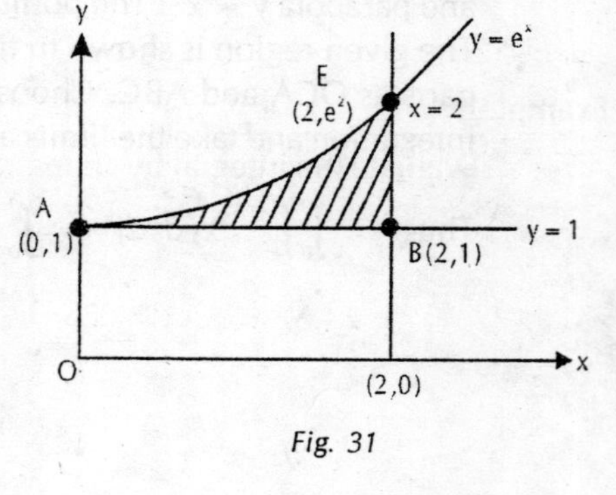

Fig. 31

from $x = \log y$ to $x = 2$ and width y varies from $y = 1$ to $y = e^2$. Thus the given integral can be written with change of order of integration as

$$\int_0^2 \int_1^{e^x} dydx = \int_1^{e^2} \int_{\log y}^2 dxdy$$

$$= \int_1^{e^2} (2 - \log y)dy = [2y - y\log y + y]_1^{e^2}$$

$$= e^2 - 3 \qquad \text{Answer}$$

Example 8

Evaluate the following double integral by changing the order of integration

$$\int_0^{\infty} \int_{-y}^{y} (y^2 - x^2)e^{-y} dxdy$$

Solution:

Here integration is done w.r.t.x first later w.r.t.y . The region of integration is shaded as shown in fig -32, and is bounded by the lines $y = 0$, $y = \infty$, $x = -y$, and $x = y$. When the order of integration is changed, consider a vertical strip whose length varies from $y = x$ to $y = \infty$ and width varies from $x = 0$ to ∞, in first quadrant only. Since the area is symmetrical to y- axis , so the required integral is twice the area in first quadrant. Hence, the given integral can be written with change of order of integration as

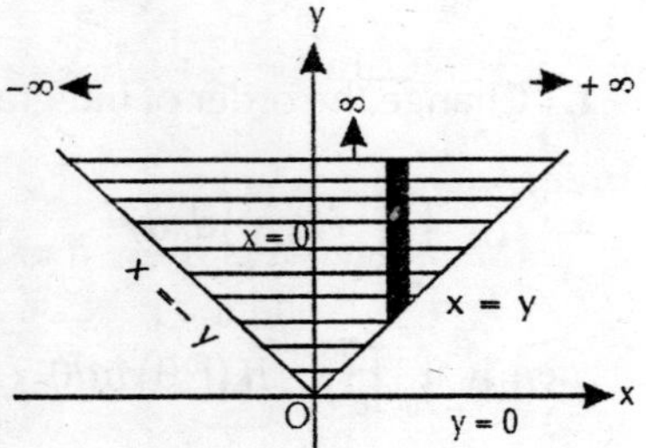

Fig. 32

$$\int_0^{\infty} \int_{-y}^{y} (y^2 - x^2)e^{-y} dxdy = 2\int_0^{\infty} \int_x^{\infty} (y^2 - x^2)e^{-y} dydx$$

$$= 2\int_0^{\infty} \left[-y^2e^{-y} - 2ye^{-y} - 2e^{-y} + x^2e^{-y}\right]_x^{\infty} dx$$

$$= 2\int_0^{\infty} \left(2xe^{-x} + 2e^{-x}\right)dx$$

$$= 4\left[-xe^{-x} - e^{-x} - e^{-x}\right]_0^{\infty} = 8 \qquad \text{Proved}$$

Example 9

Evaluate the integral by changing its order

$$I = \int_0^{\infty} \int_x^{\infty} \frac{e^{-y}}{y} dydx$$

Solution :

The region of integration is bounded by $x = 0, x = \infty$, $x = y$ and $y = \infty$. To change the order of integration, choose an elementary strip parallel to x - axis. For this, x varies from $x = 0$ to $x = y$ and y varies from $y = 0$ to $y = \infty$. Thus, we have

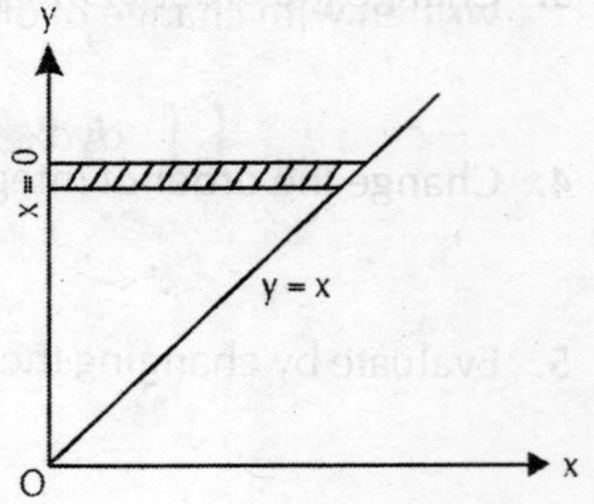

Fig. 32

$$I = \int_0^\infty \int_0^y \frac{e^{-y}}{y} dydx = \int_0^\infty \frac{e^{-y}}{y} [x]_0^y dy$$

$$= \int_0^\infty e^{-y} dy = \left[-e^{-y}\right]_0^\infty = 1$$

Exercise 7.4

1. Change the order of integration of the following integrals:

(i) $\int_0^a \int_0^y f(x,y)dxdy$ (ii) $\int_0^4 \int_x^{2\sqrt{x}} f(x,y)dydx$

(iii) $\int_0^{\frac{\pi}{2}} \int_0^{2a\cos\theta} rf(r,\theta)drd\theta$ (iv) $\int_0^a \int_0^{\frac{b}{b+x}} f(x,y)dydx$

2. Evaluate the following integrals by changing the order of integration :

(i) $\int_0^1 \int_0^{\sqrt{1-x^2}} y^2 dydx$ (ii) $\int_0^a \int_{\sqrt{ax}}^a \frac{y^2 dydx}{\sqrt{y^4 - a^2x^2}}$

(iii) $\int_0^1 \int_{4y}^4 e^{x^2} dxdy$ (iv) $\int_0^{2a} \int_{\frac{x^2}{4a}}^{3a-x} (x^2 + y^2) dydx$

(v) $\int_0^a \int_0^x \frac{\cos y}{\sqrt{(a-x)(a-y)}} dydx$ (vi) $\int_0^1 \int_1^{x^2} \frac{x^2}{y} dydx$

(vii) $\int_0^\infty \int_{-y}^y (y^2 - x^2) e^{-y} dxdy$ (viii) $\int_0^{4a} \int_{\frac{x^2}{4a}}^{2\sqrt{ax}} dydx$

(ix) $\int_0^{\frac{a}{\sqrt{2}}} \int_x^{\sqrt{a^2-x^2}} \log(x^2 + y^2) dydx$, $a > 0$ (x) $\int_0^{\frac{\pi}{2}} \int_0^{2a\cos\theta} r\sin\theta drd\theta$

3. Change the order of integration of $\int_0^1\int_{x^2}^{x}(x^2+y^2)^{\frac{-1}{2}}\,dydx$

4. Change the order of integration of $\int_0^a\int_{a-\sqrt{a^2-y^2}}^{a+\sqrt{a^2-y^2}} xydxdy$ and hence find its value.

5. Evaluate by changing the order of integration of $\int_0^5\int_0^{x^2} x(x^2+y^2)\,dydx$

Answers

1. (i) $\int_0^a\int_x^a f(x,y)\,dydx$ (ii) $\int_0^4\int_{\frac{y^2}{4}}^{y} f(x,y)\,dxdy$ (iii) $\int_0^{2a}\int_0^{\cos^{-1}\left(\frac{r}{a}\right)} f(r,\theta)\,rd\theta dr$

(iv) $\int_0^{\frac{b}{(a+b)}}\int_0^a f(x,y)\,dxdy+\int_{\frac{b}{(a+b)}}^{1}\int_0^{\frac{b(1-y)}{y}} f(x,y)\,dxdy$

2. (i) $\dfrac{\pi}{16}$ (ii) $\dfrac{\pi a^2}{6}$ (iii) $\dfrac{1}{8}\left(e^{16}-1\right)$

(iv) $\dfrac{314}{35}a^4$ (v) 2sina (vi) $\dfrac{2}{9}(24\log 2-7)$

(vii) 8 (viii) $\dfrac{16}{3}a^2$ (ix) $\dfrac{\pi a^2}{4}\left(\log a-\dfrac{1}{2}\right)$

(x) $\dfrac{2a^2}{3}$

3. $\int_0^1\int_y^{\sqrt{y}}\left(x^2+y^2\right)^{\frac{-1}{2}}\,dxdy$ 4. $\int_0^{2a}\int_0^{\sqrt{a^2-(x-a)^2}} xydydx\,,\ \dfrac{2}{3}a^4$ 5. $\dfrac{29}{24}\cdot 5^6$

7.8 Application of Double Integral

(a) **Area of a plane Region :** The area A of a plane regular region D is given by a two - fold iterated integral

$$A = \iint_D ds = \iint_D dxdy = \int_{x=a}^{b} \int_{y=y_1(x)}^{y_2(x)} dydx = \int_{y=c}^{d} \int_{x=x_1(y)}^{x_2(y)} dxdy$$

(b) **Mass contained in a plane Region :** Let $f(x, y) > 0$ be the surface density (mass per unit area) of a given plane region D. Then the amount of mass M contained in the plane region D is given by

$$M = \iint_D f(x,y)\,dxdy$$

(c) **Centre of Gravity (centroid) of a plane Region :** The co - ordinates (x_c, y_c) of centroid of a plane region D with surface density $f(x, y)$ and mass M are given by

$$x_c = \frac{\iint_D x \cdot f(x,y)\,dxdy}{M}, \quad y_c = \frac{\iint_D y \cdot f(x,y)\,dxdy}{M}$$

(d) **Moment of Inertia of plane Region :** Moment of inertia of a plane region D with surface density $f(x, y)$, relative to x -axis , y- axis and origin O are respectively given by

$$I_{xx} = \iint_D y^2 f(x,y)\,dxdy$$

$$I_{yy} = \iint_D x^2 f(x,y)\,dxdy$$

$$I_0 = I_{xx} + I_{yy} = \iint_D (x^2 + y^2) f(x,y)\,dxdy$$

Note that I_0 is also called as polar moment of inertia.

(e) **Volume under a surface of a soilid as a Double Integral :** Let $z = f(x, y) > 0$ be the equation of a surface . Let the curve c be the boundary of the plane domain D in the xy - plane. Further, let v be the volume of a solid Q, under the surface $z = f(x, y)$ and above xy -plane and a cylindrical surface whose generator are parallel to the z - axis, while the directrix Q is C (see fig - 34)

Then the double integral of $f(x, y)$ taken over D gives the volume V under the surface $z = f(x, y)$.

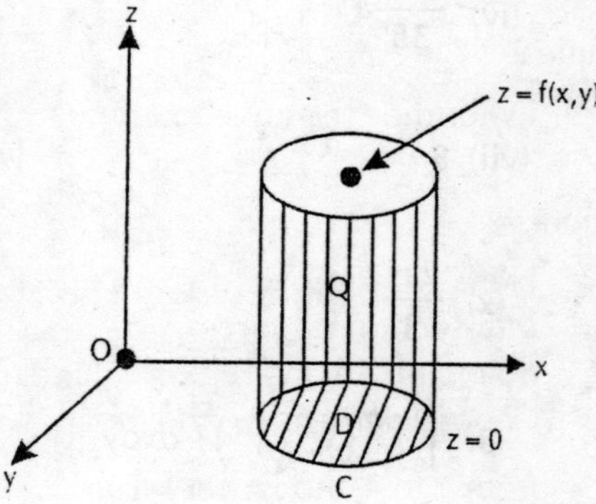

Fig. 34

$$V = \iint_D f(x,y)\,dxdy$$

Illustrative Examples

Example 1

Find the area bounded by the curves $y^2 = x^3$ and $x^2 = y^3$

Solution :

The point of intersection of curves $y^2 = x^3$ and $x^2 = y^3$ are (0, 0) and (1, 1). So,

Area $A = \iint_D dxdy = \iint_D dydx$

Where, the region D can be covered by varying y from $y = x^{\frac{3}{2}}$ to $y = x^{\frac{2}{3}}$ and x varies from 0 to 1 as shown in fig - 35.

Thus,

$$A = \int_{x=0}^{1}\int_{y=x^{\frac{3}{2}}}^{x^{\frac{2}{3}}} dydx$$

$$= \int_0^1 [y]_{x^{\frac{3}{2}}}^{x^{\frac{2}{3}}} dx = \int_0^1\left(x^{\frac{2}{3}} - x^{\frac{3}{2}}\right)dx$$

$$= \left[\frac{3}{5}x^{\frac{5}{3}} - \frac{2}{5}x^{\frac{5}{2}}\right]_0^1 = \frac{3}{5} - \frac{2}{5} = \frac{1}{5}$$

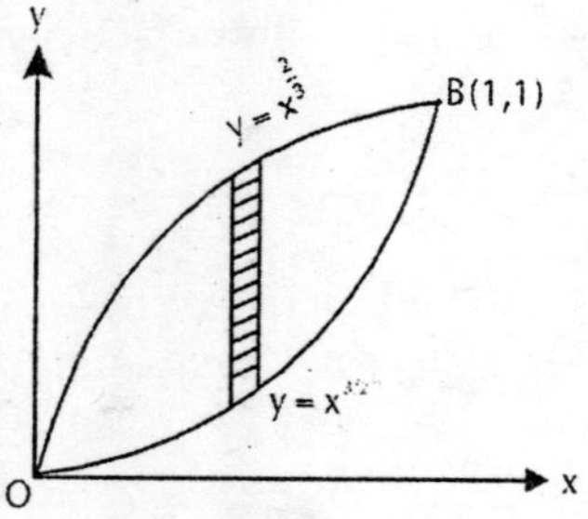

Fig. 35

Example 2

Determine the area bounded by the curves $x^2 = 4y$, $xy = 2$ and $y = 4$.

Solution:

Here $x^2 = 4y$ is a parabola about y - axis . Points of intersection of curves $x^2 = 4y$ and $xy = 2$ is (2, 1) in first quadrant. So, the region of integration is shown in fig - 36. Choose elementary strip parallel is to x - axis. For this y varies from 1 to 4 and x varies from $2/y$ to $2\sqrt{y}$. Hence

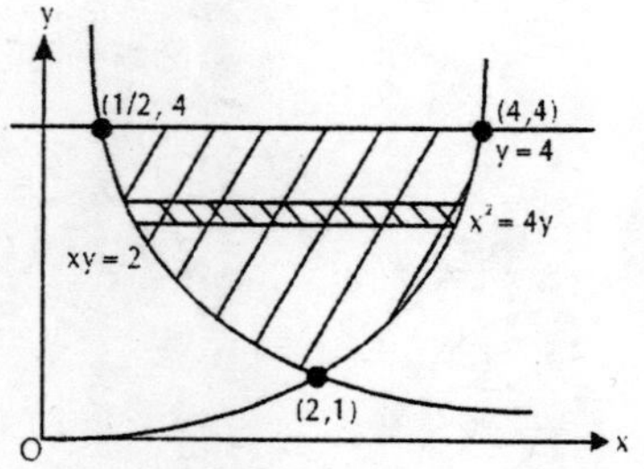

Fig. 36

Area $A = \int_{y=1}^{4}\int_{x=\frac{2}{y}}^{2\sqrt{y}} dxdy$

$$= \int_1^4 [x]_{\frac{2}{y}}^{2\sqrt{y}} dy$$

$$= \int_1^4\left(2y^{\frac{1}{2}} - \frac{2}{y}\right)dy = \left[\frac{4}{3}y^{\frac{3}{2}} - 2\log y\right]_1^4 = \frac{28}{3} - 2\log 4 \quad \text{Answer}$$

Example 3

Find the mass, co – ordinates of centre of gravity and moments of inertia I_{xx}, I_{yy} and I_o of the plane region D having mass density $(x + y)$ and bounded by the parabola $x = y - y^2$ and the straight line $x + y = 0$.

Solution :

The given region D is shown in fig -37 by OAB, where O (0, 0), A (0, 1)and B (–2, 2). The region D can be covered with x varying from $-y$ to $(y - y^2)$ and y from 0 to 2. Therefore, the mass M contained in region D is

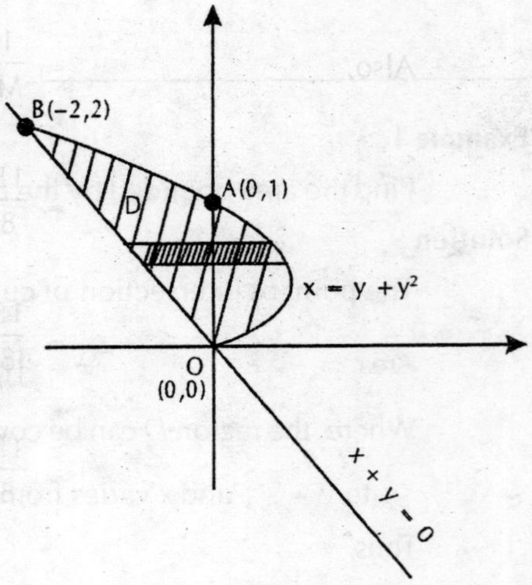

Fig. 37

$$M = \iint_D f(x,y)\,dxdy$$

$$= \int_0^2 \int_{-y}^{y-y^2} (x+y)\,dxdy$$

$$= \int_0^2 \left[\frac{x^2}{2} + xy\right]_{-y}^{y-y^2} dy = \int_0^2 \left(\frac{1}{2}y^4 - 2y^3 + 2y^2\right) dy$$

$$= \left[\frac{1}{2}\frac{y^5}{5} - \frac{1}{2}y^4 + \frac{2}{3}y^3\right]_0^2 = \frac{8}{15} \qquad ...(1)$$

Again, let (x_c, y_c) be the co-ordinates of C.G. of D, then

$$x_c = \frac{1}{M}\iint_D x f(x,y)\,dxdy$$

$$= \frac{15}{8}\int_0^2 \int_{-y}^{y-y^2} x(x+y)\,dxdy \qquad \left[\because M = \frac{8}{15}\right]$$

$$= \frac{15}{8}\int_0^2 \left[\frac{x^3}{3} + y\frac{x^2}{2}\right]_{-y}^{y-y^2} dy$$

$$= \frac{15}{8}\int_0^2 \frac{1}{6}\left[4y^3 - 12y^4 + 9y^5 - 2y^6\right] dy$$

$$= \frac{15}{8}\cdot\frac{1}{6}\left[\frac{4y^4}{4} - \frac{12}{5}y^5 + \frac{3}{2}y^6 - \frac{2}{7}y^7\right]_0^2 = -\frac{73}{7} \qquad \text{Answer}$$

Also,
$$y_c = \frac{1}{M}\int_0^2\int_{-y}^{y-y^2} y(x+y)\,dxdy$$

$$= \frac{15}{8}\int_0^2\left[\frac{yx^2}{2}+y^2x\right]_{-y}^{y-y^2} dy$$

$$= \frac{15}{8}\int_0^2\frac{1}{2}\left(y^5-3y^4+\frac{5}{2}y^3\right)dy$$

$$= \frac{15}{8}\left[\frac{y^6}{6}-\frac{3y^5}{5}+\frac{5}{8}y^4\right]_0^2 = \frac{15}{16}\times\frac{22}{15} = \frac{11}{8}$$

$$\therefore \quad (x_c, y_c) = \left(-\frac{73}{7},\frac{11}{8}\right) \quad \ldots(2)$$

Finally, moments of inertia

$$I_{xx} = \int_0^2\int_{-y}^{y-y^2} y^2(x+y)\,dxdy = \int_0^2\left[\frac{x^2y^2}{2}+xy^3\right]_{-y}^{y-y^2} dy$$

$$= \int_0^2\left(\frac{y^6}{2}-2y^5+2y^4\right)dy = \left[\frac{y^7}{14}-\frac{1}{3}y^6+\frac{2}{5}y^5\right]_0^2$$

$$= \frac{64}{105} \quad \ldots(3)$$

$$I_{yy} = \int_0^2\int_{-y}^{y-y^2} x^2(x+y)\,dxdy = \int_0^2\left[\frac{x^4}{4}+\frac{x^3y}{3}\right]_{-y}^{y-y^2}$$

$$= \frac{1}{12}\int_0^2\left(3y^8-16y^7+30y^6-24y^5+8y^4\right)dy$$

$$= \frac{1}{12}\left[\frac{3y^9}{9}-2y^8+\frac{30}{7}y^7-4y^6+\frac{8}{5}y^5\right]_0^2 = \frac{256}{105} \quad \ldots(4)$$

$$I_0 = I_{xx}+I_{yy} = \frac{64}{105}+\frac{256}{105} = \frac{320}{105}$$

$$= \frac{64}{21} \quad \text{Answer} \quad \ldots(5)$$

Example 4

Find the mass, centroid and moments of inertia relative to x – axis, y – axis and origin of a rectangle $0 \leq x \leq 4$, $0 \leq y \leq 2$ having mass density xy.

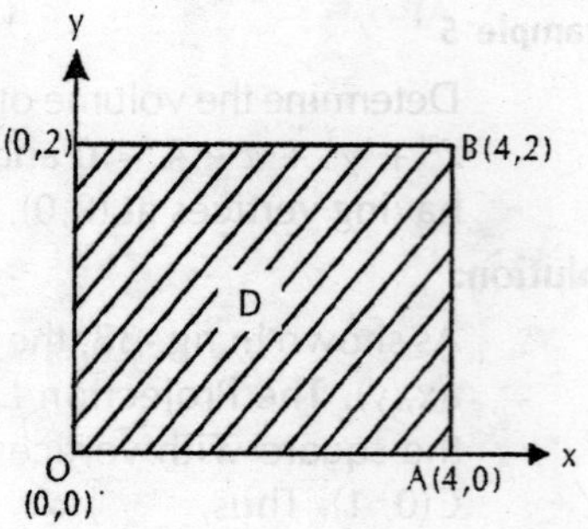

Fig. 38

Solution:

The region D has been shown in fig – 38. The mass density is given xy so,

The mass $$M = \iint_D f(x,y)\,dxdy$$

$$= \int_0^4 \int_0^2 xy\,dxdy$$

$$= \left[\frac{x^2}{2}\right]_0^4 \left[\frac{y^2}{2}\right]_0^2 = 8 \times 2 = 16 \qquad ...(1)$$

Again, let (x_c, y_c) be the co – ordinates of centroid, then,

$$x_c = \frac{1}{M}\iint_D xf(x,y)\,dxdy = \frac{1}{16}\int_0^4 \int_0^2 x^2y\,dxdy$$

$$= \frac{1}{16}\left[\frac{x^3}{3}\right]_0^4 \left[\frac{y^2}{2}\right]_0^2 = \frac{1}{16} \times \frac{64}{3} \times 2 = \frac{8}{3}$$

and $$y_c = \frac{1}{M}\iint_D yf(x,y)\,dxdy = \frac{1}{16}\int_0^4 \int_0^2 y^2x\,dxdy$$

$$= \frac{1}{16}\left[\frac{y^3}{3}\right]_0^2 \left[\frac{x^2}{2}\right]_0^4 = \frac{1}{16} \times \frac{8}{3} \times 8 = \frac{4}{3}$$

Hence, $\left(\frac{8}{3}, \frac{4}{3}\right)$ be required centroid of region D. ...(2)

Finally, moments of inertia

$$I_{xx} = \iint_D y^2 f(x,y)\,dxdy = \int_0^4 \int_0^2 xy^3\,dxdy = \left[\frac{x^2}{2}\right]_0^4 \left[\frac{y^4}{4}\right]_0^2 = 32 \qquad ...(3)$$

Similarly, $$I_{yy} = \iint_D x^2 f(x,y)\,dxdy = \int_0^4 \int_0^2 x^3y\,dxdy = \left[\frac{x^4}{4}\right]_0^4 \left[\frac{y^2}{2}\right]_0^2 = 128 \qquad ...(4)$$

And $$I_0 = I_{xx} + I_{yy} = 32 + 128 = 160 \qquad ...(5)$$

Example 5

Determine the volume of the space below the paraboloid $x^2 + y^2 + z - 4 = 0$ and above the square in *xy*-plane having vertices at (0,0), (0, 1), (1, 0), (1, 1).

Solution:

As shown in fig.-38, the top surface is $z = 4 - x^2 - y^2 = f(x, y)$. The Projection *D* of this surface on xy-plane is the square with vertices *O*(0, 0), *A*(1, 0), *B*(1, 1) and *C*(0, 1). Thus,

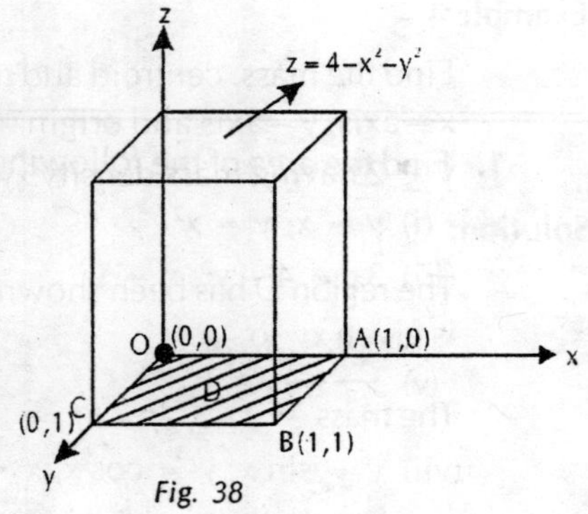

Fig. 38

$$\text{volume} \quad V = \iint_D f(x,y)dx\,dy$$

$$= \int_0^1\int_0^1 (4 - x^2 - y^2)dx\,dy = \int_0^1\left[4y - x^2y - \frac{y^3}{3}\right]_0^1 dx$$

$$= \int_0^1\left(\frac{11}{3} - x^2\right)dx = \left[\frac{11}{3}x - \frac{x^3}{3}\right]_0^1 = \frac{10}{3} \quad \text{Answer}$$

Example 6

Find the volume of the tetrahedron in space cut from the first octant by the plane $6x + 3y + 2z = 6$.

Solution:

The tetrahedron is bounded below by xy-plane and bounded above by the surface

$$z = \frac{6 - 6x - 3y}{2} = f(x, y).$$

The projection *D* of surface *f*(*x*, *y*) onto *xy*-plane is a triangle *OAB* bounded by the lines $x = 0$, $y = 0$ and $6x + 3y = 6$ as shown in fig.-39. This region *D* can be covered by varying y from 0 to $2 - 2x$, and x from 0 to 1. Thus required volume

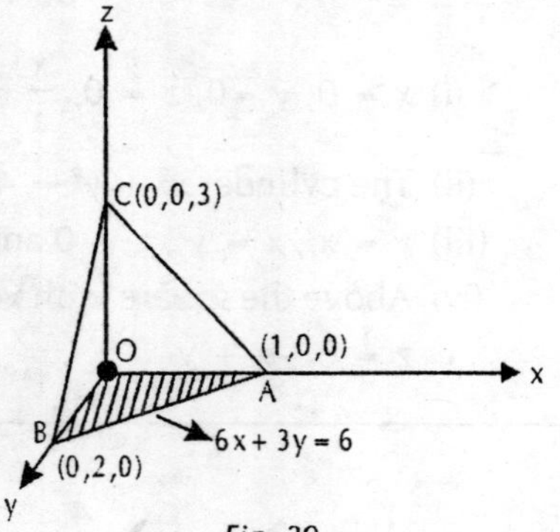

Fig. 39

$$V = \iint_D f(x,y)dy\,dx$$

$$= \int_0^1\int_0^{2-2x} \frac{6 - 6x - 3y}{2}dy\,dx$$

$$= \int_0^1\left[(3 - 3x)y - \frac{3}{4}y^2\right]_0^{2-2x} dx = 3\int_0^1(1 - x^2)dx = 1 \quad \text{Answer}$$

Exercise 7.5

1. Find the area of the following domains in xy-plane bounded by the indicative curves :

(i) $y = x, y = x^2$ (ii) $y = 3x - x^2, y = x$

(iii) $3x = 4 - y^2, x = y^2$ (iv) $y = x^2, y = x + 2$

(v) $\frac{x^2}{a^2}+\frac{y^2}{b^2}=1$ (vi) $y^2 = 4ax, x + y = 3a$

(vii) $y = \sin x, y = \cos x, x = 0$ (viii) $3y^2 = 25x$ and $5x^2 = 9y$

2. Find the mass contained in the ellipse $\frac{x^2}{a^2}+\frac{y^2}{b^2}=1$ with surface density $(x + y)^2$.

3. Find the centroid of the region in first quadrant bounded by $\left(\frac{x}{a}\right)^{\frac{2}{3}}+\left(\frac{y}{b}\right)^{\frac{2}{3}}=1, x = 0, y = 0$ and having surface density kxy where k is constant.

4. Determine the Co-ordinates of the centre of gravity of the plane region bounded by $y = \log x, y = 1, y = 0$ and $x = 0$.

5. A thin plate of uniform thickness and constant density ρ covers the region of xy-plane and is bounded by $y = x^2$ and $y = x + 2$. Find the mass M and moment of inertia about y-axis.

6. By means of double integral, find the volume V of the following solids bounded by :

(i) $x = 0, y = 0, z = 0, \frac{x}{a}+\frac{y}{b}+\frac{z}{c}=1$

(ii) The cylinder $x^2 + y^2 = 4$ and the planes $y + z = 4$ and $z = 0$

(iii) $y = x^2, x = y^2, z = 0$ and $z = 12 + y - x^2$

(iv) Above the square with vertices at (0, 0), (2, 0), (0, 2), (2, 2) and under the plane $z = 8 - x + y$

Answers

1. (i) $\frac{1}{6}$ (ii) $\frac{4}{3}$ (iii) $\frac{16}{9}$ (iv) $\frac{9}{2}$

(v) πab (vi) $\frac{10}{3}a^2$ (vii) $\sqrt{2}-1$ (viii) 5

2. $\frac{\pi ab}{4}(a^2+b^2)$ **3.** $\left(\frac{128}{429}a, \frac{128}{429}b\right)$ **4.** $\left(\frac{e+1}{4}, \frac{1}{e-1}\right)$ **5.** $M=\frac{9}{2}\rho,\ I_y=\frac{63}{20}\rho$

6. (i) $\frac{abc}{6}$ (ii) 16π (iii) $\frac{549}{144}$ (iv) 32

7.9 Triple Integral

Let $f(x, y, z)$ be a function of three independent variables x, y and z defined at every point of a 3-Dim region V.

Divide the region V into n elementary volumes $\delta v_1, \delta v_2 ..., \delta v_n$ and let (x_r, y_r, z_r) be any point inside the δv_r. Find the sum

$$\sum_{r=1}^{n} f(x_r, y_r, z_r)\delta v_r$$

Then $$\iiint_V f(x,y,z)dv = \lim_{\substack{n\to\infty \\ \delta v_r \to 0}} \sum_{r=1}^{n} f(x_r, y_r, z_r)\delta v_r \qquad ...(1)$$

To extend definition of repeated integrals for triple integral, first integrate $f(x, y, z)$ w.r.t. z keeping x and y as constant between the limits, in general, depending upon x and y. This would reduce $f(x, y, z)$ to a function of x and y only.

Thus, let $$\phi(x, y) = \int_{z=z_1(x,y)}^{z_2(x,y)} f(x,y,z)dz \qquad ...(2)$$

Now, use the method of double integral to integrate $\phi(x, y)$ w.r.t. x and y.

Hence, $$\iiint_V f(x,y,z)dv = \int_{x=a}^{b}\int_{y=y_1(x)}^{y_2(x)}\int_{z=z_1(x,y)}^{z_2(x,y)} f(x,y,z)dx\, dy\, dz \qquad ...(3)$$

Now solve as usual.

Note:

(i) If $f(x, y, z) = 1$, then $\iiint_V dv$ gives the volume.

(ii) Technique of change of variables remains applicable for triple integral and is similar to the technique used for double integral.

(iii) Mass of the solid, $M = \iiint_V \rho dv$, where ρ is density function.

(iv) Moment of Inertia of a solid: Let $\rho(x, y, z)$ be the density of the substance then moment of inertia of a solid relative to x, y, z axes respectively is denoted and defined by

$$I_{xx} = \iiint_V (y^2 + z^2)\,\rho(x,y,z)\, dx\, dy\, dz$$

$$I_{yy} = \iiint_V (x^2 + z^2)\,\rho(x,y,z)\, dx\, dy\, dz$$

and $$I_{zz} = \iiint_V (x^2 + y^2)\,\rho(x,y,z)\, dx\, dy\, dz$$

(v) Centre of gravity of a solid (x_c, y_c, z_c) : Let $\rho(x, y, z)$ be the density function then

$$x_c = \frac{\iiint_x x\rho(x,y,z)\,dv}{\iiint_V \rho\,dv} = \frac{1}{M}\iiint_V x\,\rho(x,y,z)\,dv$$

Similarly, $$y_c = \frac{1}{M}\iiint_V y\,\rho(x,y,z)\,dv$$

and $$z_c = \frac{1}{M}\iiint_V z\,\rho(x,y,\dot{z})\,dv$$

Illustrative Examples

Example 1

Evaluate the integral $\int_0^1\int_0^{1-x}\int_0^{1-x-y} \frac{dxdydz}{(x+y+z+1)^3}$

Solution:

Let $$I = \int_0^1\int_0^{1-x}\int_0^{1-x-y} \frac{dxdydz}{(x+y+z+1)^3}$$

$$= \int_0^1\int_0^{1-x}\left[-\frac{1}{2(x+y+z+1)^2}\right]_0^{1-x-y} dxdy$$

$$= -\frac{1}{2}\int_0^1\int_0^{1-x}\left(\frac{1}{4}-\frac{1}{(x+y+1)^2}\right)dxdy$$

$$= -\frac{1}{2}\int_0^1\left[\frac{1}{4}y+\frac{1}{(x+y+1)}\right]_0^{1-x} dx$$

$$= -\frac{1}{2}\int_0^1\left[\frac{1}{4}(1-x)+\frac{1}{2}-\frac{1}{x+1}\right]dx = -\frac{1}{2}\int_0^1\left(\frac{3}{4}-\frac{x}{4}-\frac{1}{x+1}\right)dx$$

$$= -\frac{1}{2}\left[\frac{3}{4}x-\frac{x^2}{8}-\log(x+1)\right]_0^1 = -\frac{1}{2}\left[\frac{3}{4}-\frac{1}{8}-\log 2\right] = \frac{1}{2}\left[\log 2-\frac{5}{8}\right]$$

Example 2:

Evaluate $I = \int_0^1 \int_0^x \int_0^{x+y} (x+y+z)\,dz\,dy\,dx$

Solution :

First integrating w.r.t. z, we obtain

$$I = \int_0^1 \int_0^x \left[(x+y)z + \frac{z^2}{2}\right]_0^{x+y} dydx$$

$$= \frac{3}{2}\int_0^1 \int_0^x (x+y)^2\,dydx = \frac{3}{2}\int_0^1 \int_0^x (x^2+y^2+2xy)\,dydx$$

Integrating now w.r.t. y, we get

$$I = \frac{3}{2}\int_0^1 \left[x^2y + \frac{y^3}{3} + xy^2\right]_0^x dx$$

$$= \frac{3}{2}\int_0^1 \frac{7}{3}x^3\,dx = \frac{7}{2}\int_0^1 x^3\,dx$$

Finally, integrating w.r.t. x, we have

$$I = \frac{7}{2}\left[\frac{x^4}{4}\right]_0^1 = \frac{7}{8} \quad \text{Answer}$$

Example 3

Find the volume bounded by the elliptic paraboloids $z = x^2 + 9y^2$ and $z = 18 - x^2 - 9y^2$

Solution :

The given surfaces intersect on the elliptic cylinder

$$x^2 + 9y^2 = z = 18 - x^2 - 9y^2$$

$$\Rightarrow \quad x^2 + 9y^2 = 9$$

$$i.e. \quad \frac{x^2}{3^2} + \frac{y^2}{1^2} = 1$$

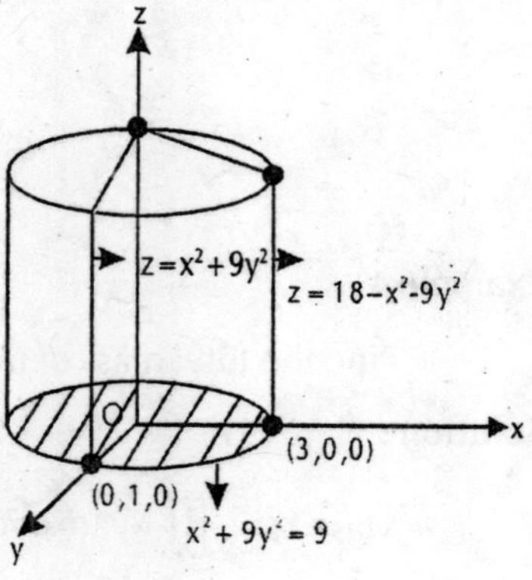

Fig. 40

which represents the projection of the body onto xy-plane as shown in fig-40. This volume can be covered as follows :

z : from $\left(x^2+9y^2\right)$ to $(18-x^2-9y^2)$

y : from $-\sqrt{\dfrac{9-x^2}{9}}$ to $\sqrt{\dfrac{9-x^2}{9}}$

x : from -3 to 3

Thus the volume v bounded by the elliptic paraboloids is given by

$$v = \int_{-3}^{3}\int_{-\sqrt{\frac{9-x^2}{9}}}^{\sqrt{\frac{9-x^2}{9}}}\int_{x^2+9y^2}^{18-x^2-9y^2} dzdydx$$

$$= \int_{-3}^{3}\int_{-\sqrt{\frac{9-x^2}{9}}}^{\sqrt{\frac{9-x^2}{9}}}\left[\left(18-x^2-9y^2\right)-\left(x^2+9y^2\right)\right]dydx$$

$$= 2\int_{-3}^{3}\int_{-\sqrt{\frac{9-x^2}{9}}}^{\sqrt{\frac{9-x^2}{9}}}\left(9-x^2-9y^2\right)dydx$$

$$= 2\int_{-3}^{3}\left[9y-x^2y-3y^3\right]_{-\sqrt{\frac{9-x^2}{9}}}^{\sqrt{\frac{9-x^2}{9}}} dx$$

$$= \frac{8}{9}\int_{-3}^{3}\left(9-x^2\right)^{\frac{3}{2}} dx, \quad \text{Now put } x = 3\cos\theta\ ,\ dx = -3\sin\theta\, d\theta$$

$$= 72\int_{0}^{\pi}\sin^4\theta d\theta = 72\int_{0}^{\pi}\left(\frac{1-\cos 2\theta}{2}\right)^2 d\theta = 27\pi$$

Example 4

Find the total mass of the cube $0 \le x \le 1, 0 \le y \le 1, 0 \le z \le 1$, with the density function xyz.

Solution :

$$\text{Mass } M = \iiint_v xyzdxdydz$$

$$= \int_0^1\int_0^1\int_0^1 xyz\ dxdydz$$

$$= \int_0^1\int_0^1\left[xy\frac{z^2}{2}\right]_0^1 dxdy = \frac{1}{2}\int_0^1\left[\frac{xy^2}{2}\right]_0^1 dx = \frac{1}{4}\left[\frac{x^2}{2}\right]_0^1 = \frac{1}{8} \quad \text{Answer}$$

Example 5

Evaluate the triple integral $I = \int_0^1 \int_y^1 \int_0^{1-x} xdzdxdy$

Solution :

We have, $$I = \int_0^1 \int_y^1 [xz]_0^{1-x} dxdy = \int_0^1 \int_y^1 (x - x^2) dxdy$$

$$= \int_0^1 \left[\frac{x^2}{2} - \frac{x^3}{3}\right]_y^1 dy = \int_0^1 \left[\frac{1}{6} - \frac{y^2}{2} + \frac{y^3}{3}\right] dy$$

$$= \left[\frac{1}{6}y - \frac{y^3}{6} + \frac{y^4}{12}\right]_0^1 = \frac{1}{12}$$

Example 6

Find the mass and centroid of the tetrahedron bounded by the coordinate plane and the plane $\frac{x}{a} + \frac{y}{b} + \frac{z}{c} = 1$

Solution :

Let ρ be the constant density of given tetrahedron which is shown in fig.-41. Then mass

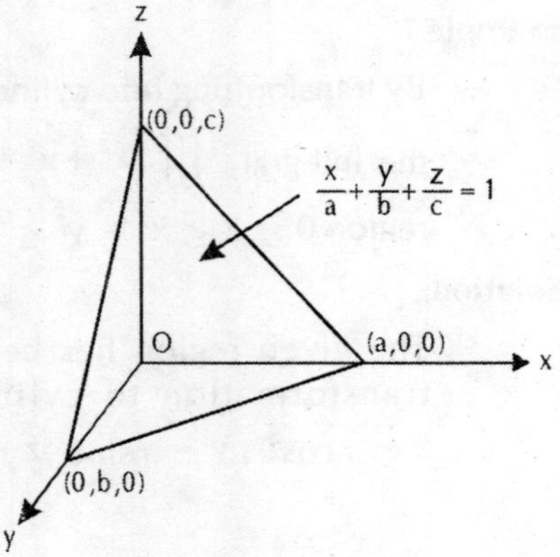

Fig. 41

$$M = \iiint_V \rho dxdydz$$

$$= \int_0^a \int_0^{b\left(1-\frac{x}{a}\right)} \int_0^{c\left(1-\frac{x}{a}-\frac{y}{b}\right)} \rho \, dxdydz$$

$$= \rho \int_0^a \int_0^{b\left(1-\frac{x}{a}\right)} c\left(1 - \frac{x}{a} - \frac{y}{b}\right) dxdy$$

$$= \rho c \int_0^a \left[\left(1 - \frac{x}{a}\right)y - \frac{y^2}{2b}\right]_0^{b\left(1-\frac{x}{a}\right)} dx$$

$$= \frac{\rho cb}{2} \int_0^a \left(1 - \frac{x}{a}\right)^2 dx = \frac{\rho cb}{2}\left[-\frac{a}{3}\left(1 - \frac{x}{a}\right)^3\right]_0^a = \frac{\rho cb}{2} \cdot \frac{a}{3} = \frac{\rho abc}{6} \quad \text{...(1)}$$

Again, Let (x_c, y_c, z_c) be the co – ordinates of centroid, then

$$x_c = \frac{1}{M}\iiint_v \rho x\,dxdydz = \frac{1}{M}\int_0^a \int_0^{b\left(1-\frac{x}{a}\right)} \int_0^{c\left(1-\frac{x}{a}-\frac{y}{b}\right)} \rho x\,dzdydx$$

$$= \frac{1}{M}\int_0^a \int_0^{b\left(1-\frac{x}{a}\right)} \rho xc\left(1-\frac{x}{a}-\frac{y}{b}\right)dydx$$

$$= \frac{\rho c}{M}\int_0^a \left[x\left(1-\frac{x}{a}\right)y - \frac{xy^2}{2b}\right]_0^{b\left(1-\frac{x}{a}\right)} dx$$

$$= \frac{\rho cb}{2M}\int_0^a x\left(1-\frac{x}{a}\right)^2 dx = \frac{\rho cb}{2M}\cdot\frac{a^2}{12}$$

$$= \frac{\rho cba^2}{24}\times\frac{6}{\rho abc}\left[by(1)\right] = \frac{a}{4}$$

Similarly, $y_c = \frac{b}{4}$ and $z_c = \frac{c}{4}$

Example 7

By transforming into cylindrical co – ordinates evaluate the integral $\iiint_v \left(x^2+y^2+z^2\right)dxdydz$ taken over the region $0 \le z \le x^2 + y^2 \le 1$.

Solution:

The given region has been shown in fig – 42. For transformation to cylindrical co-ordinates, put $x = r\cos\theta$, $y = r\sin\theta$, $z = z$

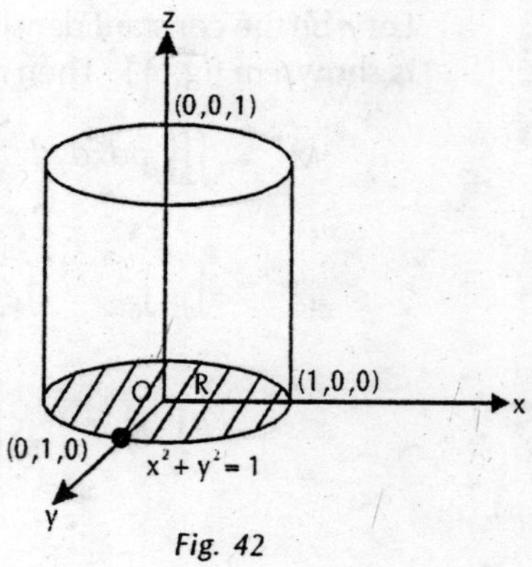

Fig. 42

Also $J = \frac{\partial(x,y,z)}{\partial(r,\theta,z)} = r$

∴ Given integral can be written as

$$\int_{z=0}^{1}\int_{\theta=0}^{2\pi}\int_{r=0}^{1}\left(r^2\cos^2\theta + r^2\sin^2\theta + z^2\right)r\,drd\theta dz$$

$$= \int_0^1\int_0^{2\pi}\left[\frac{r^2}{4}+\frac{z^2r^2}{2}\right]_0^1 d\theta dz = \int_0^1\left[\frac{1}{4}+\frac{z^2}{2}\right]\left[\theta\right]_0^{2\pi} dz$$

$$= 2\pi\left[\frac{z}{4}+\frac{z^3}{6}\right]_0^1 = 2\pi\left[\frac{1}{4}+\frac{1}{6}\right] = \frac{5\pi}{6}$$

Example 8:

Evaluate $\iiint_v xyz\left(x^2+y^2+z^2\right)^{\frac{n}{2}} dxdydz$ taken over the positive octant of the sphere

$x^2+y^2+z^2 = b^2$, provided $n + 5 > 0$

Solution :

Introducing spherical polar co - ordinates as $x = r\sin\theta\cos\phi$, $y = r\sin\theta\sin\phi$ and $z = r\cos\theta$

Also,
$$J = \frac{\partial(x,y,z)}{\partial(r,\theta,\phi)} = r^2\sin\theta$$

Substituting above transformation, given integral becomes

$$I = \int_{\phi=0}^{\frac{\pi}{2}}\int_{\theta=0}^{\frac{\pi}{2}}\int_{r=0}^{b} r^3\sin^2\theta\cos\theta\sin\phi\cos\phi\cdot r^n\cdot r^2\sin\theta drd\theta d\phi$$

Note that given region is hemi sphere of base as circle $x^2+y^2 = b^2$

So, for this θ varies 0 to $\frac{\pi}{2}$, ϕ varies 0 to $\frac{\pi}{2}$ and r varies from 0 to b.

Hence,
$$I = \int_0^b r^{n+5}dr\int_0^{\frac{\pi}{2}}\sin^3\theta\cos\theta d\theta\int_0^{\frac{\pi}{2}}\cos\phi\sin\phi d\phi$$

$$= \left[\frac{r^{n+6}}{n+6}\right]_0^b\left[\frac{\sin^4\theta}{4}\right]_0^{\frac{\pi}{2}}\left[\frac{\sin^2\phi}{2}\right]_0^{\frac{\pi}{2}}$$

$$= \frac{b^{n+6}}{n+6}\cdot\frac{1}{4}\cdot\frac{1}{2} = \frac{b^{n+6}}{8(n+6)}$$ provided $n + 5 > 0$ Answer

Example 9

Evaluate $\iiint\left(\frac{1-x-y-z}{xyz}\right)^{\frac{1}{2}} dxdydz$ taken over the volume bounded by the planes $x = 0$, $y = 0$, $z = 0$ and $x + y + z = 1$.

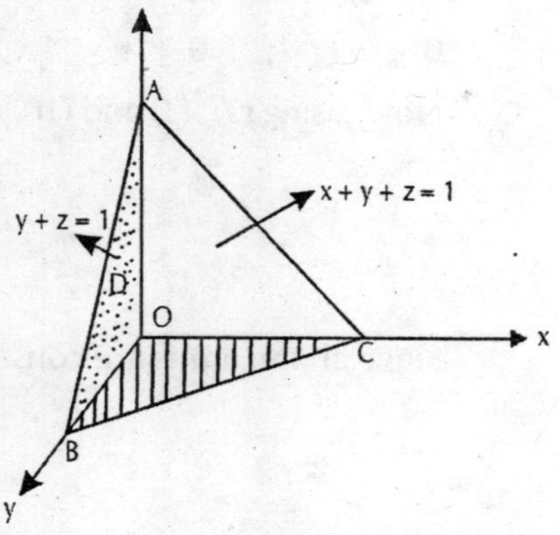

Fig. 43

Solution :

The region of integration is shown in fig.-43. Now put

$$x + y + z = u,\ y + z = uv,$$
$$z = uvw \qquad ...(1)$$

$$\Rightarrow \qquad x = u(1-v),\ y = uv(1-w),\ z = uvw \qquad ...(2)$$

Also,

$$J = \frac{\partial(x,y,z)}{\partial(u,v,w)} = \begin{vmatrix} \frac{\partial x}{\partial u} & \frac{\partial x}{\partial v} & \frac{\partial x}{\partial w} \\ \frac{\partial y}{\partial u} & \frac{\partial y}{\partial v} & \frac{\partial y}{\partial w} \\ \frac{\partial z}{\partial u} & \frac{\partial z}{\partial v} & \frac{\partial z}{\partial w} \end{vmatrix}$$

$$= \begin{vmatrix} 1-v & -u & 0 \\ v(1-w) & u(1-w) & -uv \\ vw & uv & uv \end{vmatrix} = u^2 v \qquad ...(3)$$

Let the triangular region OAB be the projection of the surface on yz – plane. Then the tetrahedral volume will be covered by the following limits

$0 \le z \le 1$ ...(4)

$0 \le y \le 1 - z$ ...(5)

$0 \le x \le 1 - z - y$ ...(6)

Now since $z = uvw$, from (4)

$0 \le uvw \le 1$ ...(7)

Since $y = uv(1 - w)$, from (5)

$0 \le uv(1 - w) \le 1 - uvw \Rightarrow 0 \le uv \le 1$ ...(8)

Since $x = u(1 - v)$, from (6)

$0 \le u(1 - v) \le 1 - uvw - uv(1 - w)$

$\Rightarrow \quad 0 \le u \le 1$ ...(9)

Using (9) in (7) and (8), the limits of integration for the new variables u, v, w will be

$0 \le u \le 1, \quad 0 \le v \le 1, \quad 0 \le w \le 1$ ...(10)

Now, using (2), (3) and (10), the given integral transforms to

$$I = \int_0^1 \int_0^1 \int_0^1 \left(\frac{1-u}{u^3 v^2 w(1-v)(1-w)} \right)^{\frac{1}{2}} u^2 v\, du\, dv\, dw$$

Since all the limits are constant, so it may be written as

$$I = \int_0^1 \sqrt{u(1-u)}du \int_0^1 \frac{dv}{\sqrt{1-v}} \int_0^1 \frac{dw}{\sqrt{w(1-w)}}$$

Put $u = \sin^2\theta,\ v = \sin^2\phi,\ w = \sin^2 t$, it becomes

$$I = \int_0^{\frac{\pi}{2}} \sin\theta\cos\theta \cdot 2\sin\theta\cos\theta d\theta \cdot \int_0^{\frac{\pi}{2}} \frac{2\sin\phi\cos\phi}{\cos\phi}d\phi \cdot \int_0^{\frac{\pi}{2}} \frac{2\sin t\cos t}{\sin t\cos t}dt$$

$$= 2\cdot\frac{1}{4}\cdot\frac{1}{2}\cdot\frac{\pi}{2}\cdot 2\cdot 1\cdot 2\cdot\frac{\pi}{4} = \frac{\pi^2}{4}$$ Answer

Example 10

Find the volume of the solid surrounded by the surface

$$\left(\frac{x}{a}\right)^{\frac{2}{3}} + \left(\frac{y}{b}\right)^{\frac{2}{3}} + \left(\frac{z}{c}\right)^{\frac{2}{3}} = 1.$$

Solution:

The desired volume $V = \iiint_V dxdydz$

Substitute

$$\frac{x}{a} = u^3,\ \frac{y}{b} = v^3,\ \frac{z}{c} = w^3, \qquad ...(1)$$

$$\Rightarrow \qquad x = au^3,\ y = bv^3,\ z = cw^3, \qquad ...(2)$$

Now
$$J = \frac{\partial(x,y,z)}{\partial(u,v,w)} = \begin{vmatrix} 3au^2 & 0 & 0 \\ 0 & 3bv^2 & 0 \\ 0 & 0 & 3cw^2 \end{vmatrix}$$

$$= 27\,abc\,u^2v^2w^2 \qquad ...(3)$$

Using (1), equation of given surface reduces to V^* (say)

$$u^2 + v^2 + w^2 = 1 \qquad ...(4)$$

which represents a sphere of radius 1 and center at origin.

Using (1),(3) the given integral transforms to

$$V = \iiint_{V^*} 27abc\,u^2v^2w^2 dudvdw \qquad ...(5)$$

where v^* is sphere $u^2 + v^2 + w^2 = 1$

To evaluate this integral, introduce again spherical polar co-ordinates as

$$u = r\sin\theta\cos\phi,\ v = r\sin\theta\sin\phi,\ w = r\cos\theta$$

also, $J = r^2 \sin\theta$ and to describe positive octant of V^*, $0 \le r \le 1, 0 \le \theta \le \frac{\pi}{2}, 0 \le \phi \le \frac{\pi}{2}$

$\therefore$ (5) becomes

$$V = 8 \times 27abc \int_0^{\frac{\pi}{2}} \int_0^{\frac{\pi}{2}} \int_0^1 r^8 \sin^5\theta \cos^2\theta \sin^2\phi \cos^2\phi \, d\phi dr d\theta$$

$$= 216\, abc \int_0^1 r^8 dr \int_0^{\frac{\pi}{2}} \sin^5\theta \cos^2\theta \, d\theta \int_0^{\frac{\pi}{2}} \sin^2\phi \cos^2\phi \, d\phi$$

$$= 216abc \left[\frac{r^9}{9}\right]_0^1 \frac{\pi}{16} \cdot \frac{8}{35 \cdot 3}$$

$$= \frac{4abc}{35} \quad \text{Answer}$$

Exercise 7.6

1. Evaluate the following triple integrals :

(i) $\int_0^2 \int_1^z \int_0^{yz} xyz \, dxdydz$

(ii) $\int_1^2 \int_y^{y^2} \int_0^{\log x} e^z dzdxdy$

(iii) $\int_0^a \int_0^x \int_0^{y+x} e^{x+y+z} dzdydx$

(iv) $\int_0^{\frac{\pi}{2}} \int_x^{\frac{\pi}{2}} \int_0^{xy} \cos\left(\frac{z}{x}\right) dzdydx$

(v) $\iiint x^2yz \, dxdydz$ over the volume bounded by $x^2 + y^2 = 9, z = 0, z = 2$

(vi) $\int_0^a \int_0^x \int_0^y xyz \, dzdydx$.

2. Find the volume of the tetrahedron bounded by co-ordinates plane and $\frac{x}{a}+\frac{y}{b}+\frac{z}{c} = 1$.

3. Compute the volume of the solid bounded by the plane $2x + 3y + 4z = 12$, xy-plane and cylinder $x^2 + y^2 = 1$.

4. Compute the volume bounded by $xy = z$, $z = 0$ and $(x-1)^2 + (y-1)^2 = 1$.

5. Evaluate $\iiint xyz\, dxdydz$ over the positive octant of the sphere $x^2+y^2+z^2 = b^2$ by transforming to spherical polar co-ordinates.

6. Use cylindrical co-ordinates, to evaluate $\iiint_v (x^2+y^2)\,dxdydz$ over the region v bounded by paraboloid $z = 9-x^2-y^2$ and the plane $z=0$.

7. Use spherical co-ordinates to evaluate

$$\int_0^1\int_0^{\sqrt{1-x^2}}\int_0^{\sqrt{1-x^2-y^2}}(1-x^2-y^2-z^2)^{\frac{-1}{2}}\,dxdydz.$$

8. Find the mass M of the tetrahedron bounded by the co-ordinate planes and the plane $\frac{x}{a}+\frac{y}{b}+\frac{z}{c} = 1$ given that the density at any point (x, y, z) is $kxyz$.

9. Find the centroid of the solid octant of the ellipsoid $\left(\frac{x}{a}\right)^2+\left(\frac{y}{b}\right)^2+\left(\frac{z}{c}\right)^2 = 1$, if the density at any point of solid varies as xyz.

10. Find the centre of gravity of a homogeneous solid sphere of radius b.

11. Find the moment of inertia about the z-axis of the solid ellipsoid inside

$$\frac{x^2}{a^2}+\frac{y^2}{b^2}+\frac{z^2}{c^2} = 1$$

12. Find the volume of the cone $0 \le z \le h(a-r)/a$.

Answers

1. (i) $\frac{7}{2}$ (ii) $\frac{47}{24}$ (iii) $\frac{e^{4a}}{8}-\frac{3}{4}e^{2a}+e^{a}-\frac{3}{8}$

(iv) $\frac{\pi}{2}-1$ (v) $\frac{648}{5}$ (vi) $\frac{a^6}{48}$

2. $\frac{|abc|}{6}$ 3. 3π 4. π

5. $\frac{b^6}{48}$ 6. $\frac{243\pi}{2}$ 7. $\frac{\pi^2}{8}$

8. $\frac{Ka^2b^2c^2}{720}$ 9. $\left(\frac{16a}{35},\frac{16b}{35},\frac{16c}{35}\right)$ 10. $x_c=0, y_c=0, z_c=\frac{3b}{8}$

11. $I = \rho abc(a^2+b^2)\frac{4\pi}{15} = \frac{1}{5}M(a^2+b^2)$ where mass $M = \frac{4}{3}\pi\rho abc$

12. $\frac{\pi a^2 h}{3}$

7.10 Gamma Function

For $n > 0$, the improper integral $\int_0^\infty e^{-x}x^{n-1}dx$ is defined as Gamma function and denoted by $\Gamma(n)$ (read as gamma n). It is also known as Eulerian Integral of second kind. Thus,

$$\Gamma(n) = \int_0^\infty e^{-x}x^{n-1}dx\,,\ n > 0 \qquad ...(1)$$

Note that Gamma function plays an important role in evaluation of definite integrals.

7.11 Properties of Gamma Function

(a) To show that $\Gamma(n+1) = n\Gamma(n)$

We have $\Gamma(n+1) = \int_0^\infty e^{-x}x^{(n+1)-1}dx$ [by (1)]

$$= \left[-e^{-x}x^n\right]_0^\infty + n\int_0^\infty e^{-x}x^{n-1}dx$$

$$= 0 + n\Gamma(n)$$

Thus, $\boxed{\Gamma(n+1) = n\Gamma(n)}$ ...(2)

(b) To show that $\Gamma(n) = (n-1)!$, where n is positive integer.

We have $\Gamma(n) = (n-1)\Gamma(n-1)$ [by (2)]

$$= (n-1)(n-2)\Gamma(n-2) \qquad \text{[by (2)]}$$

$$= (n-1)(n-2)(n-3)...3.2.\Gamma(2) \qquad \text{[by repeated application of (2)]}$$

$$= (n-1)(n-2)(n-3)...3.2.1.\Gamma(1)$$

$$= (n-1)!\Gamma(1) = (n-1)!\int_0^\infty e^{-x}dx \qquad \text{[by (1)]}$$

$$= (n-1)!\left[-e^{-x}\right]_0^\infty = (n-1)!$$

Thus $\boxed{\Gamma(n) = (n-1)!\text{, n is positive integer}}$ ...(3)

Remark 1:

The formula $\Gamma(n+1) = n\Gamma(n)$ is called as recurrence formula for Gamma Function.

Remark 2:

Note that $\boxed{\Gamma(1) = 1}$ as Find out in (b).

Remark 3:

The Gamma Function can be generalised to n < 0 by the use of recurrence formula in the form $\Gamma(n) = \dfrac{\Gamma(n+1)}{n}$ the process is known as analytic continuation.

(c) To show that, $\Gamma\left(\frac{1}{2}\right) = \sqrt{\pi}$

Put $n = \frac{1}{2}$ in (1), we obtain

$$\Gamma\left(\frac{1}{2}\right) = \int_0^\infty e^{-x} x^{-\frac{1}{2}}\, dx$$

Putting $x = v^2 \Rightarrow dx = 2v\,dv$, we get

$$\Gamma\left(\frac{1}{2}\right) = 2\int_0^\infty e^{-v^2}\, dv \qquad ...(4)$$

Writing u in place of v in (4), we obtain

$$\Gamma\left(\frac{1}{2}\right) = 2\int_0^\infty e^{-u^2}\, du \qquad ...(5)$$

Multiplying (4) and (5), we get

$$\left[\Gamma\left(\frac{1}{2}\right)\right]^2 = 4\int_0^\infty \int_0^\infty e^{-(u^2+v^2)} du\, dv$$

Now let $u = r\cos\theta$, $v = r\sin\theta$ then $u^2 + v^2 = r^2$ and $\theta = \tan^{-1}\left(\frac{v}{u}\right)$. Also $du\,dv = r dr\, d\theta$ and r varies from 0 to ∞; θ varies from 0 to $\frac{\pi}{2}$.

$$\therefore \quad \left[\Gamma\left(\frac{1}{2}\right)\right]^2 = 4\int_0^{\frac{\pi}{2}} \int_0^\infty e^{-r^2} r\, dr\, d\theta$$

$$= 4\int_0^{\frac{\pi}{2}} \left[-\frac{1}{2}e^{-r^2}\right]_0^\infty d\theta = 2\int_0^{\frac{\pi}{2}} d\theta = \pi$$

Hence, $$\boxed{\Gamma\left(\frac{1}{2}\right) = \sqrt{\pi}}$$ ...(6)

(d) To show that $\Gamma(0) = \infty$

From (2), we have $\Gamma(n) = \dfrac{\Gamma(n+1)}{n}$

As $n \to 0$, $\Gamma(0) = \lim\limits_{n\to 0}\dfrac{\Gamma(1)}{n} = \lim\limits_{n\to 0}\dfrac{1}{n} = \infty$

Thus, $$\boxed{\Gamma(0) = \infty}$$ ...(7)

Further note that $\Gamma(-1), \Gamma(-2), \Gamma(-3)$ etc. are also undefined. Hence Gamma function is continuous for any $n > 0$ and is discontinuous at $n = 0, -1, -2, \ldots$.

Thus, $\Gamma(n)$ is defined for all n, except for zero and negative integers.

(e) To show that $\Gamma(n+1) = (m+1)^{n+1}(-1)^n \int_0^1 x^m (\log x)^n \, dx$

where n is a positive integer and $m > -1$.

We have $\int_0^1 x^m (\log x)^n \, dx$

Put $x = e^{-y}$ then $dx = -e^{-y}\,dy = -x\,dy$

$$\therefore \quad \int_0^1 x^m (\log x)^n \, dx = \int_0^\infty e^{-my}(-y)^n e^{-y} dy = (-1)^n \int_0^\infty y^n e^{-(m+1)y} dy$$

Put $(m+1)y = u$ so, $dy = \dfrac{du}{m+1}$

$$\therefore \quad \int_0^1 x^m (\log x)^n \, dx = (-1)^n \int_0^\infty \frac{u^n}{(m+1)^n}\cdot e^{-u}\cdot\frac{du}{(m+1)}$$

$$= \frac{(-1)^n}{(m+1)^{n+1}}\int_0^\infty e^{-u}\cdot u^{(n+1)-1} du$$

$$= \frac{(-1)^n}{(m+1)^{n+1}}\Gamma(n+1) \qquad \text{[by (1)]}$$

Thus, $$\boxed{\Gamma(n+1) = (m+1)^{n+1}(-1)^n \int_0^1 x^m (\log x)^n \, dx}$$ ...(8)

where n is a positive integer and $m > -1$.

7.12 Simple Transformations of Gamma Function

(i) To prove that $\frac{\Gamma(n)}{k^n} = \int_0^\infty e^{-ky} y^{n-1} dy$

By definition of Gamma function, we have

$$\Gamma(n) = \int_0^\infty e^{-x} x^{n-1} dx \ , n > 0 \quad ...(1)$$

Putting $x = ky$, so that $dx = kdy$

also, when $x = 0, y = 0$ and when $x \to \infty, y \to \infty$

$$\therefore \quad \Gamma(n) = \int_0^\infty e^{-ky} (ky)^{n-1} kdy$$

$$= k^n \int_0^\infty e^{-ky} y^{n-1} dy$$

$$\Rightarrow \quad \boxed{\frac{\Gamma(n)}{k^n} = \int_0^\infty e^{-ky} y^{n-1} dy}$$

(ii) To prove that $\Gamma(n) = 2\int_0^\infty e^{-x^2} x^{2n-1} dx$

Put $x = y^2$ in (1), so that $dx = 2y\, dy$ and

$$\Gamma(n) = \int_0^\infty e^{-y^2} (y^2)^{n-1} 2y\, dy$$

$$= 2\int_0^\infty e^{-y^2} y^{2n-1} dy$$

$$= 2\int_0^\infty e^{-x^2} x^{2n-1} dx \qquad \left[\text{since } \int_a^b f(y)dy = \int_a^b f(x)dx\right]$$

Thus, $\Gamma(n) = 2\int_0^\infty e^{-x^2} x^{2n-1} dx$

(iii) To prove that $n\Gamma(n) = \int_0^\infty e^{-x^{\frac{1}{n}}} dx$, $n > 0$.

Put $x = y^{\frac{1}{n}} \Rightarrow x^n = y \Rightarrow nx^{n-1} dx = dy$

then from (1), we obtain

$$\Gamma(n) = \int_0^\infty e^{-y^{\frac{1}{n}}} \cdot \left(y^{\frac{1}{n}}\right)^{n-1} \frac{dy}{ny^{1-\frac{1}{n}}}$$

$$= \frac{1}{n}\int_0^\infty e^{-y^{\frac{1}{n}}}\,dy = \frac{1}{n}\int_0^\infty e^{-x^{\frac{1}{n}}}\,dx$$

Thus $$\boxed{n\Gamma(n) = \int_0^\infty e^{-x^{\frac{1}{n}}}\,dx}$$

(iv) To prove that $\Gamma(n) = \int_0^1 \left(\log\frac{1}{x}\right)^{n-1} dx,\ n > 0$

Put $$x = \log\frac{1}{y} \Rightarrow y = e^{-x} \Rightarrow dy = -e^{-x}dx$$

$$\Rightarrow \quad dy = -y\,dx$$

Then from (1), we obtain

$$\Gamma(n) = -\int_1^0 e^{-\left(\log\frac{1}{y}\right)}\cdot\left(\log\frac{1}{y}\right)^{n-1}\frac{dy}{y}$$

$$= \int_0^1 \left(\log\frac{1}{y}\right)^{n-1} dy$$

Thus $$\boxed{\Gamma(n) = \int_0^1 \left(\log\frac{1}{x}\right)^{n-1} dx}$$

Illustrative Examples

Example 1

Evaluate $\int_0^\infty x^{\frac{1}{4}} e^{-\sqrt{x}}\,dx$

Solution:

We have $$I = \int_0^\infty e^{-\sqrt{x}} x^{\frac{1}{4}}\,dx$$

Now put $$\sqrt{x} = u \Rightarrow x = u^2 \Rightarrow dx = 2u\,du$$

$$\therefore \quad I = \int_0^\infty e^{-u}\left(u^2\right)^{\frac{1}{4}} 2u\,du = 2\int_0^\infty e^{-u}u^{\frac{3}{2}}\,du$$

$$= = 2\int_0^\infty e^{-u} u^{\frac{5}{2}-1} du = 2\Gamma\left(\frac{5}{2}\right)$$

$$= 2\cdot\frac{3}{2}\Gamma\left(\frac{3}{2}\right) = 2\cdot\frac{3}{2}\cdot\frac{1}{2}\Gamma\left(\frac{1}{2}\right) \qquad [\because \Gamma(n+1) = n\Gamma(n)]$$

$$= \frac{3}{2}\sqrt{\pi} \qquad \text{Answer} \qquad \left[\because \Gamma\left(\frac{1}{2}\right) = \sqrt{\pi}\right]$$

Example 2

Evaluate $\int_0^\infty x^{n-1} e^{-h^2x^2} dx$

Solution:

We have $\qquad I = \int_0^\infty x^{n-1} e^{-h^2x^2} dx$

Putting $\qquad h^2 x^2 = y$, so that $h^2 2x dx = dy$

$$\therefore \qquad dx = \frac{1}{2}\frac{dy}{h^2 x} = \frac{1}{2}\frac{dy}{h\sqrt{y}}$$

$$\therefore \qquad I = \int_0^\infty e^{-y} \cdot \left(\frac{\sqrt{y}}{h}\right)^{n-1} \frac{1}{2}\frac{dy}{h\sqrt{y}}$$

$$= \frac{1}{2h^n}\int_0^\infty e^{-y} y^{\frac{n-2}{2}} dy$$

$$= \frac{1}{2h^n}\int_0^\infty e^{-y} y^{\frac{n}{2}-1} dy$$

$$= \frac{1}{2h^n}\Gamma\left(\frac{n}{2}\right) \qquad \text{Amswer} \qquad \text{[By definition of Gamma function]}$$

Example 3

Evaluate $\int_0^\infty x^6 e^{-2x} dx$

Solution :

We have $\qquad I = \int_0^\infty e^{-2x} x^6 dx$

Putting $\qquad 2x = y \Rightarrow dx = \frac{dy}{2}$, then

$$I = \int_0^\infty e^{-y} \left(\frac{y}{2}\right)^6 \tfrac{1}{2} dy = \frac{1}{2^7}\int_0^\infty e^{-y} y^{7-1}\, dy \quad = \frac{1}{2^7}\Gamma(7)$$

$$\Rightarrow \qquad \sqrt{\pi} = 2\int_0^\infty e^{-y^2} dy$$

$$= \frac{1}{2^7}\Gamma(7) \qquad \text{[By definition]}$$

$$= \frac{1}{2^7} 6! = \frac{4^5}{8} \qquad \text{Answer} \qquad \left[\because \Gamma(n+1) = n!, n > 0\right]$$

Example 4

Prove that $\int_0^\infty e^{(2ax - x^2)} dx = \frac{1}{2}\sqrt{\pi} \cdot e^{a^2}$

Solution:

We have

$$I = \int_0^\infty e^{(2ax - x^2)} dx = \int_0^\infty e^{a^2 - (x^2 - 2ax + a^2)} dx$$

$$= \int_0^\infty e^{a^2 - (x-a)^2} dx = e^{a^2} \int_0^\infty e^{-(x-a)^2} dx$$

Putting $(x - a) = y \Rightarrow dx = dy$

$$\therefore \qquad I = e^{a^2} \int_0^\infty e^{-y^2} dy \qquad \text{...(1)}$$

Now by the definition of gamma function

$$\Gamma(n) = \int_0^\infty e^{-u} u^{n-1} du \;,\; n > 0$$

Put $n = \frac{1}{2}$, we obtain

$$\Gamma\left(\frac{1}{2}\right) = \int_0^\infty e^{-u} u^{\frac{-1}{2}} du$$

Now Put $u = y^2 \Rightarrow du = 2y\, dy$, we get

$$\Gamma\left(\frac{1}{2}\right) = \int_0^\infty e^{-y^2} (y^2)^{\frac{-1}{2}} \cdot 2y dy \qquad \left[\because \Gamma\left(\frac{1}{2}\right) = \sqrt{\pi}\right]$$

$$= \sqrt{\pi} = 2\int_0^\infty e^{-y^2} dy$$

$$\Rightarrow \qquad \int_0^\infty e^{-y^2}dy = \frac{\sqrt{\pi}}{2} \qquad \text{...(2)}$$

using (2), (1) becomes

$$I = e^{a^2}\cdot\frac{\sqrt{\pi}}{2} = \frac{1}{2}\sqrt{\pi}\cdot e^{a^2} \qquad \text{Proved}$$

Example 5

Evaluate $\int_0^\infty x^{\frac{-3}{2}}\left(1-e^{-x}\right)dx$

Solution:

Let
$$I = \int_0^\infty x^{\frac{-3}{2}}\left(1-e^{-x}\right)dx$$

$$\Rightarrow \qquad I = \left[\left(1-e^{-x}\right)\frac{x^{\frac{-1}{2}}}{\left(\frac{-1}{2}\right)}\right]_0^\infty - \int_0^\infty e^{-x}\cdot\frac{x^{\frac{-1}{2}}}{\left(\frac{-1}{2}\right)}dx$$

$$= 0 + 2\int_0^\infty e^{-x}x^{\frac{-1}{2}}dx$$

$$= 2\int_0^\infty e^{-x}x^{\frac{1}{2}-1}dx = 2\Gamma\left(\frac{1}{2}\right) \qquad \text{Answer} \qquad \text{[By definition]}$$

$$= 2\sqrt{\pi}$$

Example 6

Prove that $\int_0^\infty \frac{x^a}{a^x}dx = \frac{\Gamma(a+1)}{(\log a)^{a+1}}, \; a > 0$

Solution:

Let
$$I = \int_0^\infty \frac{x^a}{a^x}dx = \int_0^\infty x^a a^{-x}dx$$

$$= \int_0^\infty x^a\left[e^{\log_e a}\right]^{-x}dx \qquad \left[\because a = e^{\log_e a}\right]$$

$$= \int_0^\infty e^{-x\log_e a}x^{(a+1)-1}dx$$

$$= \frac{\Gamma(a+1)}{(\log_e a)^{a+1}} \qquad \text{Proved} \qquad \left[\because \int_0^\infty e^{-kx}x^{n-1}dx = \frac{\Gamma(n)}{k^n}\right]$$

Example 7

Evaluate using gamma function, $\int_0^\infty a^{-bx^2} dx$

Solution:

We have $\quad I = \int_0^\infty a^{-bx^2} dx$

$$= \int_0^\infty \left(e^{\log_e a}\right)^{-bx^2} dx \qquad [\because a = e^{\log_e a}]$$

$$= \int_0^\infty e^{-bx^2 \log_e a} dx \qquad ...(1)$$

Putting $\quad (b \log a)x^2 = y$, so that

$$dx = \frac{dy}{2\sqrt{(b \log a) y}}$$

From (1),

$$I = \int_0^\infty e^{-y} \cdot \frac{dy}{2\sqrt{(b \log a)} \cdot \sqrt{y}} = \frac{1}{2\sqrt{b \log a}} \int_0^\infty e^{-y} \cdot y^{\frac{1}{2}-1} dy$$

$$= \frac{1}{2\sqrt{b \log a}} \Gamma\left(\frac{1}{2}\right) \qquad \text{[By definition]}$$

$$= \frac{\sqrt{\pi}}{2\sqrt{b \log a}}$$

Example 8

Evaluate $\int_1^0 x^5 \left(\log \frac{1}{x}\right)^3 dx$ using gamma function.

Solution:

Let $\quad I = \int_1^0 x^5 \left(\log \frac{1}{x}\right)^3 dx$

Putting $\quad x = e^{\frac{-y}{5}} \Rightarrow \log\left(\frac{1}{x}\right) = \frac{y}{5}$

$$\therefore \quad I = \int_0^\infty e^{-y} \left(\frac{y}{5}\right)^3 dy = \frac{1}{125} \int_0^\infty e^{-y} y^{4-1} dy$$

$$= \frac{1}{125}\Gamma(4) = \frac{3!}{125} = \frac{6}{125} \quad \text{Answer}$$

Example 9

With certain limitation on the value of *a, b, m* and *n*, show that

$$\int_0^\infty \int_0^\infty e^{-(ax^2+by^2)} x^{2m-1} y^{2n-1} dxdy = \frac{\Gamma(m)\Gamma(n)}{4a^m b^n}$$

Solution: Given that

$$I = \int_0^\infty \int_0^\infty e^{-(ax^2+by^2)} x^{2m-1} y^{2n-1} dxdy$$

$$= \int_0^\infty e^{-ax^2} x^{2m-1} dx \times \int_0^\infty e^{-by^2} y^{2n-1} dy$$

$$= I_1 \times I_2 \quad \ldots(1)$$

Where $I_1 = \int_0^\infty e^{-ax^2} x^{2m-1} dx$ and $I_2 = \int_0^\infty e^{-by^2} y^{2n-1} dy$

To calculate I_1, we put $ax^2 = u$

$$\Rightarrow \quad x = \left(\frac{u}{a}\right)^{\frac{1}{2}}$$

$$\Rightarrow \quad dx = \frac{du}{2\sqrt{au}}$$

$$\therefore \quad I_1 = \int_0^\infty e^{-u} \left(\frac{u}{a}\right)^{\frac{2m-1}{2}} \frac{du}{2\sqrt{au}}$$

$$= \frac{1}{2a^m} \int_0^\infty e^{-u} u^{m-1} du = \frac{1}{2a^m}\Gamma(m), \; m > 0, a > 0 \quad \ldots(2)$$

Similarly, $$I_2 = \frac{1}{2b^n}\Gamma(n) \; ; \; n > 0, b > 0 \quad \ldots(3)$$

Using (2) and (3), (1) becomes

$$I = \frac{\Gamma(m)}{2a^m} \times \frac{\Gamma(n)}{2b^n}$$

$$\Rightarrow \quad I = \frac{\Gamma(m)\Gamma(n)}{4a^m b^n} \quad \text{Hence proved}$$

Where $a > 0, b > 0, m > 0, n > 0$

7.13 Beta Function

The Beta function is denoted and defined as

$$B(m,n) = \int_0^1 x^{m-1}(1-x)^{n-1}\,dx \qquad \text{...(1)}$$

Where *m,n* are positive numbers, integer or fractional. This is also known as Eulerian integral of first kind.

7.14. Simple Properties of Beta Function

(i) To show that: $B(m,n) = B(n,m)$ **[Symmetry]**

We have $B(m,n) = \int_0^1 x^{m-1}(1-x)^{n-1}\,dx$

$$= \int_0^1 (1-x)^{m-1}[1-(1-x)]^{n-1}\,dx \qquad \left[\because \int_0^a f(x)dx = \int_0^a f(a-x)dx\right]$$

$$= \int_0^1 (1-x)^{m-1} x^{n-1}\,dx = \int_0^1 x^{n-1}(1-x)^{m-1}\,dx$$

$$= B(n,m)$$

Thus, $B(m,n) = B(n,m)$

(ii) To show that: $\int_0^a x^{m-1}(a-x)^{n-1}\,dx = a^{m+n-1}B(m,n)$.

From (1), we have $\int_0^1 x^{m-1}(1-x)^{n-1}\,dx = B(m,n)$

Putting $x = \dfrac{y}{a} \Rightarrow dx = \dfrac{dy}{a}$ and y varies from 0 to a.

$$\int_0^a \left(\frac{y}{a}\right)^{m-1}\left(1-\frac{y}{a}\right)^{n-1}\frac{dy}{a} = B(m,n)$$

$$\Rightarrow \frac{1}{a^{m+n-1}}\int_0^a y^{m-1}(a-y)^{n-1}\,dy = B(m,n)$$

$$\Rightarrow \boxed{\int_0^a x^{m-1}(a-x)^{n-1}\,dx = a^{m+n-1}B(m,n)}$$

(iii) To show that $B(m,n) = \int_0^1 \dfrac{x^{m-1}+x^{n-1}}{(1+x)^{m+n}}\,dx$

We know that $B(m,n) = \int_0^1 x^{m-1}(1-x)^{n-1}\,dx$

Putting $x = \dfrac{1}{y+1}$ so that $dx = -\dfrac{dy}{(1+y)^2}$ and y varies from ∞ to 0.

$$\therefore \quad B(m, n) = \int_{\infty}^{0}\left(\frac{1}{1+y}\right)^{m-1}\left[1-\frac{1}{1+y}\right]^{n-1}\cdot(-)\frac{dy}{(1+y)^2}$$

$$= \int_0^{\infty}\frac{1}{(1+y)^{m-1}}\cdot\frac{y^{n-1}}{(1+y)^{n-1}}\cdot\frac{dy}{(1+y)^2}$$

$$\Rightarrow \quad \boxed{B(m,n) = \int_0^{\infty}\frac{y^{n-1}}{(1+y)^{m+n}}dy} \quad ...(1)$$

(In this integral m and n may be changed, by the virtue of symmetry of the function:)

Again, (1) may be written as

$$B(m, n) = \int_0^{\infty}\frac{y^{n-1}}{(1+y)^{m+n}}dy = \int_0^{1}\frac{y^{n-1}}{(1+y)^{m+n}}dy + \int_1^{\infty}\frac{y^{n-1}}{(1+y)^{m+n}}dy \quad ...(2)$$

In order to solve second integral on R.H.S., put

$$y = \frac{1}{x}, \text{ so that } dy = -\frac{1}{x^2}dx \text{ and } x \text{ varies from 1 to 0.}$$

$$\therefore \quad \int_1^{\infty}\frac{y^{n-1}}{(1+y)^{m+n}}dy = \int_1^{0}\left(\frac{1}{x}\right)^{n-1}\frac{x^{m+n}}{(1+x)^{m+n}}\left(-\frac{1}{x^2}\right)dx = \int_0^{1}\frac{x^{m-1}}{(1+x)^{m+n}}dx \quad ...(3)$$

using (3), (2) becomes

$$B(m, n) = \int_0^{1}\frac{y^{n-1}}{(1+y)^{m+n}}dy + \int_0^{1}\frac{x^{m-1}}{(1+x)^{m+n}}dx$$

$$= \int_0^{1}\frac{x^{n-1}}{(1+x)^{m+n}}dx + \int_0^{1}\frac{x^{m-1}}{(1+x)^{m+n}}dx \quad \left[\because \int_0^{a}f(x)dx = \int_0^{a}f(y)dy\right]$$

$$= \int_0^{1}\frac{x^{m-1}+x^{n-1}}{(1+x)^{m+n}}dx$$

Hence, $\boxed{B(m,n) = \int_0^{1}\frac{x^{m-1}+x^{n-1}}{(1+x)^{m+n}}dx}$

(iv) To show that $B(m, n) = 2\int_0^{\frac{\pi}{2}}\sin^{2m-1}\theta\cos^{2n-1}\theta d\theta$

We have $B(m, n) = \int_0^{1}x^{m-1}(1-x)^{n-1}dx$

Putting $\quad x = \sin^2\theta \Rightarrow dx = 2\sin\theta\cos\theta d\theta$ and θ varies from 0 to $\frac{\pi}{2}$

$$\therefore \quad B(m, n) = \int_0^{\frac{\pi}{2}} (\sin^2\theta)^{m-1}(1-\sin^2\theta)^{n-1} 2\sin\theta\cos\theta d\theta$$

$$= 2\int_0^{\frac{\pi}{2}} \sin^{2m-1}\theta\cos^{2n-1}\theta d\theta$$

Thus $\quad \boxed{B(m,n) = 2\int_0^{\frac{\pi}{2}} \sin^{2m-1}\theta \cdot \cos^{2n-1}\theta \cdot d\theta}$

Particular case:

When $m = \frac{1}{2}, n = \frac{1}{2}$, above expression gives,

$$B\left(\frac{1}{2}, \frac{1}{2}\right) = 2\int_0^{\frac{\pi}{2}} d\theta = 2\cdot\frac{\pi}{2} = \pi$$

(v) Show that $(a-b)^{m+n-1} B(m, n) = \int_b^a (x-b)^{m-1}(a-x)^{n-1} dx$

We know that $\quad B(m, n) = \int_0^1 y^{m-1}(1-y)^{n-1} dy$

Let $\quad y = \frac{x-b}{a-b} \Rightarrow dy = \frac{dx}{a-b}$ and y varies from b to a.

$$\therefore \quad B(m, n) = \int_b^a \left(\frac{x-b}{a-b}\right)^{m-1}\left(1-\frac{x-b}{a-b}\right)^{n-1} \frac{dx}{a-b}$$

$$= \frac{1}{(a-b)^{m+n-1}} \int_b^a (x-b)^{m-1}(a-x)^{n-1} dx$$

$$\therefore \quad \int_b^a (x-b)^{m-1}(a-x)^{n-1} dx = (a-b)^{m+n-1} B(m,n)$$

Thus, $\boxed{\int_b^a (x-b)^{m-1}(a-x)^{n-1} dx = (a-b)^{m+n-1} B(m,n)}$

7.15 Relation between Beta and Gamma Function

To prove that $\quad B(m, n) = \frac{\Gamma(m)\Gamma(n)}{\Gamma(m+n)}, m > 0, n > 0$

Proof :

we know that

$$\frac{\Gamma(n)}{z^n} = \int_0^\infty e^{-zy}y^{n-1}dy \qquad ...(1)$$

or,

$$\Gamma(n) = \int_0^\infty z^n e^{-zy}y^{n-1}dy \qquad ...(2)$$

Also, $\Gamma(m) = \int_0^\infty e^{-y}y^{m-1}dy \qquad ...(3)$

Now multiply (2) by z^{m-1} e^{-z} on both sides, we get

$$\Gamma(n)\cdot e^{-z}\cdot z^{m-1} = \int_0^\infty z^{m+n-1}e^{-zy}y^{n-1}e^{-z}dy$$

$$= \int_0^\infty z^{m+n-1}e^{-(y+1)z}y^{n-1}dy$$

Integrating both the sides w.r.t. z from 0 to ∞, we get

$$\Gamma(n)\int_0^\infty e^{-z}z^{m-1}dz = \int_0^\infty\left[\int_0^\infty z^{m+n-1}e^{-(y+1)z}dz\right]y^{n-1}dy$$

$$\Rightarrow \qquad \Gamma(n)\Gamma(m) = \int_0^\infty \frac{\Gamma(m+n)}{(y+1)^{m+n}}\cdot y^{n-1}dy \qquad \text{[by the property (1) and (3)]}$$

$$\Gamma(m)\Gamma(n) = \Gamma(m+n)\int_0^\infty \frac{y^{n-1}}{(1+y)^{m+n}}dy$$

$$= \Gamma(m+n)\cdot B(m,n) \qquad \text{[By (1) of (iii) of 13]}$$

Thus,

$$\boxed{B(m,n) = \frac{\Gamma(m)\Gamma(n)}{\Gamma(m+n)}},\ m > 0,\ n > 0$$

Deduction (i) $\qquad \Gamma(n)\Gamma(1-n) = \dfrac{\pi}{\sin n\pi}$ where $0 < n < 1$

Proof:

We know that

$$\frac{\Gamma(m)\Gamma(n)}{\Gamma(m+n)} = B(m,n) = \int_0^\infty \frac{y^{n-1}}{(1+y)^{m+n}}dy$$

choose $\qquad m + n = 1$, so that $m = (1-n)$

$$\therefore \qquad \frac{\Gamma(1-n)\Gamma(n)}{\Gamma(1)} = \int_0^\infty \frac{y^{n-1}}{1+y}dy,\ 0 < n < 1$$

$$\Rightarrow \quad \boxed{\Gamma(n)\Gamma(1-n) = \frac{\pi}{\sin n\pi}} \qquad \left[\because \Gamma(1) = 1, \int_0^\infty \frac{y^{n-1}}{1+y}dy = \frac{\pi}{\sin n\pi}\right]$$

Where $0 < n < 1$

Deduction(ii)To prove that $\Gamma\left(\frac{1}{2}\right) = \sqrt{\pi}$

Proof:

Since $\Gamma(n)\Gamma(1-n) = \frac{\pi}{\sin n\pi}$

put $n = \frac{1}{2}$, we obtain

$$\Gamma\left(\frac{1}{2}\right)\Gamma\left(\frac{1}{2}\right) = \frac{\pi}{\sin\frac{\pi}{2}}$$

$$\Rightarrow \quad \left[\Gamma\left(\frac{1}{2}\right)\right]^2 = \pi$$

$$\Rightarrow \quad \Gamma\left(\frac{1}{2}\right) = \sqrt{\pi}$$

Deduction(iii) To prove that

$$\int_0^{\frac{\pi}{2}} \cos^m\theta \sin^n\theta \, d\theta = \frac{\Gamma\left(\frac{m+1}{2}\right)\Gamma\left(\frac{n+1}{2}\right)}{2\Gamma\left(\frac{m+n+2}{2}\right)}, m > -1, n > -1$$

Proof:

Let $I = \int_0^{\frac{\pi}{2}} \cos^m\theta \sin^n\theta \, d\theta$

put $\sin^2\theta = x$

$$\Rightarrow \quad 2\sin\theta\cos\theta \, d\theta = dx$$

$$\Rightarrow \quad d\theta = \frac{dx}{2\sqrt{x}\cdot\sqrt{1-x}}$$

also, when $\theta = 0, x = 0$ and when $\theta = \frac{\pi}{2}, x = 1$

$$\therefore \quad I = \int_0^1 (1-x)^{\frac{m}{2}} \cdot x^{\frac{n}{2}} \frac{dx}{2\sqrt{x}\cdot\sqrt{(1-x)}}$$

$$= \frac{1}{2}\int_0^1 x^{\frac{n-1}{2}} \cdot (1-x)^{\frac{m-1}{2}}\, dx$$

$$= \frac{1}{2}\int_0^1 x^{\frac{n+1}{2}-1} (1-x)^{\frac{m+1}{2}-1}\, dx$$

$$= \frac{1}{2} B\left(\frac{n+1}{2}, \frac{m+1}{2}\right)$$

$$= \frac{1}{2} \frac{\Gamma\left(\frac{m+1}{2}\right)\Gamma\left(\frac{n+1}{2}\right)}{\Gamma\left(\frac{m+n+2}{2}\right)} \qquad \left[\because B(m,n) = \frac{\Gamma(m)\Gamma(n)}{\Gamma(m+n)}\right]$$

Illustrative Examples

Example 1

Evaluate the integral $\int_0^1 x^4(1-\sqrt{x})^5 dx$

Solution:

Let $\sqrt{x} = y \Rightarrow x = y^2 \Rightarrow dx = 2y\, dy$

$\therefore$ given integral becomes

$$\int_0^1 y^8(1-y)^5 \cdot 2y dy = 2\int_0^1 y^{(10-1)}(1-y)^{6-1} dy$$

$$= 2\, B(10, 6) = 2\frac{\Gamma(10)\Gamma(6)}{\Gamma(16)} = \frac{2 \cdot 9!5!}{15!} = \frac{1}{15015} \quad \text{Answer}$$

Example2

Evaluate $\int_0^1 (1-x^3)^{\frac{-1}{2}} dx$

Solution:

Let $x^3 = y \Rightarrow x = y^{1/3} \Rightarrow dx = \frac{1}{3} y^{\frac{-2}{3}} dy$

$\therefore$ given integral becomes

$$\int_0^1 (1-x^3)^{\frac{-1}{2}} dx = \int_0^1 (1-y)^{\frac{-1}{2}} \cdot \frac{1}{3} y^{\frac{-2}{3}} dy$$

$$= \frac{1}{3}\int_0^1 y^{\frac{1}{3}-1}(1-y)^{\frac{1}{2}-1} dy = \frac{1}{3}B\left(\frac{1}{3},\frac{1}{2}\right) = \frac{1}{3}\frac{\Gamma\left(\frac{1}{3}\right)\Gamma\left(\frac{1}{2}\right)}{\Gamma\left(\frac{5}{6}\right)} \quad \text{Answer}$$

Example 3

Express $\int_0^1 y^m(1-y^p)^n dy$ in terms of Beta function and hence evaluate $\int_0^1 y^5(1-y^3)^{10} dy$.

Solution:

Let $y^p = z \Rightarrow y = z^{1/p} \Rightarrow dy = \frac{1}{p} z^{\frac{1}{p}-1} dz$

$$\therefore \quad \int_0^1 y^m(1-y^p)^n dy = \int_0^1 z^{\frac{m}{p}}(1-z)^n \cdot \frac{1}{p} z^{\frac{1}{p}-1} dz$$

$$= \frac{1}{p}\int_0^1 z^{\frac{m+1}{p}-1}(1-z)^{\overline{n+1}-1} dz = \frac{1}{p}B\left(\frac{m+1}{p}, n+1\right) \quad \text{...(1)}$$

Putting $m = 5, p = 3, n = 10$ in (1), we obtain

$$\int_0^1 y^5(1-y^3)^{10} dy = \frac{1}{3}B(2,11)$$

$$= \frac{1}{3}\frac{\Gamma(2)\Gamma(11)}{\Gamma(13)} = \frac{1}{3}\frac{1.\Gamma(1)\Gamma(11)}{12\cdot 11\cdot \Gamma(11)}$$

$$= \frac{1}{3\times 12\times 11} \quad \left[\because \Gamma(n+1) = n\Gamma(n)\right]$$

$$= \frac{1}{396} \quad \text{Answer}$$

Example 4

Evaluate $\int_0^\infty \frac{x^8(1-x^6)}{(1+x)^{24}} dx$

Solution :

Let $I = \int_0^\infty \frac{x^8(1-x^6)}{(1+x)^{24}} dx = \int_0^\infty \frac{x^8}{(1+x)^{24}} dx - \int_0^\infty \frac{x^{14}}{(1+x)^{24}} dx$

$$= \int_0^\infty \frac{x^{9-1}}{(1+x)^{9+15}} dx - \int_0^\infty \frac{x^{15-1}}{(1+x)^{15+9}} dx$$

$$= B(9, 15) - B(15, 9)$$

$$= 0 \qquad \text{Answer} \qquad [\because B(m,n) = B(n,m)]$$

Example 5

Prove that $\int_0^{\frac{\pi}{2}} \left(1 - \frac{\sin^2\theta}{2}\right)^{\frac{-1}{2}} d\theta = \frac{\left[\Gamma\left(\frac{1}{4}\right)\right]^2}{4\sqrt{\pi}}$

Solution:

Let $I = \int_0^{\frac{\pi}{2}} \left(1 - \frac{\sin^2\theta}{2}\right)^{\frac{-1}{2}} d\theta = \sqrt{2}\int_0^{\frac{\pi}{2}} \frac{d\theta}{\sqrt{2-\sin^2\theta}}$

$$= \sqrt{2}\int_0^{\frac{\pi}{2}} \frac{d\theta}{\sqrt{1+\cos^2\theta}}$$

Put $\cos\theta = t \Rightarrow -\sin\theta d\theta = dt \Rightarrow d\theta = \frac{dt}{-\sqrt{1-t^2}}$

$\therefore$ $I = -\sqrt{2}\int_1^0 \frac{dt}{\sqrt{1+t^2}\cdot\sqrt{1-t^2}} = \sqrt{2}\int_0^1 \frac{dt}{\sqrt{1-t^4}}$

Again put $t^2 = \sin\theta \Rightarrow 2t\,dt = \cos\theta d\theta$

$\therefore$ $I = \sqrt{2}\int_0^{\frac{\pi}{2}} \frac{\cos\theta d\theta}{2\sqrt{\sin\theta}\cdot\sqrt{1-\sin^2\theta}} = \frac{1}{\sqrt{2}}\int_0^{\frac{\pi}{2}} \sin^{\frac{-1}{2}}\theta\cos^0\theta d\theta$

$$= \frac{1}{2\sqrt{2}} \frac{\Gamma\left(\frac{1}{4}\right)\Gamma\left(\frac{1}{2}\right)}{\Gamma\left(\frac{3}{4}\right)} = \frac{1}{2\sqrt{2}} \frac{\left[\Gamma\left(\frac{1}{4}\right)\right]^2 \pi}{\Gamma\left(\frac{1}{4}\right)\Gamma\left(\frac{3}{4}\right)} \qquad \ldots(1)$$

Since $\Gamma\left(\frac{1}{4}\right)\Gamma\left(\frac{3}{4}\right) = \Gamma\left(\frac{1}{4}\right)\Gamma\left(1-\frac{1}{4}\right) = \frac{\pi}{\sin\frac{\pi}{4}} = \pi\sqrt{2}$

$\therefore$ (1) gives, $I = \frac{1}{2\sqrt{2}} \frac{\left[\Gamma\left(\frac{1}{4}\right)\right]^2 \sqrt{\pi}}{\pi\sqrt{2}} = \frac{\left[\Gamma\left(\frac{1}{4}\right)\right]^2}{4\sqrt{\pi}}$ Proved

Example 6

Evaluate $\int_0^\infty \frac{xdx}{1+x^6}$

Solution :

Let $I = \int_0^\infty \frac{xdx}{1+x^6}$

Putting $x^6 = y \Rightarrow x = y^{1/6} \Rightarrow dx = \frac{1}{6} y^{\frac{-5}{6}} dy$

$$\therefore \quad I = \int_0^\infty \frac{y^{\frac{1}{6}}}{(1+y)} \frac{1}{6} y^{\frac{-5}{6}} dy = \frac{1}{6}\int_0^\infty \frac{y^{\frac{-2}{3}}}{(1+y)} dy$$

$$= \frac{1}{6}\int_0^\infty \frac{y^{\frac{1}{3}-1}}{(1+y)^{\frac{1}{3}+\frac{2}{3}}} dy = \frac{1}{6} B\left(\frac{1}{3}, \frac{2}{3}\right)$$

$$= \frac{1}{6} \frac{\Gamma\left(\frac{1}{3}\right)\Gamma\left(\frac{2}{3}\right)}{\Gamma\left(\frac{1}{3}+\frac{2}{3}\right)} = \frac{1}{6} \frac{\Gamma\left(\frac{1}{3}\right)\Gamma\left(1-\frac{1}{3}\right)}{\Gamma(1)}$$

$$= \frac{1}{6} \frac{\pi}{\sin\frac{\pi}{3}} \qquad \left[\because \Gamma(n)\Gamma(1-n) = \frac{\pi}{\sin n\pi} \text{ and } \Gamma(1) = 1\right]$$

$$= \frac{\pi}{3\sqrt{3}}$$

Example 7

Prove that $\int_0^{\frac{\pi}{2}} \sqrt{\sin\theta} d\theta \times \int_0^{\frac{\pi}{2}} \frac{d\theta}{\sqrt{\sin\theta}} = \pi$

Solution:

We know that

$$\int_0^{\frac{\pi}{2}} \sin^n\theta \cos^0\theta d\theta = \frac{\Gamma\left(\frac{n+1}{2}\right)\Gamma\left(\frac{1}{2}\right)}{2\Gamma\left(\frac{n+2}{2}\right)} \qquad ...(1)$$

Now, let $$I = \int_0^{\frac{\pi}{2}} \sqrt{\sin\theta}\,d\theta \times \int_0^{\frac{\pi}{2}} \frac{d\theta}{\sqrt{\sin\theta}}$$

$$= \int_0^{\frac{\pi}{2}} \sin^{\frac{1}{2}}\theta\cos^0\theta\,d\theta \times \int_0^{\frac{\pi}{2}} \sin^{-\frac{1}{2}}\theta\cos^0\theta\,d\theta$$

$$= \frac{\Gamma\left(\frac{3}{4}\right)\Gamma\left(\frac{1}{2}\right)}{2\Gamma\left(\frac{5}{4}\right)} \times \frac{\Gamma\left(\frac{1}{4}\right)\Gamma\left(\frac{1}{2}\right)}{2\Gamma\left(\frac{3}{4}\right)} \qquad \text{[By (1)]}$$

$$= \frac{\pi\,\Gamma\left(\frac{1}{4}\right)}{4\,\Gamma\left(\frac{5}{4}\right)} = \frac{\pi\,\Gamma\left(\frac{1}{4}\right)}{4\cdot\frac{1}{4}\Gamma\left(\frac{1}{4}\right)} = \pi \qquad \text{Proved} \qquad \left[\because \Gamma(n+1) = n\Gamma(n)\right]$$

Example 8

Show that $\int_0^{\infty} \frac{x^2dx}{(1+x^4)^3} = \frac{5\pi\sqrt{2}}{128}$

Solution:

Put $$x = \sqrt{\tan\theta} \Rightarrow dx = \frac{1}{2\sqrt{\tan\theta}}\cdot\sec^2\theta\,d\theta$$

$$\therefore \int_0^{\infty} \frac{x^2dx}{(1+x^4)^3} = \int_0^{\frac{\pi}{2}} \frac{\tan\theta\cdot\frac{1}{2}(\tan\theta)^{\frac{-1}{2}}\sec^2\theta}{\left(1+\tan^2\theta\right)^3}d\theta$$

$$= \frac{1}{2}\int_0^{\frac{\pi}{2}} (\tan\theta)^{\frac{1}{2}}(\sec\theta)^{-4}\,d\theta = \frac{1}{2}\int_0^{\frac{\pi}{2}} \sin^{\frac{1}{2}}\theta\cos^{\frac{7}{2}}\theta\,d\theta$$

$$= \frac{1}{2}\cdot\frac{1}{2}\frac{\Gamma\left(\frac{3}{4}\right)\Gamma\left(\frac{9}{4}\right)}{\Gamma(3)}$$

$$= \frac{1}{4\cdot 2!}\Gamma\left(\frac{3}{4}\right)\cdot\frac{5}{4}\cdot\frac{1}{4}\Gamma\left(\frac{1}{4}\right)$$

$$= \frac{5}{128}\Gamma\left(\frac{1}{4}\right)\Gamma\left(1-\frac{1}{4}\right) = \frac{5}{128}\frac{\pi}{\sin\frac{\pi}{4}} = \frac{5\pi\sqrt{2}}{128} \qquad \text{Proved}$$

Example 9

Evaluate $\int_0^\infty e^{-x^{\frac{1}{3}}}dx$

Solution:

Let $x^{1/3} = y \Rightarrow x = y^3 \Rightarrow dx = 3y^2 dy$

$$\therefore \quad \int_0^\infty e^{-x^{\frac{1}{3}}}dx = \int_0^\infty e^{-y}\cdot 3y^2 dy$$

$$= 3\int_0^\infty e^{-y}y^{3-1}dy = 3\Gamma(3)$$

$$= 3\cdot 2! = 6 \quad \text{Answer}$$

7.16 Duplication Formula

To show that $\Gamma(p)\Gamma\left(p+\frac{1}{2}\right) = \frac{\sqrt{\pi}}{2^{2p-1}}\Gamma(2p)$ where $p > 0$

Proof:

We know that

$$\int_0^{\frac{\pi}{2}} \sin^m\theta\cos^n\theta d\theta = \frac{\Gamma\left(\frac{m+1}{2}\right)\Gamma\left(\frac{n+1}{2}\right)}{2\Gamma\left(\frac{m+n+2}{2}\right)} \quad \text{...(1)}$$

Now putting $n = m$ in (1), we obtain

$$\frac{\left[\Gamma\left(\frac{m+1}{2}\right)\right]^2}{2\Gamma(m+1)} = \int_0^{\frac{\pi}{2}}(\sin\theta\cos\theta)^m d\theta$$

$$= \frac{1}{2^m}\int_0^{\frac{\pi}{2}}(2\sin\theta\cos\theta)^m d\theta = \frac{1}{2^m}\int_0^{\frac{\pi}{2}}(\sin 2\theta)^m d\theta$$

Again putting $2\theta = \phi \Rightarrow d\theta = \frac{d\phi}{2}$, we get

$$\frac{\left[\Gamma\left(\frac{m+1}{2}\right)\right]^2}{2\Gamma(m+1)} = \frac{1}{2^m}\int_0^\pi \sin^m\phi\cdot\frac{d\phi}{2}$$

$$= \frac{1}{2^m}\int_0^{\frac{\pi}{2}} \sin^m \phi \cos^0 \phi \, d\phi$$

$$= \frac{1}{2^m}\frac{\Gamma\left(\frac{m+1}{2}\right)\Gamma\left(\frac{1}{2}\right)}{2\Gamma\left(\frac{m+2}{2}\right)} \qquad \text{[by (1)]}$$

$$\Rightarrow \qquad \frac{\Gamma\left(\frac{m+1}{2}\right)}{\Gamma(m+1)} = \frac{\sqrt{\pi}}{2^m}\frac{1}{\Gamma\left(\frac{m+2}{2}\right)}$$

Let ,

$$\frac{m+1}{2} = p \quad \Rightarrow \quad m = 2p-1, p > 0$$

$$\therefore \qquad \frac{\Gamma(p)}{\Gamma(2p)} = \frac{\sqrt{\pi}}{2^{2p-1}}\cdot\frac{1}{\Gamma\left(\frac{2p+1}{2}\right)}$$

$$\Rightarrow \qquad \boxed{\Gamma(p)\Gamma\left(p+\frac{1}{2}\right) = \frac{\sqrt{\pi}}{2^{2p-1}}\cdot\Gamma(2p)} \quad \text{where } p > 0$$

which is known as Duplication formula

Deduction (i) To prove that

$$2^m\Gamma\left(\frac{m+1}{2}\right)\Gamma\left(\frac{m+2}{2}\right) = \sqrt{\pi}\,\Gamma(m+1)$$

for all real values of m

Proof:

Put $2p-1 = m$ in Duplication formula, we get

$$\Gamma\left(\frac{m+1}{2}\right)\Gamma\left(\frac{m+2}{2}\right) = \frac{\sqrt{\pi}}{2^m}\Gamma(m+1)$$

$$\Rightarrow \qquad 2^m\Gamma\left(\frac{m+1}{2}\right)\Gamma\left(\frac{m+2}{2}\right) = \sqrt{\pi}\,\Gamma(m+1) \quad \text{(proved)}$$

Deduction (ii)To prove that

$$\Gamma\left(m+\frac{1}{2}\right) = \frac{(2m)!}{2^m . m!}\sqrt{\pi} \text{ , where } m > 0$$

Proof:

We have

$$\frac{\Gamma(2m)}{\Gamma(m)} = \frac{(2m-1)!}{(m-1)!} = \frac{2m\cdot(2m-1)!}{2m\cdot(m-1)!} = \frac{(2m)!}{2\cdot m!} \qquad ...(1)$$

Now using Duplication formula, we have

$$\Gamma\left(m+\frac{1}{2}\right) = \frac{\sqrt{\pi}}{2^{2m-1}}\frac{\Gamma(2m)}{\Gamma(m)}$$

$$= \frac{\sqrt{\pi}}{2^{2m-1}}\cdot\frac{(2m)!}{2\cdot(m)!} \qquad \text{[by (1)]}$$

$$= \frac{\sqrt{\pi}}{2^{m}}\frac{(2m)!}{m!} \qquad \text{Proved}$$

7.17 To Show That

$$\Gamma\left(\frac{1}{n}\right)\Gamma\left(\frac{2}{n}\right)\Gamma\left(\frac{3}{n}\right)...\Gamma\left(\frac{n-1}{n}\right) = \frac{(2\pi)^{\frac{(n-1)}{2}}}{\sqrt{n}}$$

where n is an integer greater than 1.

Proof: Let

$$I = \Gamma\left(\frac{1}{n}\right)\Gamma\left(\frac{2}{n}\right)\Gamma\left(\frac{3}{n}\right)...\Gamma\left(\frac{n-1}{n}\right) \qquad ...(1)$$

Writing (1) in reverse order, we have

$$I = \Gamma\left(\frac{n-1}{n}\right)\Gamma\left(\frac{n-2}{n}\right)...\Gamma\left(\frac{2}{n}\right)\cdot\Gamma\left(\frac{1}{n}\right)$$

or,

$$I = \Gamma\left(1-\frac{1}{n}\right)\Gamma\left(1-\frac{2}{n}\right)...\Gamma\left(1-\frac{n-2}{n}\right)\Gamma\left(1-\frac{n-1}{n}\right) \qquad ...(2)$$

Multiplying (1) and (2), we obtain

$$I^2 = \left\{\Gamma\left(\frac{1}{n}\right)\Gamma\left(1-\frac{1}{n}\right)\right\}\left\{\Gamma\left(\frac{2}{n}\right)\Gamma\left(1-\frac{2}{n}\right)\right\}...\left\{\Gamma\left(\frac{n-2}{n}\right)\Gamma\left(1-\frac{n-2}{n}\right)\right\}\cdot\left\{\Gamma\left(\frac{n-1}{n}\right)\Gamma\left(1-\frac{n-1}{n}\right)\right\}$$

$$= \frac{\pi}{\sin\frac{\pi}{n}}\cdot\frac{\pi}{\sin\frac{2\pi}{n}}...\frac{\pi}{\sin\frac{(n-2)\pi}{n}}\cdot\frac{\pi}{\sin\frac{(n-1)\pi}{n}} \qquad \left[\text{since}\,\Gamma(n)\Gamma(1-n) = \frac{\pi}{\sin n\pi}\right]$$

$$= \frac{\pi^{n-1}}{\sin\left(\frac{\pi}{n}\right)\sin\left(\frac{2\pi}{n}\right)...\sin\frac{(n-1)\pi}{n}} \qquad ...(3)$$

Now using trigonometrical identity,

$$\frac{\sin n\theta}{\sin\theta} = 2^{n-1}\sin\left(\theta+\frac{\pi}{n}\right)\sin\left(\theta+\frac{2\pi}{n}\right)...\sin\left(\theta+\frac{(n-1)\pi}{n}\right)$$

taking limit $\theta \to 0$, we obtain

$$n = 2^{n-1}\sin\left(\frac{\pi}{n}\right)\sin\left(\frac{2\pi}{n}\right)...\sin\frac{(n-1)\pi}{n}$$

$$\Rightarrow \sin\left(\frac{\pi}{n}\right)\sin\left(\frac{2\pi}{n}\right)...\sin\frac{(n-1)\pi}{n} = \frac{n}{2^{n-1}} \qquad ...(4)$$

Using (4), equation (3) becomes

$$I^2 = \frac{\pi^{n-1}\cdot 2^{n-1}}{n}$$

$$\Rightarrow \qquad I = \frac{(2\pi)^{\frac{n-1}{2}}}{\sqrt{n}} \qquad \text{(proved)}$$

Illustrative Examples

Example 1

Show that $B(m+1, n) = \frac{m}{m+n}B(m,n)$

Solution:

We know that $B(m, n) = \frac{\Gamma(m)\Gamma(n)}{\Gamma(m+n)}$

$$\therefore \quad B(m+1, n) = \frac{\Gamma(m+1)\Gamma(n)}{\Gamma(m+n+1)} = \frac{m\Gamma(m)\Gamma(n)}{(m+n)\cdot\Gamma(m+n)}$$

$$= \frac{m}{m+n}B(m,n)$$

Example 2

Prove that $B(n,n) = 2^{1-2n} B\left(n,\frac{1}{2}\right)$

Solution :

We have
$$B\left(n,\frac{1}{2}\right) = \frac{\Gamma(n)\Gamma\left(\frac{1}{2}\right)}{\Gamma\left(n+\frac{1}{2}\right)} = \frac{\Gamma(n)\Gamma(n)\Gamma\left(\frac{1}{2}\right)}{\Gamma(n)\Gamma\left(n+\frac{1}{2}\right)}$$

$$= \frac{\Gamma(n)\Gamma(n)\sqrt{\pi}}{\sqrt{\pi}\,\Gamma(2n)\cdot 2^{1-2n}}$$ [By Duplication formula]

$$= \frac{B(n,n)}{2^{1-2n}}$$

$$\Rightarrow \quad B(n,n) = 2^{1-2n} B\left(n,\frac{1}{2}\right)$$

Example 3

Prove that
$$\Gamma\left(\frac{1}{6}\right) = \frac{\sqrt{3}}{2^{\frac{1}{3}}\sqrt{\pi}}\left[\Gamma\left(\frac{1}{3}\right)\right]^2$$

Solution:

By Duplication formula, we have
$$\Gamma(m)\Gamma\left(m+\frac{1}{2}\right) = \frac{\sqrt{\pi}}{2^{2m-1}}\Gamma(2m)$$

Putting $m = \frac{1}{6}$, we get
$$\Gamma\left(\frac{1}{6}\right)\Gamma\left(\frac{2}{3}\right) = \frac{\sqrt{\pi}\,\Gamma\left(\frac{1}{3}\right)}{2^{\frac{-2}{3}}}$$

$$\therefore \quad \Gamma\left(\frac{1}{6}\right) = \frac{\sqrt{\pi}\,\Gamma\left(\frac{1}{3}\right)}{2^{\frac{-2}{3}}\Gamma\left(\frac{2}{3}\right)} \qquad \text{...(1)}$$

Again, we know that $\Gamma(n)\Gamma(1-n) = \dfrac{\pi}{\sin n\pi}$

Putting $n = \dfrac{1}{3}$, we obtain

$$\Gamma\left(\frac{1}{3}\right)\Gamma\left(\frac{2}{3}\right) = \frac{\pi}{\sin\frac{\pi}{3}} = \frac{2\pi}{\sqrt{3}}$$

$$\therefore \quad \Gamma\left(\frac{2}{3}\right) = \frac{2\pi}{\sqrt{3}\,\Gamma\left(\frac{1}{3}\right)} \qquad \ldots(2)$$

Using (2), (1) becomes

$$\Gamma\left(\frac{1}{6}\right) = \frac{\sqrt{\pi}\,\Gamma\left(\frac{1}{3}\right)}{2^{\frac{-2}{3}}} \cdot \frac{\sqrt{3}\,\Gamma\left(\frac{1}{3}\right)}{2\pi} = \frac{\sqrt{3}}{2^{\frac{1}{3}}\sqrt{\pi}}\left[\Gamma\left(\frac{1}{3}\right)\right]^2 \qquad \text{Proved}$$

Example 4

Show that $\int_{-\infty}^{\infty} \cos\left(\frac{\pi}{2}x^2\right)dx = 1$

Solution:

Let $$I = \int_{-\infty}^{\infty} \cos\left(\frac{\pi}{2}x^2\right)dx = 2\int_0^{\infty} \cos\left(\frac{\pi}{2}x^2\right)dx \qquad \ldots(1)$$

Now substituting $\dfrac{\pi x^2}{2} = y \Rightarrow x = \sqrt{\dfrac{2}{\pi}}\,y^{\frac{1}{2}}$

$$\therefore \quad dx = \frac{1}{2}\left(\sqrt{\frac{2}{\pi}}\right)y^{\frac{-1}{2}}dy$$

$$\therefore \quad I = 2\cdot\frac{1}{2}\sqrt{\frac{2}{\pi}}\int_0^{\infty} y^{\frac{-1}{2}}\cos y\,dy = \sqrt{\frac{2}{\pi}}\int_0^{\infty} y^{\frac{-1}{2}}\cos y\,dy$$

$$= \text{Real part of } \sqrt{\frac{2}{\pi}}\int_0^{\infty} y^{\frac{-1}{2}}e^{-iy}\,dy \qquad \left[\because e^{-i\theta} = \cos\theta - i\sin\theta\right]$$

$$= \text{R.P. of } \sqrt{\frac{2}{\pi}}\int_0^{\infty} e^{-iy}y^{\frac{1}{2}-1}\,dy$$

$$= \text{R.P. of } \sqrt{\frac{2}{\pi}}\frac{\Gamma\left(\frac{1}{2}\right)}{\sqrt{e^{i\pi/2}}} = R.P.of\sqrt{\frac{2}{\pi}}\frac{\sqrt{\pi}}{\left(\cos\frac{\pi}{2}+i\sin\frac{\pi}{2}\right)^{\frac{1}{2}}}$$

$$= \text{R.P. of } \sqrt{2}\left(\cos\frac{\pi}{2}+i\sin\frac{\pi}{2}\right)^{\frac{-1}{2}}$$

$$= \text{R.P. of } \sqrt{2}\left(\cos\frac{\pi}{4}-i\sin\frac{\pi}{4}\right)$$ [by Demoivre's theorem]

$$= \sqrt{2}\cos\frac{\pi}{4} = \sqrt{2}\cdot\frac{1}{\sqrt{2}} = 1$$ Proved

Example 5

Calculate $\int_0^\infty \cos\left(\lambda^2 x^2\right)dx$

Solution:

Let $I = \int_0^\infty \cos\left(\lambda^2 x^2\right)dx$

Putting $x^2 = z$

$\Rightarrow$ $dx = \frac{1}{2}z^{\frac{-1}{2}}dz$

$\therefore$ $I = \int_0^\infty \cos\lambda^2 z\cdot\frac{1}{2}z^{\frac{-1}{2}}dz = \frac{1}{2}\int_0^\infty z^{\frac{1}{2}-1}\cos\lambda^2 z\,dz$

$$= \text{R.P. of } \frac{1}{2}\int_0^\infty e^{-i\lambda^2 z}z^{\frac{1}{2}-1}dz$$

$$= \text{R.P. of } \frac{1}{2}\frac{\Gamma\left(\frac{1}{2}\right)}{\left(i\lambda^2\right)^{\frac{1}{2}}} = R.P.of\ \frac{\sqrt{\pi}}{2\lambda}\left(\cos\frac{\pi}{2}+i\sin\frac{\pi}{2}\right)^{\frac{-1}{2}}$$

$$= \frac{\sqrt{\pi}}{2\lambda}\cos\frac{\pi}{4} = \frac{1}{2\lambda}\sqrt{\frac{\pi}{2}}$$ Proved

Example 6

Evaluate $\int_0^1 \log\Gamma(y)\,dy$

Solution:

Let $$I = \int_0^1 \log\Gamma(y)\,dy \qquad ...(1)$$

or, $$I = \int_0^1 \log\Gamma(1-y)\,dy \qquad ...(2) \qquad \left[\because \int_0^a f(x)dx = \int_0^a f(a-x)dx\right]$$

Adding (1) and (2), we get

$$2I = \int_0^1 [\log\Gamma(y) + \log\Gamma(1-y)]dy$$

$$= \int_0^1 \log[\Gamma(y)\Gamma(1-y)]dy$$

$$= \int_0^1 \log\left(\frac{\pi}{\sin\pi y}\right)dy \qquad \left[\because \Gamma(n)\Gamma(1-n) = \frac{\pi}{\sin n\pi}\right]$$

$$= \int_0^1 (\log\pi - \log\sin\pi y)dy$$

$$= \log\pi[y]_0^1 - \int_0^\pi \log\sin z \cdot \frac{1}{\pi}dz \qquad \left[for\,2^{nd}\,\text{integral, put}\,\pi y = z\right]$$

$$= \log\pi - \frac{2}{\pi}\int_0^{\frac{\pi}{2}} \log\sin z\,dz$$

$$= \log\pi - \frac{2}{\pi}\left(-\frac{\pi}{2}\log 2\right) = \log\pi + \log 2 = \log 2\pi$$

$$\therefore \qquad I = \frac{1}{2}\log 2\pi \qquad \text{Answer}$$

Example 7

Prove that $\Gamma\left(\frac{3}{2}-y\right)\Gamma\left(\frac{3}{2}+y\right) = \left(\frac{1}{4}-y^2\right)\pi\sec\pi y$, provided $-1 < 2y < 1$.

Solution:

$$\text{L.H.S.} = \Gamma\left(\frac{3}{2}-y\right)\Gamma\left(\frac{3}{2}+y\right)$$

$$= \left(\frac{1}{2}-y\right)\Gamma\left(\frac{1}{2}-y\right)\cdot\left(\frac{1}{2}+y\right)\Gamma\left(\frac{1}{2}+y\right)$$

$$= \left(\frac{1}{4}-y^2\right)\Gamma\left(\frac{1}{2}-y\right)\Gamma\left(\frac{1}{2}+y\right) = \left(\frac{1}{4}-y^2\right)\Gamma\left(\frac{1-2y}{2}\right)\Gamma\left(1-\frac{1-2y}{2}\right)$$

$$= \left(\frac{1}{4}-y^2\right)\frac{\pi}{\sin\left(\frac{1-2y}{2}\right)\pi} = \left(\frac{1}{4}-y^2\right)\frac{\pi}{\sin\left(\frac{\pi}{2}-\pi y\right)}$$

$$= \left(\frac{1}{4}-y^2\right)\frac{\pi}{\cos \pi y} = \left(\frac{1}{4}-y^2\right)\pi \sec \pi y$$ **Proved**

= R.H.S.

Example 8

Show that $\int_0^{\frac{\pi}{2}} \tan^n \theta \, d\theta = \frac{\pi}{2}\sec\frac{n\pi}{2}, -1 < n < 1$

Solution:

Let $I = \int_0^{\frac{\pi}{2}} \tan^n \theta \, d\theta = \int_0^{\frac{\pi}{2}} \sin^n \theta \cos^{-n} \theta \, d\theta$

$$= \frac{\Gamma\left(\frac{n+1}{2}\right)\Gamma\left(\frac{1-n}{2}\right)}{2\Gamma(1)} = \frac{1}{2}\Gamma\left(\frac{n+1}{2}\right)\Gamma\left(1-\frac{n+1}{2}\right)$$

$$= \frac{1}{2}\frac{\pi}{\sin\left(\frac{n+1}{2}\right)\pi} = \frac{1}{2}\frac{\pi}{\sin\left(\frac{\pi}{2}+\frac{n\pi}{2}\right)} = \frac{1}{2}\frac{\pi}{\cos\frac{n\pi}{2}}$$

$$= \frac{1}{2}\sec\frac{n\pi}{2}$$ **Proved**

Exercise 7.7

1. Evaluate the following integrals :

(a) $\int_0^\infty \sqrt{x}\, e^{-x^3} dx$ (b) $\int_0^\infty (8-x^3)^{-1/3} dx$

(c) $\int_0^\infty e^{-x^2} x^{-1/2} dx \int_0^\infty x^2 e^{-x^4} dx$ (d) $\int_0^\infty x^6 e^{-2x} dx$

(e) $\int_0^\infty \frac{e^{-pt}}{\sqrt{t}}\,dt$ (f) $\int_0^\infty \frac{x\,dx}{1+x^6}$

(g) $\int_0^3 \frac{dt}{\sqrt{3t-t^2}}$ (h) $\int_0^1 \frac{dt}{\sqrt{-\log t}}$

2. Prove that $\int_0^1 t^m(\log t)^n\,dt = \frac{(-1)^n \cdot n!}{(m+1)^{n+1}}$ where n is a positive integer and $m > -1$.

3. show that $\int_0^\infty \frac{t^{m-1}}{(a+bt)^{m+n}}\,dt = \frac{B(m,n)}{a^n b^m}$, where m, n, a and b are positive integers.

4. Prove that $\int_0^\pi \sqrt{\frac{\sin\theta}{(5+3\cos\theta)^n}}\,d\theta = \frac{\left[\Gamma\left(\frac{3}{4}\right)\right]^2}{2\sqrt{2\pi}}$

5. Evaluate the integrals :

(i) $\int_0^1 \frac{dx}{(1-x^n)^{1/n}}$ (ii) $\int_0^1 x^m(1-x^m)^p\,dx$

(iii) $\int_0^1 \frac{dt}{(1-t^n)^{\frac{1}{2}}}$ (iv) $\int_0^\infty x^m e^{-ax^n}\,dx$

6. Evaluate (i) $\Gamma\left(-\frac{5}{2}\right)$ (ii) $\Gamma\left(-\frac{15}{2}\right)$

7. Evaluate the following integrals :

(i) $\int_0^1 \sqrt{1-x^4}\,dx$ (ii) $\int_0^1 \frac{x^{m-1}}{(1+ax)(1-x)^m}\,dx$

(iii) $\int_0^{\frac{\pi}{2}} \sin^2 x\,dx \times \int_0^{\frac{\pi}{2}} \sin^{q+1} x\,dx$

8. Show that $\Gamma(0.1)\Gamma(0.2)\Gamma(0.3)\ldots\Gamma(0.9) = \frac{(2\pi)^{\frac{9}{2}}}{\sqrt{10}}$

9. Prove that $\Gamma\left(\frac{3}{2}-m\right)\Gamma\left(\frac{3}{2}+m\right) = \left(\frac{1}{4}-m^2\right)\cdot \pi \sec m\pi$, where $-1 < 2m < 1$.

10. Prove that $\int_0^1 \frac{x^{m-1}(1-x)^{n-1}}{(x+a)^{m+n}}\,dx = \frac{B(m,n)}{a^n(1+a)^m}; a,m,n > 0$

11. Prove that $\int_0^\infty e^{-x^{\frac{1}{m}}}\,dx = m\Gamma(m)$

12. Show that $1\cdot 3\cdot 5\cdot \ldots.(2m-1) = \frac{2^m\Gamma\left(m+\frac{1}{2}\right)}{\sqrt{\pi}}$

13. Evaluate the following :

(i) $\int_0^1 \left(\log\frac{1}{x}\right)^{m-1} dx, m > 1$ (ii) $\int_0^1 \frac{dy}{\sqrt{y\log\frac{1}{y}}}$

(iii) $\int_0^1 (y\log y)^3\,dy$ (iv) $\int_0^\infty \frac{y^a}{a^y}\,dy$

(v) $\int_0^1 y^{n-1}\left[\log\left(\frac{1}{y}\right)\right]^{m-1} dy$

14. Express $\int_0^1 y^m(1-y^n)^p\,dy$ in terms of Gamma function and then calculate $\int_0^1 y^5(1-y^3)^{10}\,dy$

15. Prove that $\int_0^{\pi/2} \sqrt{\cot\theta}\,d\theta = \frac{1}{2}\Gamma\left(\frac{1}{4}\right)\Gamma\left(\frac{3}{4}\right)$

16. Prove that $\int_0^{\pi/2} \left(\sqrt{\tan\theta}+\sqrt{\sec\theta}\right) d\theta = \frac{1}{2}\Gamma\left(\frac{1}{4}\right)\Gamma\left(\frac{3}{4}\right) + \sqrt{\pi}\,\Gamma\left(\frac{3}{4}\right)$

Answers

1. (i) $\frac{\sqrt{\pi}}{3}$ (ii) $\frac{2\pi}{3\sqrt{3}}$ (iii) $\frac{\pi}{4\sqrt{2}}$ (iv) $\frac{45}{8}$

(v) $\sqrt{\frac{\pi}{p}}, p > 0$ (vi) $\frac{\pi}{3\sqrt{3}}$ (vii) π (Viii) $\sqrt{\pi}$

5. (i) $\frac{\pi}{n}\sin\frac{\pi}{n}$ (ii) $\frac{\Gamma(p+1)\,\Gamma\left(\frac{m+1}{n}\right)}{n\Gamma\left(p+1+\frac{m+1}{n}\right)}$ (iii) $\frac{\sqrt{\pi}\,\Gamma\left(\frac{1}{n}\right)}{\pi\,\Gamma\left(\frac{1}{2}+\frac{1}{n}\right)}$

(iv) $\frac{1}{n\,a^{(m+1)/n}}\Gamma\left(\frac{m+1}{n}\right)$

6. (i) $\frac{-8}{15}\sqrt{\pi}$ (ii) $\frac{\sqrt{\pi}2^8}{1\cdot 3\cdot 5\cdot 7\ldots 15}$

7. (i) $\frac{[\Gamma(1/4)]^2}{6\sqrt{2\pi}}$ (ii) $\frac{1}{(1+a)^m}\cdot\frac{\pi}{\sin m\pi}$ (iii) $\frac{\pi}{2(q+1)}$

13. (i) $\overline{|(n)}$ (ii) $\sqrt{2\pi}$ (iii) $-\frac{3}{128}$ (iv) $\frac{1}{(\log a)^{a+1}}\overline{|a+1}$ (v) $\frac{1}{n^m}\overline{|m}$

7.18 Use Of Beta, Gamma Functions To Solve Triple Integrals

As earlier, we have studied the method of solution of triple integrals but it was time consuming and lengthy. So, there arose a need to study some simpler and easier method for solving the triple integrals. For this purpose, we may take help of Beta and Gamma Functions.

Dirichlet's Integral

If v is a region bounded by $x \geq 0, y \geq 0, z \geq 0$ and $x+y+z \leq 1$, then

$$\iiint_v x^{l-1}y^{m-1}z^{n-1}dx\,dy\,dz = \frac{\Gamma(l)\Gamma(m)\Gamma(n)}{\Gamma(l+m+n+1)}$$

Proof : The triple integral may be written as

$$I = \int_0^1\int_0^{1-x}\int_0^{1-x-y} x^{l-1}y^{m-1}z^{n-1}dx\,dy\,dz$$

$$= \int_0^1\int_0^{1-x} x^{l-1}y^{m-1}\left[\frac{z^n}{n}\right]_0^{1-x-y} dx\,dy$$

$$= \frac{1}{n}\int_0^1\int_0^{1-x} x^{l-1}y^{m-1}(1-x-y)^n dx\,dy$$

Put $y = (1-x)t$, so that $dy = (1-x)dt$, we obtain

$$I = \frac{1}{n}\int_0^1\int_0^1 x^{l-1}[(1-x)t]^{m-1}[1-x-(1-x)t]^n(1-x)dx\,dt$$

$$= \frac{1}{n}\int_0^1\int_0^1 x^{l-1}(1-x)^{m+n}t^{m-1}(1-t)^n dx\,dt$$

$$= \frac{1}{n}\int_0^1 x^{l-1}(1-x)^{\overline{m+n+1}-1}dx \cdot \int_0^1 t^{m-1}(1-t)^{\overline{n+1}-1}dt$$

$$= \frac{1}{n} B(l, m+n+1)\cdot B(m, n+1)$$

[By definition of Beta function]

$$= \frac{1}{n}\frac{\Gamma(l)\Gamma(m+n+1)}{\Gamma(l+m+n+1)}\cdot\frac{\Gamma(m)\Gamma(n+1)}{\Gamma(m+n+1)} \qquad \left[\therefore\ B(m,n) = \frac{\Gamma(m)\Gamma(n)}{\Gamma(m+n)}\right]$$

$$= \frac{1}{n}\frac{\Gamma(l)\,\Gamma(m)\cdot n\Gamma(n)}{\Gamma(l+m+n+1)} \qquad [\therefore\ \Gamma(n+1) = n\Gamma(n)]$$

$$= \frac{\Gamma(l)\,\Gamma(m)\,\Gamma(n)}{\Gamma(l+m+n+1)}$$ Proved

This integral is known as Dirichlet's Integral and is very useful in evaluating multiple integrals.

Remark :

(i) The above result can be generalized for more than 3 variable, *i.e.*, if region is $x_i \geq 0, i = 1, 2 \ldots n$ and $x_1 + x_2 + \ldots + x_n \leq 1$, then

$$\int_1\int_2\ldots\int_n x_1^{p_1-1}\,x_2^{p_2-1}\,x_3^{p_3-1}\ldots x_n^{p_n-1}\,dx_1\,dx_2\ldots dx_n$$

$$= \frac{\Gamma(p_1)\Gamma(p_2)\ldots\Gamma(p_n)}{\Gamma(p_1+p_2+\ldots+p_n+1)}$$

(ii) If the region of integration is $x \geq 0, y \geq 0, z \geq 0$ and $x+y+z \leq h$, then modified result will be

$$\iiint_V x^{l-1} y^{m-1} z^{n-1} dx\, dy\, dz = h^{l+m+n} \frac{\Gamma(l)\Gamma(m)\Gamma(n)}{\Gamma(l+m+n+1)}$$

Also, as discussed in remark (i), this result can be generalized for more than 3 variables.

(iii) If the region of integration is $x \geq 0, y \geq 0, z \geq 0$ and $h_1 \leq x+y+z \leq h_2$, then

$$\iiint_V x^{l-1} y^{m-1} z^{n-1} dx\, dy\, dz = \frac{\Gamma(l)\Gamma(m)\Gamma(n)}{\Gamma(l+m+n)} \int_{h_1}^{h_2} f(u) \cdot u^{l+m+n-1} du$$

Above result can also be generalized for more than 3 variables. This is known as Liouville's extension of Dirichlet's Integral.

Illustrative Examples

Example 1

Find the volume of the solid surrounded by the surface

$$\left(\frac{x}{a}\right)^{\frac{2}{3}} + \left(\frac{y}{b}\right)^{\frac{2}{3}} + \left(\frac{z}{c}\right)^{\frac{2}{3}} = 1$$

Solution :

The given surface is

$$\left(\frac{x}{a}\right)^{\frac{2}{3}} + \left(\frac{y}{b}\right)^{\frac{2}{3}} + \left(\frac{z}{c}\right)^{\frac{2}{3}} = 1 \quad \text{...(1)}$$

Now put $\left(\frac{x}{a}\right)^{\frac{2}{3}} = u, \left(\frac{y}{b}\right)^{\frac{2}{3}} = v, \left(\frac{z}{c}\right)^{\frac{2}{3}} = w$ ...(2)

$\Rightarrow$ $x = a\,u^{3/2},\ y = b\,v^{3/2},\ c = c\,w^{3/2}$

$\Rightarrow$ $dx = \frac{3}{2} a\,u^{1/2} du,\ dy = \frac{3}{2} b\,v^{1/2} dv,\ dz = \frac{3}{2} c\,w^{1/2} dw$ ...(3)

Using above substitution, surface (1) becomes

$u + v + w = 1$ ...(4)

$\therefore$ volume $v = 8 \times$ volume bounded by co-ordinate planes and surface in first octant.

$$= 8\iiint dx\,dy\,dz$$

$$= 8\iiint \frac{27}{8} abc\, u^{1/2}\, v^{1/2} w^{1/2}\, du\, dv\, dw \qquad \text{by (3)}$$

$$= 27abc \iiint u^{\frac{3}{2}-1}\, v^{\frac{3}{2}-1}\, w^{\frac{3}{2}-1}\, du\, dv\, dw$$

where $u \geq 0,\ v \geq 0,\ w \geq 0$ and $u+v+w \leq 1$

$$= 27abc \frac{\Gamma\left(\frac{3}{2}\right)\Gamma\left(\frac{3}{2}\right)\Gamma\left(\frac{3}{2}\right)}{\Gamma\left(\frac{3}{2}+\frac{3}{2}+\frac{3}{2}+1\right)} \qquad \text{[by Dirichlet's th}^{\text{m}}\text{.]}$$

$$= 27abc \frac{\frac{1}{2}\sqrt{\pi}\cdot\frac{1}{2}\sqrt{\pi}\cdot\frac{1}{2}\sqrt{\pi}}{\frac{9}{2}\cdot\frac{7}{2}\cdot\frac{5}{2}\cdot\frac{3}{2}\cdot\frac{1}{2}\sqrt{\pi}} = \frac{4}{35}\pi abc \qquad \text{Answer}$$

Example 2

Find the mass of an octant of the ellipsoid $\frac{x^2}{a^2}+\frac{y^2}{b^2}+\frac{z^2}{c^2}=1$, the density at any point being $\rho = \kappa xyz$

Solution :

Let us consider the first octant for calculations.

Mass $\quad M = \iiint \rho\, dx\, dy\, dz = \kappa \iiint xyz\, dx\, dy\, dz \qquad ...(1)$

Now the given surface is $\frac{x^2}{a^2}+\frac{y^2}{b^2}+\frac{z^2}{c^2}=1$

Put $\frac{x^2}{a^2}=u,\ \frac{y^2}{b^2}=v,\ \frac{z^2}{c^2}=w$

$\Rightarrow \quad x = au^{1/2},\ y = bv^{1/2},\ z = cw^{1/2}$

$\Rightarrow \quad dx = \frac{a}{2}u^{-1/2}du,\ \ dy = \frac{b}{2}v^{-1/2}dv,\ dz = \frac{c}{2}w^{-1/2}dw$

using these, (1) becomes

$$M = \kappa \iiint \frac{a^2b^2c^2}{8}\, du\, dv\, dw$$

where $u \ge 0,\ v \ge 0,\ w \ge 0$ and $u+v+w \le 1$

$$\therefore \quad M = \frac{\kappa\, a^2b^2c^2}{8}\iiint u^{1-1}\, v^{1-1}\, w^{1-1}\, du\, dv\, dw$$

$$= \frac{\kappa\, a^2b^2c^2}{8}\frac{\Gamma(1)\,\Gamma(1)\,\Gamma(1)}{\Gamma(1+1+1+1)} \qquad \text{[by Dirichlet's integral]}$$

$$= \frac{\kappa\, a^2b^2c^2}{8}\cdot\frac{1}{3!} = \frac{\kappa\, a^2b^2c^2}{48} \qquad \text{Answer}$$

Example 3

Show that $\iint_R x^{l-1}y^{m-1}\, dx\, dy = \dfrac{\Gamma(l).\Gamma(m)}{\Gamma(l+m+1)}h^{l+m}$ where R is the domain $x \ge 0,\ y \ge 0$ and $x+y \le h$

Solution :

The given integral is

$$I = \iint_{R'} (x)^{l-1}(y)^{m-1}dx\ dy \quad \text{where } x \ge 0, y \ge 0, x+y \le h$$

Put $\quad x = hx,\ y = hy,$ then

$$I = \iint_{R'} (hx)^{l-1}(hy)^{m-1}h^2\, dx\, dy$$

where R' is $x \ge 0, y \ge 0$ and $x+y \le 1$

$$= \int_0^1\int_0^{1-x} x^{l-1}y^{m-1}\cdot h^{l+m}\, dx\, dy$$

$$= h^{l+m}\int_0^1 x^{l-1}\left[\frac{y^m}{m}\right]_0^{1-x} dx \qquad = \frac{h^{l+m}}{m}\int_0^1 x^{l-1}(1-x)^{\overline{m+1-1}}dx$$

$$= \frac{h^{l+m}}{m}B(l, m+1) \qquad = \frac{h^{l+m}}{m}\frac{\Gamma(l)\Gamma(m+1)}{\Gamma(l+m+1)}$$

$$= \frac{h^{l+m}}{m}\frac{\Gamma(l)\cdot m\Gamma(m)}{\Gamma(l+m+1)} \qquad = h^{l+m}\frac{\Gamma(l)\,\Gamma(m)}{\Gamma(l+m+1)} \qquad \text{Proved}$$

Example 4

Prove that $\iiint \dfrac{dx\, dy\, dz}{(1+x+y+z)^3} = \dfrac{1}{2}\log 2 - \dfrac{5}{16}$ the integral being taken throughout the volume bounded by the planes $x = 0,\ y = 0,\ z = 0$ and $x + y + z = 1$.

Solution :

The region of integration is $x \geq 0, y \geq 0, z \geq 0$ and $0 < x + y + z < 1$, then by Liouville's theorem

$$I = \iiint \frac{x^{l-1} y^{l-1} z^{l-1}}{(1+x+y+z)^3} \, dx\, dy\, dz$$

$$= \frac{\Gamma(1)\,\Gamma(1)\,\Gamma(1)}{\Gamma(1+1+1)} \int_0^1 \frac{u^{1+1+1-1}}{(1+u)^3} du$$

$$= \frac{1}{2}\int_0^1 \frac{u^2}{(1+u)^3} du = \frac{1}{2}\int_0^1 \left[\frac{1}{u+1} - \frac{2}{(u+1)^2} - \frac{1}{(u+1)^3}\right] du$$

$$= \frac{1}{2}\left[\log(u+1) + \frac{2}{u+1} + \frac{1}{2(u+1)^2}\right]_0^1$$

$$= \frac{1}{2}\log 2 - \frac{5}{16} \qquad \text{Proved}$$

Example 5

Evaluate $\iiint e^{x+y+z}\, dx\, dy\, dz$ taken over the positive oc tant such that $x + y + z \leq 1$

Solution :

The region of integration is given as $x \geq 0, y \geq 0, z \geq 0$ and $0 < x + y + z \leq 1$, then by Liouville's theorem

$$\iiint e^{x+y+z} \cdot x^{l-1} y^{l-1} z^{l-1} dx\, dy\, dz = \frac{\Gamma(1)\,\Gamma(1)\,\Gamma(1)}{\Gamma(1+1+1)} \int_0^1 e^u \cdot u^{3-1}\, du$$

$$= \frac{1}{2}\int_0^1 e^u \cdot u^2 du = \frac{1}{2}\left\{\left[e^u \cdot u^2\right]_0^1 - \int_0^1 e^u \cdot 2u\, du\right\}$$

$$= \frac{e}{2} - \left\{(e^u \cdot u)_0^1 - \int_0^1 e^u \cdot du\right\} = \frac{e}{2} - e + e - 1$$

$$= \frac{e-2}{2} \qquad \text{Answer}$$

Example 6

Evaluate $\iiint \log(x+y+z) dx\, dy\, dz$ taken over all positive values of x, y, z subjected to $x + y + z \leq 1$.

Solution :

The region of integration is given as $x \geq 0, y \geq 0, z \geq 0$ and $0 < x + y + z \leq 1$, then by Liouville's theorem